T0319618

STATISTICS FOR SCIENTISTS AND ENGINEERS

STATISTICS FOR SCIENTISTS AND ENGINEERS

RAMALINGAM SHANMUGAM
RAJAN CHATTAMVELLI

Published by John Wiley & Sons, Inc., Hoboken, New Jersey
Published simultaneously in Canada

For general information on our other products and services or for technical support, please contact our Customer Care Department within the United States at (800) 762-2974, outside the United States at (317) 572-3993 or fax (317) 572-4002.

Wiley also publishes its books in a variety of electronic formats. Some content that appears in print may not be available in electronic formats. For more information about Wiley products, visit our web site at www.wiley.com.

Library of Congress Cataloging-in-Publication Data:

Shanmugam, Ramalingam.
  Statistics for scientists and engineers / Ramalingam Shanmugam, Rajan Chattamvelli.
      pages cm
  Includes bibliographical references and index.
  ISBN 978-1-118-22896-8 (cloth)
1.   Mathematical statistics. I. Chattamvelli, Rajan. II. Title.
  QA276.S456 2015
  519.5–dc23

                                                    2015004898

Printed in the United States of America
10 9 8 7 6 5 4 3 2 1

CONTENTS

7   Continuous Distributions   **255**

PREFACE

This is an intermediate-level textbook on probability and mathematical statistics. This book can be used for one-semester graduate courses in engineering statistics. Prerequisites include a course in college algebra, differential and integral calculus, and some linear algebra concepts. Exposure to computer programming is not a must but will be useful in Chapters 1 and 5. It can also be used as a supplementary reading material for mathematical statistics, probability, and statistical methods. Researchers in various disciplines will also find the book very helpful as a ready reference.

AUDIENCE

This book is mainly aimed at three reader groups: (i) Advanced undergraduate and beginning graduate students in engineering and physical sciences and management. (ii) Scientists and engineers who require skill sets to model practical problems in a statistical setting. (iii) Researchers in mathematical statistics, engineering, and related fields who need to make their knowledge up-to-date. Some of the discussions in Chapters 1, 9, and 11 are also of use to computer professionals. It is especially suited to readers who look for nonconventional concepts, problems, and applications. Some of the chapters (especially Chapters 2–8) can also be used by students in other disciplines. With this audience in mind, we have prepared a textbook with the hope that it will serve as a gateway to the wonderful subject of statistics. Readers will appreciate the content, organization, and coverage of the topics.

PURPOSE

This book discusses the theoretical framework needed to build, analyze, and interpret various models. Our aim is to present the basic concepts in an intuitively appealing and easily understood way. This book will help the readers to choose the correct model or distinguish among various choices that best captures the data or solve the problem at hand. As this book lays a strong foundation of the subject, interested readers will be able to pursue advanced modeling and analysis with ease.

APPROACH

Theoretical concepts are developed or explained in a step-by-step and easy-to-understand manner. This is followed by practical examples. Some of the difficult concepts are exemplified using multiple examples drawn from different fields. Exercises are chosen to test the understanding of concepts. Extensive bibliography appears at the end of the book.

MAIN FEATURES

Most important feature of the book is the large number of worked out examples from a variety of fields. These are self-explanatory and easily grasped by students. These are drawn from medical sciences and various disciplines of engineering, and so on. In addition, extensive exercises are provided at the end of each chapter. Some novel methods to find the mean deviation (MD) of discrete and continuous distributions are introduced in Chapters 6, 7, and 9.

MATHEMATICS LEVEL

This book is ideal for those who have done at least one course in college algebra and calculus. Sum and product notations given in Chapter 1 are used in Chapters 6 and 8. Set theory concepts are sparingly used in Chapter 5. Basic trigonometric concepts are used in Chapters 7 and 11. Differential calculus concepts are needed to solve some of the problems in Chapters 10 and 11, especially in finding the Jacobian of transformations. Integral calculus is used extensively in Chapters 7, 10, and 11. Some concepts on matrices and linear algebra are needed in Chapters 10 and 11.

COVERAGE

The book starts with an introductory chapter that fills the gap for those readers who do not have all the prerequisites. Chapter 1 introduces the basic concepts, including several notations used throughout the book. Important among them are the notations for combinations, summation, and products. It briefly discusses the scales of measurement and gives examples of various types of data. The summation notation is extensively discussed, and its variants such as nested sums, fractional steps, symmetric sums, summation over sets, and loop unrolling are thoroughly discussed. The product notation is discussed next and its applications to evaluating powers of the form x^n where x and n are both large are given. Rising and falling factorial notation is briefly discussed. These are extremely helpful for scientists and working engineers in various fields. Data discretization and data transformations are also introduced. The chapter ends with a discussion of testing for normality of data. This chapter can be skipped for senior-level courses. Working engineers and professionals may need to skim through this chapter, as it contains a few useful concepts of immense practical value.

Chapter 2 discusses the measures of location. These are essential tools for anyone working with numeric data. All important measures such as arithmetic mean, geometric and harmonic means, median, and mode are discussed. Some updating formulas for the means are also given. Important among them are the updating formula for weighted mean, geometric mean and harmonic means, and trimmed means, as well as updating formulas for origin and scale changed data and windowed data. The sample median, mode, quartiles, and percentiles are also explained.

Popular measures of spread appear in Chapter 3. A categorization of spread measures helps the reader to distinguish between various measures. These include linear and nonlinear measures, pivotal and pivot-less measures, additive and nonadditive measures, absolute and relative measures, distance-based measures, and so on. The sample range and its advantages and applications are discussed. An illuminating discussion of the "degrees-of-freedom" concept appears in page 3–13. A summary table gives a comparison of various spread measures (pp. 3–8). The average absolute deviation (AAD) (also called *sample mean absolute deviation*) and its properties are discussed next. Sample variance and standard deviation are the most frequently used measures of spread. These are discussed, and some updating formulae for sample variance are derived. This is followed by the formula for pooling sample variance and covariance, which forms the basis for a divide-and-conquer algorithm. Some bounds on the sample standard deviation in terms of sample range are given. The chapter ends with a discussion of the coefficient of variation and Gini coefficient.

Chapter 4 discusses measures of skewness and kurtosis. Absolute versus relative measures of skewness are discussed, followed by various categories of skewness measures such as location and scale-based measures, quartile-based measures, moment-based measures, measures that utilize inverse of distribution functions, and measures that utilize L-moments. Pearson's and Bowley's measures are given and their ranges are discussed. Coefficient of quartile deviation and its properties are discussed. The range of values of various measures is summarized into a table. This is followed by a discussion of the measures of kurtosis. The kurtosis of other statistical distributions is compared with that of a standard normal with kurtosis coefficient 3, which is derived. A brief discussion of skewness–kurtosis bounds and L-kurtosis appear next. This chapter ends with a discussion of spectral kurtosis and multivariate kurtosis (which may be skipped in undergraduate courses).

Fundamentals of probability theory are built from the ground up in Chapter 5. As solving some probability problems is a challenge to those without adequate mathematical skills, a majority of this chapter develops the tools and techniques needed to solve a variety of problems. The chapter starts with a discussion of various ways to express probability. Converting repeating and nonrepeating decimal numbers into fractional form p/q is given in algorithmic form. Sample spaces are defined and illustrated using various problems. These are then used to derive the probability of various events. This is followed by building the mathematical background using set theory and Venn diagrams. A discussion of event categories appears next–simple and compound events, mutually exclusive events, dependent and independent events, and so on. Discrete and continuous events are exemplified as well as various laws of events—commutative, associative, distributive laws; the law of total probability; and De'Morgan's laws. Basic counting principle is introduced and illustrated using numerous examples from various fields. This is followed by a lengthy discussion of the tools and techniques such as permutation and combination, cyclic permutation, complete enumeration, trees, principle of inclusion and exclusion, recurrence relations, derangements, urn models, and partitions. Probability measure and space are defined and illustrated. The do-little principle of probability and its applications are discussed. The axiomatic, frequency, and other approaches to probability are given. The chapter ends with a discussion of Bayes theorem for conditional probability and illustrates its use in various problems.

Chapter 6 on discrete distributions builds the concepts by starting with the binomial theorem. As the probabilities of theoretical distributions sum to one, some of them can be easily obtained by putting particular values in the binomial expansion. A novel method to easily find the MD of discrete distributions is introduced in this chapter (Section 6.3, pp. 6–6). Important properties of distributions are succinctly summarized. These include tail probabilities, moments and location measures, dispersion measures, generating functions, and recurrence relations. It is shown that the rate of convergence of binomial distribution to the Poisson law is quadratic in p and linear in n (pp. 6–37). This provides new insight into the classical textbook rule that "binomial tends to the Poisson law when $n \to \infty$ and $p \to 0$ such that np remains a constant." Analogous results are obtained for the limiting behavior of negative binomial distributions. Distribution of the difference of successes and failures in Bernoulli trials are obtained in simple form. Other distributions discussed include geometric, Poisson, hypergeometric, negative hypergeometric, logarithmic, beta binomial, and multinomial distributions. Researchers in various fields will find Chapters 6 and 7 to be of immense value.

Chapter 7 introduces important continuous distributions that are often encountered in practical applications. A general method to find the MD of continuous distributions is derived in page 7–4, which is very impressive as it immensely reduces the arithmetic work. This helpful result is extensively used throughout the chapter. A relation between variance of continuous distributions and tail areas is derived. Alternate parametrizations of some distributions are given. List of distributions include uniform (rectangular), exponential, beta-I, beta-II, gamma, arc-sine, cosine, normal, Cauchy, central χ^2 and chi-, Student's t, Snedecor's F, inverse Gaussian, log-normal, Pareto, Laplace, Weibull, Rayleigh, Maxwell, and Fisher's Z distributions. Important results are summarized and several algorithms for tail areas are discussed. These results are used in subsequent chapters.

Mathematical expectation is discussed in Chapter 8. Expectation and variance using distribution functions are discussed next. Expectation of functions of random variables appears in page 8–20. Properties of expected values, expectation of functions of random variables, variance, covariance, moments, and so on appear next. This is followed by a discussion of conditional expectation, which is used to derive the mean of mixture distributions. Several important results such as expressions for variance (pp. 8–47) and expectation of functions of random variables (pp. 8–50) are summarized. These are needed in subsequent chapters.

Chapter 9 on generating functions gives a brief introduction to various generating functions used in statistics. This includes probability generating function, moment generating function, cumulant generating function, and characteristic functions. These are derived for several distributions and their inter-relationships are illustrated with examples. Two novel generating functions are introduced in this chapter–first one to generate the cumulative distribution function (CDF-GF) in Section 9.3 (pp. 9–10) and second one to generate MD (MD-GF) (Section 9.4, pp. 9–11). Factorial moment generating functions and its relationship to Stirling numbers are briefly mentioned. This chapter is strongly coupled with Chapters 6–8. Readers in prior chapters may want to refer to the results in this chapter as and when needed.

Functions of random variables are discussed in Chapter 10. These are used in deriving distributions of related statistics. This chapter discusses various techniques such as method of distribution functions, Jacobian method, probabilistic methods, and area-based methods, and it also discusses distribution of absolute values of symmetric random variables, distribution of $F(x)$ and $F^{-1}(x)$, and so on. These results are applied to find the MD of continuous distributions using a simple integral of $t\,dt/f(F^{-1}(t))$ from lower limit to $F(\mu)$. Other topics discussed include distribution of squares, square roots, reciprocals, sums, products, quotients, integer part, and fractional part of continuous random variables. Distributions of trigonometric and transcendental functions are also discussed. The chapter ends with a discussion of various transformations of normal variables.

Joint distributions are briefly discussed in Chapter 11 and some applications are given. Marginal and conditional distributions are discussed and illustrated with various examples. The concept of the Jacobian is introduced in Section 11.2 in page 11–7. Derivation of joint distributions in a bivariate setup is given in Section 11.2.1, pp. 11–9. An immensely useful summary table of 2D transformations appears in page 11–15 for ready reference. Various polar transformations such as plane polar, spherical polar, and toroidal polar and its inverses are discussed, and a summary table is given in page 11–28. A good understanding of integration is absolutely essential to grasp some of the examples in this chapter.

Working professionals will find the book to be very handy and immensely useful. Some of the materials in this book were developed during Dr. Ramalingam Shanmugam's teaching of engineering statistics in the University of Colorado and second author's teaching at Frederick University, Cyprus. Any suggestions or comments for improvement are welcome. Please mail them to the first author at rs25@swt.edu.

The first author would like to thank Wiley editorial staff, especially Ms. Kari Capone, Ms. Amy Henderson, and others in Wiley for tremendous help during the entire production work and their patience. The second author would like to thank his brothers C.V. Santosh Kumar and C.V. Vijayan for all the help and encouragements. The first author dedicates this book to his wife srimathi Malarvizhi Shanmugam, and the second author dedicates it to his late grandfather Mr. Keyath Kunjikannan Nair.

RAM SHANMUGAM
RAJAN CHATTAMVELLI
San Marcos, TX
Thanjavur, India
November, 2014

ABOUT THE COMPANION WEBSITE

This book is accompanied by a companion website:

www.wiley.com/go/shanmugam/statistics

The website includes:

- Solutions Manual available to Instructors.

1

DESCRIPTIVE STATISTICS

After finishing the chapter, students will be able to

- Explain the meaning and uses of statistics
- Describe the standard scales of measurement
- Interpret various summation ($\sum$) and product ($\prod$) notations
- Apply different types of data transformations
- Distinguish various data discretization algorithms (DDAs)

1.1 INTRODUCTION

Statistics has borrowed several ideas and notations from various other fields. This section summarizes some important concepts and notations that will be used in this book. As examples, the factorial, permutation and combination, summation, product (including rising and falling factorials), and other mathematical notations including set-theoretic operators are discussed in this chapter. These notations are extensively used in descriptive and inferential statistics. A discussion on the most commonly used scales of measurement in this chapter gives better insight into the types of data most often encountered. Various techniques to transform these data into any desired range are also given. Data discretization techniques to categorize continuous data are exemplified. These notations, tools, and techniques are immensely useful to better grasp the rest of the chapters.

Statistics for Scientists and Engineers, First Edition. Ramalingam Shanmugam and Rajan Chattamvelli.
© 2015 John Wiley & Sons, Inc. Published 2015 by John Wiley & Sons, Inc.

Most of the statistical analyses use a sample of data values. These data are either collected from a target population, obtained from field trials, or generated randomly. They are then either summarized (using summary measures) or subjected to one or more analyses. Popular summarization measures include location measures, spread measures, skewness and kurtosis measures, dependency measures, and other measures. Chapter 2 introduces the most common location measures. These include arithmetic, geometric, and harmonic means; median; and the mode. Trimmed versions of these measures are obtained by dropping data values at either or both of the extremes. As a special case, as the geometric and harmonic means are defined only for positive data values ($ >\$0$), an analyst may drop all zero and negative data values to obtain left-trimmed versions of them. The trimmed means, weighted means, and their updating formula are given. The sample median, mode, quartiles, and percentiles are also explained.

Dispersion measures are discussed in Chapter 3. A categorization of spread measures helps the reader to distinguish between linear and nonlinear measures, pivotal and pivotless measures, additive and nonadditive measures, absolute and relative measures, and distance-based measures. The most frequently used spread measures are the sample range, variance, or standard deviation. These are discussed and some updating formulae for sample variance are derived. The sample variance and covariance can be computed recursively using a divide-and-conquer strategy by dividing the sample into two subsamples and pooling the corresponding variance or covariance of subsamples in a sensible and orderly way. The coefficient of variation and Gini coefficient are also discussed.

Popular measures of skewness and kurtosis are discussed in Chapter 4. Different types of skewness measures such as location and scale-based measures, quartile-based measures, moment-based measures, measures that utilize inverse of distribution functions, or L-moments are discussed. The measures of kurtosis discussed are skewness–kurtosis bounds, L-kurtosis, and spectral kurtosis. Chapter 5 discusses probability theory with emphasis on solving problems from various fields. Essential tools and techniques such as permutation and combination, cyclic permutation, complete enumeration, trees, principle of inclusion and exclusion, recurrence relations, derangements, urn models, and partitions used to solve probability problems are discussed, and conditional probability and Bayes theorem are introduced.

Discrete distributions and their properties are discussed in Chapter 6. A novel method to easily find the mean deviation (MD) of any discrete distribution is introduced in this chapter, and its practical use is illustrated. It is shown that the rate of convergence of binomial distribution to the Poisson [307,308] law is quadratic in p and linear in n. Distribution of the absolute value of the difference between successes and failures in independent Bernoulli trials is obtained in simple form. Commonly encountered continuous distributions are discussed in Chapter 7. An impressive method of immense practical value to find the mean deviation of continuous distributions is derived and is extensively used throughout the chapter. Variance of continuous distributions is shown to be related to the tail areas. Mathematical expectation and its properties are discussed in Chapter 8. An introduction to various generating functions used in statistics appears in Chapter 9. Two new families of generating functions (for generating cumulative distribution functions (CDF-GF) and mean deviation (MD-GF)) are introduced in this chapter.

Functions of random variables appear in Chapter 10. Different techniques such as method of distribution functions, Jacobian method, probabilistic methods, and area-based methods are discussed. Distribution of squares, square roots, reciprocals, sums, products, quotients, integer part, and fractional part of continuous random variables and distributions of trigonometric and transcendental functions are also discussed. Joint, marginal, and conditional distributions are discussed in Chapter 11. The concept of Jacobians is introduced, and an immensely useful summary table of 2D transformations is given. Various polar transformations such as plane polar, spherical polar, and toroidal polar and their inverses are also discussed and summarized.

1.2 STATISTICS AS A SCIENTIFIC DISCIPLINE

Statistics has its origin in describing (collecting, organizing, and interpreting) numeric data collected on subjects of interest. This branch of statistics is called *descriptive statistics*. It deals with finite random samples drawn from the totality of all elements concerned (which is assumed to be large and is called the *population* (pp. 1–11)). It uses numerical measures (such as mean, variance, and correlation) and graphical techniques to summarize information in a concise and comprehensible form. These are intended for communication, interpretation, or subsequent processing by humans or machines. *Inferential statistics* is concerned with making inferences about the parameters or on the functional form of a population (defined in the following) using small random samples drawn from it. A great majority of inferential statistics do make assumptions on the form of the underlying density, on the parameters, or on the data range. Data for descriptive and inferential statistics are often numeric. Large samples are typically used in descriptive statistics than in inferential statistics. For example, scatterplots and other visualization tools require more data points than those used in statistical quality control or testing statistical hypotheses.

Definition 1.1 Statistics is a branch of scientific discipline that deals with the systematic collection, tabulation, summarization, classification, analysis and modeling of data, extracting summary information from numeric data, and drawing potentially useful conclusions from past or observed data, or verifying experimental hypotheses.

This definition does not cover every branch of statistics, as the subject continues to diversify into various applied sciences. For example, statistical quality control is used to check process deviations to see if they are well within the tolerance limits or if they fall beyond predefined levels. Testing of statistical hypotheses involve well-defined experimental steps that use tabulated values of a test statistic to draw reasonable conclusions (accepting or rejecting a research hypothesis) about an unknown population parameter that describes some characteristic of the distribution.

Similarly, engineering applications of statistics include searching for potentially useful patterns and trends in large data collections using regression [304] models, hierarchical (tree) models, cluster analysis, partial least squares, and so on. The least-squares principle finds applications in neural networks (NNs), digital signal processing, data compression, and many engineering fields.

Population census was conducted long ago in 3340 BC in Egypt [en.wikipedia.org/wiki/census]. William-I of England conducted a complete census of adults and households in AD 1086. Regular population census started in the United States in 1790 and in the United Kingdom in 1801 at 10-year periods. Numeric data on birth, death, and marriages were collected in several countries of Western Europe subsequently. This branch of statistics that deals with vital (life-connected) data is known as *vital statistics*. Study of numeric data on education, housing, and social welfare came to be known as *social statistics*. Economic statistics deals with data analysis on unemployment, economic indicators (consumer price index of essential commodities, purchasing power of people in various strata, etc.), industrial production, import, and export. Experimental statistics deals with data analysis or comparison techniques in field experiments using samples or simulated trials. Agricultural statistics is a related field that deals with analysis of yield in field experimentation. Data analysis techniques that utilize data spread across a frame-of-reference (such as the Earth's surface or deep space) are known as *spatial statistics*. They involve data of categorical or quantitative types along with a spatial frame of reference that serves as a window. Medical statistics is a mixture of the above that deals with summary measures, experimental designs, sampling, predictions, classifications, and clustering, to name a few. Mathematical statistics comprises an umbrella of properties of random variables, sampling distributions, expectations, estimation, and inference (see inferential statistics). Nonparametric statistics have minimal assumptions on the data—on the data range, distribution function, or the location and scale parameters of the population. Statistical techniques are also applied in various other fields such as insurance (actuarial statistics), engineering (engineering statistics), psychology, and education.

1.2.1 Scales of Measurement

The most popular scales of measurement are nominal, ordinal, interval, and ratio (NOIR) scales originated by Stevens [265]. A great majority of common data during the 1940s belonged to one of these four categories. These are called *standard data scales*. These scales were well known in statistical analysis much before 1946 as categorical and quantitative scales. The nominal and ordinal data are together called *categorical*, *nonmetric*, or *qualitative* data because they are labeled using a finite alphabet of distinct and consistent symbols. These need not be numbers because we can use letters (in various languages such as English and Greek), enumerated constants, or strings to label them. The interval and ratio type of data are together called *quantitative* data. These are always numeric (either integers or floating point

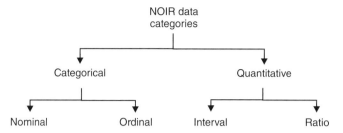

Figure 1.1 The NOIR scale of measurement.

numbers with integer and fractional parts). Most statisticians are familiar with categorical (nominal and ordinal) and quantitative (interval and ratio) (or C&Q for short) data scales[1], as the majority of statistical methods works on them. Different statistical techniques are used with C&Q data, as discussed in subsequent chapters.

Text data are built on a nominal or ordinal set of characters [53]. One example is the binary string data that uses a sequence of 0's and 1's (Figure 1.1). Majority of text data use alphabetic characters of natural languages that are ordinal data. Subsets of alphabetic characters or special characters can also be used to encode text data as in genomics and bioinformatics. Audio, video, animation, and multimedia (AVAM) data are always compressed when stored in digital form. These use interval or ratio type of data at the basic level.

Spatial data are stored with reference to a coordinate frame, a surface, or a map. It consists of quantitative or categorical data that are tied to unique points in the corresponding frame. One example is GIS data that use the Earth's surface as the base. Earth-centric data can be represented in multiple ways (using (longitude, latitude, altitude) coordinates, using GPS coordinates, etc.).

Temporal data are time dependent. One common example is the stock market index, which varies continuously during the trading time. Spatiotemporal data are spatial data that vary over time. Examples are data generated by tornadoes, hurricanes, and other natural disasters that persist for some time, sea surface temperatures (which vary during 24-hour intervals with maximum occurring when the Sun has passed over the particular sea), HTTP connection requests generated across the world (which subsides down after midnight at particular regions), and so on. The extended data types can be structured or unstructured, compressed or uncompressed. Newer data types and standards are also being devised to store data collected using special hardware devices, wireless devices, and satellites. These data are transmitted as radio waves in analog form, which are then converted into digital form. High-quality compression standards that use quantization and least squares principle are used for efficient storage of this type of data. Newer applications of statistics are also emerging in extended data analysis. For instance, statistical measures computed

---

[1]quantitative data are also called *metric data*

from summary information extracted from extended data are being used in information retrieval using latent semantic analysis [53, 294]. Well-developed statistical techniques also exist for spatial and temporal data analysis.

1.3 THE NOIR SCALE

This section briefly reviews the standard scales of measurement, which we call the NOIR typology. A basic understanding of this concept is helpful to students and researchers in statistics and other fields, who are faced with the data collection task. Data can be collected using questionnaires, web forms, or machines or special devices. A precise granularity level is important for choosing each variable and to extract the maximum information from subjects that generate the data.

1.3.1 The Nominal Scale

It is a categorization of a variable into a set of mutually exclusive (distinct) values. A most common example is the human blood group $= \{A, B, O, AB\}$. There is no logical order among the blood groups. Hence, we cannot say that a person with blood group "A" is better or worse than a person with blood group "B" (Owing to the advances being made in the mapping and analysis of human genes, it is now possible to identify humans who are more susceptible to certain illnesses using their blood type. Someday, it may be possible to logically order the blood groups conditionally on various disease categories.). They are just distinguishing labels given to persons based on a surface marker on the red blood cells. By convention, they are denoted by capital letters of the English alphabet. Another common nominal variable is the rhesus factor (Rh-F) coded as $\{+, -\}$. The labels $+$ and $-$ are chosen historically, but it could be any two distinct symbols, or numbers such as $\{1, 0\}$.

Definition 1.2 A variable that takes a value among a finite set of mutually exclusive codes that have no logical order among themselves is called a nominal variable, and the data that it generates is called *nominal data*.

Some nominal data are always coded numerically. Examples are binary support vector machines (SVMs) whose class labels are always coded as $(-1, +1)$, dummy variables in linear programming (coded as 0 or 1). Decision trees (DTs) and NNs use any type of numeric or nonnumeric codes for nominal data. Pearson's χ^2-statistic based goodness of fit tests use mutually exclusive and collectively exhaustive nominal categories that are numerically coded. It uses the contingency coefficient (Section 1.12, pp. 1–20) to find the correlation between two numerically coded nominal variables.

1.3.2 The Ordinal Scale

Some of the nominal variables can be ordered using the values they take on two or more subjects. Examples are the severity of an accident $= \{mild, severe,$

critical, deadly}, patient-type = {child, teenager, youth, adult, senile}, and various ratings = {poor, fair, good, excellent}.

Definition 1.3 Nominal data that can be logically ordered are called *ordinal data* and the corresponding variable is called an *ordinal variable*.

One of the relational operators is used to order them. For instance, poor < fair < good < excellent. Sometimes, this ordering is done literally. An example is the seasons = {spring, summer, autumn, winter}. These seasons repeat among themselves. Hence, it is called a *cyclic ordinal variable*. Other examples are the days of the week, months of the year, letters of an alphabet, and so on, which are literally ordered ordinal data. We could categorize ordinal data as alphabetically ordered, numerically ordered, and literally ordered; each of them can be cyclically ordered if there is a natural succession among them. Alphabetically ordered ordinal data use the precedence order among the letters of a natural language (e.g., English and Japanese) to decide which comes first.

Numerically coded ordinal data can be compared using relational operators. Hence, the median and mode are appropriate measures of location for them. We could also use the quartiles, percentiles, contingency coefficient, and so on.

1.3.3 The Interval Scale

Some of the ordinal data can be coded as integers or functions thereof (as in date data type).

Definition 1.4 If numerically coded ordinal data have the property that the differences between any two values represent equal difference in the amount of the characteristic measured, it is called an *interval data*.

In other words, all ordinal data with well-defined intervals are interval data. One common example is the date data type. A characteristic property of interval data is that there is no natural zero. The natural zero need not always be the zero point. This can differ among variables. As an example, the fever is measured in Fahrenheit or centigrade scales. The normal [305] human body temperature is 98.4°F. If a person has higher body temperature, we say that there is fever. Thus, the origin for fever is 98.4°F, as it is the cutoff limit.

1.3.4 The Ratio Scale

Interval data with a clear definition of zero are called *ratio data*. Thus, both the differences between data values and their ratios are meaningful. Common examples are the height and weight. If a father is 5 feet 4 inches and his sibling is 2 feet 8 inches, then the father is twice as tall as the sibling, and the same relationship holds irrespective of the unit of measurement. Hence even if we change to the metric system, the same relationship will hold. Other examples are the price of an article, speed of vehicles, capacity of disks or pen drives, and so on. The zero point for some ratio variables may never be materialized. For instance, consider the task of classifying a patient as overfat or underfat. This is done using the body mass index (BMI) measure.

The BMI is defined as follows:

$$BMI = (\text{weight in Kg.})/(\text{height in meter squared}). \qquad (1.1)$$

This is a ratio variable. Typical cutoff limit varies from country to country, but a BMI of 25 is a common cutoff limit for adults. If the person's BMI is above 25, we say that he or she is overfat. The BMI can never be zero (not even for newborns). This means that a zero BMI is hypothetical. As the square of a number between zero and one is less than itself, the cutoff will vary for infants shorter than 1 meter.

1.4 POPULATION VERSUS SAMPLE

Most statistical procedures are based on a random sample drawn from a "population" of interest. The meaning of statistical population is slightly different from the literal meaning of population. Literally, it denotes a group of living organisms that are often large in size. A statistical population can have temporary or permanent existence. It can be small in size. As examples, the set of all HTTP requests on the web on any day is a statistical population that is large in size. However, majority of HTTP requests last at most a few seconds. The set of GPS satellites in orbit is another population that is small in size, as also the set of atomic powered interplanetary spacecrafts. Each element of a population is assumed to be measurable on a set of variables. These variables often follow well-defined statistical laws.

Definition 1.5 The totality of all elements of interest in a study is called the *population.*

A statistical population may comprise animate or inanimate objects, symbols, entities, and so on. It may or may not be finite. It can be confined to a specific location or could be spread around a known or unknown locality. The first step in obtaining a sample is identifying the correct population. Hence, the population is often defined unambiguously by the experimenter for each research study. The study of elements of the entire population is called a *census*.

Illustration 1 Consider a study to correlate the marks obtained by students in a science subject to ownership of a computer or iPad and the use of the Internet. The population in this study is the totality of all students in that subject who owns a computer or iPad with Internet connectivity. This can be confined to a university, a country, a geographic region, and so on. Thus, the population could vary depending on the spread of the subjects.

Illustration 2 Toy manufacturers make toys for kids in specific age groups. In a study to find out toys that are injurious to kids in 1–5 age group, the population is the set of all toys for this age group. As measurements are taken on subjects, the true population is the totality of kids in the above-mentioned age group who use these toys.

Definition 1.6 A random sample is a true representative subset of a population, which is much smaller in size and each element of which generates recordable and meaningful data.

By the true representative, we mean that each and every element of the population has an equal chance of being included in our random sample. If the population is of finite size, sampling with replacement will ensure that the chance of drawing a sample from the population remains the same. The method in which a sampled item from a population is not replaced (back into the population) before the next item is drawn is called *sampling* without replacement. This does not matter for infinite populations. If the population size is finite and small, sampling without replacement results in a nonrandom sample.

Illustration 3 A foundation offered five scholarships to students in a college who secured distinction in their final exam. If there are 25 eligible students, we need to select a random sample of five students from among them. To preclude the possibility of a selected student receiving two scholarships, we need to do a sampling without replacement.

Random samples are much easier to work with because they are much smaller in size. Owing to some asymptotic properties of sample estimates, we often restrict the random sample size between 30 and a few hundred. Bigger sample sizes give better results in some statistical procedures. A researcher decides upon an optimal sample size using the cost of sampling an item, population characteristics, number of unknown parameters, sampling distribution of statistics, and so on. For instance, if the population has distinct data clusters, a technique called *stratified sampling* can be used to select a random sample from each cluster based on the cluster size. Sampling of elements from populations is an extensively studied field called *sampling theory* in statistics.

1.4.1 Parameter Versus Statistic

A parameter describes the population of interest. A population can have zero or more parameters. Consider the Cauchy distribution $f(x) = \frac{1}{\pi} \frac{1}{(1+x^2)}$ for $-\infty < x < \infty$. It has no parameters, although it describes a population[2]. Parameters are values that characterize the population. They may be a part of the functional form or the range of values assumed. For instance, a left-truncated Poisson distribution with truncation point k has the functional form

$$g(x; \lambda, k) = f(x; \lambda) / \left[1 - \sum_{j=0}^{k} P(j) \right] = e^{-\lambda} \lambda^x / \left[x! \left(1 - \sum_{j=0}^{k} P(j) \right) \right], \quad (1.2)$$

where $P(j) = e^{-\lambda} \lambda^j / j!$ Here k is a range parameter. The location (central value) and spread (scale) are the most important characteristics of a population. They may either be described as functions of the same set of parameters (as in χ^2 distribution with $\mu = n, \sigma^2 = 2n$, where n is the degrees of freedom; or the Poisson law with $\mu = \sigma^2 = \lambda$)

---

[2]The more general form of Cauchy's distribution is $f(x; a, b) = \frac{K}{\pi} \frac{1}{a^2 + (x-b)^2}$, with two parameters a and b.

or by separate parameters as in the univariate normal law $N(\mu, \sigma^2)$ with mean μ and variance σ^2; the Laplace law which has PDF

$$f(x; a, b) = (1/2b) \exp (-|x - a|/b), \quad b > 0, \quad -\infty < a < \infty, \tag{1.3}$$

with $\mu = a$ and $\sigma^2 = 2b^2$ or the logistic law with CDF

$$F(x; a, b) = 1/[1 + \exp (-(x - a)/b)] \text{ with } \mu = a \text{ and } \sigma^2 = \pi^2 b^2/3. \tag{1.4}$$

For several distributions, the mean and variance are complex functions of the parameters. As examples, the BINO(n, p) distribution has mean $\mu = np$ and variance $= \mu * q = npq$. Similarly, the noncentral chi-square distribution has mean $\mu = n + \lambda$ and variance $\sigma^2 = 2(\mu + \lambda) = 2n + 4\lambda$. The shape of the distribution can be very sensitive to the parameters. One example is the beta distribution BETA-I(p, q) with mean $\mu = p/(p + q)$, which is asymmetric in the parameters p and q and variance

$$\sigma^2 = \mu \, q/(p + q + 1) = pq/[(p + q)(p + q + 1)], \tag{1.5}$$

which is symmetric in p and q. If p is increased or decreased by keeping q fixed, the distribution changes shape rapidly. The logarithmic distribution

$$f(x; p, q) = pq^x/x, \quad 0 < q < 1, \quad x = 1, 2,... \tag{1.6}$$

has $\mu = pq/(1 - q)$ and $\sigma^2 = \mu^2(\frac{1}{pq} - 1)$. This discussion shows that various parameters contribute differently to the mean and variance in particular and to other moments in general. This complexity can be used as a measure of the parameter dependence on the shape of distributions. This also affects the asymptotic convergence behavior of various distributions to other standard distributions. When there exist three or more parameters, they may all contribute as functions to the location and scale. For example, the noncentral F distribution NCF(m, n, λ) has mean $\mu = \frac{n}{n-2} \frac{m+\lambda}{m}$, for $n > 2$, which is linear in λ, and nonlinear in the degrees of freedom parameters. If the population parameters are unknown, they are estimated from a sample drawn from that population.

Definition 1.7 A well-defined function of the sample values is called a *statistic*.

By our definition, a statistic does not involve the unknown population parameters, but it could involve the sample size or any function of it. For example, if X is a random sample of size n from a population with sample mean $\bar{x}$, then $n - \bar{x}$ is a statistic as it involves the sample values and sample size. The unknown population parameters are estimated using a statistic (or a function of it, which is also a statistic if it does not involve unknowns). We have used the word "well-defined function" in our definition of statistic. This includes not only arithmetic and other mathematical functions but also special functions such as minimum and maximum of sample values and integer part of data values (ceil and floor functions in computer programming).

1.5   COMBINATION NOTATION

The combination notation is used in several statistical distributions. Examples include the binomial, negative binomial, hypergeometric, and beta-binomial distributions. We denote it as $\binom{n}{x}$, which is read as "n choose x." Other notations include ^nC_x and $C(n, x)$. This represents the combination of n things taken x at a time. Symbolically,

$$\binom{n}{x} = n!/[x! * (n - x)!] \text{ where } \binom{n}{0} = \binom{n}{n} = 1, \binom{n}{1} = n, \qquad (1.7)$$

where $n!$ is pronounced either as "n factorial" or as "factorial n" and abbreviated as "fact n." By convention $0! = 1! = 1$. As the numerator and denominator involve products of integers, it can be evaluated in multiple ways. Write the $n!$ in the numerator as $n * (n - 1)!$, and the $x!$ in the denominator as $x * (x - 1)!$ to get

$$\binom{n}{x} = \frac{n}{x} * \binom{n - 1}{x - 1}. \qquad (1.8)$$

As it represents the number of ways to select x items out of n distinct items, $\binom{n}{x}$ is always an integer when n and x are integers. Formula (1.8) can result in approximations (for example, $\binom{5}{3}$ is evaluated as $(5/3) * (4/2) * 3 = 9.99$, instead of 10, owing to floating point truncations (5/3 is truncated to 1.6666). A solution is to use Pascal's identity

$$\binom{n}{x} = \binom{n - 1}{x} + \binom{n - 1}{x - 1}. \qquad (1.9)$$

As this involves only addition, it will always give an integer as the final result. A related notation used in negative binomial and negative hypergeometric distributions is $\binom{-n}{x} = (-1)^x \binom{n+x-1}{x}$.

1.6   SUMMATION NOTATION

The ubiquitous $\sum$ notation is extremely useful to express functions of sample values, random variables, complicated sequences, series, and so on in concise form. This section introduces several summation notations that are extensively used in the present and subsequent chapters. A good grasp of various summation notations is essential for students and practitioners of statistics and for people in many other disciplines such as algorithmics, numerical methods, digital signal processing, and parallel computing. It is extensively used in probability distributions (more so in discrete than in continuous distributions), sampling distributions, mathematical probability and expectations, generating functions, design of experiments, regression and correlation, contingency tables, and order statistics, to name a few. There are many variants to the summation notation. All of them starts with the Greek capital symbol $\sum$

(which is read as *sum*). Each $\sum$ has an associated index variable (indexvar)[3] and has an implied meaning of an iteration or enumeration of the expression that follows it (summand) over the range of the implicit indexvar (which is assumed to be an integer and is usually denoted by i, j, k, p, q, or r with or without subscripts). Thus, all of the following expressions are equivalent:

$$\sum_{j=1}^{n} X_j = \sum_{r=1}^{n} X_r = \sum_{s} X_s = X_1 + X_2 + \cdots + X_n. \tag{1.10}$$

The indexvar could also be explicit when the set notation is used for iteration. The subscript of $\sum$ denotes the initial values, conditions, or initializations, and the superscript denotes the terminal (final) values or conditions. The subscript, superscript, or both can also be missing, as in the above-mentioned example, if they can be inferred from the context.

In the great majority of applications, the indexvar varies from low values to high values. However, there are a few applications in which this can either be in the reverse (high to low) or two-way varying. As an example of reverse summing, consider the problem of accumulating binomial right-tail probabilities (survival function) until it accumulates to say c. This is mathematically expressed as

$$\sum_{x=k}^{n} \binom{n}{x} p^x q^{n-x} = c. \tag{1.11}$$

The indexvar here is chosen as x to indicate that we are summing probabilities. This can be easily accumulated by starting the summation at $x = n$ and iterating backward ($x = n - 1, x = n - 2, \cdots , x = k$) until the desired sum is accumulated (this can also be expressed in terms of the incomplete beta function; see page 6–34). Similarly, infinite Poisson-weighted distributions such as noncentral χ^2 and noncentral beta need to evaluate the Poisson weights $e^{-\lambda} \lambda^x / x!$ for large λ values. If the computations are carried out in single precision arithmetic, the first Poisson term $e^{-\lambda} \lambda^0 / 0! = e^{-\lambda}$ will result in a memory underflow (a machine will misinterpret it as zero) for $\lambda > 104$ and a memory overflow for $\lambda > 183.805$ (see [47], pp. 231–232 for details). This leads to error propagation, as each subsequent term is evaluated iteratively. As the mode of the Poisson distribution is λ, we could start the computations at λ and iterate in both directions [49, 51, 52]. This will allow one to compute such PDF and CDF for much higher λ values. Alternatively, we could use double precision arithmetic.

1.6.1   Nested Sums

Nested sums and iterations are usually denoted by multiple $\sum$ symbols[4]. For instance, $\sum_j \sum_k$ denotes a nested sum over two index variables j and k. When multiple $\sum$ occur

---

[3] also called *dummy variable* or *running variable*, but dummy variable has another meaning in linear programming and decision theory.
[4] If there are only two sums as $\sum_j \sum_k$, it is called a *double sum*.

as a block, they are evaluated from *right to left*. In other words, the leftmost indexvar is fixed and right indexvars are varied in the summand until they are complete, thereby accumulating the partialsum. Then, the leftmost indexvar is incremented using the step size, and the process is continued. One common example is Pearson's χ^2 statistic:

$$U = \sum_{j=1}^{r} \sum_{k=1}^{c} \frac{(o_{jk} - e_{jk})^2}{e_{jk}}, \qquad (1.12)$$

where o_{jk} are the observed and e_{jk} are the expected frequencies in an $r \times c$ table. The order in which nested sums are found is sometimes important (as in matrix multi-plications). In the above-mentioned sum, we iterate first over the (inner) k variable, followed by the (outer) j variable. Thus, j is set to its lowest value ($=1$), and k is varied over its full range (1 to c). Then, j is incremented by 1 and k is again varied over its full range, and so on. If the expression to be evaluated is symmetric in the dummy variables (as in equation 1.12), the order of summation can be interchanged. The summand may sometimes be tightly coupled with the indexvars. Consider the minimization criterion used in K-means clustering as

$$E = \sum_{j=1}^{k} \sum_{i=1}^{|C_j|} ||x_i^{(j)} - c_j||^2, \qquad (1.13)$$

where $|C_j|$ is the number of data points in cluster j, c_j's are the cluster centroids, k is the number of clusters, and $x_i^{(j)}$ denotes ith data value in jth cluster. Here, the summand is tightly coupled with the outer indexvar. Thus, we cannot easily interchange the summations.

When there are several nested summations in one block, each of them must be assigned a *unique* index variable. The variable names do not actually matter, as they are dummy indexvars. Any constant multiplier(s) that does not depend upon the indexvars could be taken outside all summations. Any multiplicand expressions that do not depend on the inner indexvars could be moved as much outside as possible such that their dependence is only on the indexvars to their left. For example, consider the sum

$$S_1 = \sum_{j=1}^{k} \sum_{i=1}^{m} c * u_i * v_j. \qquad (1.14)$$

As v_j is independent of i and c is a constant, we could rewrite it as

$$S_1 = c * \sum_{j=1}^{k} v_j * \left(\sum_{i=1}^{m} u_i \right) \qquad (1.15)$$

(see Exercise 1.51). Similarly,

$$S_2 = \sum_{k} (a * c_k + b * d_k) = a * \sum_{k} c_k + b * \sum_{k} d_k. \qquad (1.16)$$

We could also combine multiple $\sum$ into a single $\sum$ if there is no scope for confusion. Thus, the above-mentioned double sum in equation (1.12) can also be written using $\sum_{j,k}$. However, multiple $\sum$ is the recommended notation, as it is easy to comprehend and useful to convert such expressions into computer programs. Moreover, the indexvars in inner sums are sometimes dependent on the outer indexvars (as in equation 1.13, where the upper limit of the inner indexvar "i" depends on the outer indexvar j. See also equations 1.30 and 1.31 in page 1–26 and 1–38 in page 1–30).

1.6.2 Increment Step Sizes

The increments of the indexvar are assumed to be in steps of 1 by default. This is true for most of the summations that we encounter in statistics and computer science. However, there exist some applications in engineering and numerical computing where the increments are fractions. If this increment is in steps of c ($\neq 1$), it is indicated at the middle of the $\sum$ symbol. The increment step c can in general be a multiple of an integer or a fraction. It can rarely grow exponentially in some applications (see Exercise 1.48 pp. 1–68).

1.6.2.1 Fractional Incremental Steps (FISs)   When the increment is a fraction, we could recode the indexvar to force it to be an integer and adjust the summand accordingly. For example, consider the summation $\sum_{j=1}^{n} f(j)$, where $f(j)$ is any arbitrary function of j (along with other parameters), the indexvar j varies in steps of $c = 0.5$, and n is an integer or half-integer >1. We could write it as $\sum_{j=1}^{2n-1} f((j+1)/2)$. Here, the indexvar has been "inflated" to vary from 1 to $2n - 1$ in steps of 1 and $(j+1)/2$ has been substituted in the summand to compensate for the inflated index. In general, if we need to increment j from u to v in steps of a proper fraction $c = 1/k$, then the indexvar is inflated to vary from u to $(v*k - u*(k-1))$, and the sum is evaluated as

$$S = \sum_{j=u}^{vk-uk+u} f(u(1 - 1/k) + j/k). \qquad (1.17)$$

As u and k are known constants, we could also write this as

$$S = \sum_{j=u}^{(v-u)k+u} f(k' + j/k), \qquad (1.18)$$

where $k' = u(1 - 1/k)$. Indeed, changing the indexvar as $i = j - u$, this could also be written in the alternative form

$$S = \sum_{j=u \ \text{step}(1/k)}^{v} f(j) = \sum_{i=0}^{(v-u)k} f(u + i/k), \qquad (1.19)$$

which is much better suited for computer implementation.

■ **EXAMPLE 1.1   Simplified sum**

Simplify $\sum_j f(j)$ for index j from 1 to 3 in steps of 1/3, using the above-mentioned technique.

Solution 1.1   The seven possible values of j are $1, 1\frac{1}{3}, 1\frac{2}{3}, 2, 2\frac{1}{3}, 2\frac{2}{3}, 3$. Here $u = 1, v = 3, c = 1/3$ so that $k = 3$. Thus, the new sum is

$$\sum_{j=0}^{(v-u)k} f(u+j/k) = \sum_{j=0}^{(3-1)*3} f(1+j/3) = \sum_{j=0}^{6} f(1+j/3). \qquad (1.20)$$

■

1.6.2.2   *Integral Incremental Steps (IISs)*   If the indexvar increments in steps of 2, we evaluate the sum $\sum_{j=1}^{n} f(j)$ with step size $c = 2$ as

$$S = \sum_{j=1\ step\ 2}^{n} f(j) = \sum_{k=1}^{\lfloor (n+1)/2 \rfloor} f(2*k-1) = \sum_{k=0}^{\lfloor (n-1)/2 \rfloor} f(2*k+1), \qquad (1.21)$$

where $\lfloor x \rfloor$ denotes the *floor operator* that returns the largest integer less than x. If c is an integer > 2, we modify the sum as

$$S = \sum_{j=1\ step\ c>2}^{n} f(j) = \sum_{k=1}^{\lfloor (n+c-1)/c \rfloor} f(1+c*(k-1)) = \sum_{k=0}^{\lfloor (n-1)/c \rfloor} f(1+c*k).$$
$$(1.22)$$

These are used in discrete signal processing and transforms. In general, if we need to increment j from u to v in steps of an integer multiple $c \geq 2$, then the index is deflated to vary from 0 to $(v-u)/c$, and the sum is evaluated as

$$step\ c \sum_{j=u}^{v} f(j) = \sum_{j=0}^{\lfloor (v-u)/c \rfloor} f(u+c*j). \qquad (1.23)$$

A special case is accumulating the sum $\sum_{j=-k}^{k} f(j)$ in steps of size c. This can be unfolded as

$$\sum_{j=-k}^{k} f(j) = \sum_{j=0}^{\lfloor 2k/c \rfloor} f(c*j-k). \qquad (1.24)$$

Equation (1.24) is valid for both c an integer and a fraction. When c is a fraction, the upper limit $\lfloor 2k/c \rfloor$ is simply blown up.

■ **EXAMPLE 1.2   Simplified symmetric sum**

If j varies in fractional steps of 1/3, simplify the following sums:– (i) $S_1 = \sum_{j=-1}^{1} 1/(1+j^2)$, (ii) $S_2 = \sum_{j=-2}^{2} \sin(\pi*j)$.

Solution 1.2 (i) Using (1.24), the first sum has upper limit $\lfloor 2 \times 1/(1/3) \rfloor = 6$, so that it can be unfolded as $\sum_{j=0}^{6} f((1/3) * j - 1) = 1/2 + 1/(1 + 4/9) + 1/(1 + 1/9) + 1 + 1/(1 + 1/9) + 1/(1 + 4/9) + 1/2 = 2 * (1/2 + 9/13 + 9/10) + 1 = 5.184615$

(ii) Using (1.24), we get $S_2 = \sum_{j=0}^{\lfloor 2 \times 2/(1/3) \rfloor} \sin(\pi * (j/3 - 2))$. This simplifies to zero as positive and negative terms cancel out.   ∎

■ **EXAMPLE 1.3   Simplified double sum**

If j varies in fractional steps of 1/4, and k varies in integer steps of 3, simplify following sums:– (i) $S_1 = \sum_{j=2}^{5} \sum_{k=1}^{10} 1/(j + k)$, (ii) $S_2 = \sum_{k=1}^{10} \sum_{j=1}^{k} 1/(j + k)$.

Solution 1.3 In Case (i), $u = 2, v = 5$, and $c = 1/4$ so that we apply FIS first to get

$$S_1 = \sum_{j=0}^{(5-2)*4} \sum_{k=1}^{10} 1/(2 + (j/4) + k) = \sum_{j=0}^{12} \sum_{k=1}^{10} 1/(2 + (j/4) + k). \tag{1.25}$$

For the inner sum (indexvar k), we have $u = 1, v = 10, k = 3$ so that $(v - u)/k = 9/3 = 3$ and

$$S_1 = \sum_{j=0}^{12} \sum_{k=0}^{3} 1/(2 + (j/4) + (1 + 3 * k)) = \sum_{j=0}^{12} \sum_{k=0}^{3} 1/[3 + 3k + j/4]. \tag{1.26}$$

In Case (ii), we first apply FIS to get

$$S_2 = \sum_{k=0}^{(10-1)/3} \sum_{j=1}^{3*k+1} 1/(1 + j + 3 * k), \tag{1.27}$$

where the inner index still increments in step size 1/4. Next, we apply IIS to indexvar k to get

$$S_2 = \sum_{k=0}^{3} \sum_{j=0}^{12*k} 1/(2 + j/4 + 3 * k). \tag{1.28}$$
                                                                                     ∎

The most frequent use of $\sum$ notation in statistics is to denote the arithmetic sum of n quantities that are distinguished only by one or more subscripts. In the following discussion, we introduce the most common summation notations.

1. Subscript fully varying summation
    Consider the summation

$$\sum_{j=1}^{n} x_j = \sum x_j = x_1 + x_2 + \cdots + x_n, \tag{1.29}$$

where x_j is the jth value of the series $X[]$. Here, each of the x_j's are either known data values or random variables. The j is called the *summation variable* or *index of summation*. This notation is used in arithmetic means of samples, of random variables, in order statistics, and in probability theory for disjoint events. Because bivariate data are arranged as a matrix and identified by a row and a column, we could extend the above-mentioned notation as $\sum_{i=1}^{m} \sum_{j=1}^{n} x_{ij}$ to denote the sum of all of the $m \times n$ entries or values. This is sometimes compactly written as $\sum_i \sum_j x_{ij}$ or as $\sum_{i,j} x_{ij}$. The summation notation is also used in variances as $s^2 = \sum_{j=1}^{n} (x_j - \bar{x})^2/(n-1)$. Consider the summation $\sum_{j=1}^{n} (x_j - c)^k$ or $\sum (x_j - c)^k$ for short. This denotes the expanded sum $(x_1 - c)^k + (x_2 - c)^k + \cdots + (x_n - c)^k$, where x_j is the jth value of the series $X[]$, and c, k are constants. As in the above-mentioned case, we could extend this to bivariate data as $\sum_{j=1}^{n} x_j y_j$, where x_j and y_j are the values of two traits generated by the jth subject. When there are many traits, we may represent them by an additional subscript, rather than by separate variables as in $\sum_{i=1}^{m} \sum_{j=1}^{n} (x_{ij} - c)^k$ for a given k.

2. Subscript partially varying summation
   This is a variant of the above-mentioned notation, where we use $\neq$ or $\geq$ to restrict one or more summation indexes. Consider the problem of summing the elements in the upper triangular portion (above the diagonal) of a 2D array. If x_{ij} denotes the (i, j)th element, this sum is given by $S = \sum_{i=1}^{n} \sum_{j=i}^{n} x_{ij}$ or equivalently $\sum_{i=1}^{n} \sum_{j \geq i} x_{ij}$. In some applications, we may have to omit particular values of the summation index as in $\sum_{i=1}^{n} \sum_{j=1; j \neq i}^{n} x_{ij}$. As another example, the sample variance can be represented as

$$s_n^2 = 1/[n(n-1)] \sum_{i=1}^{n} \sum_{j>i}^{n} (x_i - x_j)^2, \tag{1.30}$$

where the inner indexvar depends on the outer indexvar. As this is symmetric in x_i and x_j, we could also write it as

$$s_n^2 = 1/[n(n-1)] \sum_{i=2}^{n} \sum_{j=1}^{i-1} (x_i - x_j)^2. \tag{1.31}$$

3. Summation over a set
   This is an extension of the above-mentioned notation. In some applications, we have distinct nonoverlapping subsets that makeup a set. We may have to accumulate some information about each of the subsets. The above-mentioned summation notation can be modified such that the summation index varies over each subset: $\sum_{j \in S_j} f(x_j)$. These sets can be specified either explicitly or implicitly using a condition. For instance, the set of all odd integers can be specified as "j is odd," where j is the summation index. Suppose that we are interested in finding the probability of an even number of heads when six coins are thrown. Let Y

denotes the event that an even number of Heads appear. The possible values of Y are $S_1 = 0, 2, 4, 6$. Here, $Y = 0$ indicates that all six coins ended up as Tails and six indicates that all of them were Heads. We know that this is solvable using the binomial distribution with $n = 6$. If p denotes the probability of a head showing up and $q = 1 - p$ denotes the probability of a tail showing up, we are interested in finding the probability

$$P(Y \text{ is even}) = \sum_{j \text{ even}} \binom{n}{j} p^j q^{n-j} = \sum_{j \in S_1} \binom{n}{j} p^j q^{n-j}. \qquad (1.32)$$

4. Function of summation variable
Consider the simple summation $\sum_{j=1}^{n} j$ or $\sum_{j=1}^{n} j^2$. Here, our summand (the quantity summed) is either j itself or a function of it. A more complicated example is the tail areas of Poisson probability defined as $P_j(\lambda) = e^{-\lambda} \lambda^j / j!$ or the binomial density $b_j(n, p) = \binom{n}{j} p^j (1 - p)^{n-j} = (1 - p)^n \binom{n}{j} (p/q)^j$, for $j = 0, 1, \cdots, n$, and $q = 1 - p$. The sum of the probabilities on the left tail is called the CDF and that on the right tail is called the *survival function*. Symbolically, $F_x(k, \lambda) = \sum_{j=0}^{k} e^{-\lambda} \lambda^j / j!$ is the CDF up to and including k, where k is a number in $(0, \infty)$. Similarly, the binomial survival function is given by $G_x(k, n, p) = 1 - F_x(k - 1, n, p) = \sum_{j=k}^{n} \binom{n}{j} p^j (1 - p)^{n-j}$. This notation will be used in Chapter 6.

5. Superscript varying summation
Consider the summation $\sum_{k=1}^{n} (x_j - c)^k$ or $\sum_k (x_j - c)^k$ for short. This denotes the expanded sum $(x_j - c)^1 + (x_j - c)^2 + \cdots + (x_j - c)^k$, where x_j, the jth value of the series X[], and c are constants. This notation is used in generating functions. In differential calculus, it denotes jth derivative as in $\frac{d^j}{dx^j} f(x) = (\frac{d}{dx})^j f(x)$, which is interpreted as applying the differential operator $(\frac{d}{dx})$ repeatedly j times. As another example, the jth derivative of cumulant generating function is

$$(\partial/\partial t)^j K_x(t) = (\partial/\partial t)^j \ln(M_x(t)) = \kappa_j + \kappa_{j+1}(t/1) + \kappa_{j+2}(t^2/(1 * 2)) + \cdots, \qquad (1.33)$$

from which by putting $t = 0$, we could separate out the jth cumulant. Putting $t = 1$ gives $(\frac{\partial}{\partial t})^j \ln(M_x(t))|_{t=1} = \sum_{r=0}^{\infty} \frac{\kappa_{j+r}}{r!}$ (see also 9.1 (pp. 9–2)). Note that the superscript may or may not mean powers. For instance, it means various *states* in stochastic processes and game theory. It could be negative in time-dependent autoregressive processes and discrete signal processing. For example, the generating function $A(z, n) = \sum_{j=1}^{n} a_j(n) z^{-j}$ denotes the autoregressive process $x(n) = -\sum_{i=1}^{n} a_i(n) x(n - i) + \sigma(n)$, where negative powers of z denote time lags to the past from the reference point. If it does not imply powers, we could enclose them in parenthesis to avoid confusion. A superscript could denote an omitted data value as in the Jackknife estimator of a parameter using a sample of size n as

$$J = n * t_n - (1/n) \sum_{j-1}^{n} t_{n-1}^{(j)}, \qquad (1.34)$$

where t_n is the estimate using all sample values and $t_{n-1}^{(j)}$ denotes the estimate without jth data value (using other n-1 values).

6. Combination summation

This type of summation may involve a combination of the above-mentioned two types. Consider the expression $M_x(t) = \sum_{j=0}^{\infty} (t^j/j!)\mu_j$, which is called the *ordinary moment generating function*. Here, the summation index appears as a subscript on μ, as superscript on the dummy variable t, and as a function $1/j!$. As another example, consider the noncentral beta distribution NCB(p, q, λ) with shape parameters (p, q), and noncentrality parameter $\lambda > 0$. The mean of this distribution is expressed as an infinite sum of Poisson-weighted central beta means as

$$\mu = \sum_{j=0}^{\infty} \frac{e^{-\lambda/2}(\lambda/2)^j}{j!} \frac{p+j}{p+q+j} \simeq 1 - \frac{q}{C}\left(1 + \frac{\lambda}{2C^2}\right), \tag{1.35}$$

where $C = p + q + \lambda/2$ [52]. Here, j appears as powers of $(\lambda/2)$ and as a function $\frac{p+j}{j!(p+q+j)}$ (see Table 8.3, pp. 8–41 in Chapter 8).

1.7 PRODUCT NOTATION

We have used the $+$ operator in the summation notation discussed earlier. There are many situations where we need to use the product ($*$) operator instead of $+$. Examples are the rising and falling factorial moments, geometric mean (Section 2.7, pp. 2–29), likelihood function in maximum likelihood estimation (MLE) of statistical parameters, multivariate distribution theory, conditional distributions, multiple and partial correlations, and some special numbers. It is also used in inverse discrete Fourier transforms, numerical interpolation, and many other engineering fields. If $x_1, x_2, \cdots, x_n$ are nonzero numbers (positive or negative), their product is denoted as $P = \prod_{j=1}^{n} x_j$. By taking logarithm of both sides, we get $\log(P) = \sum_{j=1}^{n} \log(x_j)$, using $\log(xy) = \log(x) + \log(y)$. It immediately follows that

$$\prod_{j=1}^{n} x_j^{b_j} = \exp\left(\sum_{j=1}^{n} b_j * \log_e(x_j)\right). \tag{1.36}$$

The objective function in geometric programming (GP) is the *posynomial* $f(x) = \sum_{t=1}^{T} c_t \prod_{j=1}^{N} x_j^{a_{jt}}$, where $c_t > 0, a_{jt}$ are real and $x_j > 0 \,\forall j$. Applying equation 1.36 immediately gives

$$f(x) = \sum_{t=1}^{T} c_t * \exp\left(\sum_{j=1}^{N} a_{jt} * \log_e(x_j)\right). \tag{1.37}$$

The multinomial probabilities are expressed as $\frac{n!}{x_1! . x_2! \cdots x_n!} p_1^{x_1} . p_2^{x_2} \ldots p_n^{x_n}$, where p'_js are probabilities that add up to 1 and '.' denotes multiplication ($*$) (see Chapter 6). This can be concisely written as $P = \frac{n!}{\prod_{j=1}^{n} x_j!} \prod_{k=1}^{n} p_k^{x_k}$. By taking log (to base e) and using equation 1.36, this becomes

$$P = \exp\left(\sum_{j=2}^{n} \log_e(j) + \sum_{k=1}^{n} x_k * \log_e(p_k) - \sum_{j=1}^{n} \sum_{k=1}^{x_j} \log_e(k)\right),$$

where we have used $\log(1) = 0$, and $\log_e(x_j!) = \sum_{k=1}^{x_j} \log_e(k)$ as x_j's are integers. This can be written as

$$\frac{n!}{x_1! . x_2! \cdots x_n!} p_1^{x_1} . p_2^{x_2} \cdots p_n^{x_n} = \exp\left(\sum_{j=2}^{n} \ln(j) + \sum_{k=1}^{n} x_k * \ln(p_k) - \sum_{j=1}^{n} \sum_{k=1}^{x_j} \ln(k)\right).$$
$$(1.38)$$

For an MLE example, let $f_x(x, \theta)$ be the PDF of a distribution from which a sample $x_1, x_2, \cdots, x_n$ of size n is drawn. Then, the likelihood function is $L(x_1, x_2, \cdots, x_n; \theta) = \prod_{j=1}^{n} f_x(x_i, \theta)$. The unknown parameters are then estimated by maximizing the likelihood or equivalently maximizing the log-likelihood.

Expressions independent of the indexvars can be taken outside all independent products. For example, consider the product $P = \prod_{j=0}^{k} \prod_{i=0}^{m} c * u_i * v_j$. As v_j is independent of i and c is a constant, we could rewrite it as

$$P = \prod_{j=0}^{k} \prod_{i=0}^{m} c * u_i * v_j = c^{k+m+2} * \prod_{j=0}^{k} v_j^{m+1} * (\prod_{i=0}^{m} u_i) \qquad (1.39)$$

(note that j varies $k+1$ times and i varies $m+1$ times). As the expression within the product[5] is symmetric in i and j, we could also write it as $P = c^{k+m+2} \prod_{i=0}^{m} u_i^{k+1} * (\prod_{j=0}^{k} v_j)$. In the particular case when $u = v$ ($u_j = v_j$), this simplifies to

$$P = c^{k+m+2} * \begin{cases} \prod_{j=0}^{k} u_j^{k+m+2} \prod_{i=k+1}^{m} u_i^{k+1} & \text{if } k < m; \\ \prod_{j=0}^{m} u_j^{k+m+2} \prod_{i=m+1}^{k} u_i^{m+1} & \text{if } m < k; . \\ \prod_{j=0}^{m} u_j^{2(m+1)} = (\prod_{j=0}^{m} u_j^2)^{(m+1)} & \text{if } m = k. \end{cases}$$

1.7.1 Evaluating Large Powers

Expressions of the form $x^n, (1-x)^n$, or λ^x occur in many probability distributions such as gamma, beta, Weibull, Pareto, power series, and Poisson distributions. In some applications, we need to compute the PDF for just a few x values for a large

---

[5]We denoted the expression within the sums as *summand*. Strictly speaking, we cannot use summand here as it is enclosed by product symbol. A better word might be *prodand*.

fixed n or large integer values of x for fixed λ. There are many ways to evaluate them. The following example first considers an efficient method when the power n is large and then explains the computational details when both x and n are large.

■ EXAMPLE 1.4

Evaluate x^n where n is a large integer.

Solution 1.4 We consider two cases depending on whether n is a power of 2 or not. (i) Let $n = 2^m$, where m is an integer. We could evaluate it as $x^{n/2} * x^{n/2}$, where each of the terms are recursively evaluated. ■

Case (ii): n is not a power of 2. If n is of the form $2^k \pm v$, where v is a small number (say 1,2, or 3), we could still utilize Case (i). As examples, $n = 15 = 2^4 - 1$, so that $x^{15} = x^{16}/x$, and $x^{67} = x^{64} * x^3$. Otherwise, we convert n into its binary representation as $n = b_k b_{k-1} \cdots b_1 b_0$, where b_0 is the least significant bit (LSB) and b_k is the most significant bit (MSB). The n and b's are connected by $n = \sum_{j=0}^{k} b_j 2^j$, where we have rearranged the summation index to match significant bits from right to left. Substitute for n to get $x^n = x^{\sum_{j=0}^{k} b_j 2^j}$. Using $x^{m+n} = x^m * x^n$, this becomes $x^{b_0} * (x^2)^{b_1} * \cdots (x^{2^k})^{b_k}$. This can be written as $\prod_{j=0}^{k} (x^{2^j})^{b_j}$. As the powers of all x's are of the form 2^k, the case (i) applies for large k. Because b'_js are binary digits (0 or 1), we need to evaluate the expressions for $b_j = 1$ only (as $(x^{2^j})^0 = 1 \forall j$). Hence, the above-mentioned product could also be expressed as $x^n = \prod_{j=0, b_j=1}^{k} (x^{2^j})^{b_j}$. Sequential algorithms convert a decimal number ($n > 0$) into its binary representation using the *repeated division by base* ($=2$) method. This generates the binary digits from LSB to MSB. We could then check each and every bit to see if it is 1 and accumulate the corresponding product term $(x^{2^j})^{b_j}$ immediately. This is especially useful in cryptography applications that work with expressions of the form $x^p - 1$ where p is a very large prime number. As fractions are converted into binary using the *repeated multiplication by base* method, the above-mentioned discussion is equally applicable to evaluate expressions involving $x^{1/n}$ too, where n is large.

In the particular case when x and n are both very large, we break x using the prime factorization theorem into $x = p_1^{n_1} p_2^{n_2} \cdots p_m^{n_m} = \prod_{i=1}^{m} p_i^{n_i}$, where p_i's are prime numbers and n_i's are integers (≥ 1). Then $x^n = (\prod_{i=1}^{m} p_i^{n_i})^n$. Substitute $n = \sum_{j=0}^{k} b_j 2^j$ to get $x^n = (\prod_{i=1}^{m} p_i^{n_i})^{\sum_{j=0}^{k} b_j 2^j}$. This can be simplified into the form

$$x^n = \prod_{i=1}^{m} (\prod_{j=0}^{k} (p_i^{2^j})^{b_j})^{n_i}, \qquad (1.40)$$

where the inner product is carried out only for $b_j = 1$. Note that $p_i^{2^{j+1}} = p_i^{2^j} * p_i^{2^j}$. Hence, these can be kept in an array and updated in each pass.

We could also combine multiple products as well as sums and products. Multiple products are, however, not of much use in engineering statistics. The index of summation and products could also be a function (say $R(j)$) of the respective indexvars as used in [157], page 27–36. An extension is to allow the step size to be any integer >1. For example, when the step size is 2, we get

$$x_{(j,2)} = x * (x - 2) * \cdots * (x - 2j + 2) = \text{step } 2 \prod_{k=0}^{j-1} (x - 2 * k). \tag{1.41}$$

1.8 RISING AND FALLING FACTORIALS

This section introduces a particular type of the product form presented earlier. These expressions are useful in finding factorial moments of discrete distributions whose PDF involves factorials or binomial coefficients. In the literature, theseare known as *Pochhammer's notation* for rising and falling factorials. This will be explored in subsequent chapters.

1. Rising Factorial Notation
   Factorial products come in two flavors. In the rising factorial, a variable is incremented successively in each iteration. This is denoted as

$$x^{(j)} = x * (x + 1) * \cdots * (x + j - 1) = \prod_{k=0}^{j-1} (x + k) = \frac{\Gamma(x + j)}{\Gamma(x)}. \tag{1.42}$$

2. Falling Factorial Notation
   In the falling factorial, a variable is decremented successively at each iteration. This is denoted as

$$x_{(j)} = x * (x - 1) * \cdots * (x - j + 1) = \prod_{k=0}^{j-1} (x - k) = \frac{x!}{(x - j)!} = j! \binom{x}{j}. \tag{1.43}$$

Writing equation (1.42) in reverse gives us the relationship $x^{(j)} = (x + j - 1)_{(j)}$. Similarly, writing equation (1.43) in reverse gives us the relationship $x_{(j)} = (x - j + 1)^{(j)}$.

1.9 MOMENTS AND CUMULANTS

Moments of a distribution are denoted by the Greek letter μ. They have great theoretical significance in statistical distribution theory. Moments can be defined about any arbitrary constant "c" as

$$\mu_n(c) = E(x - c)^n = \begin{cases} \sum_x (x - c)^n f(x), & \text{for discrete distributions} \\ \int_x (x - c)^n f(x) dx, & \text{for continuous distributions.} \end{cases} \tag{1.44}$$

If $c = 0$ in this definition, we get raw moments, and when $c = \mu_1 = \mu$, we get the central moments.

The corresponding sample moments are analogously defined as $m_k(c) = \sum_{j=1}^{n} (x_j - c)^k f(x_j)$. The population moments may not always exist, but the sample moments will always exist. For example, the mean of a standard Cauchy distribution does not exist, however, the sample mean exists. The moment generating function (MGF) provides a convenient method to express population moments. It is defined as

$$M_x(t) = E(e^{tx}) = 1 + \frac{t}{1!}\mu_1 + \frac{t^2}{2!}\mu_2 + . \tag{1.45}$$

The logarithm of the MGF is called cumulant generating function (KGF). Symbolically,

$$\log(M_x(t)) = K_x(t) = \kappa_1 t + \kappa_2 t^2/2! + \cdots + \kappa_r t^r/r! + .. \tag{1.46}$$

See also equation (1.33) in pp. 1–31. L-moment is an extension that is discussed in Ref. 145. A thorough discussion of generating functions appears in Chapter 9.

1.10   DATA TRANSFORMATIONS

Data transformation is used in various statistical analyses. It is especially useful in hand computations when the numbers involved are too large or too small. Computing summary measures such as mean and variance of large numbers can be simplified using linear transformation techniques discussed in the following. Similarly, as the test statistic in analysis of variance (ANOVA) computations involves the ratio of sums of squares, a change of scale transformation is applicable. If the spread (variance) of data are too large or too small, an appropriate change of scale transformation can ease the visualization of data.

1.10.1   Change of Origin

As the name implies, this shifts all data points linearly (by subtracting or adding a constant from each sample value) as $Y = X - c$. The constant c (positive or negative) is preferably an integer when sample values are all integers.

■ EXAMPLE 1.5   Reservoir inflow

The amount of water inflow into a reservoir during 6 hours in cubic feet is $X = \{286, 254, 242, 247, 255, 270\}$. Apply the change of origin method and find the mean.

Solution 1.5 We subtract 240 (chosen arbitrarily) from each observation to get $x_i' = x_i - 240$ as $X' = \{46, 14, 2, 7, 15, 30\}$, from which $\sum_i x_i' = 114$. The mean of X' is $\bar{x}' = 114/6 = 19$. The mean of the original data is $240 + 19 = 259 = \bar{x}$. ∎

It is trivial to prove that the range of original data is preserved by a change of origin transformation because $(x_n - k) - (x_1 - k) = x_n - x_1$ and $(x_n + k) - (x_1 + k) = x_n - x_1$.

1.10.2 Change of Scale

This technique divides each large observation by the same constant (>1). This is useful when numbers are large and has high variability. Examples are family income, total insured amounts, annual insurance premiums, defaulting loan amounts, advertising expenses in various media and regions, and so on. Let $x_1, x_2, ..x_n$ be "n" sample values and c be a *nonzero* constant. Define $y_i = x_i/c$. If c is less than the minimum of the observations, each of the y_i's are greater than 1, if x_i's are positive. Similarly, if c is greater than the maximum of the observations, each y_i's are less than 1. For values of c between minimum and maximum of the sample, we get values on the real line (positive real line if all x_i's are positive). If all values are small fractions, we may multiply by a constant to scale them up.

■ EXAMPLE 1.6   Change of scale to find the mean

Error measurements of a device are as follows. Scale the data and find the mean:-
$X = \{0.001, 0.006, 0.0095, 0.015, 0.03\}$.

Solution 1.6 Choose $c = 1000$ and scale using $y_i = cx_i$ to get $Y = \{1, 6, 9.5, 15, 30\}$. This is an example of decimal scaling in which the decimal point is moved by multiplying/dividing by a power of 10. The mean of Y is $\bar{y} = 61.5/5 = 12.3$ and thus the mean of X is $\bar{x} = 0.0123$. ■

1.10.3 Change of Origin and Scale

This is the most frequently used technique to standardize data values. Depending on the constants used to change the origin and scale, a variety of transformed intervals can be obtained.

Theorem 1.1 A sample in the range (a, b) can be transformed to a new interval (c, d) by the transformation $y = c + [(d - c)/(b - a)] * (x - a)$.

Proof: By putting $x = a$ in the expression gives $y = c$. Putting $x = b$ gives $y = c + [(d - c)/(b - a)] * (b - a) = c + (d - c) = d$. As $(x - a)/(b - a)$ and $(y - c)/(d - c)$ both map points in the respective intervals to the $[0,1]$ range, all intermediate values in (a, b) get mapped to a value in (c, d) range. This proves the result. ■

■ EXAMPLE 1.7

Amount of fluoride (in milligrams) in drinking water collected from six places are [60, 90, 118, 150, 165, 170]. Transform the data to the range [10, 60].

Solution 1.7 Here $a = 60, b = 170, c = 10, d = 60$. Thus $(d - c)/(b - a) = (60 - 10)/(170 - 60) = 5/11$. Hence, the required transformation is $y_i = 10 + (5/11)(x_i - 60)$. Substitute each successive value of x to get $Y = [10, 23.6364, 36.3636, 50.9091, 57.7273, 60]$. ∎

Corollary 1 Prove that the sample x in the range (a,b) can be transformed to a new interval (c,d) by the transformation

$$y_i = \frac{1}{(b - a)}[(d - c) * x_i + (bc - ad)]. \tag{1.47}$$

Proof: Write $c + [(d - c)/(b - a)] * (x_i - a)$ as $c + (x_i - a) * [(d - c)/(b - a)]$. Distribute $(x_i - a)$ as two products to get $[c + x_i * (d - c)/(b - a) - a * (d - c)/(b - a)]$. Take $(b\text{-}a)$ as the common denominator. The first and third expressions simplify to $(bc - ac - ad + ac)/(b - a)$. Cancel out "$ac$" to get $(bc - ad)/(b - a)$. Combine with the second expression and take the common denominator outside to get the required result. ∎

1.10.4   Min–Max Transformation

This transformation is used to map any sample values to the interval $[0, 1]$, $[-1, +1]$, and so on.

Theorem 1.2 Any numeric variable x in the interval (x_{min}, x_{max}) can be transformed to a new interval $[0, 1]$ by the transformation $y = (x - x_{min})/R$, where $R = (x_{max} - x_{min})$ is the range and x_{min}, x_{max} are the minimum and maximum of the sample values.

Proof: Substituting $x = x_{min}$ gives $y = 0$ and $x = x_{max}$ gives $y = 1$. Hence, the transformed values are mapped to $[0, 1]$. ∎

▣ **EXAMPLE 1.8**

Transform the fluoride data in page 1–38 into the $[0,1]$ range.

Solution 1.8 As the minimum is 60 and maximum is 170, the required transformation is $y = (x - 60)/(170 - 60)$. This gives $Y = [0, 0.2727, 0.5273, 0.8182, 0.9545, 1]$. ∎

Lemma 1 A sample x in any finite range can be mapped to the interval $[-1, +1]$ by a simple change of origin and scale transformation.

Proof: Consider the transformation $y = 2 * [(x - x_{min})/R] - 1$, where $R = (x_{max} - x_{min})$. When $x = x_{min}, y$ becomes -1 and when $x = x_{max}, y = +1$. All intermediate values are mapped to points within the interval $[-1, +1]$. Thus the result. This holds even if x_{min} is negative. ∎

■ EXAMPLE 1.9

Transform the data [34, 43, 55, 62, 68, 74] to the interval $[-1, +1]$.

Solution 1.9 Here, the minimum is 34 and maximum is 74. Thus, the range is 40. Using Lemma 1, we get the transformed data as $y = [2 * (x - 34)/40] - 1 = [-1, -0.55, 0.05, 0.40, 0.70, +1]$. ■

Theorem 1.3 A sample x in any finite range with at least two elements can be mapped to the range $Y = [-k, +k]$ by the transformation

$$y_i = \frac{k}{R}[2 * x_i - (x_{min} + x_{max})], \quad \text{where} \quad R = (x_{max} - x_{min}). \tag{1.48}$$

Proof: Putting $x = x_{min}$ gives $y = \frac{k}{R}[2 * x_{min} - (x_{min} + x_{max})] = \frac{k}{R}[x_{min} - x_{max}] = -k$ because $R = (x_{max} - x_{min})$. Putting $x = x_{max}$ gives $y = \frac{k}{R}[2 * x_{max} - (x_{min} + x_{max})] = +k$. All intermediate values are mapped to the interval $(-k, +k)$. For instance $x_c = (x_{min} + x_{max})/2$ gets mapped to 0. This proves the result. ■

■ EXAMPLE 1.10

Transform the above-mentioned data into the intervals $[-0.5, +0.5]$, and $[-3, +3]$.

Solution 1.10 Here $k = 0.5$, so that $k/R = 0.5/(74 - 34) = 0.0125, x_{min} + x_{max} = 74 + 34 = 108$, giving the transformation $y = 0.0125 * (2 * x - 108)$. Resulting y vector is $[-0.5, -0.275, 0.025, 0.20, 0.35, 0.5]$. For the $[-3, +3]$ range, $k = 3$ and $k/R = 3/40 = 0.075$, giving the transformation $y = 0.075 * (2 * x - 108)$. Resulting values are $[-3, -1.65, 0.15, 1.2, 2.1, 3]$. ■

Remark 1 In the particular case when the sample has just two elements, they are mapped exactly to $-k$ and $+k$ respectively.

Proof: Consider a sample (x, y) of size two. Rearrange them such that $x_{min} = x, x_{max} = y$ or vice versa. Substituting in equation 1.3, x_{min} and x_{max} gets mapped to $-k$ and $+k$, respectively. ■

■ EXAMPLE 1.11

The incomes of six families are [34,000, 43,000, 55,000, 62,000, 68,000, 74,000]. Transform the data to the interval $[-1, +1]$.

Solution 1.11 We will choose $c = 10,000$ and divide each value by c to get $X' = [3.4, 4.3, 5.5, 6.2, 6.8, 7.4]$. Here $R = (7.4 - 3.4) = 4, k = 1$ and $(x_{min} + x_{max}) = 3.4 + 7.4 = 10.8$, giving the transformation $y = 0.25 * (2 * x - 10.8)$ where x varies over the original data values. The resulting y vector is $[-1, -0.55, 0.05, 0.40, 0.70, +1]$. ■

1.10.5 Nonlinear Transformations

Linear data transformations may be insufficient in some engineering applications. The popular nonlinear transformations are square-root transformation, trigonometric and hyperbolic transformations, logarithmic and exponential transformations, power transformations, and polynomial transformations. These transformations are used either to stabilize the variance of the data or to bring the data into one of the well-known distributional form.

1.10.6 Standard Normalization

This transformation is so called because it is extensively used in statistics to standardize arbitrary scores. Here, the origin is changed using the mean of the sample, and the scale is changed using the standard deviation of the sample. Symbolically $y_i = (x_i - \bar{x})/s$, where s is the standard deviation. The resulting values of y are called z-scores and will almost always lie in the interval $[-3, +3]$. A disadvantage of this transformation is that it uses the mean and variance that need a single pass through the data. If standard normalization is applied manually, a quick check can be carried out as follows. If the sum of the z-scores is nonzero, it is an indication that either the calculation is wrong or error has propagated. Ideally, we expect the sum of the z-scores to be less than a small number (say <0.00001).

■ **EXAMPLE 1.12 Shear strength of bonded joints**

The shear strength of bonded joints (in MPa) are $X=(22, 30, 81, 26, 44, 29, 61, 35)$. Apply the standard normalization.

Solution 1.12 The sum of the data is 328, from which the mean is 41. Sum of squares is 16,344 so that the variance is 413.7142857 (and s is 20.33996769). Thus, the transformation $y=(x-41)/20.33996769$, which gives $Y=(-0.93412, -0.54081, 1.96657, -0.737464, 0.14749, -0.58997, 0.983286, -0.294986)$. ■

■ **EXAMPLE 1.13 Carbon nanoparticles**

The amount of carbon particles in a nano-device is $X = \{32, 148, 21, 940, 36, 182, 39, 276, 14, 260, 43, 769, 25, 313, 25, 312\}$. Compute the z-scores.

Solution 1.13 As the numbers are large, apply a change of scale transformation. Divide each data by 10,000 to get $S'=\{0.32148, 0.2194, 0.36182, 0.39276, 0.1426, 0.43760, 0.25322, 0.25312\}$. The mean of scaled data is 0.29775, and variance is $s^2 = 0.009634$, from which the standard deviation is obtained as 0.098152944. The corresponding z-scores are easily found as $[00.24176, -0.79824, 0.65276, 0.967979, -1.580696, 1.42482, -0.4536797, -0.454698]$. The corresponding z-scores for S are also the same, which can be verified using the transformation $z = (x - 29775)/9815.294$. ■

1.11   DATA DISCRETIZATION

As the name implies, discretization (also known as *binning*) is the process of categorizing a continuous variable (called *source variable*) measured in the *interval* or *ratio scale* of NOIR typology into a small number of groups (called *bins*) with minimal loss of information.

Definition 1.8 DDAs divide the global range of a continuous attribute into nonoverlapping and piece-wise continuous intervals in an optimal way, where each continuous interval is assigned a categorical label.

Univariate DDA has only one source variable with well-defined logical boundaries (upper and lower). Continuous periodic data are discretized using a technique called *sampling using Shannon's law*. We will consider only aperiodic functions in the rest of the chapter. These boundaries can also be $\mp\infty$. Nothing is assumed on the distribution of the source variable—it can be uniformly distributed over its range or can follow one of the other statistical laws.

1.12   CATEGORIZATION OF DATA DISCRETIZATION

The DDA can be classified into the following categories—(i) supervised, semisupervised, or unsupervised; (ii) global or local, and (iii) static or dynamic. Supervised DDA explores the class information (category labels) in the data intensively. Entropy-based binning and purity-based binning are supervised algorithms. Static DDA discretizes each attribute independently without regard to attribute interactions. Dynamic DDA on the other hand searches for all attributes simultaneously and takes care of attribute interactions. In the following discussion, we will use a simple parenthesis '(' to denote an open interval and a '[' to denote a closed interval. We have a choice of either keeping the right margin open, except for the last bin, or keeping the left margin open, except for the first bin.

1.12.1   Equal Interval Binning (EIB)

This method is also called equal width binning (EWB). It is the simplest unsupervised DDA, as it does not use the class label information of training data. Moreover, it does not require data sorting. Only inputs to this algorithm are the *minimum* x_1 and *maximum* x_n of n observations, and a user-supplied constant $k(\geq 2)$ that represents the number of bins. This minimum and maximum can be found without data sorting, either using a single iteration over the data or using a recursive divide-and-conquer strategy. The range $R = x_n - x_1$ is then divided into k equal width intervals (say $S = R/k$), so that $x_n = x_1 + k * S$. The ith interval is then given by $(x_1 + (i-1) * S, x_1 + i * S)$ for $i = 1, 2, \cdots, k$, where boundaries are properly taken care of. A disadvantage of EIB is that it is sensitive to data outliers on both sides. A simple solution is to use percentiles of the data X, say P_5 and P_{95}, as the minimum and maximum and use it in the range calculation, so that $S = (P_{95} - P_5)/k$. The leftmost and rightmost intervals can finally be made unequal widths as $[x_1, P_5 + S)$

and $[P_{95} - S, x_n]$. In this case, the intervals 1 and k are unequal and all others are equal width. This does not matter as we assign categorical labels to these intervals.

It is well known that excessive fat intake and a compulsion to over-eat are the major contributing factors in the pathogenesis of obesityx. Sedentary life styles and fatty food eating habits make many people overfat in developed countries, resulting in a negatively skewed BMI distribution. The BMI of most adults varies between 17 and 35. This of course is country specific. An ideal BMI value is a key indicator of the overall health and fitness of an individual. Too low or too high BMI values quite often indicate ill-health. Very low values can be due to immunity-related illnesses and anemia. Some genetic disorders and addiction to fatty food can result in very high BMI values (there are less than a dozen genetic markers that increase the BMI and contribute to obesity). Those with BMI values between 30 and 35 are called *obese*. Those above 35 BMI are called *morbidly obese*, for which surgical options (bariatric surgery) are available. Management of obesity is important in adolescence as it could lead to heart problems in later life.

Because the normal BMI range for adults is a narrow interval, we expect a smaller fraction of the people to fall in this range than in the other ranges. An EWD will probably produce wrong results, as the "normal" range is too narrow. An EFB could give better results if the population were naturally divided equally among under-fats, normals, and over-fats. In addition, as females are in general shorter than males, the BMI distribution for males and females is different. Thus, the proportion of females in our sample could impact the binning boundary. Note that the BMI-based categorization of an individual into the three body-fatness groups is very explicit as there are no overlaps. On the contrary, consider binning a group of people into {diabetic, nondiabetic} categories based on the BMI value. All overfat people are not diabetic, and there are few underfat diabetic patients too. Thus, the classes have high overlaps. The extent of the overlap determines the error rate in binning. The entropy-based binning algorithms can give good binning in this kind of situations if the class labels of the training data are exactly known.

Instead of fixing the lower and upper cutoff at P_5 and P_{95}, we could arbitrarily choose two percentiles, which are often taken symmetrically (this is not necessary if the distribution is highly skewed). If P_l and P_{100-l} are the lower and upper percentiles, the above-mentioned formula becomes

$$[x_1, (1 - 1/k)P_l + (1/k)P_{100-l}), (P_l + (i - 1)(P_{100-l} - P_l)/k, P_l + i(P_{100-l} - P_l)/k)$$

for $i = 1, 2, \cdots, k - 1$ and $[(1 - 1/k)P_{100-l} + (1/k)P_l, x_n]$. This is recommended only when the data size is large. For smaller data sizes, an outlier test may be carried out to individually remove them one by one and then use the remaining data for binning. Multiple continuous attributes are discretized one at a time (simultaneous discretization algorithms that care for attribute interactions (correlations) are also reported in the literature). This can also be done in parallel, as the majority of computation time is spent in assigning the correct bin to each data value. Because nothing is assumed about the data distribution, other than the range, this method is called *blind binning*. If the class labels (categories) of the data are already known, we could estimate the error rate (discussed in the following) by comparing the actual and predicted class labels. Obviously, the error rate is maximum with EWB when compared to other DDA. Thus, the real question is whether we should compromise on the high predictive accuracy attainable by supervised DDA at the expense of extra computations over the simplicity of unsupervised algorithms such as EIB.

EXAMPLE 1.14

The body mass index (BMI) of 15 patients is as follows. Discretize the data using EIB with (i) $k = 3$, (ii) $k = 4$, and (iii) $k = 7$ bins. $X = \{26.2, 25.6, 25.1, 23.3, 23.7, 23.4, 29.7, 28.5, 25.2, 21.4, 28.3, 33.4, 27.8, 24.4, 25.9\}$.

Solution 1.14 Here, the minimum is 21.4 and maximum is 33.4, so that the range is 12. For case (i), we need to divide the range into three equal widths, so that $S = R/k = 12/3 = 4$. The bins are $b1 = [21.4, 25.4), b2 = [25.4, 29.4)$, and $b3 = [29.4, 33.4]$. If the labels are $U = underweight$, $N = normal$, and $O = overweight$, the new data are $Y = \{N, N, U, U, U, U, O, N, U, U, N, O, N, U, N\}$. For case (ii), we get $S = R/k = 12/4 = 3$. The bins are $b1 = [21.4, 24.4), b2 = [24.4, 27.4), b3 = [27.4, 30.4)$, and $b4 = [30.4, 33.4]$. Let the labels be U = underweight, N = normal, O = over weight, and H = Heavy. Note that the "Normal" category has lost its significance because the normal[6] BMI is in the range (25–26). Discretized data are $Y = \{N, N, N, U, U, U, O, O, N, U, O, H, O, N, N\}$. For case (iii), $S = R/k = 12/7 = 1.7143$. The bins are $A = [21.40, 23.114), B = [23.114, 24.829), C = [24.829, 26.543), D = [26.5435, 28.257), E = [28.257, 29.971), F = [29.971, 31.686), G = [31.686, 33.4]$. The discretized data becomes $\{C, C, C, B, B, B, E, E, C, A, E, G, D, B, C\}$. Note that the class $F = [29.971, 31.686)$ has zero frequency. This is a common problem when narrow range data are discretized into a large number of bins. ∎

1.12.2 Equal Frequency Binning (EFB)

This method divides the total range such that each subinterval has more or less the same number of data items. As this obviously requires some knowledge about the data distribution, EFB is in general computationally more complex than EIB. If data are known to be approximately uniform, we could first apply the EIB and perturb the boundary, if necessary, to get the EFB. The results obtained by EIB and EFB are often different, except in particular cases (when data are uniformly distributed, when $(n = 2, k = 2), (n = 3, k = 2)$, etc.). If k is a power of 2, we could apply the median finding algorithm repeatedly using the divide-and-conquer principle to easily get the bin boundaries. Otherwise, we sort the data values and pick out the bin boundaries using the following algorithm. If the number of data points is large and k is small,

---

[6]Another categorization is *normal* = (18 − 25), *overweight* = (26 − 30), obese = (30 − 35), morbid ≥ 35.

the frequency of each class will be more or less equal. A problem with this binning is the duplicate values that could get split across the boundary of two adjacent classes. Consider for example discretizing $X = \{1, 1, 2, 2, 2, 3\}$ into $k = 2$ bins. As $n = 6$, we would split it into $b1 = \{1, 1, 2\}$ and $b2 = \{2, 2, 3\}$. Here 2 appears in both the bins. By our algorithm, each value in $b1$ is assigned one label (say X), and all values in $b2$ are assigned another label (say Y), so that the discretized data becomes $\{X, X, X, Y, Y, Y\}$. If the EIB algorithm is used, we have $R = 3 - 1 = 2$ and $S = R/2 = 1$ so that we either get $b1 = [1, 2)$ and $b2 = [2, 3]$ or $b1 = [1, 2]$ and $b2 = (2, 3]$. In the first case, the discretized data becomes $\{X, X, Y, Y, Y, Y\}$, and in the second case, we get $\{X, X, X, X, X, Y\}$. This example shows that EIB and EFB could give totally different results. Eliminating all duplicates solves the problem because the data range will remain the same after a duplicates deletion (but it could result in a smaller n if there were at least one pair of duplicates). Above data without duplicates is $X = \{1, 2, 3\}$ with new $n = 3$. For $k = 2$ bins, we get the boundaries as $b1 = [1, 1.5)$ and $b2 = [1.5, 3]$. Discretized data becomes $\{X, X, Y, Y, Y, Y\}$.

◼ **EXAMPLE 1.15   Discretize BMI data using EFB**

> Discretize the 15 BMI data in Example 1.14 using EFB with (i) $k = 3$ and (ii) $k = 7$ bins.

> **Solution 1.15** Data in sorted order is $X = \{21.4, 23.3, 23.4, 23.7, 24.4, 25.1, 25.2,$ $25.6, 25.9, 26.2, 27.8, 28.3, 28.5, 29.7, 33.4\}$. Here $n = 15, S = 15/3 = 5$. For case (i), we need to assign the same label to all values in $(X[1 + (i - 1) *$ $S], X[1 + i * S])$ so that $b1 = (X[1], X[5]) = [21.4, 24.4], b2 = (X[6], X[10]) =$ $[25.1, 26.2]$, and $b3 = (X[11], X[15]) = [27.8, 28.3, 28.5, 29.7, 33, 33.4]$. The binned original data (unsorted) using labels (U, N, O) is $Y = \{N, N, N, U, U, U, O, O, N,$ $U, O, O, O, U, N\}$. ◼

For case (ii), we have $k = 7, s = [15/7] = 2$ so that the binned data is $Y = \{OO\ UUUU\ OO\ UU\ OOO\ UO\}$ or $Y = \{O\ UUUUU\ OO\ UU\ OOO\ UO\}$.

Note that the class labels of test data are not utilized in any of the above-mentioned unsupervised DDA. This results in loss of classification information (when DDA is used in engineering context). Other unsupervised DDAs include Holte's 1R algorithm [125] that constrains each bin to have at least m prespecified data instances of a majority class and Kerber's Chi–Merge algorithm [153]. The following algorithm adjusts the boundaries to decrease entropy at each interval.

1.12.3   Entropy-Based Discretization (EBD)

These are hierarchical discretization methods that maximize Shannon's entropy in the resulting discretized space or minimize entropy to control the number of intervals induced in the continuous space. As EBD considers the class labels of the data, it is a supervised learning algorithm. The EBD induces a binary tree in the data by recursively splitting it using a fixed attribute at each level. If the data are unsorted, each value is considered one-by-one as a pivot for a possible split. Let $T = x[j]$ be the current pivot. Then, we split the data into two bins as $b1 = (x[i] < T] \forall$ (i) and $b2 = (x[i] \geq T \forall$ (i). We assume that each of the bins contains representative data items of each of the classes. In other words, if there are k classes, we assume that at least one of the data items in each bin will belong to one of the classes. Sometimes, this assumption may not hold, as our classes become more and more pure. The entropy for this split is calculated as $\frac{|S1|}{|S|} * \text{Ent}(S1) + \frac{|S2|}{|S|} * \text{Ent}(S2)$, where $|S1|$ is the number of elements in bin $b1$ and $|S|$ the total number of data items under current consideration. The entropy is calculated using all of the classes as $\text{Ent}(S_i) = -\sum_{j=1}^{k} P(c_j) * \log_2 P(c_j)$, where k is the number of classes and $P(c_j)$ the fraction of items belonging to class C_j in the respective subset S_i. As log $(0) = -\infty$, irrespective of the base of the logarithm, we will drop those classes not represented in the bins (or combine the corresponding bins with its neighbors). The information gain resulting from the split at T is found as the difference between the entropies before and after split:

$$\text{InfGain}(S, T) = \text{Ent}(S) - \left[\frac{|S1|}{|S|} \text{Ent}(S1) + \frac{|S2|}{|S|} \text{Ent}(S2) \right], \qquad (1.49)$$

where $\text{Ent}(S) = -\sum_{j=1}^{k} P'(c_j) * \log_2 P'(c_j)$ (here $P'(c_j)$ is the fraction of items belonging to class C_j in the original set before split).

Those intervals with entropy 0 or with only one data value are kept. Others are split recursively. The splitting is stopped using the MDL principle when $\text{InfGain}(S, T) < \delta =$

$$(\log_2(n - 1) + \log_2(3^k - 2) - [k * \text{Ent}(S) - k_1 * \text{Ent}(S_1) - k_2 * \text{Ent}(S_2)])/n, \qquad (1.50)$$

where n is the size of data and k the total number of classes. Alternately, the difference between the entropy of the parent and maximum of the child node entropies is computed at each step, and iterations are terminated if this difference is small. As stated earlier, further splitting is continued only for *impure* intervals (if an interval is totally pure, then all values in it belong to the same class and its entropy is 0). If the class labels are highly correlated with an (increasing or decreasing) sort order of one or more attributes, we could considerably speedup the above-mentioned algorithm by taking T at the boundary of each class. As an example, consider discretizing medical patients as high-blood pressure (HBP) (C_1) and low-blood pressure (C_2) groups. Majority of HBP patients are also overfat. Thus, the BMI and HBP are highly correlated. If we sort the data in increasing order of BMI, there will be some overlap along the class boundary (on occasion there could also be some outliers in both classes).

■ EXAMPLE 1.16   Discretize BMI data using EBD

Discretize the 15 BMI data in Example 1.14 (pp. 1–14) using EBD.

Solution 1.16 As EBD is a supervised learning algorithm, we will label[7] the data as $X = \{26.2(O), 25.6(N), 25.1(N), 23.3(U), 23.7(U), 23.4(U), 29.7(O), 28.5(O), 25.2(N), 21.4(U), 28.3(O), 33.4(O), 27.8(O), 24.4(U), 25.9(N)\}$. There are 6 O's, 4 N's, and 5 U's in the original data. The entropy before split is $-6/15 * \log_2 (6/15) - 4/15 * \log_2(4/15) - 5/15 * \log_2(5/15) = -0.4 * (-1.3219) - 0.26667 * (-1.90689) - 0.33333 * (-1.58496) = 0.52877 + 0.5085 + 0.52832 = 1.565596$. The set S_1 contains all sample values <26.2 and S_2 contains all sample values ≥ 26.2. For convenience, we represent only the class labels as $S_1 = \{N, N, U, U, U, N, U, U, N\}$ and $S_2 = \{O, O, O, O, O, O\}$. Here, S_1 contains only four N's and five U's; and S_2 contains only six O's. The corresponding entropies are easily computed as $\text{Ent}(S_1) = -(4/9) * \log_2(4/9) - (5/9) * \log_2(5/9) = -0.444444 * (-1.169925) - 0.555556 * (-0.8479969) = 0.5199667 + 0.4711094 = 0.991076$, and $\text{Ent}(S_2) = 0.0$. Information gain for this split is given by $\text{Ent}(S) - \left[\frac{|S1|}{|S|} * \text{Ent}(S1) + \frac{|S2|}{|S|} * \text{Ent}(S2) \right] = 1.565596 - (9/16) * 0.991076 - (6/16) * 0.0 = 0.97095$. Iterations are continued using each of the subsequent values as split points. The results are summarized in Table 1.1. The maximum information gain 0.97095 occurs for split point 26.2. Hence, we will keep the set S_2 intact and recursively split S_1. Proceeding as above, we get the entropies as 25.60 (0.31976), 25.10 (0.99108), 23.30 (0.10219), 23.70 (0.37888), 23.40 (0.22479), 25.20 (0.55773), 21.40 (0.00000), 24.40 (0.59000), and 25.90 (0.14269). The optimal split point is 25.1 with maximum value

TABLE 1.1 Computation of Entropies

Pivot	S_1	S_2	$\text{Ent}(S_1)$	$\text{Ent}(S_2)$	InfoGain
26.2	4 N, 5 U	6 O	0.99108	0.00000	**0.97095**
25.6	2 N, 5 U	2 N, 6 O	0.86312	0.81128	0.73012
25.1	5 U	4 N, 6 O	0.00000	0.97095	0.91830
23.3	1 U	4 N, 4 U, 6 O	0.00000	1.55666	0.11272
23.7	3 U	4 N, 2 U, 6 O	0.00000	1.45915	0.39828
23.4	2 U	4 N, 3 U, 6 O	0.00000	1.52623	0.24286
29.7	4 N, 5 U, 4 O	2 O	1.57662	0.00000	0.19919
28.5	4 N, 5 U, 3 O	3 O	1.55459	0.00000	0.32193
25.2	1 N, 5 U	3 N, 6 O	0.65002	0.91830	0.75461
21.4	0 N, 0 U	4 N, 5 U, 6 O	0.00000	1.56560	0.00000
28.3	4 N, 5 U, 2 O	4 O	1.49492	0.00000	0.46932
33.4	4 N, 5 U, 5O	1 O	1.57741	0.00000	0.09335
27.8	4 N, 5 U, 1 O	5 O	1.36096	0.00000	0.65829
24.4	4 U	4N, 1U, 6O	0.00000	1.32218	0.59600
25.9	3 N, 5 U, 0 O	1 N, 6 O	0.95443	0.59167	0.78045

U = Underfat, n = Normal, and O = Overfat.

[7]See another categorization at the footnote of page 1–48

0.99108. Hence, the three intervals are U $=$ BMI < 25.1, N $= (25.1, 26.2)$, and O $=$ BMI ≥ 26.2. ∎

1.12.4 Error in Discretization

If the true class labels of data to be discretized are known apriori, we could estimate the error in discretizing as follows. Let there be k original classes. Denote original data by X and discretized data by Y. For class C_j, let n_j be the total number of items in X. If all of them are correctly classified in Y, the error rate for class j is zero. Let n_{p_j} be the number of correctly classified items of C_j and n_{q_j} be the wrongly classified number of items, so that $n_j = n_{p_j} + n_{q_j}$. Note that in the case of just two classes, all n_{q_j} items will belong to the other class. However, if there are >2 classes, n_{q_j} will contain all items belonging to $\overline{C}_j$. Then, the error rate for class j is n_{q_j}/n_j. The error rate for the entire data is obtained by summing over all of the classes as $\epsilon = \sum_{j=1}^{k} n_{q_j}/n_j$. The minimum of ϵ occurs when all items are correctly classified with minimum value 0. The maximum occurs when all items are incorrectly classified with maximum value 1. When there are no class overlaps (ϵ is very small), the predictive accuracy is maximum. The information extracted by the discretization process can be used to classify new data instances. As the DDA returns a set of disjoint, piece-wise continuous set of intervals, these intervals define the boundaries for various classes. We could divide the available data into a training set and a test set. The training set can then be used to construct the class boundaries (bins). These bin boundaries are then put to use in discretizing test data (in a classification context). In this sense, the DDA is a semisupervised learning model.

1.13 TESTING FOR NORMALITY

Several statistical procedures such as ANOVA tests and t-tests assume normality of data. Similarly, the error terms are assumed to be normally distributed with zero mean in linear regression models. There are two categories of normality tests— (i) visual displays and (ii) numerical tests.

1.13.1 Graphical Methods for Normality Checking

Visual displays (also known as *graphical methods*) use one or more graphs or diagrams to visually display the data distribution. They can be drawn as overlapping diagrams with a normal distribution for reference comparison. If the data have distinct mode, the normal curve with the same variance as the data is drawn so as to align the normal mode with the data mode. If the data mode is not unique, the normal curve uses the means for alignment. Note that the theoretical normal curve extends from $-\infty$ to ∞, whereas the sample data are always in a finite range. Hence, we will look for alignment with a normal curve at the central part of the data distribution rather than at the tails (away from the location measure).

The popular graphical methods include histograms, frequency polygons and curves, box plots, Quantile-Quantile (Q-Q) plots, Moran plots, ogive curves, and dot plots. These can be categorized into two types:– (i) isochronous graphical methods produce shapes that resemble a normal curve. Examples are the histograms, frequency curves. These graphs can be used as quick tests to check for normality. (ii) Nonisochronous graphical methods produce particular shapes or patterns that do not have direct resemblance to a normal curve but are similar to the shapes obtained for data from a normal distribution. For instance, if points lie close to a straight line in a Q–Q plot, it is an indication of normality of data. This is due to the fact that quantiles of any two identical distributions when plotted along the X and Y axes gives rise to a straight line plot. Checking for symmetry using a box plot is easy, but checking for normality is more involved.

1.13.2 Ogive Plots

Ogives are graphical plots of cumulative distribution functions $F(x)$ or survival probabilities $G(x) = 1 - F(x)$. They are of two types, called less-than ogive (positive ogive) and more-than ogive (negative ogive). Mathematically, we plot $(x, F(x))$ for each sample value arranged in increasing order for the less-than ogive and plot $(x, 1.0 - F(x))$ for the more-than ogive. The positive ogive is more popular among data analysts because several software packages support only this option. In addition, the positive ogive passes through the origin, whereas the negative ogive touches the y-axis at $y = 1$. If both of them are plotted in the same graph, they will intersect at the median. If the intersection point is less than the normal median and more toward the origin, it is an indication that the sample has come from a right-skewed distribution. On the other hand, an intersection point away from the normal median indicates that the sample has come from a left-skewed distribution. Similar reasoning holds for left- and right-truncated distributions. The ogive curves exhibit anti-symmetry (lower left-tail and upper right-tail both tails off similarly for data from symmetric distributions). Thus, the tail shapes can throw some light on whether the parent population is symmetric or not. Note that a variate is symmetric around a constant μ if $F_x(\mu - x_k) + F_x(\mu + x_k) = 1 \; \forall \, k > 0$. In terms of density functions, this can be expressed as $f_x(\mu - x_k) = f_x(\mu + x_k)$. A standard normal ogive may be superimposed on an ogive obtained from standardized empirical data to check for deviations from normality. If the original data are normal, both ogives will almost overlap. However, as linear combinations of normal variates are normally distributed, these methods cannot reveal whether the original data are linear combinations of normal laws or purely normal. Nevertheless, these methods are less technical than the numerical tests.

1.13.3 P–P and Q–Q Plots

A probability–probability (P–P) plot is another method to check if a sample has come from a known distribution. We plot the CDF of the standardized variate along the X-axis and the corresponding cumulative probabilities from a theoretical distribution along the Y-axis. Mathematically, we plot $(F(x_j - \bar{x})/s, G(x))$ for each sample value arranged in increasing order, where $G(x)$ is the CDF of the hypothesized distribution from which the sample came. In other words, it compares the empirical cumulative distribution function of a variate (say along the X-axis) with the CDF of a theoretical distribution (say along the Y-axis). Thus, both of the axes are calibrated from 0 to 1.0, starting with the origin. A reference line with slope 45° is also drawn in the positive quadrant along the stretch of the data. If the sample is indeed drawn from the hypothesized population, the data points will clutter around the reference line. Any major scattering away from the reference line indicates that the hypothesis is wrong. The Q–Q plot is very similar, except that we plot the quantiles of the data along the X-axis and quantiles of theoretical distribution along the Y-axis. A skewness–kurtosis plot can also be used if multiple samples are available. This plot uses the X-axis for skewness and the Y-axis for kurtosis or vice versa. As the normal distribution has skewness 0 and kurtosis 3, the sample values must clutter around the point with coordinates (0,3) (or (3,0) if skewness is plotted along the Y-axis) if the parent population is normal. Several researchers have modeled the skewness–kurtosis relationships empirically.

Using the above-mentioned techniques, even beginning practitioners and analysts can easily be trained to check normality. All of the positive ogive plot, P–P plot and Q–Q plot pass through the origin. An advantage of ogive plot and Q–Q plot is that they can be used to check if data came from any theoretical distributions and not only for normal populations. As an example, if data are known to come from a student-t distribution, we could plot the data quantiles along the X-axis and quantiles of student's t along the Y-axis.

1.13.4   Stem-and-Leaf Plots

If all data values are integers with a fixed range (say they have two or three digits), one could also use the stem-and-leaf (S&L) plot to check if data are approximately normal (this method, however, cannot distinguish between continuous and discrete distributions. For example, the S&L plot of data from a binomial distribution with a p close to 0.5 and large n will resemble that from a normal law). They are unsuitable for higher dimensional data. This method depends on the user's familiarity with the normal law too. Most statistics textbooks give the figure of only the standard normal law $N(0, 1)$. As the dispersion parameter $\sigma^2 > 0$, the normal curve can take a variety of shapes. Hence unless a normal distribution is superimposed on the observed data, slight deviations from normality are difficult to judge. In addition, the success also depends on the class width chosen for some of these plots. If a histogram is prepared with a small class interval, some of the classes may be empty (there may not be any data points falling in this range, so that their frequency counts are zeros). This is more likely to occur in classes toward the tails, especially when data contain outliers (exceptions do exist as in the case of U-shaped distributions). Thus, a trial and error method with many class widths may be needed to reasonably conclude that the data are indeed drawn from a normal law.

Box plots are more appropriate to check for outliers than for normality. It uses the five-number summary of a sample, namely, the (minimum, Q_1, median= Q_2, Q_3, maximum). If the data are symmetrically distributed, the Q_1 and Q_3 are equidistant from the median ($Q_2 - Q_1 = Q_3 - Q_2$). This is easy to catch if the scale of the graph is large enough (so that the boxes are long). In addition, the mean and median should coincide for symmetric distributions. As the box is drawn from Q_1 to Q_3, the mean and median should bisect the box area (they must be approximately at the center of the box; considering any sampling errors). The difference $Q_3 - Q_1$ is called *IQR*. All observations that fall below $-1.5*$IQR of Q_1 and above $1.5*$IQR of Q_3 are considered to be outliers.

1.13.5 Numerical Methods for Normality Testing

Numerical tests are more reliable as they can catch all kinds of normality violations (normality may be violated due to dispersion, skewness or kurtosis, or a combination of these. In a normal curve, 68.26% of the frequencies lie in $\mu \mp \sigma$, 95.44% of the frequencies lie in $\mu \mp 2\sigma$, and 99.74% of the frequencies lie in $\mu \mp 3\sigma$ (see Chapter 8). Location and spread measures computed from the data cannot in general reveal if the parent population is normal or not. As the mean, median, and mode coincide for symmetric distributions, these measures can quite often reveal symmetry for large samples. If the sample size is small, the above-mentioned measures may be in proximity (close-by) even for asymmetric distributions. A symmetry test and a skewness measure can jointly be used to check for normality. As discussed in the following, there are many symmetric distributions with the same skewness. Hence, this method cannot always guarantee the normality of a population. Similarly, linear combination of several symmetric distributions is known to be symmetric. If data are known to be asymmetric, the inverse Gaussian distribution IG(μ, λ) with PDF $f(x; \mu, \lambda) = (\lambda/2\pi x^3)^{\frac{1}{2}} \exp\{-\frac{\lambda}{2\mu^2 x}(x - \mu)^2\}$ is the preferred choice for data modeling and fitting. Table 1.2 summarizes popular normality testing using graphical and analytical methods.

A data plot can reveal any possible lack of symmetry for 2D samples. This is more difficult to visualize when dimensionality is more than three. In addition, if the variables (in 2D or more) are measured in different units, one may have to do a data transformation to concisely visualize the data. Any slight departures from symmetry may not be apparent in such situations. This is more challenging when the sample size is small. Suppose we have data from two asymmetric distributions. Deciding whether one is more asymmetric than the other is harder when they are mirror image

TABLE 1.2 Normality Testing Using Graphical and Analytical Methods

Type	Graphical	Numeric
Descriptive	Box plot and stem-and-leaf plot	P–P plots
Inferential	Q–Q plots,	Kolmogorov & Smirnov test, Lillioforos test, Shapiro–Wilks, Anderson–Darling and Jarque–Bera tests

asymmetric—one is skewed to the left and the other is skewed to the right. Setting aside the geometric intuition behind skewness as evinced through graphical plots and displays, a numeric score derived from the data can certainly help to understand the amount of lack of symmetry. If such a measure takes positive and negative values, we could even distinguish between left-skewed and right-skewed distributions. Several skewness measures have been reported for this purpose. These are interpreted as measures of lack of symmetry because increasing values indicate how far they are away from symmetry. See Refs 60, 134, and 170 for other normality tests.

1.14   SUMMARY

This chapter introduced different data types encountered in statistical analysis. Some notations to better understand statistics in particular and mathematical sciences in general are given below. Most students are familiar with the summation and product notations. However, these can sometimes be intricate and often tricky solutions exist to simplify them. Some of the concepts such as sum and product notations, data discretization, and transformation may be skipped depending on the level of the course. Readers who are already familiar with summation and product notations, combinations, and so on can have a bird's eye view of the respective sections. Equations that are unfamiliar or tedious can be skipped in the first reading as these are meant only to familiarize the reader with various notations. See Ref. 298 for an unsupervised and Ref. 174 for a Bayesian data discretization algorithm.

EXERCISES

1.1 Mark as True or False

a) Interval data have no natural zero point

b) Quartile differences are interval data

c) The mode of a sample can coincide with the minimum of the sample

d) The median is meaningless for ordinal data

e) A scale of proportionality exist among values of numeric ordinal data

f) The entropy of a set can be negative

g) All arithmetic operations are allowed on numeric nominal variables

h) Data discretization works only for unlabeled data.

1.2 What are the main branches of statistics? How does sample size differ among these branches?

1.3 Give examples of nominal and ordinal data. What are some restrictions on coding these types of data?

1.4 Distinguish between categorical and quantitative data. What are some statistical procedures that use each of them? Which encapsulates more information?

1.5 Distinguish between standard and extended data types. Identify some numeric measures used in each of the standard data types.

1.6 What type of variable is each of the following: (i) BMI, (ii) Systolic blood pressure, (iii) Earthquake intensity, (iv) consumer price index, (v) GRE scores.

1.7 Can you apply the change of origin technique to find the median of a sample containing very large numbers? mode of same sample?

1.8 Consider the alphabet of any natural language. What type of data are these?

1.9 Which means are easier to evaluate for distributions that have $\binom{n}{x}$ in the probability mass function?.

1.10 What data are the basic building blocks of text encoded data?

1.11 Define parameter and statistic. Can a statistic take arbitrarily large values?

1.12 Distinguish between population and sample. Give examples of enumerable populations.

1.13 What are some problems encountered in computing the Poisson PDF for large parameter values?

1.14 What are some situations in which the summation variables in a double sum can be interchanged?.

1.15 Give an example situation where the index variable is varied from high to low values.

1.16 What is a nested sum? How are they evaluated? What are the possible simplifications in a nested sum evaluation?

1.17 Give an example of a summation over a set. Give examples of subscript varying and superscript varying summations.

1.18 Give examples of double summations where the inner indexvar is dependent on the outer indexvar.

1.19 What is the most appropriate indexvar to model thermal conductivity problems?

1.20 In what situations can you exchange the indexvars in a double sum?

1.21 Give an example where the indexvar increments in fractions.

1.22 Give an example where the indexvar is varied from high to low values. Is it possible to convert such summations in the low to high indexvar values using an index transformation?

1.23 What type of summation will you use in very large matrix multiplication problems where each matrix is decomposed into several submatrices of appropriate order?

1.24 Give examples of summations in which the upper limit for the indexvar is known only at run-time.

1.25 Distinguish between supervised and unsupervised data discretization algorithms.

1.26 If X_{mxm} is a square matrix, use the $\sum$ notation to find the sum of each of the following:–(i) diagonal elements, (ii) tridiagonal elements (main diagonal plus adjacent diagonals), and (iii) lower triangular elements (including the diagonal).

1.27 The number of hours that a battery can be continuously operated in different devices after a 30-minute recharge is given below. Transform the data into the intervals $[-0.5, +0.5], [-1, +1]$. $X = \{32, 19, 24, 31, 20, 27\}$.

1.28 Using equation 1.30 in page 1–30, prove that $s_n^2 = (1/[2n(n-1)]) * \sum_{i=1}^{n} \sum_{j \neq i=1}^{n} (x_i - x_j)^2$.

1.29 Express $(\sum_{j=1}^{n} x_j)^2$ in terms of $\sum_{j=1}^{n} x_j^2$ and $\sum_{j \neq k; j,k=1}^{n} x_j * x_k$. Use it to express $\exp\left(-(\sum_{j=1}^{n} x_j)^2\right)$ as a product.

1.30 If the indexvar increments in steps of 2, evaluate the sum $\sum_{j=0}^{n} f(j)$ using unit incrementing indexvar.

1.31 If the indexvar increments in steps of c, evaluate the sum $\sum_{j=-k}^{k} f(j)$ using unit incrementing indexvar.

1.32 If Gini diversity index is defined as $D^2 = (1/n^2) \sum_{i=1}^{n} \sum_{j \neq i=1}^{n} (x_i - x_j)^2$, prove that $D^2 \leq 2s_n^2$.

1.33 What is a data requirement for using the entropy-based discretization? In what situations is it best?

1.34 The following data gives the marks scored by 10 students in engineering statistics. Discretize the data using EIB and EFB. $X = \{56, 62, 68, 73, 75, 78, 81, 88, 90, 93\}$. Transform the data to $[-3, +3]$ range using min–max transformation. Obtain the z-scores and compare with min–max transformed data.

1.35 The power of a discrete signal measured at 2N+1 points is given by $P = 1/(2N + 1) \sum_{n=-N}^{N} x[n]$,

where $x[n]$ is the signal value recorded at time $t = n$. Rewrite the expression where n varies from 0 to $2N$. What is the power when the signals are either compressed using $y[n] = x[kn]$ for $k > 1$ or expanded using $y[n] = x[\text{ln }]$ for $0 < l < 1$.

1.36 Describe how you can discretize discontinuous data (with gaps in between)? Which algorithm is best in such cases?

1.37 Discretize the data $X = \{56, 62, 68, 73, 75, 78, 81, 88, 90, 93\}$ using EPB if the labels are $D = (50, 60), C = [60, 75), B = [75, 90), A = [90, 100]$.

1.38 What is data discretization? What are some of its applications in engineering? Describe how you will discretize if the range (spread) of values is too large.

1.39 Transform the above-mentioned data to the $[-1, +1]$ and $[-3, +3]$ ranges using min–max transformation (pp. 1–48) and compare the results using the z-score transformation.

1.40 Can the minimum data value in EIB (pp. 1–45) be negative? Can both the minimum and maximum be negative?

1.41 How will the EFB (pp. 1–48) divide n data items into k intervals if n is not a multiple of k?

1.42 The first-order Bragg reflection of X-ray at different angles through a crystal gave the nanometer wavelengths as $\{0.0795, 0.0841, 0.0790, 0.0844, 0.0842, 0.0840\}$.

Transform the data into the intervals $[-0.5, +0.5]$ and $[-1, +1]$.

1.43 The resistance of an electronic circuit was measured using five different components as $\{5.2\Omega, 4.9\Omega, 5.12\Omega, 4.95\Omega, 5.1\Omega\}$. Transform the data to $[-1, +1]$ range. Convert data into z-scores.

1.44 Dielectric strength (kV/mm) of some thermoplastics is given below. Discretize the data into three intervals using EFB and EIB. Transform the data to the intervals (i) $[1,5]$, (ii) $[-1, +1]$. $X = \{15.6, 19.5, 17.2, 18.1, 17.6, 15.3, 18.0, 16.8, 16.4, 19.0\}$.

1.45 The number of hours that a battery can be continuously operated in different devices after a 30-minute recharge is given below. Transform the data into the interval

$[-1, +1]$ and $[-3, +3]$ ranges. $X = \{32, 19, 24, 31, 20, 27\}$.

1.46 A plastic polymer thread is subjected to an elongation stress test to see how much it can be stretched before it breaks. Elongation at break point is expressed as a percentage of its original length as $X = \{9.2\%, 6.7\%, 15.3\%, 18.0\%, 11.6\%, 10.8\%, 7.7\%, 16.1\%, 8.5\%, 12.0\%\}$. Transform the data to the $[-3, +3]$ range.

1.47 Soluble dissolvents (in mg/L) in drinking water are measured at different places in a city. $X = \{560, 458, 490, 525, 482, 554, 499, 538, 540, 507, 481, 513\}$. Standardize the data. Will you prefer the change of origin, change of scale, or both transformations? Transform the data to the $[-1, +1]$ range.?

1.48 If the index varies in powers of b (in steps of b^j), prove that $\sum_{j=0}^{n} f(j) = \sum_{k=0}^{\log_b(n)} f(b^k)$.

1.49 Prove that $\prod_{j=1}^{n}(1 - \frac{1}{j+1}) = \frac{1}{n+1}$ and $\prod_{j=2}^{n}(1 - \frac{2}{j+1}) = \frac{2}{n(n+1)}$.

1.50 Consider an expression for echo delay estimation in audio echo cancellation algorithms $\hat{\psi}(\text{lag}) = \frac{1}{k} \sum_{j=-\text{lag}}^{k} X[j] * X[j + \text{lag}]$. Use loop rerolling technique to express it in terms of an indexvar that is always positive.

1.51 Rewrite the summation $\sum_{k=1}^{n} \sum_{j=1}^{l} \sum_{i=1}^{m} c * u_i * v_{j+k} * w_k$ by taking terms independent of indexvars outside the summations.

1.52 Simplify $S_3 = \sum_{k=-10}^{10} \sum_{j=-2}^{2} 1/(j^2 + k^2)$ if k varies in steps of 2, j varies in steps of $1/4$.

1.53 Prove that $\prod_{j=1}^{n}(\frac{j}{2j+1}) = 2^n (n!)^2/(2n + 1)!$.

2

MEASURES OF LOCATION

After finishing the chapter, students will be able to

- Distinguish between location and scale population parameters
- Describe important measures of location (central tendency)
- Understand trimmed mean and weighted mean
- Comprehend Quartiles, Deciles, and Percentiles
- Use data transformations to compute various measures
- Apply updating formula for arithmetic, geometric, and harmonic means
- Prudently choose the correct measure for each situation

2.1 MEANING OF LOCATION MEASURE

The literal meaning of "location" is a place or point of interest with respect to (wrt) a frame of reference. In statistics, a location indicates a single point (for univariate data) that best describes the data at hand.

Definition 2.1 A well-defined function of the sample values that purports to summarize the locational information of data into a concise number is called a measure of location or central tendency.

The concept of location is applicable to a sample as well as to a population. Population locations are indicated by parameters (described below). For example,

Statistics for Scientists and Engineers, First Edition. Ramalingam Shanmugam and Rajan Chattamvelli.
© 2015 John Wiley & Sons, Inc. Published 2015 by John Wiley & Sons, Inc.

a parameter θ is called a location parameter if the functional form of the PDF is $f(x \mp \theta)$. Here, θ is a nonzero real number. Sample locations are measured by functions of sample values that return a real number within the range of the sample. It need not coincide with the sample data (i.e., x-value) for a sample drawn from a discrete distribution. These are also called measures of central tendency.

2.1.1 Categorization of Location Measures

Many meaningful functions of sample values can be used as sample location measures. Such a measure is expected to locate the central part of the data. Naturally, a measure that uses each and every sample value is more meaningful in engineering applications. The arithmetic mean (simply called mean), geometric mean, and harmonic mean (HM) belong to this category. Trimmed versions of them remove a small amount of extreme observations, and compute the value for the rest of the data. Weighted version of them give different importance to different data. The mean need not always coincide with one of the data values. A medoid is that data value that is closest to the mean in a distance sense. Medoids for large samples need not be unique (as there could exist multiple data points at equal distance from the mean). As it depends on the mean, it also belongs to the above category. Yet other types of measures that use the frequency of data rather than data values are available. One example is the mode that locates the data value with maximum frequency. This is more meaningful for grouped data. The sample median uses the count of data values to divide the total frequency into two equal parts. An extension of this concept uses quartiles, deciles, and percentiles that are useful when the data size is large. Among these measures, a change of origin transformation is meaningful to the arithmetic mean only, and a change of scale transformation is applicable to all the three means. These are discussed in subsequent sections.

2.2 MEASURES OF CENTRAL TENDENCY

Statistical distributions come in various shapes. Some of them are always symmetric around a real number for univariate distributions (or a vector for multivariate distributions), which can be zero or nonzero. Examples include the standard normal, standard Cauchy, and Student's T distributions (symmetric about 0), general normal distribution $N(\mu, \sigma^2)$, which is symmetric about μ, and general Cauchy distributions. Examples of asymmetric distributions include the exponential, beta and gamma distributions, F distribution, Pareto distribution, and so on. Some of these distributions are symmetric for particular parameter values though. For instance, the 2-parameter beta distribution BETA-I (a, b) is symmetric when the parameters are equal $(a = b)$, and the binomial distribution BINO (n, p) is symmetric when $p = 1/2, \forall n$. As mentioned below, a great majority of statistical distributions are asymmetric. Most of the symmetric distributions are of continuous type.

The "central tendency" measures the location of symmetry of symmetric distributions, and the center of gravity of asymmetric distributions. We call it a *location measure* because they can locate the approximate centering of the distribution along the real line (univariate case). The most commonly used measures of location are the arithmetic mean, median, and the mode. Among them, the arithmetic mean is a linear measure as it uses the sum of the data values in the numerator. Geometric and HMs are nonlinear measures (geometric mean is log-linear as shown below). A change of origin transformation (e.g., using the mean as the pivot) can be used to align the location measures of different distributions. Arithmetic, geometric, and HMs and the median always lie between the minimum and maximum of the sample values (for $n \geq 2$), while the mode may get aligned with the extremes.

2.3 ARITHMETIC MEAN

The mean of a population is denoted by the Greek letter μ, and the corresponding sample mean is denoted by $\bar{x}$ (or $\bar{x}_n$ where n is the sample size). We define it as

$$\mu = \begin{cases} \sum_{k=-\infty}^{\infty} x_k\, p_k & \text{if } X \text{ is discrete;} \\ \int_{x=-\infty}^{\infty} x f(x) dx & \text{if } X \text{ is continuous.} \end{cases}$$

The summation or integration needs to be carried out only throughout the range of the respective random variable (as the PDF is defined to be zero outside the range). This represents the weighted average of all possible values of a random variable with the corresponding probabilities as weights. The mean is the first moment because it is obtained by putting $j = 1$ in

$$\mu_j = \begin{cases} \sum_{k=-\infty}^{\infty} x_k^j\, p_k & \text{if } X \text{ is discrete;} \\ \int_{-\infty}^{\infty} x^j f(x) dx & \text{if } X \text{ is continuous.} \end{cases}$$

The simple (arithmetic[1]) mean of a sample of size n is defined as the sum of the sample values divided by the sample size. Symbolically

$$\bar{x}_n = (x_1 + x_2 + \cdots + x_n)/n = \sum_{j=1}^{n} x_j/n. \tag{2.1}$$

where the subscript n on the left hand side (LHS) denotes the sample size, and on x_n denotes the nth data value. We write it as $\bar{x}$ when no ambiguity is present. Duplicate values, if any, are counted distinctly in finding the mean. It is evident from equation (2.1) that the mean of a sample need not coincide with one of the sample values for $n > 1$ (median for odd sample size, and mode will always coincide with a sample value). Distributing the constant with each of the sample values results in

$$\bar{x}_n = (x_1/n + x_2/n + \cdots + x_n/n). \tag{2.2}$$

---

[1] In statistical parlance, "mean" or "average" always denote the "arithmetic mean." It is also called average value, although we reserve this term to mathematical expectation (Chapter 8).

This shows that the sample mean gives equal weights or importance to each sample data item. If a sample contains several zeros, all of them are counted in the above definition. Subtract $\bar{x}_n$ from both sides of equation (2.1), and write $\bar{x}_n$ on the right-hand side (RHS) as n terms each of which is $\bar{x}_n/n$ to get

$$0 = \bar{x}_n - \bar{x}_n = (x_1/n - \bar{x}_n/n) + (x_2/n - \bar{x}_n/n) + \cdots + (x_n/n - \bar{x}_n/n). \qquad (2.3)$$

Take $(1/n)$ as common factor from RHS, and write the rest of the terms using the summation notation. This gives

$$(1/n) * \sum_{i=1}^{n} (x_i - \bar{x}_n) = 0. \qquad (2.4)$$

As $(1/n)$ is a constant, this means that the sum of the deviations of sample values from its mean is always zero. This can also be stated as follows:

Lemma 1 If $\sum_{i=1}^{n}(x_i - c) = 0$ for a sample, then $c = \bar{x}_n$.

Proof: Apply the summation to each individual term in the bracket to get $\sum_{i=1}^{n} x_i - \sum_{i=1}^{n} c$. From the definition of $\bar{x}_n$, we have $\sum_{i=1}^{n} x_i = n * \bar{x}_n$. As the summand in the second term is a constant, $\sum_{i=1}^{n} c = n * c$. Substitute in the above to get $n * \bar{x}_n - n * c = 0$, or equivalently $n * (\bar{x}_n - c) = 0$. As n being the sample size is nonzero, the only possibility is that $c = \bar{x}_n$. This result will be used in subsequent chapters. ∎

The sample mean is the most extensively used location measure due to its desirable properties in inferential statistics. As the mean utilizes each and every observation in a sample, it rapidly converges to the population mean as $n \to \infty$. The arithmetic mean is not an appropriate measure of central tendency when nominal variables are coded numerically. However, the mean is meaningful in one situation—when a dichotomous nominal (i.e., binary) variable is coded as 0 and 1, the mean gives the *proportion* of items that are coded as 1. As we cannot compare nominal data, the median p. 54 also is meaningless. The mode p. 58 is the most appropriate measure of central tendency for nominal data.

2.3.1   Updating Formula For Sample Mean

As mentioned above, the mean of a sample can be found if the sum of the observations and the sample size are known. All of the sample values may not be readily available in some scientific and industrial applications. As an example, suppose the data come from sensors installed in a large factory. Several industries and factories have a multitude of sensors such as temperature (heat), light, pressure, humidity (moisture), gas, and chemical sensors installed at various strategic points. In addition, some specialized industries such as chip design factories, DVD, and floppy disk manufacturing plants measure dust and microparticle suspension in the air to ensure that they do not get deposited into sensitive chip components, circuits, or platters. Smoke and radiation sensors are more important in space stations. Similarly, some pharmaceutical companies have microbe sensors on the machine parts that manufacture some medicines. A high concentration of microbes in the ingredients could be lethal to patients if it contaminates just a few of the tablets or capsules manufactured[2]. Each of the sensors can have variations in terms of calibration. For example, there are separate heat sensors for air, water, liquids (different liquids boil at different temperatures; it slightly differs for the same liquid in the presence of various solvents, or combinations of them; the boiling point also depends on the altitude), chemicals, and surface temperatures. This will vary from factory to factory. While air temperature, smoke, and humidity sensors are more important in textile factories, pressure and temperature sensors are more important in robotic factories. As another example, *hydroponics farms* are closed (air-tight) laboratories in which plants are grown in sand or water tubes or containers. The light, nutrient concentrations, and temperature sensors are the most important, followed by water and microbe concentrations in hydroponics farms. These can be continuously monitored using various sensors.

In all of the above cases, we wish to continuously check process deviations using quality control charts or statistical models that heavily depend on the sample mean. In such situations, we could find the mean of already available data, and iteratively update the mean when new data items are received from various sensors. This is called online updating. Suppose we have a sample of size n with mean $\bar{x}_n$. If an additional observation x_{n+1} is added to our sample, the new mean becomes

$$\bar{x}_{n+1} = (x_1 + x_2 + \cdots + x_n + x_{n+1})/(n + 1) = \sum_{i=1}^{n+1} x_i/(n + 1). \tag{2.5}$$

Multiply numerator and denominator by n, and separate out the last term x_{n+1} to get

$$\bar{x}_{n+1} = [n/(n + 1)] * (x_1 + x_2 + \cdots + x_n)/n + (x_{n+1}/(n + 1)). \tag{2.6}$$

Replace $(x_1 + x_2 + \cdots + x_n)/n$ by $\bar{x}_n$, to get

$$\bar{x}_{n+1} = [n/(n + 1)] * \bar{x}_n + x_{n+1}/(n + 1). \tag{2.7}$$

Take $1/(n+1)$ as a common factor, and write this as

$$\bar{x}_{n+1} = [n\bar{x}_n + x_{n+1}]/(n + 1). \tag{2.8}$$

Add and subtract $\bar{x}_n/(n + 1)$ on the RHS, and combine $n\bar{x}_n/(n + 1) + \bar{x}_n/(n + 1)$ as $(n + 1)\bar{x}_n/(n + 1) = \bar{x}_n$, to get the alternate form

$$\bar{x}_{n+1} = \bar{x}_n + (x_{n+1} - \bar{x}_n)/(n + 1). \tag{2.9}$$

---

[2]Most pharmaceutical companies have quality control specialists who sample the produced medicines on a periodic basis and checks for contaminations.

Each newly received data item is used only once in the updating formula. Note that the correction term $(x_{n+1} - \bar{x}_n)/(n + 1)$ can be positive or negative depending on whether the new data item x_{n+1} is $>$ or $< \bar{x}_n$. In the particular case when $x_{n+1} = \bar{x}_n$, the mean is unchanged. This provides a recursive algorithm for arithmetic mean [2].

▉ EXAMPLE 2.1   Find mean by updating formula

Thickness of paint layer applied on straight locations using nylon brush is dependent on the paint viscosity and smoothness of the surface. Paint-layer tends to be thicker on harsh surfaces than smooth ones. A sample surface of size $1'' \times 1''$ is test-painted, and the layer thickness (in mm) after drying is noted down at 10 random spots. Use the updating formula (2.9) to compute the mean paint thickness.
$$X = \{0.26, 0.51, 0.39, 0.27, 0.44, 0.58, 0.34, 0.29, 0.4, 0.53\}$$

Solution 2.1   Form a sequence of pairs $(x_i, \bar{x}_i)$ where $\bar{x}_i$ is the mean of all data until the current one. We get (0.26, 0.26), (0.51, 0.385), (0.39, 0.38667), (0.27, 0.3575), (0.44, 0.374), (0.58, 0.4083), (0.34, 0.39857), (0.29, 0.385), (0.4, 0.386667), (0.53, 0.401) as the values. The second value in the last pair is the mean $\bar{x}_n$.   ▉

In some applications, we have the mean of subsamples already available. As examples, the mean marks of two or more classes in the same college, the mean yield of two or more plots in an agricultural experimentation, the mean purchase amount of day-time and night-time customers to an online store, and the average sales amount in two consecutive time periods (days, months, years, etc.) all record multiple means for different samples. These separately computed means could be combined, irrespective of their individual sample sizes, using the following theorem.

Theorem 2.1   If $\bar{x}_1$ and $\bar{x}_2$ are the means of two samples of sizes n_1 and n_2, respectively, the mean of the combined sample is given by $\bar{x} = (n_1\bar{x}_1 + n_2\bar{x}_2)/(n_1 + n_2)$.

Proof: The $n_1\bar{x}_1$ and $n_2\bar{x}_2$ in the RHS represent the sum of the observations of the first and second sample, respectively, so that their sum is the grand total of all observations. By dividing this total by $(n_1 + n_2)$ gives the grand mean on the LHS. This result can be extended to any number of samples as follows: ▉

Corollary 1   If $\bar{x}_i, i = 1, 2, .. m$ are the means of m samples of sizes $n_1, n_2, \dots, n_m$, respectively, the mean of the combined sample is given by $\bar{x} = (n_1\bar{x}_1 + n_2\bar{x}_2 + \cdots + n_m\bar{x}_m)/(n_1 + n_2 + \cdots + n_m)$.

As $E(\bar{x}_n) = \mu$, the sample mean is used as an unbiased estimator of the unknown population mean μ. This has two interpretations. If repeated random samples of small size n are drawn from a population, the mean of these samples will clutter around the population mean μ. On the other hand, the mean of a sample of size n converges in probability to the population mean as $n \to \infty$. Equivalence of both these statements can be understood from the above lemma, where $N = (n_1 + n_2 + \cdots + n_m) \to \infty$ with each of the $n_i's$ being equal, and m is large.

⬛ **EXAMPLE 2.2 Combined mean**

Two trucks work continuously to transport passenger luggage from an airport to a terminal. If the mean weight (in tonnes) transported in 10 trips by truck-1 is 58, and 12 trips of truck-2 is 46, what is the mean weight transported by these two trucks combined?

Solution 2.2 Here, $\bar{x}_1 = 58$, $\bar{x}_2 = 46$. Hence, $\bar{x} = (10 * 58 + 12 * 46)/(10 + 12) = 1132/22 = 51.4545$ tonnes. ∎

Corollary 2 If an existing observation x_n is removed from a sample of size n with mean $\bar{x}_n$, the new mean is given by $(n\bar{x}_n - x_n)/(n - 1)$.

Corollary 3 If m observations with mean $\bar{x}_m$ are removed from a sample of size n with mean $\bar{x}$, the new mean is given by $\bar{x}_{new} = (n\bar{x} - m\bar{x}_m)/(n - m)$.

2.3.2 Sample Mean Using Change of Origin and Scale

The change of origin technique is useful to compute the mean when the sample values are large. If the variables are transformed as $y_i = x_i - c$, the means are related as $\bar{y}_n = \bar{x}_n - c$. In this case, the updating formula becomes $\bar{y}_n = [(n - 1) * \bar{y}_{n-1} + x_n - c]/n$. This can also be written as

$$\bar{y}_n = (1 - 1/n)\,\bar{y}_{n-1} + (x_n - c)/n. \tag{2.10}$$

The change of scale transformation $Y = c * X$ gives $\bar{y}_n = c * \bar{x}_n$. We could simultaneously apply the change of origin and scale transformation to the data as $z_i = (x_i - c)/d$. The means are then related as $\bar{z}_n = (\bar{x}_n - c)/d$. The updating formula then becomes

$$\bar{z}_n = [(1 - 1/n) * \bar{z}_{n-1} + (x_n - c)/(nd)]. \tag{2.11}$$

The above equation is quite useful in iteratively computing the mean when the data values are large and have large variance. As an example, microparticle sensors have limited range (or visibility) to maintain correct accuracy. If the range is 1cm^3 (theoretically, it is a sphere of appropriate radius (if they are setup above ground) or a semi-sphere (if they are mounted on walls or flat surfaces) such that there are no empty regions between adjacent sensors) around its sensing point, the number of microparticles in it could be very large, which could vary depending on the air current. Similarly, smoke sensors installed in rooms or buildings near the road or highway sides, or inside vehicles on the road have a cutoff threshold for the number of carbon particles. If this number is beyond the threshold, it is flagged as *smoke* from fire. If it is below the threshold, it is assumed as engine exhausts or cigarette smoke, and so on. The numbers used in all these situations are large in magnitude and have large variance. However, we need to only accumulate the values for a suitable time window. A smoke detector is least concerned with the number of carbon particles it encountered 2 min ago. Its window is very small, perhaps 1–3 s. The window size of microparticle sensors could vary depending on the air-current—if air circulates fast, the window is a few milliseconds, and if it circulates slowly, it could be 1 or 2 s. This could also vary among sensors installed in other media such as water, liquids, or chemicals. If the window size is d, the general updating formula given above becomes

$$\bar{z}_t = \bar{z}_{t-1} + (x_t - x_{t-d}) / d, \text{ for } t = d + 1, d + 2, \ldots \quad (2.12)$$

This is called the 'window mean' as it simply accumulates the mean of the most recently seen d data values.

▉ EXAMPLE 2.3   Mean updating

The mean of the number of particles received in a sensor for 6 s is 1600. If two new counts (970 and 1830) are recorded in subsequent seconds, find the new mean using updating formula (2.9) in page 47.

Solution 2.3 Our updating formula is $\bar{x}_{n+1} = \bar{x}_n + (x_{n+1} - \bar{x}_n)/(n + 1)$. We are given that $n = 6$ (as the particles are counted in intervals of 1 s), $\bar{x}_n = 1600$. For $x_{n+1} = 970$, the correction term is $\delta_{n+1} = (x_{n+1} - \bar{x}_n)/(n + 1) = (970 - 1600)/7 = -90$. Substitute $\bar{x}_{n+1} = \bar{x}_n + \delta_{n+1}$ to get the new mean as $1600 - 90 = 1510$. The new correction term is $\delta_{n+2} = (x_{n+2} - \bar{x}_{n+1})/(n + 2) = (1830 - 1510)/8 = 40$. Substitute in $\bar{x}_{n+2} = \bar{x}_{n+1} + \delta_{n+2}$ to get the new mean as $1510 + 40 = 1550$.   ∎

2.3.3   Trimmed Mean

Data outliers have a major influence on the arithmetic mean, as they are given equal importance as other data values. A solution is to delete extreme observations from the low and high end of a sample (of sufficiently large size) and compute the mean of the rest of the data. These are called trimmed means. They can be left-trimmed (only low end data are discarded), right-trimmed (only high end data are discarded), or simply trimmed (from both the ends). It is symmetrically trimmed if an equal number of observations are discarded from both the ends. Using the summation notation introduced in Chapter 1, this becomes

$$\bar{x}_m^t = (x_{(k+1)} + x_{(k+2)} + \cdots + x_{(n-k)})/(n - 2k) = \frac{1}{(n - 2k)} \sum_{i=k+1}^{n-k} x_{(i)}. \quad (2.13)$$

where $\bar{x}_m^t$ denotes that this is the trimmed mean of $m = n - 2k$ data values, and $x_{(i)}$ is the ith order statistic. This definition uses a count (k) to truncate data values from both the ends. A cutoff threshold can also be used to discard data values from either or both the ends of a rearranged sample. In fact, an entire sample need not be sorted (arranged in increasing or decreasing order) to find the trimmed mean.

2.3.4   Weighted Mean

Each observation (sample value) is weighted by $1/n$ in the simple mean (see equation (2.2)). The weighted mean is an extension in which we multiply (or divide) each observation by an appropriate nonzero weight. If $w_1, w_2, \ldots, w_n$ are the weights associated with $x_1, x_2, \ldots, x_n$, respectively, the weighted mean is given by

$$\bar{x}_n(\mathbf{w_n}) = (w_1 x_1 + w_2 x_2 + \cdots + w_n x_n)/(w_1 + w_2 + \cdots + w_n) = \sum_{i=1}^{n} w_i x_i / \sum_{i=1}^{n} w_i.$$

Weighted mean assigns different importance to different sample observations. For example, if the data were collected over a time window (as in supermarket sales), more recent transactions must be highly weighted than distant ones to the past. Similarly in some medical studies in which the age of a patient is correlated with the outcome of an experiment, patients in various age groups may be weighted differently. We denote the weighted mean by $\bar{x}_n(\mathbf{w})$ (or $\bar{x}_n(\mathbf{w_n})$) to distinguish it from simple mean, and to indicate that the weights are the parameters. Different weightings may be used on the same sample. When all the weights are equal, the weighted mean reduces to the arithmetic mean.

2.3.5 Mean of Grouped Data

The mean of grouped data is obtained from the above by replacing w_i's with corresponding class frequencies f_i's as

$$\bar{x}_n = \sum_{i=1}^{n} f_i x_i / F \quad \text{where } F = \sum_{i=1}^{n} f_i. \tag{2.14}$$

Here, f_i are the frequencies and x_i is the middle point of the respective class. It is assumed that there are no open classes (such as $x < 5$ or $x > 100$) at the extremes. In such cases, the median is more appropriate. Each of the class widths are assumed to be equal in equation (2.14). A Shepperd's correction may be applied to get more accurate results. This is desirable because the middle value of a class is used to compute the mean (and higher order moments) under the assumption that the entire frequency falling in a class is concentrated at or around the middle value. This warrants a correction to compensate for the distribution of data throughout the class. There is no correction for the first moment μ_1. For μ_2, the correction term is $h^2/2$ so that the corrected term is $\mu_2 - h^2/2$ where h is the class width. If there are a large number of classes and some of the adjacent classes have relatively very low frequencies, they may be combined to reduce the computation.

2.3.6 Updating Formula for Weighted Sample Mean

An updating formula could also be developed for the weighted mean as follows. Start with equation (2.14) for $n + 1$ as

$$\bar{x}_{n+1}(\mathbf{w_{n+1}}) = \sum_{i=1}^{n} w_i x_i / \sum_{i=1}^{n+1} w_i + w_{n+1} x_{n+1} / \sum_{i=1}^{n+1} w_i. \tag{2.15}$$

Multiply and divide the first term on the RHS by $\sum_{i=1}^{n} w_i$, and then replace $\sum_{i=1}^{n} w_i x_i / \sum_{i=1}^{n} w_i$ by $\bar{x}_n(\mathbf{w_n})$ to get

$$\bar{x}_{n+1}(\mathbf{w_{n+1}}) = \left(\sum_{i=1}^{n} w_i / \sum_{i=1}^{n+1} w_i \right) \bar{x}_n(\mathbf{w_n}) + w_{n+1} x_{n+1} / \left(\sum_{i=1}^{n+1} w_i \right). \qquad (2.16)$$

Add and subtract $w_{n+1}\bar{x}_n(\mathbf{w_n}) / \sum_{i=1}^{n+1} w_i$ on the RHS, then take $\sum_{i=1}^{n+1} w_i$ as a common factor from first two terms, and cancel out $\sum_{i=1}^{n+1} w_i$ from numerator and denominator of the first term to get

$$\bar{x}_{n+1}(\mathbf{w_{n+1}}) = \bar{x}_n(\mathbf{w_n}) + \left[w_{n+1} / \sum_{i=1}^{n+1} w_i \right] (x_{n+1} - \bar{x}_n(\mathbf{w_n})). \qquad (2.17)$$

In terms of the mean of the weights, this becomes $\bar{x}_{n+1}(\mathbf{w_{n+1}}) = \bar{x}_n(\mathbf{w_n}) + \frac{w_{n+1}}{(n+1)\bar{w}_{n+1}}(x_{n+1} - \bar{x}_n(\mathbf{w_n}))$. When $x_{n+1} = \bar{x}_n(\mathbf{w_n})$, the weighted mean will remain the same, irrespective of the weight assigned to the new sample data item.

■ EXAMPLE 2.4   Calories burned while exercising

Calories burned on a treadmill by a person depends on many things including speed of the belt, age, and physical stature. Table 2.1 gives the calories burned and speed on treadmill of 16 visitors to a health club. Find the weighted mean using equation (2.17).

Solution 2.4 Calculations are shown in Table 2.1. Weighted mean is computed directly to check the computations. The last entry in the last column gives the weighted mean as 8.0282.   ■

TABLE 2.1   Weighted Mean Example: Calories Burned on Threadmill

c	v	$c * v$	(2.17)	Direct	c	v	$c * v$	(2.17)	Direct
6.4	7.60	48.64	6.4000	6.4000	7.5	8.4	63.00	7.8105	7.8105
8.3	8.20	68.06	7.3861	7.3861	9.1	13.0	118.3	8.0068	8.0068
7.2	7.40	53.28	7.3267	7.3267	6.6	7.0	46.20	7.9002	7.9002
9.7	10.00	97.00	8.0416	8.0416	8.4	10.0	84.0	7.9490	7.9490
8.9	9.00	80.10	8.2246	8.2246	7.5	8.3	62.25	7.9154	7.9154
6.9	8.00	55.20	8.0135	8.0135	6.7	7.8	52.26	7.8354	7.8354
8.0	7.80	62.40	8.0117	8.0117	9.85	14.0	137.90	8.0482	8.0482
6.3	6.00	37.80	7.8513	7.8513	7.7	8.1	62.37	8.0282	8.0282

First column is the calories burned per minute while exercising on a treadmill. Second column gives the speed of walking/jogging in miles/hour. Third column is the product. Fourth column gives the weighted mean using equation (2.17). Fifth column is direct calculation using equation (2.14). Subsequent columns repeat the data.

2.3.7   Advantages of Mean

The AM can be computed even if data contain many zeros. In addition, it possesses some desirable statistical properties in other fields of statistics such as testing of hypotheses and inferences. It is meaningful for ordinal or higher scales of measurements that are numerically coded. There is one particular case of nominal data for which the mean is meaningful. If the nominal data are coded as either 0 or 1, the mean will give the relative frequency of sample values that are coded as 1. As a simple example, suppose the sex of patients to a clinic are coded as $0 =$ Female, $1 =$ Male. If 120 patients visit the clinic on a particular day, we could find the mean of these values to find out what proportion of them were males. This is due to the fact that we have coded Males as "1." What if we want to find out the proportion of females only? One solution is to subtract the males' proportion from 1 to get the female proportion (as the proportions for males and females add up to 1). The mean also has an interpretation as the *balancing point* (center of gravity) of a simple or weighted sample (see below). This implies that if one were to use a single number between the minimum and maximum of the sample values as a representative of the sample, the sample mean seems to be the most appropriate value to use.

Some of the advantages of mean are summarized below:

1. The mean is easy to compute.
2. It lends itself to further arithmetic treatment.
3. It is always unique (whereas mode of a sample need not be unique).
4. It can easily be updated (when data are added or deleted).

As the mean is a linear function of the sample values, we could deal with missing values as follows: (i) find the grand mean $\bar{x}_g$ by omitting all missing observations; (ii) replace each missing value by $\bar{x}_g$ and find the new mean $\bar{x}$.

2.3.8   Properties of The Mean

The mean satisfies many interesting properties. For example, the mean places itself in-between the extremes of observations in such a way that the sum of the deviations of observations (from it) to its left and to its right are equally balanced in terms of their magnitudes. This is proved in the following theorem.

Theorem 2.2 For any sample of size $n > 1$, the sum of the deviations of observations from the mean $\sum_{j=1}^{n}(x_j - \bar{x}_n)$ is zero.

Proof: This is already proved in equation (2.4) (p. 46). This can be extended to the weighted mean as follows: ∎

📋 **EXAMPLE 2.5**   Verify $\sum_{i=1}^{n}(x_i - \bar{x}_n) = 0$

Hexavalent chromium is a toxic chemical found in the metropolitan areas. Data in Table 2.2 gives the levels in nanogram per cubic meter for 10 different places. Compute the mean and verify whether $\sum_{i=1}^{n}(x_i - \bar{x}_n) = 0$.

TABLE 2.2   Hexavalent Chromium Levels

1	2	3	4	5	6	7	8	9	10	Sum
0.95	1.26	0.63	0.80	0.57	0.34	0.29	0.71	1.17	0.94	7.66
0.18	0.49	−0.14	0.03	−0.20	−0.43	−0.48	−0.06	0.40	0.17	0.00

Solution 2.5 The sum of the numbers is 7.66, from which the mean is found as 0.766. The second row of Table 2.2 gives the deviations of data from the mean. The last column is the sum of the deviations, which is obviously zero.

∎

EXAMPLE 2.6   AM coinciding with a data value

If the arithmetic mean of n data values coincide exactly with one of the data values (say x_k), then x_k must be the AM of the other $(n-1)$ data values.

Solution 2.6 Let there be n data values with mean $\bar{x}_n$. Then we have $n * \bar{x}_n = \sum_{i=1}^{n} x_i$. Without loss of generality, assume that the coinciding data value is x_k so that $\bar{x}_n = x_k$ and the LHS becomes nx_k. Cancel one x_k term from LHS and RHS. The multiplier on the LHS becomes $(n-1)$. What remains on the RHS is the sum of the data values less x_k. Divide both sides by $(n-1)$ to get $x_k = \bar{x}_{n-1}$. As k is arbitrary, the result follows. This result is easy to extend to GM and HM (see Exercise 2.10, p. 65).   ∎

In analysis of variance procedures, we encounter within group variances which are measured around the means of each group $\bar{x}_{i.} = \frac{1}{n}\sum_j x_{ij}$, and between group variances which are measured around overall mean $\bar{x}_{..} = \frac{1}{nk}\sum_i \sum_j x_{ij}$. Note that a "." in these expressions fixes a variable. Thus $\bar{x}_{..}$ is the mean that is averaged around all values of i and j, whereas $\bar{x}_{i.}$ is the mean that is averaged around all j values.

2.4   MEDIAN

The population median is that value below which 50% of the values fall. In other words, the median divides the total frequency (area under the distribution) into exactly equal parts. Analogous definition holds for the sample median. It is most appropriate when all sample values are different. It can be easily found if the sample values are arranged in sorted order (in ascending or descending order). The complexity of sequential data sorting is $O(n \log n)$ where n is the size of the data. Parallel sorting techniques can improve this to $O(n)$. Still, it may be time consuming to sort an entire data set, just to find the median when the data size is too large. However, efficient algorithms are available to locate approximate median without data sorting [2, 17].

The median of a sample is unique for odd sample size (middle element at $[(n+1)/2]$th position, or $x_{(n+1)/2}$). When the sample size is even, we take the arithmetic mean of the middle values (at $(n/2)$th and $(n/2 + 1)$th positions) as the median. Symbolically:

$$\text{Median} = \begin{cases} x_{(n+1)/2} & \text{if } n \text{ is odd;} \\ 0.5(x_{(n/2)} + x_{(n/2)+1}) & \text{if } n \text{ is even} \end{cases}$$

■ EXAMPLE 2.7   Median finding

Find the median of (5,2,8,4,7) and (5,2,8,9,4,7).

Solution 2.7 Here, the number of observations is odd. The sorted data set is (2,4,5,7,8). The middle element is 5, which is the median. In the second case, the number of observations is even. The sorted data set is (2,4,5,7,8,9). The middle elements are 5 and 7. The mean of these middle elements is $(5+7)/2 = 6$, which is the median.   ■

Trimmed median is meaningful when the trimming occurs at either of the extremes. If data values are discarded at the low end, the trimmed median moves to the right and vice versa. When an equal number of data values are discarded from both ends, the median will remain the same.

2.4.1   Median of Grouped Data

Finding the median of grouped data is more difficult, as we need to first locate the median class. It is found in two steps as follows:

1. Find the class to which the median belongs
2. Compute it as Median $= L + c * (n/2 - M)/f$ where L is the lower limit of the median class, c is the fixed class width, n is the sample size, M is the cumulative frequency up to median class, and f is the frequency in the median class.

Theorem 2.3 The expected absolute departure of a random variable is minimum when it is taken around the median (i.e., $E|X - c|$ is minimum when c is the median (expected values are discussed in Chapter 8)).

Proof: Let X be discrete. By definition, $E|X - c| = \sum_{x_i < c} (c - x_i)f(x) + \sum_{x_i > c} (x_i - c)f(x)$. Perturb the constant c by a small amount δc so that $c = c - \delta c$. The net change is then $\Delta = -\sum_{x_i < c} (\delta c)f(x) + \sum_{x_i > c} (\delta c)f(x)$. Taking the constant δc outside the summation, we get $\Delta = \delta c[\sum_{x_i > c} f(x) - \sum_{x_i < c} f(x)]$. If c is the median, then the expression in the square brackets is zero (because the median divides the total frequency into equal parts). Thus, the result. If X is continuous, we could write

$$|X - c| - |X - M| = \begin{cases} c - M & \text{for } x < c; \\ 2(X - c) + c - M & \text{for } c \leq x \leq M; \\ M - c & \text{for } X > M. \end{cases}$$

whereas the mean balances the data above and below it in terms of the magnitudes of observations, the median balances the frequency (count) of data above and below it, irrespective of their magnitudes (here we are assuming that the median for even sample size is the mean of the middle (sorted) sample values). Thus the median can be found iteratively using an indicator function. Define an indicator function $I(x_j) = 1$ if $x_j < \text{Median}$ and $I(x_j) = 0$ otherwise. Summing results in $\sum_{x_j} I(x_j) = n/2$ if n is even; and $(n-1)/2$ if n is odd (because $I(x_j)$ is zero at $x_j = \text{Median}$). Then the median can be defined as

$$\text{Median} = \text{maximum } x_j \text{ such that } \sum_{x_j} I(x_j) = \begin{cases} n/2 & \text{if } n \text{ is even;} \\ (n-1)/2 & \text{if } n \text{ is odd.} \end{cases}$$

As $I(x_j)$ is defined in terms of the median, we start with a guess value (say M_0) and evaluate the LHS. If it is less than the RHS, it means that our guess value was short of the true median. We increment our guess value M_0 by a small amount, and repeat the above procedure (checking $I(x_j) = 0$ values and changing perhaps some of them to 1) until equality holds. If LHS sum is greater than the RHS value, we keep on decrementing our guess value M_0 by a small amount (checking $I(x_j) = 1$ values and changing perhaps some of them to 0) until equality holds. This is easy to parallelize, and can be extended to find quartiles (discussed below).   ∎

2.4.1.1 Advantages of Median   The sample median is least influenced by extreme observations (for n>2). It can be approximated graphically using ogive curves. Median is better than the mean for skewed data. The median can be found even for open-ended data.

Finding the median of a sample of size $n \geq 4$ is computationally more involved than finding the mean. If the data are unsorted, we may require multiple passes through the data to locate the median. The nature of the sample size n (whether it is odd or even) should be known to compute the sample median, whereas this is immaterial to compute the mean and mode. The Theorem 2.1 (p. 48) allows us to use a divide and conquer strategy to find the mean of large samples by finding the mean of subsamples, but such a strategy will not in general work for finding the median.

Sample median is used as smoothing filters in digital image processing. It is also used in data clustering algorithms (*k*-median algorithm). The data item nearest to the mean (if mean does not coincide with a sample item) is called the *medoid*. This nearness can be quantified using a distance metric. The medoid is not unique in the univariate case if it is equally distant from the nearest data points on both the sides of it.

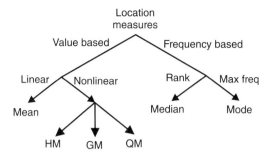

Figure 2.1   Location measures.

This is more of a problem in the multivariate case, in the presence of correlation, for which Mahalanobis distance metric is the most appropriate choice. Medoid is used in *k*-medoid algorithm of data clustering. Location measures are summarized in Figure 2.1.

2.5   QUARTILES AND PERCENTILES

Quantiles is a common name for quartiles (which divide the total frequency into four equal parts), deciles (which divide the total frequency into 10 equal parts), and percentiles (see Figure 2.2). They can be considered as generalizations of the median. As the quartiles divide the total frequency into four equal parts, there are three of them. The first quartile is denoted by Q_1. It is that value of x below which one-fourth of the frequency lie. The second quartile is the same as the median ($Q_2 = $ median). The third quartile Q_3 is that value of x below which three-fourth of the frequency lie (or above which one-fourth of the frequency lie). Deciles divide the total frequency into one-tenth parts. Percentiles are those values of x that divide the total frequency into units of (1/100). Thus, 25th percentile = first quartile. The five parameters $[x_{(1)}, Q_1, Q_2 = M, Q_3, x_{(n)}]$ is called the *five-number summary* of a sample, where $x_{(1)}$ and $x_{(n)}$ are the minimum and maximum of the sample.

The quartiles of grouped data are found using $Q_k = L + c * (kN/4 - M)/f$ where $k = 1$ for Q_1, and $k = 3$ for Q_3. Here, L is the lower limit of the respective class, c is the fixed class interval, N is the total frequency, M is the cumulative frequency up to the respective class, and f is the frequency in the respective class. This formula can be generalized to find the percentiles as $P_k = L + c * (kN/100 - M)/f$,

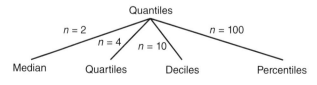

Figure 2.2   Quantiles.

where $k \in \{1, 2, 3, .., 99\}$. A quantile function $Q(u)$ is analogously defined as $Q(u) = \inf \{x : F(x) \geq u\}$, where $u \in (0, 1)$. They can easily be found for continuous distributions (find $x : x = F^{-1}(u)$). The inequality may not strictly hold for discrete distributions due to uneven split of probabilities.

Trimmed quantiles are obtained by trimming data values from either or both the ends of a sample. They are useful when outliers are present. When an equal number of data values are discarded from both extremes, the median will remain the same, but other quantiles will move uniformly toward the median.

2.6   MODE

Mode of a sample is that data item which occurs most frequently. The corresponding value is called *modal value*. If each data item is unique, any of the observations can be taken as the mode. Hence, it is most appropriate when some sample values are repeated. A population with two or more modes is called multi-modal. The mode of grouped data can be found using a two-step procedure:

1. Find the class to which the mode belongs.
2. Compute the mode using the formula Mode $= L + c * \delta_u / (\delta_l + \delta_u)$ where L is the lower limit of modal class, δ_l is the difference in frequency between modal class and the class below it, and δ_u is the difference in frequency between the next class above it and the modal class. For bivariate and higher samples, we could define conditional mode by fixing (conditioning) some of the variables. However, the existence of unique conditional modes does not necessarily mean that the mode for the entire sample is unique. As an example, in a class of students, there could exist multiple values for height or weight but it is rare to have two or more students with the same *height and weight* unless the sample is too large.

2.6.1   Advantages of Mode

As the mode is located along the maximum frequency, it is easy to find irrespective of whether the data are symmetric or skewed. Mode can be found even when the data are open ended. Some other advantages are summarized below: (1) Mode can be approximated graphically, which is useful for skewed distributions and in multivariate case. (2) Mode is not influenced by outliers. (3) The modal value coincides with a sample observation (whereas the median for even sample size and mean need not coincide with sample values). (4) Mode is the most appropriate measure for categorical data. The biggest disadvantage of mode is that it need not be unique. It can coincide with the minimum or maximum of the sample (which is not possible for mean for $n \geq 2$, although it could happen for median). When large number of data items are missing or have default values, the mode can wrongly get located at the missing value. Mode utilizes only the value of most frequently occurring observation (max frequency counts) in contrast to the mean that utilizes actual values of every item in a sample.

An approximate relation exists for samples from bell-shaped distributions as (mean–mode) $\simeq$ 3(mean–median). For right–skewed distributions (mean $\leq$ median $\leq$ mode). This is called the mean–median–mode inequality [18, 19].

2.7 GEOMETRIC MEAN

The Geometric Mean (GM) of n nonzero numbers is defined as

$$\text{GM} = (x_1 x_2 .. x_n)^{1/n} = \left(\prod_{i=1}^{n} x_i \right)^{1/n}, \qquad (2.18)$$

where $\prod$ denotes the product of the observations. We will denote GM by $\hat{x}_n$ to distinguish it from $\bar{x}_n$. If none of the observations are zeros, we could take the logarithm of both sides of equation (2.18) to get $\log(\text{GM}) = (1/n) \sum_{i=1}^{n} \log(x_i)$. This shows that $\log(\text{GM})$ is the arithmetic mean in the "log-space." Because the logarithm is defined only for positive argument, this summary measure is meaningful only when all observations are positive (if at least one observation is 0, the product will itself be zero. The usual practice in such situations is to omit all zeros, and find the GM of the remaining values).

The GM coincides with the sample observations when all observations are equal. Hence, it is most appropriate when the *products* of several positive numbers combine together to produce a resulting quantity as in rates of changes, exchange rates, inflation rates, compound interests, population growth, and so on. Other examples are successive discounts; price or stock market increases and decreases; successive size changes (enlargements or contractions) of images, graphics; successive volume changes; and so on. It is used in image enhancement applications to smooth low contrast images by taking the GM of the surrounding pixels.

Some of the rates of changes can be positive or negative. Is the geometric mean defined for negative numbers? Theoretically No!, because the nth root of a negative number is imaginary. However, if there is an even number of negatives, the product of them will be a positive number. As an example, if $X = \{-3, -2, 2, 3\}$, the product of data values is $+36$. Hence, it looks like we could define the GM when negative numbers occur in pairs. However, this is not true, because it loses the significance as a measure of central tendency, which may wrongly get located toward the positive values). This implies that even in rates of changes involving negative numbers, we should opt for the arithmetic mean. As GM inherently involves the product of individual observations, the change of origin technique is not useful. However, the change of scale transformation $Y = c * X$ provides the relationship $\hat{y}_n = c * \hat{x}_n$. The GM for grouped data is given by $\left(\prod_{i=1}^{n} (f_i x_i) \right)^{1/F}$, where $F = \sum_{j=1}^{n} f_j$ is the total frequency.

Trimmed GM is meaningful when outliers or zero values are present. A left-trimmed GM is appropriate when data contain several zeros.

2.7.1   Updating Formula For Geometric Mean

As in the case of AM, there are situations in which we need to update an already found GM using newly arrived data values. When all sample values are non-negative, an updating formula for GM can easily be derived as

$$\log(\hat{x}_n) = \left(1 - \frac{1}{n}\right) \log(\hat{x}_{n-1}) + \log(x_n)/n. \tag{2.19}$$

where the logarithm is to any base. By taking $1/n$ as a common factor, and denoting log of GM by $\hat{z}$ this could also be written as

$$\hat{z}_n = (1 - 1/n)\,\hat{z}_{n-1} + \log(x_n)/n. \tag{2.20}$$

The successive values can be evaluated iteratively by starting with $\log(\hat{x}_1) = \log(x_1), \log(\hat{x}_2) = \frac{1}{2}\log(\hat{x}_1) + \frac{1}{2}\log(x_2), \log(\hat{x}_3) = \frac{2}{3}\log(\hat{x}_2) + \log(x_3)/3$, and so on. The iterations are stopped when $\log(\hat{x}_n)$ is reached. By taking the anti-logarithm we get the required result. When the variables are transformed using the change of scale transformation $Y = c * X$, these iterations are carried out in y_i and at the end, the GM of Y is multiplied by c to get the GM of X, as shown below. Alternatively, we could add a constant $\log(c)$ to the recurrence (2.19) to iteratively update $\log(\hat{x}_i)$. As the GM involves the product of nonzero observations, weighted GM is meaningless. However, there is one situation where weighting by exponentiation is useful. Consider $GM_w = (x_1^{f_1} x_2^{f_2} .. x_n^{f_n})^{1/N}$, where $f_1, f_2, \ldots, f_n$ are nonzero real numbers, which serves as the weights. If $f_j > 1$, then x_j's are scaled up if $x_j > 1$ and scaled down if $0 < x_j < 1$ (as the powers of a fraction are less than the fraction itself). In the particular case, when $f_j = -x_j$, we get $GM_w = (x_1^{-x_1} x_2^{-x_2} .. x_n^{-x_n})^{1/N}$. Taking log of both sides we get $\log(GM_w) = (-1/N) \sum_{k=1}^{n} x_k * \log(x_k)$.

Lemma 2 Prove that the GM of change of scale transformed data (y-variable) is given by $GM(Y) = c * GM(X)$, where $y_i = cx_i$, and c is a constant (positive or negative).

Proof: $GM(y) = (cx_1.cx_2.. cx_n)^{1/n} = c(\prod_{i=1}^{n} x_i)^{1/n} = c * GM(x)$ because $(c^n)^{1/n} = c$.

∎

▣   EXAMPLE 2.8   Geometric mean for shear strength

Find the geometric mean for the shear strength $X = (32, 80, 56, 75, 69, 26, 44, 50)$ using equation (2.19).

Solution 2.8 As the numbers are large, we will apply the above lemma to compute the GM (product of the original numbers is $42{,}435{,}993{,}600{,}000 = 4.24E{+}13$, which is too big for single precision) by dividing each number by 10 to get

TABLE 2.3 Recursive Computation of Geometric Mean

$X[i]$	$X[i]/10$	$\log(X[i]/10)$	$\ln(GM)$	GM
32	3.2	1.16315081	1.16315081	3.200000000
80	8.0	2.079441542	1.621296176	5.059644256
56	5.6	1.722766598	1.65511965	5.233706096
75	7.5	2.014903021	1.745065492	5.726276491
69	6.9	1.931521412	1.782356676	5.943847650
26	2.6	0.955511445	1.644549138	5.178674512
44	4.4	1.481604541	1.621271338	5.059518589
50	5.0	1.609437912	1.61979216	5.052040191

The fourth column is computed using equation (2.19) as $\log(\hat{x}_n) = \left(1 - \frac{1}{n}\right) \log(\hat{x}_{n-1}) + \log(x_n)/n$.

$X = \{3.2, 8.0, 5.6, 7.5, 6.9, 2.6, 4.4, 5.0\}$. Here, $n = 8$. The calculations are shown in Table 2.3, where the last column contains the successive GM. The GM of scaled data is 5.05204019. Hence using Lemma 3, the GM of original data is 5.05204019*10 = 50.5204019. As $\ln(4.24359936E+13)/8 = 3.92237725$ and $\exp(3.92237725) = 50.5204019$, we get the same result directly. ∎

2.8 HARMONIC MEAN

If all the observations are nonzero, the reciprocal of the arithmetic mean of the reciprocals of observations is known as HM. For ungrouped data, it is defined as HM = $n/\sum_{i=1}^{n}(1/x_i)$. The HM is used when (nonzero) numbers combine via reciprocals as in the case of finding the mean speed of vehicles that go the same distance (not for the same duration). The HM for grouped data is given by $F/\sum_{i=1}^{n}(f_i/x_i)$, where $F = \sum_{j=1}^{n} f_j$.

We will denote it by x_n to distinguish it from $\underline{x}_n$ and $\bar{x}_n$. A simple inequality exists between the three popular means as: (AM ≥ GM ≥ HM). When each of the sample values are weighted using the same set of weights, this identity is preserved [20]. HM finds applications in clustering (k-HMs algorithm). The F-score used in text mining is the HM of precision and recall [2].

2.8.1 Updating Formula For Harmonic Mean

If the sample values arrive successively, we may have to update the HM from an already found value. The updating formula for HM is easily derived as

$$\underline{x}_n = n/[(n-1)/\underline{x}_{n-1} + 1/x_n], \tag{2.21}$$

or in terms of the HM() notation as $HM(X_n) = n/[(n-1)/HM(X_{n-1}) + 1/(x_n)]$. Dividing numerator and denominator by n, and rearranging gives

$$1/HM(X_n) = [(1 - 1/n)/HM(X_{n-1}) + 1/(nx_n)], \qquad (2.22)$$

where $HM(X_j)$ denotes the harmonic mean of $x_1, ..., x_j$, and $HM(X_1) = x_1$. Denoting the reciprocal of the HM by RHM, we could rewrite it in the easy-to-remember form

$$RHM(X_n) = [(1 - 1/n) * RHM(X_{n-1}) + 1/(n * x_n)]. \qquad (2.23)$$

As in the case of GM, the change of origin transformation is meaningless. The change of scale transformation $Y = c * X$ for HM gives $HM(y) = c*HM(x)$.

Lemma 3 If $y_i = c * x_i$, where c is a constant, prove that the HM of transformed data (y-variable) is given by $HM(Y) = c * HM(X)$.

Proof: $HM(y) = n/\sum_{i=1}^{n} 1/y_i = n/\sum_{i=1}^{n} 1/(cx_i)$. Taking the constant c to the numerator, this becomes $c * n/\sum_{i=1}^{n} 1/x_i = c * HM(x)$. ∎

Trimmed HM is meaningful when outliers or zero values are present. A left-trimmed HM is appropriate when data contain several zeros.

■ EXAMPLE 2.9   Harmonic mean finding

Find the HM for the data in Example 2–32 (pp. 2–24) using equation in page 60.

Solution 2.9 The calculations are shown in Table 2.4. Column 2 gives the reciprocal of observations. These sum to 0.170622041. Column 3 gives the RHS of equation (2.22), the reciprocals of which are given in column 4. We could verify our result by direct substitution as $n/(\sum 1/x_i) = 8/0.170622041 = 46.88726$, which agrees with the last entry in column 4. ∎

TABLE 2.4   Recursive Computation of Harmonic Mean

$X[i]$	$1/X[i]$	$\dfrac{(1-1/n)}{HM(X_{n-1})} + \dfrac{1}{nx_n}$	HM
32	0.031250000	0.031250000	32.00000000
80	0.012500000	0.021875000	45.71428571
56	0.017857143	0.020535714	48.69565217
75	0.013333333	0.018735119	53.37569500
69	0.014492754	0.017886646	55.90763087
26	0.038461538	0.021315795	46.91356872
44	0.022727273	0.021517434	46.47394202
50	0.020000000	0.021327755	46.88726001

The fourth column is computed using equation (2.22) as $HM(X_n) = 1/\left[\frac{(1-1/n)}{HM(X_{n-1})} + \frac{1}{nx_n}\right]$.

2.9   WHICH MEASURE TO USE?

Three types of means (AM, GM, and HM), along with the median and mode, serve as measures of location. In addition, the quadratic mean is defined as $\sqrt{\sum_i x_i^2/n}$. A question that analysts face is "which measure is the most appropriate?". This depends both on the nature of the data (qualitative or quantitative, positive or negative) and the application at hand. The mode is the only appropriate measure for numerically coded nominal or ordinal data. For interval and ratio type data, the median is better than the mean and mode if the distribution is skewed. As the medoid coincides with a sample observation, it is preferred when all data values are integers and arithmetic operations involve differences between data values and the medoid (as in clustering). The arithmetic mean is to be preferred when the numbers combine *additively* to produce a resultant value. Examples are consumption of materials or power, quantities measured on a scale such as heights, weights, thickness, and temperatures. The geometric mean is better suited when several nonzero numbers combine *multiplicatively* to produce a resultant value (or equivalently, the logarithm of several nonzero numbers combine additively). This includes rates of changes such as successive discounts; time-dependent growth; successive size changes (enlargements or contractions) of images, graphics; successive volume changes, power and voltage changes and so on. The HM is preferred when *reciprocals* of several nonzero numbers combine *additively* to produce a resultant value. Examples are electrical resistance or capacitance in parallel circuits, average speed of vehicles for the same distance, and so on. Quadratic mean is better suited when squares of several numbers combine additively as in squared Euclidean distances.

2.10   SUMMARY

Several popular measures of location are introduced and exemplified which are useful to compare several groups. The measures of location portray the central location of varying data values. They are sometimes called "sample statistics" which are substitutes for their population counterparts. The mean is vulnerable to unusually low or high values (which are recognized as outliers) in an uneven manner. In such situations, the median should be used. When the data size is large or there is a need to identify more often repeating value in the data, the mode is preferable over the mean or median. If neither the median nor the mode resolves the issue of uneven influences exerted by the outliers on the measures of location, the weighted mean could be chosen as a remedy.

Updating formula for some of them are also presented. These are useful in online computations, where new data arrive continuously. Important properties of these location measures are also discussed. This allows an analyst to choose the most appropriate measure of central tendency for the data at their hand [21]. Median finding algorithms are discussed in References 17, 22, and 23. See Reference 20 for a discussion of AM, GM, and HM inequalities for weighted data, and References 24, 25 for new measures of central tendency and variability. A discussion of visualizing of location measures can be found in References 26–28.

EXERCISES

2.1 Mark as True or False

a) Quartiles divide a data set into three equal parts

b) Third quartiles lies between 7th and 8th decile

c) The mode of a sample can never be the minimum of the sample

d) The median balances the frequency count of data above and below it

e) Geometric mean of data containing at least one zero is zero

f) Every sample observation contributes to the mode

g) Mode of a sample is always a sample value

h) Duplicate data values are counted distinctly in finding the mean.

2.2 The tuition fees of 230 graduate schools per semester are given in Table 2.5, where the count column indicates the number of schools charging the fee on the left. Find the mean and median of tuition fees.

TABLE 2.5   Median of Grouped Data

Fees	Count	Fees	Count	Fees	Count
0K–4K	2	8K–10K	30	14K–16K	44
4K–6K	7	10K–12K	65	16K–18K	10
6K–8K	18	12K–14K	51	>18K	3

2.3 What is a medoid? What are its uses? How can it be used to measure data spread?

2.4 To which of the location measures does a medoid converge to as the sample size is increased?

2.5 For which of the following measures is the change of origin technique useful? (a) arithmetic mean (b) geometric mean (c) harmonic mean (d) median

2.6 In what situations is the mode most appropriate, and most inappropriate? What information is needed to update the mode using new data?

2.7 Which of the following is most appropriate as a measure of location in finding the average distance of vehicles that travel the same duration?

(a) arithmetic mean (b) geometric mean (c) harmonic mean (d) all of them

2.8 Which location measure is most appropriate for the following data? (i) growth of visitors to a web site

(ii) amount of money in a compound interest account (iii) electric current in a parallel circuit (iv) debt of a company.

2.9 If the GM of n data values coincide exactly with one of the data values (say x_k), then prove x_k is the GM of the other $(n - 1)$ data values.

2.10 If the HM of n data values coincide exactly with one of the data values (say x_k), then x_k must be the HM of the other $(n - 1)$ data values.

2.11 What is the most commonly used location measure? What are its advantages over others?

2.12 What are some uses of sample median? What is a medoid? Where is it used?

2.13 What is the least stable measure of central tendency? (a) arithmetic mean (b) geometric mean (c) harmonic mean (d) mode

2.14 Prove that the sum of the deviations of sample values from the sample mean is zero. What is the corresponding population equivalent?

2.15 What is trimmed arithmetic mean? Give formula for trimmed geometric and trimmed harmonic means.

2.16 In what situations is the geometric mean most appropriate? What are some data restrictions on computing it?

2.17 If $\bar{x}_{n-2}^t$ denotes the 1-trimmed mean after deleting the smallest and largest observation in a sample, prove that $(1 - 2/n) * \bar{x}_{n-2}^t = \bar{x}_n - (x_{(1)} + x_{(n)})/n$.

2.18 Describe situations where trimmed mean and median are useful for grouped data. Can you find trimmed mean without complete data sorting?

2.19 Give examples of some situations where the harmonic mean is the most appropriate location measure. When is it most inappropriate?

2.20 Can you always find the GM and HM for standardized data $Y = (X - \bar{x})/s$, where $\bar{x}$ is the sample mean and s is the standard deviation?

2.21 Find the mean of n observations that are in arithmetic progression with first term k and common difference d.

2.22 What is the first step in computing (i) the mode? (ii) the median of raw data and grouped data?

2.23 What are the two situations in which the mean is the same as the sample value?

2.24 When is the change of origin useful in computing the mean?

2.25 The percentage of seeds that germinate from eight different plots are given below: {98.2, 92.7, 89.3, 94.4, 95.0, 83.1, 90.6, 96.1}. Which location measure is most appropriate? Find its value.

2.26 The first-order Bragg reflection of X-ray at different angles through a crystal gave the wavelengths (in nanometers) as {0.0795, 0.0841, 0.0790, 0.0844, 0.0842, 0.0840}. Use the change of scale technique to find the mean and the median.

2.27 The number of hours that a battery can be continuously operated in different devices after a 30 min recharge is given below. Find the median and mean. $X = \{32, 19, 24, 31, 20, 27\}$.

2.28 The resistance of an electronic circuit was measured using five different components as $\{5.2, 4.9\ 5.12, 4.95, 5.1\}$. Find the mean and median. Convert data to z-scores.

2.29 What is trimmed median? If extreme data values are removed from both ends of a sample, does the trimmed median differ from the original median?

2.30 Prove that $(\bar{x}_n - \mu) = \frac{1}{n}\sum_{j=1}^{n}(x_j - \mu) = (\text{Median} - \mu) + \frac{1}{n}\sum_{j=1}^{n}(x_j - \text{Median})$, where $\bar{x}_n$ is the mean of a sample of size n.

2.31 If data values are discarded from the low end of a sorted sample, the trimmed median —(a) moves to the left (b) moves to the right (c) remains the same (d) is unpredictable

2.32 The expected absolute departure of a random variable is minimum when it is taken around the—(a) mean (b) median (c) mode (d) both (a) and (b).

2.33 A plastic polymer thread is subjected to an elongation stress test to see how much it can be stretched before it breaks. Let $X = \{9.2\ 6.7\ 15.3\ 18.0\ 11.6\ 10.8\ 7.7\ 16.1\ 8.5\ 12.0\}$ denote the break point length in cm. (i) Find the mean and the median.

2.34 Soluble dissolvents (in milligram/liter) in drinking water are measured at different places in a city. Find the mean and median, and standardize the data where $X = \{560, 458, 490, 525, 482, 554, 499, 538, 540, 507, 481, 513\}$ is the amount of dissolvent in mg/L.

2.35 Should the complete sample be sorted to compute the trimmed mean using the formula $\bar{x}_m^t = \frac{1}{(n-2k)}\sum_{i=k+1}^{n-k}x_{(i)}$? If not, how much sorting is required?

3

MEASURES OF SPREAD

After finishing the chapter, students will be able to

- Describe popular measures of spread
- Understand range and inter-quartile range
- Understand variance and standard deviation
- Comprehend the Coefficient of Variation
- Apply the above concepts to practical problems

3.1 NEED FOR A SPREAD MEASURE

The prime task in many statistical analyses is to summarize the location and variability of data. One or more concise measures are used for this purpose. These are real numbers for univariate samples, and a vector or matrix for bivariate and higher dimensional samples. Chapter 2 introduced several location measures for this purpose. If repeated samples are drawn from a univariate population, they can lie anywhere within the range (min, max). As shown below, this depends on the shape of the distribution. If the parent population is unimodal (with a clear peak), a great majority of sample values will fall close to the mode. As the mean and mode coincide for symmetric unimodal distributions, we expect most of the data points to fall within the vicinity of the mean for such distributions. On the other hand, if the distribution is uniform, there is an equal chance for any new data item to fall anywhere within

Statistics for Scientists and Engineers, First Edition. Ramalingam Shanmugam and Rajan Chattamvelli.
© 2015 John Wiley & Sons, Inc. Published 2015 by John Wiley & Sons, Inc.

the range. The number of data points that fall in the vicinity of the mean or mode depends more on how fast the distributions tail-off in both directions. We expect less data points around the mean if the tailing-offs are slow, than otherwise. Thus, there are likely to be more data points in the close proximity of the mean for leptokurtic distributions (defined in Chapter 4) when sample size is large.

This shows that a location measure alone is insufficient to fully understand a data distribution. Assume that we have somehow found the mean (average) of a population. In repeated sampling from that population, why do some data points fall above the mean, and some others fall below? Can we predict with some confidence how far from a location measure (e.g., the mean) are the new data values likely to lie? What is the probability that a randomly chosen new data value will fall above the mean or two standard deviations away (in both directions) from the mean? These types of queries can be answered using spread measures discussed below. As the median divides the total frequency into two equal halves, we know that there is a 50–50 chance that a new sample value will be above or below the median. Hence for symmetric uni-modal distributions, we expect that there is an equal chance for new data values to fall above or below the mean too. However, to quantify "how far from a location measure (such as the mean) they are likely to lie," we need well-defined measures. These are called measures of dispersion or spread (we will use "measures of spread," "dispersion measure," or measure of variability synonymously).

Definition 3.1 A univariate dispersion measure concisely summarizes the extent of spread or variability of data in a sample of size $n \geq 2$ using a well-defined statistic, with a minimum value of zero indicating that there is no spread; and an increasingly positive value indicating the extent of spread of observations.

As the zero value is well-defined, this is a ratio measure. Increasing values of it indicate that the sample values are more spread-out over its range. The extent of this spread depends on whether the measure is linear or nonlinear. As shown below, some of the dispersion measures (such as the variance) are upper-bounded by the square of the range. As in the case of location measures, these are also applicable to sample and population. A population parameter is called a scale-parameter if the density function takes the form $(1/\theta) f(x/\theta)$.

There are two situations in which a univariate sample measure of spread can be zero—(i) if the sample contains just one item ($n = 1$), (ii) if all sample values coincide. In case (i) there is no spread as the sample is a singleton. This is symbolically written as $s_1^2 = 0$, where s_n^2 denotes the sample variance (defined in p. 77). Note that if the sample variance uses ($n - 1$) in the denominator, we get a zero ($1 - 1$) in the denominator. Hence, the sample variance is undefined (it is not zero) if ($n - 1$) is used in the denominator, and is zero if n is used. In case (ii), all the sample observations are the same ($x_i = x_j = c \ \forall i, j$). This is as equal as having a singleton sample. The mean in this case is c, so that each of the deviation terms in the numerator of s_n^2 is zero. The sample range and mean absolute deviation are both zero as well (as the minimum and maximum are both c).

3.1.1 Categorization of Dispersion Measures

Sample range, inter-quartile range (IQR), mean absolute deviations (from the mean or median), sample variance, and standard deviation are the most commonly used measures of spread. While the sample variance additively combines the squared deviations of sample values from its mean, the mean absolute deviation combines the absolute values of deviations additively, and the range-based measures (sample range, IQR, etc.) combine the appropriate extremes of sample values linearly. All dispersion-measures quantify the spread of data into a positive numeric scale. They are not affected by a change of origin transformation (as the entire data are translated linearly by this transformation). All of the measures defined below are affected by a change of scale transformation. There are many ways to categorize the measures of spread (see Table 3.1)—(i) linear and nonlinear measures, (ii) pivotal measures and pivot-less measures, (iii) measures that utilize sample size and those that do not use sample size, and (iv) additive and nonadditive measures.

1. Linear and Nonlinear Measures
   Linear measures combine sample values as simple linear functions or their deviations from pivotal values. For instance, the sample range is a linear function of the first and last sample values as $R = (x_{(n)} - x_{(1)})$ (see Section 3.2, p. 71). The mean deviation $\frac{1}{N} \sum_{i=1}^{n} |x_i - \bar{x}_n|$ (where $N = n - 1$, see Section 3.5, p. 76), on the other hand, is a linear function of deviations measured from the mean $\bar{x}_n$. Nonlinear measures combine sample values nonlinearly (as square-roots, squares, or higher powers). Nonlinear measures are the preferred choice in some applications because they often inflate (blow-up) the deviations so that the computed value is larger than those obtained from linear measures.

2. Pivotal Measures and Pivot-less Measures
   Some of the dispersion measures use a location measure as a pivot to quantify the spread (see below). Recall from Chapter 2 that some of the location measures (such as the means) are expressible as a function of the sample values.

TABLE 3.1 Categorization of Dispersion Measures

Measure Name	Additive	Linear	Pivotal	Absolute	Uses Size n	Distance Based
Range	No	Yes	No	Yes	No	Yes
IQR	No	Yes	No	Yes	No	Yes
AAD	No	Yes	Yes	Yes	Yes	No
Variance	Yes	No	Yes	Yes	Yes	Yes
CV	No	No	Yes	No	Yes	No

AAD = average absolute deviation, CV = coefficient of variation uses sample standard deviation s, which in turn uses a pivot. Variance can be considered as squared Euclidean distance between sample values and a vector of all $\bar{x}_n$.

By expanding such location measures as a function of the sample values, it is possible to obtain those spread measures without an explicit location measure. Nevertheless, this criterion allows us to distinguish some spread measures from the others. Exceptions are the range, IQR, and quartile deviation (QD) that do not use a location measure as a pivot.

3. Measures that Utilize the Sample Size

The variance, coefficient of variation (CV), and mean deviations discussed below fully utilize each and every sample value (and thus the sample size n). On the contrary, the sample range utilizes only the minimum and maximum; and IQR utilizes only two of the sample values denoted by Q_1 and Q_3. The range does not distinguish between multi-modal distributions, skewed distributions, and peaked distributions. Hence, the range and IQR are called minimax measures.

4. Additive and Nonadditive Measures

Additive measures are those that can be found by divide-and-conquer (D&C) method without further information. In other words, suppose we divide a sample into two subsamples and find the measure values from these subsamples. If we could combine these values obtained independently from the subsamples without additional information to find the corresponding measure for the entire sample, then it is called additive. In the case of sample range, we need extra information to find the range of the original sample. If a sample S is divided into two subsamples S_1 and S_2, and we find the ranges r_1 and r_2, we cannot find the range of the original sample S unless the subsample minimums and maximums are both known. Sometimes the subsamples may be such that all elements in one of them is less than (or greater than) all elements in the other. If such overlap information about subsamples is known, we could sometimes find the range using the minimum of lower subsample and maximum of the upper subsample. However, the overlap can occur in many ways—(i) S_1 completely subsumes S_2, (ii) S_2 completely subsumes S_1, (iii) minimum of S_2 lies between minimum and maximum of S_1, or vice versa, and (iv) minimum of S_2 is greater than maximum of S_1 or vice versa. In this case, we could obtain $\text{Range}(S) = \max(\max(S_1), \max(S_2)) - \min(\min(S_1), \min(S_2))$. Suppose the subsamples S_1 and S_2 are nonoverlapping, and additionally we know that elements in S_1 are all less than the elements in S_2. In this particular case, we could find the range as $\text{Range}(S) = \max(S_2) - \min(S_1)$. Similar arguments hold when the minimum element of S_1 is greater than the maximum of S_2, in which case the roles of S_1 and S_2 simply get swapped and we obtain Range $(S) = \max(S_1) - \min(S_2)$. Variance is an additive measure.

5. Absolute and Relative Measures

Some of the dispersion measures are absolute. They are expressed in the same unit as that of the observations. Examples of absolute dispersion measures are the range, QD, mean deviations, variance, and standard deviation. Variance, being the average of the squared deviations of observations from their mean, is expressed in the unit squared. Relative measures, on the other hand, do not depend on a unit. Examples are the coefficient of dispersion and CV. Absolute measures are easy to convert into relative measures. Simply find the unit in which they are expressed and divide by another measure (usually one of the location measures) expressed in the same unit. Sample standard deviation (s), being the positive square-root of variance (s^2), has the same unit as the data. Hence, we could divide s by any of the location measures (mean, median, or mode) to get a relative measure. As the standard deviation uses the sample mean as pivotal measure to take the deviations, it is customary to use the mean in the denominator to get a relative measure $s/\bar{x}$. Owing to the possibility of $\bar{x}$ becoming zero (resulting in a very large value) this measure is defined only for $\bar{x} \neq 0$. This measure called the CV (p. 82) can also be expressed as a percentage. As it is dimensionless, it can be used to compare the variability of data measured in different units. For instance, data collected from different geographical regions that have different currencies (dollars, euro, yen, etc.) can be compared without worrying about the currency exchange rates or conversions.

6. Distance-based Measures

Some of the dispersion measures can be cast in distance metric form. As an example, we can interpret the univariate sample range as either the Manhattan distance between x_n and x_1 as $|x_n - x_1|$, or as the Euclidean distance as $[(x_n - x_1)^{1/2}]^2$. The sample variance in the univariate case is the squared Euclidean distance $\frac{1}{N}(X - \overline{X}_n)'(X - \overline{X}_n)$, where X is the data vector of size n, and $\overline{X}_n$ is an n-vector in which each element is the sample mean $\bar{x}_n$ (i.e.,

$$\overline{X}_n' = \{\overbrace{\bar{x}_n, \bar{x}_n, \ldots, \bar{x}_n}^{n \text{ values}}\}).$$ Here, N denotes the appropriate divisor used (either $N = n - 1$, or $N = n$; see discussion below). Similarly, the average absolute deviation (AAD) can be written as $\text{AAD} = \frac{1}{N}\sum_{i=1}^{n}|x_i - \bar{x}_n|$. This could also be written in vector form in which each component is $\sqrt{|x_i - \bar{x}_n|}$. The analogue in the multivariate case is the Mahalanobis distance $(X - \overline{X}_n)'S^{-1}(X - \overline{X}_n)$, where S is the pooled sample variance–covariance matrix.

3.2 RANGE

The sample range can throw more insight into the inherent variability in a population. Suppose repeated samples are taken from a population and the range is updated each time. If it does not vary very much, it is an indication that we have captured most of the variability into the sample. As an example, if the range of temperatures in 24 h for two cities are the same, we cannot conclude that both cities have the same weather because one city, say on the sea-front, might have cooled faster at night whereas another city in mid-plains might have cooled slower. If we have the additional information that the mean temperature during the 24-h period was almost the same, we could have a better perception regarding the weather at the two cities. Thus, a measure of location along with a spread measure can describe the nature of our data in a better way than either of them alone. As temperatures increase and decrease gradually, we can conclude that the weather is more or less the same. As shown below, even this cannot fully describe the data if skewness and kurtosis are also present.

Definition 3.2 Range of a sample is the difference between the largest and smallest observation of the sample. Symbolically, if $X = \{x_1, x_2, ..x_n\}$ are the "n" sample values that are arranged in increasing order,

$$R = (x_{(n)} - x_{(1)}) = \max(X) - \min(X). \tag{3.1}$$

The range is zero in only one particular case—when all of the sample values are the same. In all other situations, it is a positive number which is an integer when the sample values are integers. Even if all data values are negative, the range is always positive as we are subtracting the minimum from the maximum. For instance, if $X = \{-11, -5, -3, -2\}$, the minimum is -11 and maximum is -2, so that the range is $\max - \min = (-2) - (-11) = 11 - 2 = 9$, where we have used the fact that maximum of negative numbers $\max(x_i : x_i < 0 \forall i) = -\min(|x_i|)$. Range is defined for interval or ratio data too. It is also meaningful for numerically coded ordinal data, if the codes are equi-spaced. Coefficient of range is defined as $\mathrm{CR} = (x_{(n)} - x_{(1)})/(x_{(1)} + x_{(n)})$, which is unit-less. If each of the sample values are positive, this measure lies in $[0,1)$. If $x_{(1)}$ is negative, this measure could take any positive value. It is assumed that $(x_{(1)} + x_{(n)})$ is nonzero.

3.2.1   Advantages of Range

The range is easy to compute and easy to interpret. We require only the smallest and largest observations of a sample to compute the range. This can be obtained in a single pass through the data (unless the data are sorted, in which case we can easily pick out the smallest and largest observations in two fetches). Range is easy to update if new data arrive continuously. For instance, suppose data are received from a traffic sensor on a continuous basis. The data may indicate either the number of vehicles in a street or locality; or the speed of a passing vehicle. As new data arrive, it is a simple matter to check if it lies above or below the minimum and maximum to decide whether the range needs to be updated. If new data are within the so far accumulated min and max, the range is unaffected. The range can be bulk-updated if old minimum and maximum are known, and several new sample values are received. Suppose a sample S_k has minimum and maximum $x_{\min}^k$ and $x_{\max}^k$. If the minimum and maximum of a new sample are $x_{\min}^{k+1}$ and $x_{\max}^{k+1}$, the new range is $\max(x_{\max}^k, x_{\max}^{k+1}) - \min(x_{\min}^k, x_{\min}^{k+1})$. Of course, we need to save the new minimum and maximum to update for subsequent iterations.

3.2.2   Disadvantage of Range

The biggest disadvantage of range is that it is extremely sensitive to outliers (on both extremes). As it does not utilize every observation of a sample, it cannot distinguish between skewed distributions that have the same range. It is not a good indicator of spread when the sample size varies. Range is not unit-less. It uses the same unit as that of the data. Thus, it is affected by a change of scale transformation. For example, if the family incomes of a sample are measured in dollars and Euros, the range will be different. It does not lend itself to further arithmetic operations (as does the sample variance).

Range is better suited for univariate data. Range of multivariate data contains too little information about the multivariate spread, especially in the presence of correlation. For example, consider a bivariate sample of say height and weight of students, or amount of two different dissolvents in drinking water. The range can measure only the difference between the individual variates X and Y.

3.2.3 Applications of Range

The sample range has lot many applications in engineering and applied sciences. It is applicable to ordinal and higher scales of measurement. It is used in quality control and process control systems. Some of the data plotting and visualization techniques use the sample range. As an example, the box-plot and range plot use the sample range. The sample range is also used in data transformations. For instance, the min–max transformation in (Section 1.9.4) uses the data range in the denominator. If the sample size is small (say 4 or 5 as in quality control applications), the range is a quite good estimate of the spread. Thus, we use average of the ranges $\overline{R} = \sum R_i/n$ in quality control charts as $(\overline{\overline{x}} \mp 3\overline{R}/(d_2\sqrt{n}))$. The mean of ranges $\overline{R}$ can indicate when a process deviates in one direction. For example, suppose a time-dependent process deviates to the "high" (or increasing) side. Even if the range remains the same, the mean of ranges will steadily increase. However, if the range deviates from both sides (either inwards or decreasing values or outwards or increasing values), the mean of ranges could remain the same.

3.3 INTER-QUARTILE RANGE (IQR)

The sample range is sensitive to outliers at both ends. This could be diminished by removing possible outliers and then computing the range of remaining data. These are called trimmed range. A generalization of it is called the IQR. We defined quartiles in Section 2.5. As the name implies, the IQR is the range of data quartiles.

Definition 3.3 The IQR is defined as $(Q_3 - Q_1)$, where Q_3 and Q_1 are the upper and lower quartiles. (Q_1 is that value below which one-fourth of the observations fall, and Q_3 is that value below which three-fourth of the observations fall, after the sample is arranged in ascending order). One half of IQR is called the QD. The *unit quantile function* is a parametrized version of it defined as $q(u) = (F^{-1}(u) - F^{-1}(1 - u))/2$, where $0 \leq u \leq 1$ and $F(x)$ denotes the cumulative distribution function. This reduces to QD for $u = 3/4$, and is negative for $u < 0.5$. It is unaffected by outliers, and provides supplementary information on the spread of observations around the center of the sample. It is used in boxplots to visually detect outliers.

3.3.1   Change of Origin and Scale Transformation for Range

Range is unaffected by a change of origin data transformation. The change of scale transformation $Y = c * X$ gives the relationship $\text{Range}(Y) = c*\text{Range}(X)$, as both extremes are scaled by the same constant. The constant c is chosen as <1 if X values are very large. This is especially useful when large data are expressed in scientific notation. In this case, dividing by 10^k is done by adjusting just the index of the number. For example, let $x = 3.6524219879E+8$. To divide x by 10^6, simply adjust E+8 to E+2 to get $x = 365.24219879$, which is the number of days in a year.

■ **EXAMPLE 3.1   Outstanding amounts on 10 bank loans**

Ten outstanding loan amounts in a bank are $X = [60{,}000, 40{,}000, 85{,}000, 37{,}000,$ $110{,}000, 280{,}000, 72{,}000, 92{,}000, 154{,}000, 81{,}000]$. Find the range and QD of the data.

Solution 3.1 As the data values are all large, we divide them by $c = 100{,}000$ to get $Y = [0.60, 0.4, .85, 0.37, 1.10, 2.80, 0.72, 0.92, 1.54, 0.81]$. The maximum and minimum values of transformed data are 2.8 and 0.37. The range of Y is $2.80 - 0.37 = 2.43$. From this the range of X is obtained by multiplying by c as $2.43 * 100{,}000 = 243{,}000$. To find the QD, we need to find Q_3 and Q_1. The first quartile is that value below which one-fourth of the data values lie. Rearranging the data in ascending order gives $Y = [0.37, 0.4, 0.60, 0.72, 0.81, 0.85, 0.92, 1.10, 1.54, 2.80]$. As there are two data values below 0.60, $Q_1 = 0.60$. Similarly, $Q_3 = 1.10$ as there are two values above it. From this we get the QD of Y as $1.10 - 0.60 = 0.50$. Multiply by 100, 000 to get the QD of X as 50,000. The quartile coefficient is $(Q_3 - Q_1)/(Q_3 + Q_1) = 50{,}000/170{,}000 = 0.294$.   ■

3.4   THE CONCEPT OF DEGREES OF FREEDOM

Degrees of freedom concept originated in data analysis. Sample variance was the most popular dispersion measure in wide use during the 19th and early 20th centuries.

The concept of degrees of freedom (DoF) is used in many branches of applied sciences. In physics and physical chemistry, it indicates the independent mode or free dimensionality in which a particle or system can move, or be oriented wrt fixed coordinate axes. In mechanical and aeronautical engineering, DoF denotes the flexibility of motion of a particle or an object in 3D. Such a particle has 6 DoF—namely: (i) up or down (heaving), (ii) left or right (swaying), (iii) forward or backward (surging), (iv) tilting up or down (pitching), (v) turning left or right wrt a plane (yawing), and (vi) tilting side-to-side (rolling). It has an entirely different interpretation in statistics, where loosely speaking, it denotes the local level of confidence left in a sample of size $n \geq 2$. If nothing has been estimated from a sample, its DoF is n. The DoF is reduced by one for each statistic (that uses all of the sample values) estimated from it. Consider the deviations $(x_1 - \bar{x}, x_2 - \bar{x}, \ldots, x_n - \bar{x})$. These deviations always sum to zero (p. 53), specifying any of the $(n - 1)$ values automatically determines the nth deviation. This is precisely the reason why we use $(n - 1)$ as the DoF of a sample from which the mean has been estimated. It may also be noted that this reduction in DoF is not a global phenomena. So, if 10 persons estimate the mean of a sample of size 15, the DoF is reduced by one for each one of them (the DoF does not become 5, but it is simply 14 for each person) under the assumption that each person's procedures or actions are independent.

Sampling distribution of the statistic $t = (\bar{x}_n - \mu)/(s/\sqrt{n})$ follows a Student's T distribution with n DoF for normal samples, where n is the sample size. Similarly, the sum of squares of n sample values drawn from a standard normal distribution has a central χ^2 distribution with n DoF. These are discussed in Chapter 11.

It is defined in terms of the mean as the pivot as in equation (3.8), which uses the sample mean $\bar{x}_n$ explicitly. Assume that the variance is computed in two steps. The first step computes the sample mean. The second step then finds the deviations of observations and finds the variance. As the mean has to be estimated from the data, some "information content" of the obtained sample is lost during this process.

For each parameter estimated from the sample, we quantify it as *a unit* loss of information. We use $(n - 1)$ in the denominator of sample variance to indicate the loss of 1 "DoF" due to the estimation of the sample mean from the data. To compensate for this loss of information, it is logical to use $(n - 1)$ as the divisor for the variance. This lead some statisticians to advocate the formula (3.8) for sample variance. However, there are many expressions for the sample variance that does not explicitly involve the sample mean. Some such formulas are given in Chapter 1, which are repeated below:

$$s_n^2 = 1/[n(n - 1)] \sum_{i=1}^{n} \sum_{j>i}^{n} (x_i - x_j)^2, \tag{3.2}$$

and

$$s_n^2 = 1/[n(n - 1)] \sum_{i=2}^{n} \sum_{j=1}^{i-1} (x_i - x_j)^2, \tag{3.3}$$

$$s_n^2 = (1/[2n(n - 1)]) * \sum_{i=1}^{n} \sum_{j\neq i=1}^{n} (x_i - x_j)^2. \tag{3.4}$$

where we have used the subscript partially varying summation notation introduced in Chapter 1. Another formula in terms of order statistic can be found in References 9 and 29.

Definition 3.4 The DoF is a concept associated with the information content of a sample that indicates a local level of confidence left in a sample as a function of the sample size. It is also applied to a statistic computed from a sample or the distribution of a parent population.

The DoF of a statistical distribution is actually a parameter. They are so-called due to an analogy with the sampling distribution of some related statistics. As examples, the Student's T distribution has a parameter which is traditionally known as "n," which is called its DoF; and Snedecor's F distribution has two parameters m,n which are called its (numerator and denominator) DoF. Other distributions that utilize the DoF concept are the χ^2 distribution, Fisher's Z distribution, Wishart distribution, and noncentral versions of these central distributions (noncentral χ^2, F, T, Z [4, 5]. Many other distributions such as the distribution of the trace of a Wishart matrix and the distribution of statistics computed from the sample variance–covariance matrix (such as the distribution of the determinant, or minimum and maximum Eigen values) can also have DoF parameter. Noncentral distributions also exist without the DoF concept. As examples, the noncentral gamma, beta, negative binomial, and hypergeometric laws have shape and scale parameters, and one or more noncentrality parameters but no DoF.

As mentioned above, 1 DoF is lost for each statistic computed from a sample. This does not mean that we must lose one DoF for each statistic. The rule is that if a statistic involves *each and every observation* of a sample, it loses 1 DoF. Thus if the mode or range of a sample is estimated, 1 DoF is not lost. But if the AM, GM, HM, variance or mean deviation, or some other statistic that utilizes each observation of a sample is estimated, 1 DoF is lost for each such estimate.

3.5   AVERAGED ABSOLUTE DEVIATION (AAD)

The AAD (also called sample mean absolute deviation (SMAD)) from the mean is defined as

$$\text{AAD} = \frac{1}{N} \sum_{i=1}^{n} |x_i - \bar{x}_n|. \tag{3.5}$$

where $N = n - 1$ if $n > 1$ and $\bar{x}_n$ is estimated from the sample (some authors use n in the denominator; this is why we have kept N which can be interpreted appropriately). As in the case of sample variance (defined below), this quantity is undefined for a sample of size 1, if $n - 1$ is used as the divisor. This is because the numerator then becomes $x_1 - \bar{x}_1 = 0$, and the expression (3.5) is of 0/0 form. But if $N = n$, the AAD is defined as zero (as the expression (3.5) becomes 0/1 = 0). Expand $\bar{x}_n$ in (3.5), and simplify to get the alternate expressions

$$\text{AAD} = \frac{1}{nN} \sum_{i=1}^{n} |(n-1)x_i - \sum_{j\neq i=1}^{n} x_j| = \frac{1}{N} \sum_{i=1}^{n} |(1 - 1/n)x_i - \frac{1}{n} \sum_{j\neq i=1}^{n} x_j|. \tag{3.6}$$

where we have used a condition on the second indexvar (p. 1–26). The corresponding population analogue is $E|X - \mu|$, where $E[\ ]$ denotes mathematical expectation

(Chapter 8). It is also called mean (absolute) deviation from the mean. As in the case of arithmetic mean, this measure uses each and every observation of the sample. It is affected by outliers, but not as much as the range. Computations can be simplified if medoid is used in place of the mean $\bar{x}_n$, resulting in AAD from the medoid. When all sample values are integers, this will ease the computations because the medoid itself being a sample value will be an integer (whereas the mean need not be an integer) so that the differences are all integers. A related statistic is median absolute deviation (AAD around the median) defined as $\sum_{i=1}^{n} |x_i - \text{Median}|/N$. Median absolute deviation around the median is the middle value of (sorted) absolute deviations of observations from the median. Symbolically,

$$\text{Median absolute deviation} = \text{Median}|x_i - M| \text{ where } M = \text{Median}(x_i). \quad (3.7)$$

Similarly, we could define the median absolute deviation around the medoid as the middle value of (sorted) absolute deviations of observations from the medoid (by replacing M by medoid in (3.7)).

3.5.1 Advantages of Averaged Absolute Deviation

As the deviation from each and every sample value is summed, it contains more information than the range. It is easy to compute as we need only the absolute deviations from the sample mean (or median).

3.5.2 Disadvantages of Averaged Absolute Deviation

It does not lend itself to further arithmetic treatment. For example, if a sample S of size n is divided into two subsamples, and the AAD of each subsample is found, it is not in general possible to combine the subsample values to find the AAD of S.

3.5.3 Change of Origin and Scale Transformation for AAD

The AAD can be found easily using the change of origin and scale transformation. Consider the change of origin transformation $Y = X + c$. Then the AAD of Y is the same as the AAD of X because each term inside the summation becomes $|y_i - \bar{y}_n| = |x_i + c - (\bar{x}_n + c)| = |x_i - \bar{x}_n|$. This holds true for averaged absolute deviation from medoid and median, as the data are simply translated. Next, consider the change of scale transformation $Y = c * X$. Each deviation term of Y is $|y_i - \bar{y}_n| = |c * x_i - c * (\bar{x}_n)| = |c| * |x_i - \bar{x}_n|$, so that $\text{AAD}(Y, N) = |c| * \text{AAD}(X, N)$.

3.6 VARIANCE AND STANDARD DEVIATION

Using the notation introduced in Section 1 (Chapter 1), we define sample variance as

$$s_n^2 = \sum_{i=1}^{n} (x_i - \bar{x}_n)^2/(n-1). \quad (3.8)$$

The variance of the population is defined as

$$\sigma^2 = \sum_{i=1}^{N} (x_i - \mu)^2 / N \tag{3.9}$$

where μ is the population mean, n is the sample size, and N is the population size. Being a sum of squares, s^2 and σ^2 are always ≥ 0. We will keep the divisor n in analogy with the sample covariance (see Chapter 8) that uses n in the denominator although $\bar{x}$ and $\bar{y}$ are estimated from the data. In addition, the sample covariance should reduce to the variance when $y_i's$ are replaced by $x_i's$ (and $\bar{y}$ is replaced by $\bar{x}$). Moreover, in the recursive algorithm for variance defined below, we assume that the variance for $n = 1$ (a single observation) is zero. This assumption is invalid if $(n-1)$ is used in the denominator.

By expanding the square, and summing the resulting terms individually, this could also be computed as

$$N * s^2 = \left(\sum_{i=1}^{n} x_i^2 - n\bar{x}^2 \right) = \sum_{i=1}^{n} x_i^2 - \left(\sum_{i=1}^{n} x_i \right)^2 / n, \tag{3.10}$$

where N is to be interpreted appropriately (as $n - 1$ or n). This has a frequency version given by

$$s_n^2 = \sum_{i=1}^{n} f_i(x_i - \bar{x}_n)^2 / F = \sum_{i=1}^{n} f_i x_i^2 / F - \left(\sum_{i=1}^{n} f_i x_i / F \right)^2, \text{ where } F = \sum_{i=1}^{n} f_i. \tag{3.11}$$

Positive square-root of variance is called the *standard deviation*. Many other formula are also available for the variance (see References 9, 30–32, which are more of theoretical interest than from a computational viewpoint.

3.6.1 Advantages of Variance

The main advantages of variance are that (i) it uses all of the sample observations, (ii) it lends itself to further arithmetic operations, (iii) distribution of sample variance is known when the population distribution is known, and (iv) it can be found without data sorting. It is well defined for univariate as well as multivariate samples (but the range and mean absolute deviation are seldom used for multivariate samples or procedures).

Theorem 3.1 The sample variance can be recursively computed as $s_{n+1}^2 = (1 - 1/n)s_n^2 + (x_{n+1} - \bar{x}_n)^2 / (n + 1)$, with initial value $s_{n=2}^2 = (x_2 - x_1)^2 / 2$.

Proof: We give a proof for the more general result for the unscaled variance. Let N denote the denominator of the sample variance for $n + 1$ DoF. Then $s_{n+1}^2 = \sum_{i=1}^{n+1} (x_i - \bar{x}_{n+1})^2 / N$, where N is the scaling factor (either n or $n + 1$). Split this into two terms to get

$$N s_{n+1}^2 = \sum_{i=1}^{n} (x_i - \bar{x}_{n+1})^2 + (x_{n+1} - \bar{x}_{n+1})^2 = (1) + (2). \text{ (say)} \tag{3.12}$$

Substitute

$$\bar{x}_{n+1} = \bar{x}_n + (x_{n+1} - \bar{x}_n)/(n + 1) \tag{3.13}$$

in (2) to get

$$[(x_{n+1} - \bar{x}_n) - (x_{n+1} - \bar{x}_n)/(n + 1)]^2 = [(x_{n+1} - \bar{x}_n)(1 - 1/(n + 1))]^2. \tag{3.14}$$

This simplifies to $n^2/(n + 1)^2 (x_{n+1} - \bar{x}_n)^2$. ∎

Next, substitute for $\bar{x}_{n+1}$ in (1) to get

$$(1) = [(x_i - \bar{x}_n) - (x_{n+1} - \bar{x}_n)/(n + 1)]^2. \tag{3.15}$$

Expand as a quadratic and sum term by term. The first term becomes $(N - 1)S_n^2$. The product term reduces to zero using equation 2.4 (p. 46) of last chapter. As the second term does not involve the index variable i, summing it n times gives the second square term as $n(x_{n+1} - \bar{x}_n)^2/(n + 1)$. Combine (1)+(2) and take $n/(n + 1)^2(x_{n+1} - \bar{x}_n)^2$ as a common factor to get

$$Ns_{n+1}^2 = (N - 1)s_n^2 + n(x_{n+1} - \bar{x}_n)^2/(n + 1). \tag{3.16}$$

Divide throughout by $N = n$ to get

$$s_{n+1}^2 = (1 - 1/n)s_n^2 + (x_{n+1} - \bar{x}_n)^2/(n + 1). \tag{3.17}$$

If $N = n + 1$ is used (instead of $N = n$ for s_{n+1}^2) as a scaling factor, the corresponding recurrence for variance becomes $(1 + \frac{1}{n})s_{n+1}^2 = s_n^2 + (x_{n+1} - \bar{x}_n)^2/(n + 1)$ with initial value $s_{n=1}^2 = 0$.

■ **EXAMPLE 3.2 Variance of chlorine in drinking water**

Chlorine in drinking water at eight locations in ml/cc are [8, 17, 12, 13, 10]. Find the variance using Theorem 3.1.

Solution 3.2 Table 3.2 gives various steps using both algorithms (that uses n in the denominator and $(n - 1)$ in the denominator). The sample means (second column) need be computed until $(n - 1)$th row. The last entry in 3rd (resp 4th) column is the variance with n (resp. $n - 1$) in the denominator. ∎

Theorem 3.2 If $(\bar{x}_1, s_1^2)$ and $(\bar{x}_2, s_2^2)$ are the arithmetic means and variances of two samples S_1 and S_2 of respective sizes n_1 and n_2, the variance of the combined sample of size $n_1 + n_2$ is given by

$$s_c^2 = \frac{1}{N_1 + N_2}\left[N_1 s_1^2 + N_2 s_2^2 + \frac{n_1 n_2}{n_1 + n_2}(\bar{x}_1 - \bar{x}_2)^2\right]. \tag{3.18}$$

where $N_1 = n_1 - 1, N_2 = n_2 - 1$.

TABLE 3.2   Recursive Calculation of Variance using Theorem 3.1

Data Values	Mean Value	Variance1 (n in Dr.)	Variance2 ($n-1$ in Dr.)
8	8.00	0.00	0.00
17	12.50	20.25	40.50
12	12.333	13.556	20.333
13	12.50	10.25	13.667
10		9.20	11.50

The second column gives successive sample means, the third column gives successive variances using n in the denominator, and the fourth column uses $(n-1)$ in the denominator of variance.

Proof: By Theorem 2.1, the combined sample mean is $\bar{x}_c = (n_1\bar{x}_1 + n_2\bar{x}_2)/(n_1 + n_2)$. The variance of the combined sample by definition is

$$s_c^2 = [1/(N_1 + N_2)] \sum_{i=1}^{n_1+n_2} (x_i - \bar{x}_c)^2. \tag{3.19}$$

Consider the expression $\sum_{i=1}^{n_1+n_2} (x_i - \bar{x}_c)^2$. Split this into two terms T_1 and T_2 over S_1 and S_2, respectively. Substitute for $\bar{x}_c$ in T_1 to get

$$T_1 = \sum_{x_i \in S_1} (x_i - (n_1\bar{x}_1 + n_2\bar{x}_2)/(n_1 + n_2))^2. \tag{3.20}$$

Take $(n_1 + n_2)^2$ outside the summation from the denominator.

$$T_1 = 1/(n_1 + n_2)^2 \sum_{x_i \in S_1} ((n_1 + n_2)x_i - (n_1\bar{x}_1 + n_2\bar{x}_2))^2. \tag{3.21}$$

Add and subtract $n_2\bar{x}_1$ inside the summation, combine $-n_1\bar{x}_1$ and $-n_2\bar{x}_1$ as $-(n_1 + n_2)\bar{x}_1$, to get $T_1 = \frac{1}{(n_1+n_2)^2}\sum_{x_i \in S_1}((n_1 + n_2)[x_i - \bar{x}_1] + n_2(\bar{x}_1 - \bar{x}_2))^2$. Expanding the square and noting that $\sum_{x_i \in S_1}(x_i - \bar{x}_1) = 0$, we get

$$T_1 = \frac{1}{(n_1 + n_2)^2} \left((n_1 + n_2)^2 \sum_{x_i \in S_1} (x_i - \bar{x}_1)^2 + n_2^2 \sum_{x_i \in S_1} (\bar{x}_1 - \bar{x}_2)^2 \right). \tag{3.22}$$

As $(\bar{x}_1 - \bar{x}_2)$ is a constant, the second expression becomes $n_1 n_2^2(\bar{x}_1 - \bar{x}_2)^2$. Also, $\sum_{x_i \in S_1}(x_i - \bar{x}_1)^2 = N_1 s_1^2$. Thus, T_1 simplifies to $N_1 s_1^2 + n_1 n_2^2(\bar{x}_1 - \bar{x}_2)^2/(n_1 + n_2)^2$. A similar reduction is possible for T_2 by adding and subtracting $n_1\bar{x}_2$ inside the summation of T_2. This gives us $T_2 = N_2 s_2^2 + n_2 n_1^2(\bar{x}_1 - \bar{x}_2)^2/(n_1 + n_2)^2$. Add T_1 and T_2, and simplify to get

$$s_c^2 = \frac{1}{N_1 + N_2}\left[N_1 s_1^2 + N_2 s_2^2 + \frac{n_1 n_2}{n_1 + n_2}(\bar{x}_1 - \bar{x}_2)^2\right]. \qquad (3.23)$$

where N_1 and N_2 are the divisors used in the respective variance (n or $n - 1$). ∎

Corollary 1 The covariance of two subsamples can be combined using the relationship

$$COV_c = \frac{1}{N_1 + N_2}\left[N_1\,COV_{c_1} + N_2\,COV_{c_2} + \frac{n_1 n_2}{n_1 + n_2}(\bar{x}_1 - \bar{x}_2)(\bar{y}_1 - \bar{y}_2)\right]. \quad (3.24)$$

where N_1 and N_2 are the scaling factors, COV_{c_1} is the covariance and $(\bar{x}_1, \bar{y}_1)$ is the mean vector of the first subsample, and COV_{c_2} is the covariance and $(\bar{x}_2, \bar{y}_2)$ is the mean vector of the second subsample.

3.6.2 Change of Origin and Scale Transformation for Variance

Theorem 3.3 If data are transformed as $y = (x - a)/c$, the variances are related as $s_y^2 = (1/c^2)s_x^2$.

Proof: The means are clearly related as $\bar{y} = (\bar{x} - a)/c$. Consider $\sum_{i=1}^{n}(y_i - \bar{y})^2 = \sum_{i=1}^{n}((x_i - a)/c - (\bar{x} - a)/c)^2 = 1/c^2 \sum_{i=1}^{n}((x_i - a) - (\bar{x} - a))^2 = 1/c^2 \sum_{i=1}^{n}(x_i - \bar{x})^2$. Divide both sides by N gives $s_y^2 = (1/c^2)s_x^2$. ∎

> Dispersion measures concisely quantify the amount of spread inherent in a sample. Some dispersion measures use a location measure as a pivot to calculate the deviations (Section 3.1.1). Both the sample range and IQR do not use a pivot. Hence, they can be used to get a preliminary estimate of the spread. Because the dispersion measures explain the inherent variability in the sample data, those measures that utilize a location measure are preferred for engineering applications. All absolute spread measures depend on the unit of measurement of the variable. In other words, they are *scale variant*. However, there are some relative measures such as the CV that does not depend on the unit of measurement. This is summarized as several theorems below (Section 3.6) in p. 82. The popular dispersion measures can be arranged according to the inherent information on the amount of spread captured by them from a sample as: range < IQR < AAD < variance. Thus, the variance and standard deviation (positive square-root of variance) contain maximum spread information in a sample.

3.6.3 Disadvantages of Variance

The main disadvantages of variance are that (i) extreme observations on either side has a large influence on variance, (ii) it is inappropriate for numeric coded ordinal data, (iii) it can result in loss of precision due to squaring when large decimal numbers are involved, and (iv) it may require 2 passes through data (in standard algorithm) although it can be computed in a single pass.

3.6.4   A Bound for Sample Standard Deviation

As the sample variance is a sum of squares, it is always ≥ 0. Several researchers have come up with bounds for variance or standard deviation. Mcleod and Henderson [33–35] gave a lower bound for the standard deviation. Shiffler and Harsha [36] provided an upper bound for the standard deviation in terms of the range. For a sample of size ≥ 3, these bounds can be combined to get a bound for the standard deviation as

$$R/(2(n-1))^{1/2} \leq s \leq 0.5 * R * (n/(n-1))^{1/2}. \tag{3.25}$$

where R is the sample range. Note that the upper bound is not tight asymptotically. As n becomes large, the quantity $(n/(n-1))^{1/2}$ will converge quickly to 1, giving $s \leq R/2$. When $n = 2$, the square-root simplifies to $\sqrt{2}$, giving $s \leq R/\sqrt{2} = 0.70711 * R$. In fact, it can be shown that for $n = 2$ the equality holds, because the variance is $(x_1 - x_2)^2/2$. A sharper upper bound can be found in Reference 37 as $s \leq R(1/2 - (n-2)/[n(n-1)])^{1/2}$ for $n \geq 2$. Write $(n-2)$ as $(n-1)-1$ to get

$$s \leq R(1/2 - 1/n + 1/[n(n-1)])^{1/2}. \tag{3.26}$$

As $1/[n(n-1)] \to 0$ when n becomes large, (3.26) becomes $s \leq R(1/2 - 1/n)^{1/2}$ for large n (say ≥ 20).

3.7   COEFFICIENT OF VARIATION

Definition 3.5 The CV of a sample is a relative ratio-measure defined as $(s/\bar{x}) \times 100$, and the corresponding population CV is $(\sigma/\mu) \times 100$. This form of CV is called the standard form, and it applies to single variables only.

3.7.1   Advantages of Coefficient of Variation

The CV is simple to understand. It is a unit-less measure whose numerical value is high when data variance is high. If the variability of two samples measured in different units are to be compared, CV is the most appropriate measure. In other words, CV allows variability of heterogeneous samples to be compared among themselves. A change of scale transformation $Y = c * X$ will not change the CV as the c will be canceled out from the numerator and denominator. It can be used to create confidence intervals. It provides caution on the sample size, normality, or departures from it.

The CV of a random sample of size 1 is zero if n is used as a divisor in the variance, and the CV is undefined if $n - 1$ is used. Symbolically, $CV(x_1) = 0$ if n is used as scaling factor. For a sample of size 2, $CV(x_1, x_2) = |x_1 - x_2|/(x_1 + x_2)$ if n is used, and $CV(x_1, x_2) = \sqrt{2}|x_1 - x_2|/(x_1 + x_2)$ if $n - 1$ is used in variance. When CV is used in the model setting, it can indicate which model better fits the data—the smaller the CV, the better the fit.

3.7.2   Disadvantages of Coefficient of Variation

As the denominator of CV contains $\bar{x}$, the arithmetic mean should be nonzero. If a variable takes both positive and negative values and the mean is very close to zero, we cannot use the CV. Thus it is most appropriate when variables are either strictly positive or strictly negative. This is a disadvantage. However, there is a catch. The CV is location variant. This means that a change of origin transformation $Y = X + c$ results in the same numerator s_y, but the denominator is shifted by c units. When CV is used to compare the variability of two samples, at least one of which has zero mean, we could choose the c carefully in such a way that the resulting means are nonzero. Alternatively, we could remove outliers from that sample which had a mean near zero with the hope of shifting it away from zero. It cannot easily be extended to the multivariate case. Another disadvantage of CV is that it can result in misleading or conflicting interpretations under some transformations such as $y = \log(x)$.

3.7.3   An Interpretation of Coefficient of Variation

As the CV is a ratio of standard deviation (which being the +ve square-root of variance is always positive) to the mean, its sign depends on the sign of $\bar{x}$ (if $\bar{x}$ is positive (negative), CV is positive (negative); if $\bar{x}$ is zero, CV is undefined). As the numerator and denominator are expressed in the same unit as the variate, the ratio is unit-less. Thus, it summarizes the dispersion of a variable as a concise and unit-less real number. Hence, it can be used across geographical boundaries, irrespective of the units in use. When two samples are being compared by the CV, a higher value may be due to a lower $\bar{x}$. Depending on whether $\bar{x} \to 0$ from below or above, the CV will tend to $\pm\infty$ for fixed variance. The CV also has an interpretation in terms of the ratio of the root mean square error (RMSE) to the mean of the dependent variable in regression models, and as the ratio of the standard error of an estimate to its estimated value [see $(\hat{\theta})/\hat{\theta}]$ in parameter estimation]. The smaller the CV, the better the goodness of fit, or the estimation procedure.

3.7.4   Change of Origin and Scale for CV

Consider the change of scale transformation first. Let $Y = c * X$, where c is nonzero. We know that $s_y = |c| * s_x$ and $\bar{y} = c * \bar{x}$. As the scaling factor c factors out from both the numerator and denominator, the CV is scale invariant ($CV_y = CV_x$). This is a desirable characteristic in some applications such as cross-national comparison of traits or attributes such as income, profits, and expenses or cross comparison of scores in different tests such as GRE and GMAT. Next, consider a change of origin transformation $Y = X + d$. In this case the standard deviation remains the same, but the mean is shifted by d units. Then, $CV_y = s_y/(\bar{x} + d) = CV_x * \bar{x}/(\bar{x} + d)$. Thus, the coefficient of variability can be increased or reduced depending on the magnitude and sign of $\bar{x}/(\bar{x} + d)$.

3.8 GINI COEFFICIENT

Gini's mean difference is a summary statistic that measures the extent of the distribution of a variable by fixing other variables. For a sample $(x_1, x_2, \ldots, x_n)$, it is defined for raw data as

$$GMD = (1/K) \sum_{i=1}^{n} \sum_{j=1}^{n} |x_i - x_j| = (2/K) \sum_{i=1}^{n} \sum_{j>i}^{n} |x_i - x_j|. \tag{3.27}$$

where $K = n(n - 1)$ if j varies from i to n or $K = n^2$ if j varies from 1 to n. For frequency distributions

$$GMD = [1/(F(F - 1))] \sum_{i=1}^{n} \sum_{j=1}^{n} f_i * f_j * |x_i - x_j|. \tag{3.28}$$

where $F = \sum_{i=1}^{n} f_i$. This has the same unit as the data. A unit-less measure can be obtained by dividing GMD by $\bar{x}$ to get the Gini coefficient of concentration as

$$GCC = (1/K) \sum_{i=1}^{n} \sum_{j=1}^{n} |x_i - x_j|/\bar{x}. \tag{3.29}$$

As it is a mean difference (see equation 3.2), it can be used to measure dispersion of highly skewed data. As the numerator is always positive, GCC can take any positive value.

3.9 SUMMARY

Popular measures of dispersion are discussed at length in this chapter. In a sense, the dispersion captures and summarizes data information. The lesser dispersion value refers to more consistency and hence reliability (see Figure 3.1). The sample dispersion measure is a surrogate of the corresponding unknown population dispersion. Commonly used dispersion measures are range, variance, QD, coefficient of variance, and absolute mean deviation (see Figure 3.2). The range $(x_n - x_1)$ is not indicative of the neighborhood of the data as the same range value could occur among smaller or larger values. Hence, the sample dispersion is preferred over the sample range. The variance is of second degree, while the mean is of first degree. The CV (which is the ratio of the standard deviation over the mean) is often used if the measure of dispersion has to be compared across samples. When the data contain one or more unusual value (which is technically called outlier), the inter-QD is selected over the variance or range. The 50th percentile (which is technically called the median) partitions the ordered data into two equal segments. The inter-QD portrays the range between the

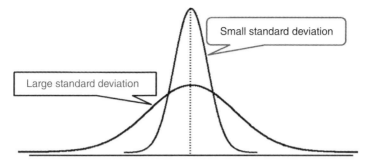

Figure 3.1 Dispersion low and high.

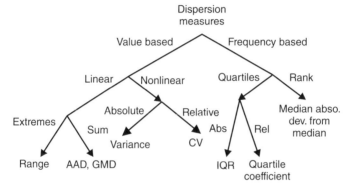

Figure 3.2 Dispersion measures. (AAD = Average Absolute Deviation from the mean, GMD = Gini's Mean Difference, CV = Coefficient of Variation, IQR = Inter-Quartile Range, Abs = Absolute, Rel = Relative.)

25th and 75th percentiles. When the data have outlier(s), the inter-QD is preferable over other measures of dispersion. Alternatively, the measure of absolute deviation is exercised when the data are skewed or have thick tails.

Every data analyst who works with numeric data will find the discussion very illuminating and easy to follow. Updating formulas that comes handy when new data arrive are presented and illustrated. A discussion of algorithms for variance can be found in References 30–32, 38–41. A measure of variance for hierarchical data appears in Reference 42. Evaluating methods of symmetry are discussed in References 43–45, and Poisson dispersion in Reference 46.

EXERCISES

3.1 Mark as True or False

   a) Dispersion is a measure of data spread

   b) Third quartile lies between 7th and 8th decile

   c) Sample range can distinguish between skewed distributions

   d) Every sample observation contributes to the range

e) Range of a sample can be negative when all data values are negative

f) A sample with large CV is less dispersed than a sample with small CV

g) CV measures the data spread irrespective of the unit of measurement.

3.2 What is the sample range when all sample observations are the same (say k)?
(a) 0 (b) k (c) ∞ (d) 1

3.3 Which of the measure additively combines the squared sample values? (a) mean absolute deviations (b) sample range (c) inter-quartile range (d) variance

3.4 What are the popular categorization of dispersion measures?

3.5 What are some desirable qualities of a good measure of dispersion?

3.6 For which of the following measures is the change of origin technique useful? (a) mean absolute deviations (b) sample range (c) inter-quartile range (d) variance

3.7 If a sample S is split into two sub-samples, can you find the range of S if the range of the subsamples are known? If not, what additional information is needed?

3.8 Identify the unit of measurement in each of the following statistics: (i) $s_n^2/\bar{x}_n$ (ii) AM/GM, where AM $= \bar{x}_n$, GM $=$ geometric mean (iii) μ_4/μ_2 if all measurements are in centimeters.

3.9 Consider a statistic defined as $s_n^d = \frac{1}{n-1}\sum_{j=1}^{n-1} d_j^2$, where $d_j = x_{j+1} - x_j$ denotes the difference between successive ordered observations. When is it minimum? What does it measure?

3.10 Prove that $s \leq R(\frac{1}{2} - \frac{1}{n})^{1/2}$ for large n, where s is the sample standard deviation and R is the range.

3.11 Why is it that a location measure alone is insufficient to fully understand a data distribution?

3.12 A city has five weight-loss clinics. Each one uses a different diet and exercise program. (i) An executive who wishes to shed the maximum body fat in shortest time should prefer a weight-loss program having
(a) low variance (b) high variance (c) negative skew (d) platykurtic
(ii) A housewife who has 15 days available should prefer a weight-loss program having_____
(a) high mean >15 (b) low variance (c) low mean and high variance (d) high mean and low variance.

3.13 If $F^{-1}(u) = \sup(x : F(x) \leq u), 0 < u < 1$ denotes the inverse of the CDF, prove that a symmetric distribution can be characterized by the condition $\alpha = F^{-1}(u) + F^{-1}(1-u) = 0$ for $u \in (0, .5)$. Prove that such distributions are symmetric around 0 if $F^{-1}(1/2) = 0$ and symmetric around μ if $F^{-1}(1/2) = \mu$. Prove that $\alpha > 0$ for positively skewed and $\alpha < 0$ for negatively skewed distributions. Can you compute this measure for frequency distributed data?

3.14 Can the QD be ever zero? If so, for what type of data? What is its interpretation?

3.15 Describe the unit quantile function. How is it useful to measure sample variance?

3.16 Critically examine the different measures for variation, and indicate their advantages and disadvantages.

3.17 In what situations is the recursive algorithm for sample variance helpful over the iterative version?

3.18 If there are two items (x_1, x_2) in a sample, prove that the mean absolute deviation around the mean and median are both equal to half the sample range.

3.19 If the sample mean and variance of two independent samples of sizes 8 and 10 are (65, 20) and (70, 32), find the variance of the combined sample.

3.20 What are the possible ranges of values for CV? How is it helpful in comparing the variability of different samples?

3.21 How is the sample standard deviation related to the range? What is the asymptotic inequality between them when sample size n is large?

3.22 If the origin and scale are transformed as $y = c * x + d$, where $c \neq 0$, prove that $CV_y = CV_x * \bar{x}/(\bar{x} + d/c)$.

3.23 How does the change of origin and scale transformation affect the sample AAD?

3.24 Consider a measure defined as $\bar{Q} = \frac{1}{4}\sum_{i=1}^{4}(Q_i - Q_{i-1})$, where $Q_0 = x_{(1)}$ is the first-order statistic, $Q_4 = x_{(n)}$ and other Q_i's are the quartiles. Can it be used to measure the dispersion? What does high values indicate?

3.25 Prove that the unit quantile function of one parameter Weibull distribution with CDF $F(x) = 1 - e^{-x^a}, x, a > 0$ is $Q(u) = [1 - \log(1 - u)]^{1/a}$. Show that the median is $\log(2)^{1/a}$.

3.26 When is the sample range meaningful for numerically coded ordinal data?

3.27 What information is needed to update the sample range using new data?

3.28 The percentage of seeds that germinate from eight different plots are as follows: {98.2, 92.7, 89.3, 94.4, 95.0, 83.1, 90.6, 96.1}. Find the coefficient of range CR.

3.29 If a quartile coefficient is defined as $qc = (Q_3 - Q_1)/(Q_3 + Q_1)$, what are the possible values? Is it absolute or relative measure?

3.30 Consider a measure defined as $[(Q_1 - p - \text{Median}) - (\text{Median} - Q_p)]/[Q_1 - p - Q_p]$, where p is a percentile. What does it measure? What are the possible values?

4

SKEWNESS AND KURTOSIS

After finishing the chapter, students will be able to

- Describe measures of Skewness
- Understand absolute and relative measures
- Comprehend Galton's, Pearson's, Bowley's, and Kelly's measures
- Interpret Pearson's and Stavig's kurtosis measures
- Describe L-kurtosis
- Understand spectral kurtosis

4.1 MEANING OF SKEWNESS

The literal meaning of "skew" is a bias, dragging, or distortion toward some particular value, group, subjects, or direction.

Definition 4.1 A measure of skewness is a numeric metric to concisely summarize the degree of asymmetry of a unimodal distribution that can be compared with other similar numbers.

A great majority of statistical distributions are asymmetric (see Table 4.1). Nevertheless, symmetrical laws such as the normal, Student's T, Laplace, Cauchy distributions, and the distribution of sample correlation coefficients are more popular.

Statistics for Scientists and Engineers, First Edition. Ramalingam Shanmugam and Rajan Chattamvelli.
© 2015 John Wiley & Sons, Inc. Published 2015 by John Wiley & Sons, Inc.

TABLE 4.1   Asymmetry of Some Statistical Distributions

Type	Discrete /Continuous	Examples
Symmetric	Discrete	Binomial($n, 1/2$), Discrete Uniform
Symmetric	Continuous	Normal, Cauchy, Student's T, Laplace, beta(p, p)
Asymmetric	Discrete	Binomial(n, p), $p \neq \frac{1}{2}$, Poisson, geometric
Asymmetric	Continuous	Exponential, Snedecor's F, gamma, beta(p, q); $p \neq q$

Some of the distributions exhibit symmetry for particular parameter values. Examples are beta(p, p), Binomial ($n, 1/2$).

Asymmetry of unimodal distributions can be due to a flat left tail or right tail. Two asymmetric mirror-image distributions (around the mode) could have exactly the same location and spread. A quantified measure of this asymmetry is needed to compare and contrast such distributions (see Reference 47 for bimodal data).

A change of origin transformation can be used to align the location measures of two distributions (or samples). Then, we could compare the spreads of the two distributions. These two measures are insufficient to fully understand the data. Two distributions could have the same location and spread, but one could tail-off slowly to the right, whereas the other could tail off slowly to the left. In other words, two asymmetric mirror-image distributions (around the mode) could have exactly the same location and spread. This is exactly the reason for studying skewness [48]. This information is useful to fit empirical distributions and in parametric analysis. In addition, possible outliers under the assumption of normality of data may turn out to be nonoutliers when the data distribution is actually skewed.

A skewed distribution may be a desired outcome in some domains. As an example, instructors in several educational institutions set the exam questions in such a way that the resulting scores reflect symmetry (the mark distribution is often unimodal with a great majority of the class having marks in the neighborhood of the mode, and a few students with scores in both the extremes—a few failures and more or less an equal number of distinctions. Owing to the heterogeneity of student population and different learning habits, a perfect bell shape is seldom achieved. Even if the instructor had a bell-shaped distribution in mind while setting the exam questions, it could end up as a skewed distribution. A mark distribution skewed to the left indicates an easy exam (or a question paper leak or mass copying), whereas a distribution skewed to the right indicates a difficult exam.

Asymmetry is the opposite of symmetry. It is literally applied to physical structures, arrangements, formations, and so on. Examples are buildings, walls, gates, paintings and pictures, arrangement of flowers and beads, formation of groups of people or ships, and so on around an axis. In statistics, it is used to describe data distributions around a central value. It is more meaningful for unimodal distributions. The uniform distribution (both discrete and continuous) exhibits a special type of symmetry. Note that the symmetry is always measured with respect to the X-axis (most software packages record the variate values along the abscissa and frequencies along the Y-axis). The following discussion is more pertinent to continuous data than discrete ones. With this convention, positive skewness (or skewed to the right) indicates a heavy and long-extending tail on the right side of the location alignment. Similarly, a negative skewness (or skewed to the left) indicates a heavy and long-extending left tail. In other words, a disproportionate amount of data falls on the left or right side of a unimodal distribution, thereby dragging or dispersing the location measures. This will be apparent only when the data size is large, and the spread is small. Consider the systolic BP of some patients. If the number of patients is small (say <12), the values may all be distinct, in which case the mode is not unique. Therefore, we may have to use grouping of data (using a class interval) to find symmetry.

The mean, median, and mode coincide for symmetric laws. In addition, the quartiles are equally distant from the median. However, for negatively skewed distribution, the mean is less than the median and mode in general (although exceptions do exist). For positively skewed distribution, the mean is greater than the median and median is greater than the mode in general. Thus, the difference between these measures can tell us if the data are symmetric or skewed [12]. It is important to remember that a skewed sample may come from a symmetric distribution and *vice versa* because of sampling errors. This error is minimal when the sample size is large.

If the distribution of marks in a series of tests of the same difficulty level moves from a positively skewed shape at the beginning to a negatively skewed shape at the end of the course, it is an indication that the teaching was effective and student participation was good.

From Table 4.1, it is evident that most of the symmetric distributions are continuous. An exception is the discrete uniform distribution, which exhibits a special type of symmetry. The binomial distribution is symmetric when $p = 0.5$, and the Poisson distribution (Chapter 6, p. 6–67) approaches symmetry for very large λ values.

4.1.1 Absolute Versus Relative Measures of Skewness

Absolute measures express the skewness in terms of the unit of measurement. As the arithmetic mean is greater than mode for positively skewed distributions, $d =$ (mean-mode) provides a quick check to see if data are skewed. This is an absolute measure because it uses the same unit as the data. Similarly $(Q_3 + Q_1) - 2 * M$ where M is the median is zero for symmetric laws and >0 for positively skewed distributions. A disadvantage of these measures is that they cannot be compared across samples. The unit can be canceled by dividing the quantity with another quantity computed from the same sample that has the same degree. For instance, Bowley's measure is obtained by dividing $(Q_3 + Q_1) - 2 * M$ by $(Q_3 - Q_1)$. This measure is zero in two situations:–(i) when data are symmetric, the average $(Q_3 + Q_1)/2$ coincides with the median M, so that the numerator is zero (ii) when all sample values are the same

(Q_3, Q_1), and M are equal. As the denominator is much greater than the numerator, the numerical value of this measure is much less than 1.

How much skewed is a skew distribution? We need a yardstick to measure the amount of departure from an otherwise symmetric law. The original skewness measures used the standard normal distribution $N(0, 1)$ as the yardstick. To quantify the amount of skewness, we could consider the standard error of skewness (SES) and measure the departure from twice the standard error. An approximate SES for a sample of size n is $\sqrt{6/n}$ (see Reference 49. If 2*SES < absolute value of skewness, we may reasonably conclude that the data are from a skewed distribution. We could also distinguish if the data are skewed to the left (skewness <0) or to the right (skewness >0). Skewness can be categorized based on the shape of both tails. A distribution which is skewed to the right may have a left tail either below a standard normal left tail, above it, partially below and partially above it, or align exactly with the normal. This gives rise to lepto-right-skewed, meso-right-skewed and platy-right-skewed distributions (where lepto-, meso- and platy- indicate the behavior of the left tail of a right-skewed distribution). The difference is subtle but important. Analogous definition holds for left-skewed distribution whose right tail may be unaligned with a normal right tail. As defined in the following, the kurtosis measures the relative concentration or amassment of probability mass toward the center (peak) of a distribution. Hence, these peculiarities are of interest among itself than from a kurtosis view point.

Note that the third-order central moments (in the numerator of some skewness measures) vanish not only for the normal law but also for any symmetric distribution. Whether the fourth cumulant vanish or not depends on the distribution. Both the skewness and kurtosis are measures of shape departures from normality.

Definition 4.2 Skewness is a numeric measure of the degree of departure of a sample of size $n > 2$ from symmetry.

The above-mentioned definition pertains to the sample, although it is defined for both the sample and the population. If a unimodal skewed distribution is superimposed on a unimodal bell-shaped distribution so as to align the peaked points exactly (location alignment), we could visualize various possible types of asymmetry in the left and right tails. A positively (respectively negatively) skewed distribution has longer tails on the right (left) side of the location alignment.

■ **EXAMPLE 4.1   Check the skewness of marks**

The mark distribution of 60 students of a class had a skewness of −0.65 at the start of the semester and +0.75 at the end. Are the marks significantly skewed?

Solution 4.1 Here $n = 60$, so that $6/n = 0.10$, SES $= 6/n^{1/2} = 0.316227766$, and 2*SES $= 0.632455532$. As the absolute values of skewness are both greater than 2*SES, we conclude that both distributions are skewed.   ■

Skewness was originally defined as a measure of asymmetry. All three location measures (the mean, median, and mode) together can throw some insight on the skewness of a data distribution. In general, mean < median < mode for negatively skewed distributions, mean = median = mode for symmetric distributions, mean > median > mode for positively skewed distributions. The variance could be exactly identical for positively and negatively skewed distributions. We need a measure of skewness to distinguish between possible asymmetries. There are many measures for this purpose [50].

4.2 CATEGORIZATION OF SKEWNESS MEASURES

1. Location and scale-based measures
   These measures combine the location and scale measures (in the numerator and denominator, respectively) to get a unit-less measure of dispersion. One popular example is Pearson's η measure [51] for a sample defined as $\eta = (\bar{x}_n - \text{mode})/s_n$. As the mode can be on the left or right of $\bar{x}_n$, the numerator can be positive, negative, or zero. The denominator being the positive square-root of variance is always positive. Hence, η can take any real value. As $\bar{x}_n$ and mode both lie in between the minimum and maximum, η is bounded by the sample range.

2. Quartile-based measures
   These measures utilize the quartiles of a distribution. The popular examples are Bowley's, Hinkley's, and Kelley's measures discussed in the following text. Another measure that utilizes the averaged deviations of percentiles from the median using a cutoff threshold can be found in Reference 43.

3. Moment-based measures
   These utilize the third central moment as $\gamma = \mu_3/\sigma^3$ for the population. They are defined for both the population and the sample. As the denominator is the cube of the standard deviations, they are also unit-less and scale invariant, that is, X and $Y = |c| * X$ have the same skewness for nonzero $c \in R$.

4. Measures that utilize inverse of distribution functions
   These measures use the inverse of theoretical distribution functions. One example is the spread function. Let $F^{-1}(u) = \sup(x : F(x) \leq u), 0 < u < 1$ denotes the inverse of the CDF. Then, a symmetric zero-centered distribution can be characterized by the condition $\alpha = F^{-1}(u) + F^{-1}(1 - u) = 0$ for $u \in (0, .5)$. Such distributions are symmetric around 0 if $F^{-1}(1/2) = 0$, and symmetric around μ if $F^{-1}(1/2) = \mu$, in which case $2\mu - \alpha = 0$. Note that $\alpha > 0$ for positively skewed and $\alpha < 0$ for negatively skewed distributions. Balanda and MacGillivray [52] used $S_F(u) = F^{-1}(0.5 + u) - F^{-1}(0.5 - u)$ for $0 \leq u \leq 0.5$ as a measure of skewness. If both F and G are unimodal and invertible, one could produce a plot of $S_F(u)/S_G(u)$ (or its inverse) to compare the relative skewness.

5. Measures that utilize L-moments
   For a real-valued random variable X with finite mean μ, the L-moment is defined as expected value of linear combination of order statistics.

The first L-moment is the same as the mean, so that $\lambda_1 = E(X) = \mu$. The second L-moment $\lambda_2 = 0.5 * [E(X_{(2:2)} - X_{(1:2)})]$ where $X_{(i:n)}$ for $i < n$ denotes the ith order statistic of a sample of size n. This is E(half range). Define $\lambda_3 = (1/3) * E(X_{(3:3)} - 2 * X_{(2:3)} + X_{(1:3)})$. The L-skewness is then defined as $\tau_3 = \lambda_3/\lambda_2$ [9, 53]. As λ_3 and λ_2 are expected values of linear combination of order statistics, τ_3 is unit-less. Analogously, $\tau_2 = \lambda_2/\lambda_1$ can be considered as a measure of dispersion.

The quartiles, deciles, and percentiles are symmetrically located from the median for all symmetric distributions. While deciles divide the total frequency (or area) into 10 equal parts, percentiles (%-tiles) divide it into 100 equal parts. We could go to finer levels into one thousand equal parts, and so on. However, these are seldom popular owing to the simple reason that sample sizes are most often <1000. If the sample size is <100, even the percentiles are not used as several percentiles could coincide. Deciles are special %-tiles, as also quartiles are special deciles. For example, $Q_2 = D_5 = P_{50}$, where Q, D, P stand for quartile, decile, and percentile, respectively. In the following discussion, $k \in \{1, 2, 3\}, j \in \{1, 2, .., 9\}$. The general formula to convert from quartile into decile is $Q_k = D_{(10/4)*k} = D_{(5/2)*k}$, provided $(5/2) * k$ is an integer. The reverse relationship is $D_j = Q_{(2/5)*j}$, provided $(2/5) * j$ is an integer (which happens to be the case for $j = 5$). Similarly $D_j = P_{10*j}$ or its reverse $P_i = D_{i/10}$ if $i/10$ is an integer. A similar relation between quartiles and percentiles is $Q_k = P_{25*k}$ or $P_i = Q_{\lceil i/25 \rceil}$. As $Q_{\lceil i/25 \rceil}$ returns the integer part (ceil operator), several percentiles can get mapped to the same quartile when the data size is small.

4.3   MEASURES OF SKEWNESS

There are many skewness measures available. They can be applied to the population or sample. Sample skewness is more important for data analysts and engineers. Most of the measures discussed in the following are sample measures. It is assumed that the sample size n is sufficiently large for the expressions involved to be meaningful. The population analogs are denoted by Greek letters and their sample counterparts by lower case English letters by convention.

4.3.1   Bowley's Skewness Measure

As skewness measures the lack of symmetry, several measures can be defined by utilizing their location relative to the distance from the median. If the distribution is symmetric, $(M - Q_1)$ and $(Q_3 - M)$ are equal. In general, $(M - P_k)$ and $(P_{100-k} - M)$ are equal where P_k denotes the percentiles and $P_{50} = M$ is the median. This property has been utilized by Harremoës [54], MacGillivray [55]. The simplest one is due to Bowley [56], who defined a skewness coefficient as

$$B_s = [(Q_3 - M) - (M - Q_1)]/(Q_3 - Q_1) = (Q_3 + Q_1 - 2M)/(Q_3 - Q_1), \quad (4.1)$$

where Q_1 and Q_3 are the lower and upper quartiles and M is the median. As the percentiles are equally distant from the median, the above-mentioned measure is zero for symmetric continuous distributions because the numerator is zero (this may not hold for discrete distributions owing to round-off errors).

The corresponding population analog is easily expressed in terms of the CDF as

$$\gamma = (F^{-1}(3/4) + F^{-1}(1/4) - 2F^{-1}(1/2))/(F^{-1}(3/4) - F^{-1}(1/4)). \qquad (4.2)$$

This measure was generalized in Reference 57, who parametrizes it in terms of an arbitrary percentile u as skewness function

$$[F^{-1}(u) + F^{-1}(1 - u) - 2F^{-1}(1/2)]/(F^{-1}(u) - F^{-1}(1 - u)), \qquad (4.3)$$

where u is normalized to the range [0,1]. This is easy to compute for continuous populations (for discrete populations, the percentiles may not align exactly on variate values). The above-mentioned measure is negative for $u < \frac{1}{2}$ (as the denominator is $-$ve). It coincides with Galton skewness measure for $u = 3/4$. Subtract and add $F^{-1}(1 - u)$ in the numerator and simplify to get the alternative form

$$1 + 2 * (F^{-1}(1 - u) - F^{-1}(1/2))/(F^{-1}(u) - F^{-1}(1 - u)). \qquad (4.4)$$

📖 **EXAMPLE 4.2   Range of Values for Bowley's Measure**

Prove that Bowley's Skewness measure lies in the interval $[-1, +1]$.

Solution 4.2 Consider $B_s = (Q_3 + Q_1 - 2M)/(Q_3 - Q_1)$. When $M = Q_1$ (highly positively skewed), the numerator simplifies as $(Q_3 + Q_1 - 2 * Q_1)$ $= (Q_3 - Q_1)$. This cancels out with the denominator giving $B_s = 1$. Similarly when $M = Q_3$ (highly negatively skewed), the numerator becomes $Q_1 - Q_3 = -(Q_3 - Q_1)$, giving $B_s = -1$. Add and subtract Q_1 in the denominator to get $B_s = (Q_3 + Q_1 - 2M)/(Q_1 + Q_3 - 2Q_1)$. Divide both the numerator and the denominator by 2 to get

$$B_s = [(Q_1 + Q_3)/2 - M]/[(Q_1 + Q_3)/2 - Q_1)]. \qquad (4.5)$$

As $Q_1 \leq Q_2 \leq Q_3$, the mean $(Q_1 + Q_3)/2$ must lie in between them. Hence, the absolute value of the numerator must be less than absolute value of the denominator. This means that $|B_s| < 1$ for all other cases. Hence, $B_s \in [-1, +1]$. ∎

TABLE 4.2   BMI of 30 Patients—Unsorted

BMI	BMI	BMI	BMI	BMI
23.3	24.6	23.4	22.9	17.7
26.2	32.2	23.0	24.4	23.7
25.6	33.6	24.0	25.2	28.6
25.1	27.4	33.0	25.2	24
23.3	24.4	29.7	24	28.5
22.1	23.5	22.3	28.4	26.3

TABLE 4.3   BMI Frequency Distribution

Class Interval	Class Middle	Frequency	Cumulative Frequency	Quartile Class
17-20	18.5	1	1	
20-23	21.5	4	5	
23-26	24.5	15	20	← 23.5
26-29	27.5	6	26	← 27.25
29-31	30.5	1	27	
≥ 31	33.5	3	30	

■ **EXAMPLE 4.3   BMI skewness calculation**

The BMI of 30 patients is given in first two columns of Table 4.2. Compute Galton's skewness coefficient.

Solution 4.3 Here $N = 30$, so that $N/4 = 7.5$. Hence, Q_1 is that value below which one-fourth of the data lie. From the sorted column (3), we find that the seventh and eighth values are both 23.3. Hence $Q_1 = 23.30$. Next, Q_3 is that value below which three-fourth of the data lie (or equivalently above which one-fourth of the data lie). From the last column, we see that eight patients have BMI $\geq$ 27.4 and there are seven patients with BMI $\geq$ 28.4. Hence $Q_3 = 27.9$. If data are grouped using a class width of 3, we get Table 4.3. Using the formula for quartiles as $Q_k = L + (N * k/4 - M) * c/f$ where $N = 30$, $L = $ lower limit of quartile class, $k = 1$ for Q_1 and 3 for Q_3, M is the cumulative frequency up to (but excluding) quartile class, c is class width ($= 3$) and f is the frequency of quartile class, we get $Q_1 = 23.5$ and $Q_3 = 27.25$. As $M = 24.45$, $(Q_3 + Q_1 - 2M)/(Q_3 - Q_1) = (50.75 - 48.9)/3.75 = 0.4933$, showing that the data are skewed to the right (Table 4.4).   ■

As the kth quartile uses the lower limit of the class where Q_k falls, frequency of that class and cumulative frequency up to the class, there is no need to sort the complete data. Fast methods exist to find any quantile when the data size is very large and one of them needs to be computed.

TABLE 4.4 Range of Skewness Measures

Name	Range	Based on
Bowley's measure	$(-1, +1)$	Quartile
Pearson's measure(η)	$(-3, +3)$	$\bar{x}, s, M$
Bowley's measure Q	$[0, \max((Q_3 - M)/2, (M - Q_1)/2)]$	Quartiles
Kelly's measure K	$(-1, +1)$	Deciles
CQD	$(0, 1)$	Quartiles

Note: M is the mode by default but could be taken as the mean or median; D_i's are the deciles.

4.3.2 Pearson's Skewness Measure

Pearson's measure of sample skewness was introduced in Section 1. It is a ratio-measure defined as $\eta = (\bar{x}_n - \text{mode})/s_n$. Nearly, bell-shaped distributions satisfy an approximate relationship $(\bar{x}\text{-mode}) \sim 3*(\bar{x}\text{-median})$. This allows us to express the above as $\eta = 3*(\bar{x}\text{-median})/s$. This is more meaningful, as the mode of a sample need not be unique. As the numerator is the difference between two location measures, η can be positive or negative. As it is divided by the standard deviation, it is unit-less. This measure returns 0 for symmetric distributions. It is <0 for negatively skewed distributions. The expected value of these statistics tends to zero when samples come from large symmetric populations. For most data, it will lie in the range $(-3, +3)$.

$$\text{Pearson's } \eta = (\bar{x} - \text{mode})/s \simeq 3 * (\bar{x} - \text{median})/s \in (-3, +3). \qquad (4.6)$$

Pearson also suggested another measure of skewness in terms of third moment of a unit normalized random variable as

$$\gamma_1(X) = E[(X - \mu)^3]/\sigma^3 = E[(X - \mu)/\sigma]^3 = E(Z^3). \qquad (4.7)$$

This can be expressed in terms of moments as μ_3/σ^3. As the standard normal distribution has skewness zero, positive values of skewness indicates a flat right tail and vice versa. Its square $\beta_1 = \gamma_1^2$ is sometimes used, under the assumption of the existence of finite second and third moments. As the numerator contains a centralized measure (with expected value zero for symmetric distributions), this measure is location invariant for unimodal distributions. As the denominator contains quantities in the same unit, it is unit-less. As the orders of the numerator and the denominator are the same, the measure is scale invariant too.

◼ EXAMPLE 4.4 Pearson's skewness calculation

Compute Pearson's skewness for the data in Table 4.2.

Solution 4.4 We find $\bar{x} = 25.49667$, and $s = 3.4455852$, Median $M = 24.45$. Substitute these values to get $\eta = 3 * (25.49667 - 24.45)/3.44558 = 3.14/3.44558 = 0.9113$, showing that the data are skewed to the right. ◼

▣ EXAMPLE 4.5   Is marks distribution bell-shaped?

The marks obtained by students in an exam have mean 70 and median 72, with a standard deviation of 8. Is the distribution of marks symmetric? If not, is it skewed to the left or right?

Solution 4.5 We find Pearson's measure of skewness as $\eta = 3*(\bar{x}\text{-median})/s = 3*(70 - 72)/8 = -6/8 = -0.75$. As this is <0, the distribution is asymmetric and is negatively skewed.   ∎

A skewness measure can be used to compare two samples drawn from distinct populations. However, as the sample statistics vary in repeated sampling from the same population, these comparisons are often vague. For example, Pearson's η coefficient has expected value (Chapter 8) zero for unimodal distributions. Suppose we take repeated samples from a uniform or U-shaped distribution. The η coefficient will vary widely in these situations (because the mode is not well defined for uniform distributions, and there are two modes for U-shaped distributions). See Reference 58 for a discussion on a quadratic-mean based skewness test.

The concept can be extended to population densities with intent to order them based on a skewness measure. One notable contribution is by van Zwet [59], who defines a partial-order among probability laws with cumulative distributions F and G as $F \leq_s G$ iff $G^{-1}(F(x))$ is convex for $x \geq k$ (or equivalently $F^{-1}(G(x))$ is concave), where k is the common point of symmetry of the distributions. This allows one to compare those distributions, the inverse of at least one of which exists. Symmetric distributions can be converted into asymmetric ones using the transformation $f(x, \lambda) = 2g(x)G(\lambda)$ where $\lambda \in \mathbb{R}$, and $G()$ denotes the CDF [60, 61].

4.3.3   Coefficient of Quartile Deviation

The coefficient of quartile deviation exclusively uses the first and third quartile

$$\text{CQD} = (Q_3 - Q_1)/(Q_3 + Q_1). \tag{4.8}$$

While Bowley's measure uses the median, CQD does not depend on the median. Hence, it is less informative. As the numerator and denominator are both linear in Q_i and measured in the same units, CQD is a unit-less ratio measure with finite range. It is always positive as both the numerator and the denominator are positive.

▣ EXAMPLE 4.6   Range of values for CQD

Prove that $\text{CQD} \in (0,1)$.

Solution 4.6 Add and subtract Q_1 in the numerator to get $(Q_3 + Q_1 - 2Q_1)$. Combine the first two terms with the denominator and write the third terms separately. Then $\text{CQD} = 1 - 2Q_1/(Q_1 + Q_3)$. Because Q_3 (being the third quartile) divides the entire data in 75%:25% ratio, it is always greater than Q_1. Thus, the ratio $Q_1/(Q_1 + Q_3)$ is always in the range (0,.5). Substituting 0 shows that CQD is always less than 1. Substituting 0.5 shows that CQD is greater than zero (see Table 4.3). Hence CQD $\in$ (0,1).   ∎

4.3.4 Other Skewness Measures

The concept of symmetry has been defined in terms of density or distribution functions in the above-mentioned discussions. Kelly's measure of skewness uses deciles and is defined as $(D_1 + D_9 - 2D_5)/(D_9 - D_1)$. The inverse Gaussian (IG)-symmetry is an analog that utilizes equality of positive and negative moments. For the IG(μ, λ) law (Chapter 7, pp. 7–64), it is easy to verify that $E(X/\mu)^{-r} = E(X/\mu)^{r+1}$, where negative index denotes inverse moments. The negative moments are defined only when $f(x) = 0$ for $x = 0$. This property is also satisfied by the log-normal law and scale mixtures of IG distributions [62].

The skewness measure defined earlier is biased. An unbiased estimator can be obtained by differently scaling it to have an expected value exactly equal to the population skewness. This is why some software packages use $n/[(n-1)(n-2)]$ $\sum_j[(x_j - \bar{x})/s]^3$ (for $n \geq 3$) as a measure of skewness. See Reference 63 for a comparison of skewness measures.

4.4   CONCEPT OF KURTOSIS

Kurtosis originated in data analysis. Some data distributions are more peaked than the standard normal law, whereas some others are less peaked. This prompted Pearson (1905) to classify distributions as leptokurtic, mesokurtic, and platykurtic. Kurtosis was originally defined using the standard normal law as a yardstick. A data distribution that has the same kurtosis as $N(0, 1)$ is called mesokurtic. Those with higher kurtosis is called leptokurtic and with lower values is called platykurtic. They are applicable to discrete and truncated data, skewed, and symmetric data that are continuous. They are more meaningful to unimodal data than rectangular data. They are less meaningful to U-shaped and other multimodal data.

Definition 4.3 Kurtosis is a measure of both the peakedness of the distribution in and around the location measure (center of mass) and a measure of the tail weights that jointly characterize the accumulation of probability mass toward the center [64].

The population analogs are denoted by Greek letters and their sample counterparts by lower case English letters by convention. Pearson's kurtosis measure for the population is denoted by β_2 and sample counterpart by b_2 (or $b_2(n)$).

4.4.1   An Interpretation of Kurtosis

Pearson's definition of kurtosis confines itself to unimodal distributions. It emphasizes the overall frequency at or around the central part (mode) of a distribution.

As theoretical distributions can take a variety of shapes depending on the parameter values, this "central part" may move to the extremes for some parameters. One example is the exponential distribution $c * \exp(-c\,x)$ or the Poisson distribution with very small λ values (say $\lambda < 0.10$) that tails off slowly to the right (the left tail of these distributions are either very short or nonexistent). Kurtosis is defined for these distributions too. Such distributions are not kurtosis comparable with others that tails off in both directions. The classical kurtosis measures how much of the probability mass is moved from the shoulders (say within $\mu \pm 2\sigma$ to $\mu \pm 3\sigma$) of a normal law to the center that results in an identical leptokurtic distribution or *vice versa* (how much mass is moved from the central part (say from $\mu \pm \sigma$) to the regions beyond, so as to get a platykurtic distribution). To quantify the amount of kurtosis, we could consider the SEK and measure the departure from twice the standard error. An approximate SEK for a sample of size n is $\sqrt{24/n}$ (see Reference 49. If $2*\text{SEK} <$ absolute value of kurtosis, we may reasonably conclude that the data are from a non-platykurtic distribution. The above-mentioned interpretation of kurtosis can be refined, resulting in a new interpretation in terms of both the tailing off behavior combined with the peakedness simultaneously [65]–[70].

As the variance is a quadratic function of the random variable (for a population) or a quadratic function of the sample values (for s^2), its second moment has power 4 (it is a biquadratic or quadratic of the random variable or sample values). In other words, the variance of the sample variance must be a function of Pearson's β_2 ($\text{Var}(s^2)/E[s^2]^2 = \beta_2 - 1$ asymptotically). As mentioned earlier, the variance is measured in the same unit as the sample values, whereas the kurtosis is a unit-less measure. If the sample size is large, it is known that $\sqrt{n}(s^2 - \sigma^2) \to N(0, (\beta_2 - 1)\sigma^4)$, which is interpreted as convergence in distribution. As $\beta_2 = 3$ for the standard normal law, the asymptotic convergence is to $N(0, 2)$. In addition, $\sqrt{n}(\log(s^2) - \log(\sigma^2)) \to N(0, (\beta_2 - 1))$ using Mann–Wald theorem [62]. It is well known that ns^2/σ^2 is distributed as χ^2_{n-1} when samples come from a normal population, so that $E(ns^2/\sigma^2)^2 \simeq 2(n-1)$. This result is used in some of the kurtosis measures defined in the following discussion.

■ **EXAMPLE 4.7   When is binomial distribution mesokurtic?**

Prove that the binomial distribution is mesokurtic when $p = \frac{1}{2}\left(1 \pm \frac{1}{\sqrt{3}}\right)$.

Solution 4.7 The coefficient of kurtosis of binomial distribution is $\beta_2 = 3 + (1 - 6pq)/npq$ (Chapter 6). For mesokurtic distributions, $\beta_2 = 3$. This means that $(1 - 6pq)/npq = 0$. As the denominator is always positive, this expression is zero when $pq = 1/6$. Write this as $p(1 - p) - 1/6 = 0$ or equivalently $6p^2 - 6p + 1 = 0$ and solve for p to get $p = \frac{1}{2} \pm \frac{\sqrt{3}}{6}$. Consider the second expression $\frac{\sqrt{3}}{6}$. Multiply numerator and denominator by $\sqrt{3}$ and cancel out 3 to get $1/2\sqrt{3}$. Substitute in the aforementioned and take (1/2) as common factor to get the condition as $p = \frac{1}{2}\left(1 \pm \frac{1}{\sqrt{3}}\right)$. The distribution is leptokurtic (respectively platykurtic) if $(1 - 6pq) >$ (respectively $<$) 0. In terms of p this becomes $p >$ (respectively $<$) $\frac{1}{2}\left(1 \pm \frac{1}{\sqrt{3}}\right)$. Equivalently, it is leptokurtic if $p < \frac{1}{2}\left(1 - \frac{1}{\sqrt{3}}\right)$ or $p > \frac{1}{2}\left(1 + \frac{1}{\sqrt{3}}\right)$ and platykurtic if $\frac{1}{2}\left(1 - \frac{1}{\sqrt{3}}\right) < p < \frac{1}{2}\left(1 + \frac{1}{\sqrt{3}}\right)$. ∎

4.4.2 Categorization of Kurtosis Measures

The kurtosis can be measured in more than one way [61]. This section gives a categorization of popular kurtosis measures.

1. Moment-based measures
   The classical kurtosis measures are moment-based and assume the existence of finite fourth moment (for the population). Most of them utilize the fourth central moment or its scale transforms. They are defined for both the population and the sample. Pearson's kurtosis is expressed for a population in terms of moments as $\beta_2 = \mu_4/\mu_2^2$. As the denominator is the fourth power of the standard deviations, they are also unit-less. Because the standard normal distribution has kurtosis 3, the quantity $\gamma_2 = \beta_2 - 3$ is widely used (see the following discussion).

2. Measures that utilize standardized variables (z-scores)
   The classical measure of Pearson's population kurtosis is defined as $\beta_2 = E[z^4]$ where $z = (x - \mu)/\sigma$. Stavig's kurtosis measure [71] is defined as $1 - E[|z|]$. Seiner Bonett used $E[g(z)]$ where

   $$g(z) = ab^{-|z|} \text{ for } 2 \leq b \leq 20, \text{ and } a[1 - |z|^b] \text{ for } 0.2 \leq b \leq 1, \quad (4.9)$$

   which gives more importance to the peak at the center for unimodal data.

3. Quantile-based measures
   These measures utilize the quantiles of a distribution. The popular ones are due to Balanda and MacGillivray [52], Groeneveld and Meeden [72] and Groeneveld [73].

4. Measures that utilize inverse of distribution functions
   These measures use the inverse of theoretical distribution functions. One example is the spread function of Balanda and MacGillivray [52]

   $$S_F(u) = F^{-1}(0.5 + u) - F^{-1}(0.5 - u) \text{ for } 0 \leq u \leq 0.5. \quad (4.10)$$

The u is called interquantile distance. If both F and G are continuous unimodal and invertible, one could produce a plot of $S_F(u)/S_G(u)$ (or its inverse) to compare the relative skewness [74].

5. Measures that utilize density crossing
   The "density crossing" is a sufficient condition to kurtosis-order two samples. Finucan [75] showed that if two distributions have the same variance, and if the frequency curves cross twice on each side of the mode, then one of them has higher kurtosis than the other.

4.4.2.1 Van Zwet ordering of kurtosis As in the case of skewness, theoretical distributions can be "kurtosis ordered" [59]. This is more meaningful for symmetric unimodal distributions. A bivariate ordering based on both the skewness and the kurtosis is more meaningful for asymmetric distributions. As various distributions have different range, they are standardized to the same range before they are ordered.

4.5 MEASURES OF KURTOSIS

Kurtosis measures are used to numerically evaluate the relative peakedness or flatness of data. The standard normal distribution can be used as a yardstick for bell-shaped data, but the concept is valid for other shapes such as J-shaped, reverse J-shaped, and cusp-shaped data. It is applicable to both the sample and the population. This has important implications in some fields. As examples, suppose that there are many weight-loss programs available. The distribution of actual weight lost, or the time spent in the program by participants can take various shapes. A negatively skewed distribution in the first case will indicate that more persons lost more weight and in the second case will indicate that participants who spent more time in the program lost more weight. A leptokurtic distribution indicates that the program was very effective in weight loss, whereas a platykurtic distribution indicates that the weight loss was gradual. Hence, people will be more attracted to a positively skewed or leptokurtic weight-loss program. Similarly consider machine servicing by various vendors or repair persons. If there are multiple shops that could do this, a client may be more interested in that service shop with a leptokurtic and positively skewed servicing time distribution. We need a standard scale to measure the amount of kurtosis.

4.5.1 Pearson's Kurtosis Measure

Using the reasoning in page 4–22, Pearson defined the population kurtosis in terms of moments as

$$\beta_2 = \mu_4/\mu_2^2 = E(X - \mu)^4/[E(X - \mu)^2]^2 = E(X - \mu)^4/\sigma^4 = E[(X - \mu)/\sigma]^4, \quad (4.11)$$

where $E()$ denotes mathematical expectation (Chapter 8) [76]. This is the fourth moment of the standardized variate $Z = (X - \mu)/\sigma$. Using $V(X) = E[X^2] - E[X]^2$ on $[(X - \mu)/\sigma]^2$, we have

$$V\{[(X - \mu)/\sigma]^2\} = E\{[(X - \mu)/\sigma]^4\} - \{E[(X - \mu)/\sigma]^2\}^2. \quad (4.12)$$

Rearranged, we get $E\{[(X - \mu)/\sigma]^4\} = V\{[(X - \mu)/\sigma]^2\} + \{E[(X - \mu)/\sigma]^2\}^2$. The second expression being the square of the variance the RHS becomes $V\{[(X - \mu)/\sigma]^2\} + \{V[(X - \mu)/\sigma]\}^2$. An interpretation of this result is that the kurtosis and variance (spread) are related through squares. As the numerator is an even function of the variate, this measure allows one to *compare the kurtosis* of asymmetric distributions. The sample kurtosis coefficient is

$$b_2 = \frac{1}{n} \sum_{j=1}^{n} (x_j - \bar{x}_n)^4 / s_n^4, \tag{4.13}$$

where s_n is the sample standard deviation. Because sums of fourth powers is always positive, $b_2 \geq 0$. As the zero-point is well defined, it is a ratio-measure with range $\in \mathbb{R}$. It is shown in the following (next page) that the kurtosis of the standard normal distribution is 3. This means that irrespective of whether the data are discrete or continuous, we could subtract 3 to get $\gamma_2 = \beta_2 - 3$ as a standardized measure of kurtosis as suggested by Fisher. Then $\gamma_2 > 0$ indicates leptokurtic and $\gamma_2 < 0$ indicates platykurtic distributions. Replacing the population quantities by the corresponding sample equivalents, we could get a *biased* estimate as

$$b_2 = \frac{\sum_j (x_j - \bar{x})^4 / n}{\left[\sum_k (x_k - \bar{x})^2 / n^2\right]} = \frac{n \sum_j (x_j - \bar{x})^4}{\left[\sum_k (x_k - \bar{x})^2\right]^2}. \tag{4.14}$$

The sample kurtosis is a biased estimator of the population kurtosis. We need to apply a different scaling factor to get the unbiased estimate. This is why some software packages use $\frac{n(n+1)}{(n-1)*(n-2)*(n-3)} \sum_j [(x_j - \bar{x}_n)/s_n]^4 - K$ where K is a correction factor $((n - 1)^2/[(n - 2)(n - 3)])$.

EXAMPLE 4.8   Classical kurtosis coefficient

Prove that the classical kurtosis coefficient measures the dispersion of $[(X - \mu)/\sigma]^2$ around its mean 1.

Solution 4.8 Replace $E[(X - \mu)/\sigma]^2$ on the RHS of equation (4.12) by $V[(X - \mu)/\sigma] + \{E[(X - \mu)/\sigma]\}^2$. As $Z = (X - \mu)/\sigma$ is a standardized variate, it has mean $E(Z) = 0$ and variance $V(Z) = 1$. If X is normally distributed, $((X - \mu)/\sigma)^2$ has a chi-square distribution having 1 DoF with mean 1 and variance 2 (Chapter 7). Substitute these values in equation (4.12) to get

$$V\{[(X - \mu)/\sigma]^2\} = E\{[(X - \mu)/\sigma]^4\} - 1. \tag{4.15}$$

This shows that $E\{[(X - \mu)/\sigma]^4\} = V\{[(X - \mu)/\sigma]^2\} + 1$ measures the dispersion of $[(X - \mu)/\sigma]^2$ about its mean 1 (which is the variance of $Z = (X - \mu)/\sigma$).

Substitute for $V\{[(X - \mu)/\sigma]^2\} = 2$ (chi-square variance with 1 DoF) shows that the kurtotis is 3 when X is normal.   ∎

The dependency of kurtosis on variance is more pronounced for symmetric distributions than for others (asymmetric and truncated distributions). The amount by which spread is reduced when kurtosis is increased depends also on whether frequency is moved from both the shoulders of a distribution to the center or only from one side (left or right) to the center, in which case the reduction of variance could be minimal. If one is interested only in the peakedness of distributions, a truncated measure that eliminates the contribution of the tail(s) may be more appropriate. The truncation point can be setup equidistant from the mode for symmetric distributions.

For the standard normal distribution, β_2 is 3. However, there exist many other distributions that also have $\beta_2 = 3$. For example, the Tukey distribution with $\lambda = 5.2$ and double gamma law with $\alpha = 0.5 * (1 + \sqrt{13})$ all have $\beta_2 = 3$, although their shapes are different [77, 78]. This is because Pearson's kurtosis measure encapsulates both the peakedness and tail weight(s) of a distribution. This is easy to understand using truncated distributions. A left-truncated distribution tails off slower than their nontruncated counterparts. Truncating a distribution at left or right α tail also increases the peak probability by $f(x)/(1 - \alpha)$ where α is the probability of truncated part.

4.5.2   Skewness–Kurtosis Bounds

Several researchers have studied the $d =$ (skewness−excess kurtosis) quantity from various perspectives. See for example References 79, and 80. Pearson obtained the bound $\beta_1^2 - \beta_2 \leq 2$ for Bernoulli distributions. This was further improved by several researchers. If the distribution is infinitely divisible, $\beta_1^2 \simeq \beta_2$, and $\beta_1^2 \leq \beta_2$ is attained for normal and Poisson distributions [81]. This property was used in Reference 82 to distinguish Poisson or normal distributions from other infinite divisible distributions. A similar quantity is $c =$ skewness/kurtosis (for kurtosis$\neq 0$), which is well behaved for symmetric distributions in general and the normal distribution in particular. See Reference 83 for some inequalities, Reference 54 in the context of minimizing information divergence under moment constraints, Reference 84 for a studentized range based test and Reference 85 for a right and left inequality order, and Reference 86 for skewness-invariant measures of kurtosis.

4.5.3   L-kurtosis

This is a generalization of kurtosis, introduced by Hosking [53], that uses L-moments (denoted by λ'_ks). For a real-valued random variable X with finite mean μ, we define the L-moment as expectation of linear combination of order statistic as follows (see page 4–10). Let $X^n_{(k)}$ denote the kth order statistic of a random sample of size n. Define $\lambda_3 = (1/3) * E(X^3_{(3)} - 2 * X^3_{(2)} + X^3_{(1)})$ and $\lambda_4 = (1/4) * E(X^4_{(4)} - 3 * X^4_{(3)} +$

$3 * X_{(2)}^4 - X_{(1)}^4$). Note that the coefficients of $X_{(k)}^j$ are the rows of Pascal's triangle with alternating signs. The L-skewness is then defined as $\tau_3 = \lambda_3/\lambda_2$. Similarly, L-kurtosis is defined as $\tau_4 = \lambda_4/\lambda_2^2$. As both are expectation of linear combination of order statistics, they are unit-less measures.

4.5.4 Spectral Kurtosis (SK)

As mentioned earlier, kurtosis can clearly distinguish between peakedness and flatness in numeric data in the interval or ratio scale. It is a quantified real number (+ve or −ve) whose magnitude represents the amount of departure of a distribution from the shoulders toward the center and tails. This property of kurtosis can be used to predict machine faults (using past data) [87], in fault diagnosis of equipments or independent parts and materials [88, 89] damage assessment of structures [90], crack detection of isotropic plates, machine diagnostics and prognostics [91], modular classification of digital signals, and so on. As an example, they can be used to warn an operator on machine overloads or wear and tear beyond a threshold. Consider an aircraft or helicopter with a fixed weight limit on the cargo and passenger compartments, respectively. If either or both of these sections exceed the weight limit, resulting in an overall overweight, the bearings sound during takeoff due to the excess weight acting down can slightly deviate from the normal takeoff sound at the same ground speed (in the case of aircraft). Similarly, by analyzing acoustic signals, one can distinguish between human footsteps from background noise (impurities) or identify submarines from whales in deep water. Seismic sensors and geophones use such signals to automatically measure the movement of objects or to distinguish between possible objects (such as vehicles, humans, other animals, or objects) and direction of movement (moving toward, away from or along a trajectory around the sensor).

Definition 4.4 Spectral Kurtosis (SK) is a ratio-measure defined in the frequency domain of a signal that reveals the deviation from Gaussianity of the spectral components with intent to separate randomly occurring signals from normal ones using cumulants.

The above-mentioned definition assumes that original data are transformed into a band-delimited frequency domain using one of the popular frequency transforms such as discrete Fourier transform (DFT), discrete wavelet transform (DWT), and so on. A simpler definition that hides the technical details is as follows:

Definition 4.5 The spectral kurtosis of a signal is the kurtosis of its sampled frequency components.

Let $x[n]$ denote a real-time discrete random process where the index n denotes the time. Let $X[m]$ denote the transformed signal in the frequency domain. Then, the SK of $x[n]$ is defined as

$$\kappa_x(m) = \kappa_4[X[m], X^*[m]]/\kappa_2[X[m], X^*[m]]^2, \qquad (4.16)$$

where κ_r is the rth order cumulant and $X^*[m]$ denotes the complex conjugate. In the case where $x[n]$ is a stationary random process, nonnull cumulants of $X[m]$ will have as many complex conjugate terms as nonconjugate terms.

4.5.5 Detecting Faults Using SK

Rotating machines typically exhibit nonstationary vibration signatures that are easy to detect using SK [92]. The popularity of SK comes from its proved effectiveness in real-time signal detection and removal of noise (impurities, harmonics, outliers, or deviants). In time-varying discrete signals, it can distinguish between constant amplitude harmonics, time-varying amplitude harmonics, and noise. In addition, it is conceptually simple and easy to compute. Its value is independent of the noise present in the input signal. This is why it has been applied in a variety of fields such as astronomy, industrial robotics, and deep-sea explorations.

Consider a healthy induction motor running at a constant speed. The harmonic components of such asynchronous machines are constant amplitude harmonics. The SK of such faulty machines can be compared with healthy ones to identify possible deviations [93]. This means that the data generated by a healthy machine is stored for future use and compared continuously with current data generated while it is presumably operating under fault (such as cracks, defunct components, or lubricant depletion) to detect any possible deviations. Owing to the heterogeneity of working conditions, a healthy data vector is used instead of a single data instances. These data are usually bandpass filtered and transformed into the frequency domain (usually using short time Fourier transform (STFT) or wavelet transforms [90] and processed in a fixed time window. The SK of both these data is found and compared to detect defects. Another application is to detect and remove (if present) radio-frequency interference (RFI) in radio astronomy and GPS. The precision of such event or object identification can be improved using multiple receivers.

SK can be used to measure the impulsiveness of signals (variation of frequencies) as a function of frequencies in a band-filtered domain [94]. They can give an indication on the most impulsive part of a vibrating signal. Kurtosis of each frequency band can be used to improve the precision in a time or frequency decomposed signal. See References 95–98 for applications to fault diagnosis.

4.5.6 Multivariate Kurtosis

The kurtosis concept has been extended to multivariate distributions by many researchers. See References 49, 99–101. The following discussion is on multivariate continuous distributions, although the concept is valid for discrete distributions. Let μ and Σ denote the mean vector and variance–covariance matrix of a multivariate distribution in $\mathbb{R}^d$. Then, the classical kurtosis measure [99] is defined as $\beta_d = E\{[(X - \mu)'\Sigma^{-1}(X - \mu)]^2\}$. As $E\{[(X - \mu)'\Sigma^{-1}(X - \mu)]\}$ is the squared Mahalanobis distance metric, this represents the second moment of Mahalanobis' squared distance. The sample analog is obtained by replacing μ by the mean vector $\overline{X}$

and Σ by the sample variance–covariance matrix. A discussion of source separation using kurtosis maximization can be found in Reference 102.

4.6 SUMMARY

This chapter discussed several measures of skewness and kurtosis. Most of the popular statistical techniques are devised for the symmetric bell-shaped data (which is technically called normal data). The skewness captures the lack of symmetry in the data trend. Kurtosis captures the tail thickness in the data trend. Financial and health data are also known to exhibit thick tailness.

Most of the popular measures of skewness and kurtosis are based on the central moments or functions of it. However, it is well known that the moments do not always determine a distribution uniquely. Several examples to support this fact are available in the literature (see References 60 and 103. This leads to skewness and kurtosis measures based on other statistics than moments. $\sqrt{\beta_1}$ and β_2 are routinely used in statistical analysis [104]. Asymptotic distributions of skewness and kurtosis coefficients are discussed in Reference 105, an application in regenerative simulation in Reference 106, and rain-drop diameter distribution in Reference 107. An application of fuzzy mean-variance-skewness to portfolio selection models can be found in Reference 108. A visualizing discussion can be found in Reference 109.

EXERCISES

4.1 Mark as True or False

a) Skewness is a measure of the lack of symmetry

b) Third moment measures the asymmetry of data

c) Zero skewness indicates symmetry around the median

d) Positive skewness indicates a long left tail

e) Every sample observation contributes to the coefficient of skewness

f) Kurtosis measures are useless in providing variance of data

g) The skewness coefficient is independent of change of scale transformation

h) Truncating data at left end increases kurtosis

i) A left-truncation of symmetric law makes it positively skewed

j) Spectral kurtosis uses frequency transformed data.

4.2 Prove that Bowley's skewness measure varies between −1 and +1.

4.3 For bell-shaped distributions prove that the skewness measures are zeros.

4.4 If Pearson's η is zero, one can infer that the___ (A) data distribution is symmetric (B) distribution is mesokurtic (C) $\bar{x}$ = mode but distribution need not be symmetric (D) distribution is bell-shaped

4.5 Show that the kurtosis of standard normal distribution is 3. Discuss how this helps in asymptotic convergence of other distributions that tend to standard normal for large parameter values.

4.6 What are some desirable qualities of a good measure of skewness?

4.7 Describe how you will check symmetry using the 5-number summary of a sample.

4.8 Arrange the following distributions according to increasing levels of kurtosis (called kurtosis ordering) (i) Student's T distribution with $n < 25$, (ii) standard Cauchy distribution, (iii) standard normal and (iv) double exponential (v) gamma distribution with parameters $(10, 2)$

4.9 Consider the distribution of marks obtained in an exam. What type of skewness is exhibited in the following situations? (i) the exam was easy for majority of students, (ii) the exam was difficult for majority of students, and (iii) questions that carry around 50 of the marks were easy questions.

4.10 What is the 5-number summary of a sample? Can you check the skewness and kurtosis of the sample using the 5-number summary?

4.11 Find the moments of gamma distribution, and obtain the measures of skewness and kurtosis.

4.12 If the sign of kurtosis statistic is positive, it indicates (a) leptokurtic distribution (b) mesokurtic distribution (c) platykurtic distribution (d) normal distribution.

4.13 What is the value of skewness for the following distributions? A) bell-shaped distribution B) continuous uniform distribution, C) symmetric triangular distribution.

4.14 What is the range of possible values of Pearson's skewness measure $\eta = (\bar{x}\text{-mode})/s$?. What does a zero value indicate? What is its expected value for bell-shaped distributions? What is a disadvantage of this measure?

4.15 What is the range of values for the standard skewness and kurtosis measures? What is the reason for defining the kurtosis measure γ_2 as $\beta_2 - 3$? Derive its value for standard normal distribution.

4.16 What does a distribution of marks in an exam skewed to the left indicate?

4.17 When is the Bowley measure and Galton measure of skewness equal?.

4.18 If a symmetric distribution is left-truncated, will the new distribution be positively or negatively skewed? Will it change dispersion?.

4.19 Does the skewness and kurtosis get affected by the change of scale transformation $Y = c * X$? Does the quantile based measures get affected?

4.20 What value does $\text{CQD} = (Q_3 - Q_1)/(Q_3 + Q_1)$ take for (i) symmetric data? (ii) for positively skewed data?

4.21 Show that the population skewness can be expressed as $\beta_1(X) = [E(X^4) - 4\mu E(X) + 6\mu_2 E(X^2) - 3\mu_4]/d^4$, where d = std.dev.

4.22 What is the range of Moore's kurtosis measure? What are the possible value ranges?

4.23 Comment on the statement "kurtosis and variance (spread) are inversely related."

4.24 What is the relation between quartiles and percentiles?

4.25 Give the mathematical expression to convert deciles into quartiles.

4.26 When is the continuous uniform distribution (CUNI(a,b)) platykurtic?

4.27 Prove that $\beta_1^2 \leq \beta_2 + 5/6$.

4.28 Consider a measure defined as $\overline{Q} = (Q_2 - Q_0)/(Q_4 - Q_2)$ where $Q_0 = x_{(1)}$ is the first-order statistic, $Q_4 = x_{(n)}$ and other Q_i's are the quartiles. Can it be used to

measure the skewness? What does high values indicate?

4.29 Prove that the kth percentile is given by $P_k = L + (N*k/100 - M)*c/f$, where L = lower limit of percentile class, M is the cumulative frequency up to (but excluding) percentile class, c is class width and f is the frequency of percentile class.

4.30 Which of the following measures uses κ_4/κ_2^2? (a) dispersion, (b) skewness, (c) kurtosis, (and d) location

4.31 Find skewness for Bragg reflection of X-ray data {0.0795, 0.0841, 0.0790, 0.0844, 0.0842, 0.0840}.

4.32 Find skewness for the seeds example data (p. 3–31) given in Chapter 3

4.33 Prove that the skewness can be increased using one-sided truncation.

5

PROBABILITY

After finishing the chapter, students will be able to

- Comprehend the concept of probability
- Explore different ways to express probability
- Understand various approaches to probability
- Grasp the meaning of events and how to assign probabilities to them
- Apply various counting rules and selection techniques
- Differentiate between dependent and independent events
- Understand conditional probability including Bayes theorem
- Practice computations of probabilities for a variety of problems

---

5.1 INTRODUCTION

Probability had its humble beginning in gambling and games of chance. The theoretical foundations of probability were laid by several 17th and 18th century mathematicians. Prominent among them are the French mathematicians Blaise Pascal (1623–1662) and Pierre de Fermat (1601–1665), Dutch astronomer Christian Huygens (1629–1695), English mathematician and physicist Isaac Newton (1642–1727), French mathematicians Abraham de Moivre (1667–1754), Pierre Simon Laplace

Statistics for Scientists and Engineers, First Edition. Ramalingam Shanmugam and Rajan Chattamvelli.
© 2015 John Wiley & Sons, Inc. Published 2015 by John Wiley & Sons, Inc.

(1749–1827), Simeon-Denis Poisson (1781–1840), German mathematician Leibnitz Gottfried (1646–1716), and so on to name a few.

Definition 5.1 Probability is a quantitative ratio capturing the possible levels of uncertainty or chance.

It is encountered in almost all applied sciences such as statistical physics, quantum mechanics, bioinformatics, and various branches of engineering. The study of probability became an essential part of statistics owing to the obvious reason that probability is deep rooted in a great majority of statistical models and procedures. For example, random sampling, frequency distributions, reliability and gaming models, estimation and inference, statistical quality control, and so on are based on the foundations of probability. Chances play a prominent role in characterizing a random sample from an unknown population. There exist many approaches to define and use probability. We begin with the most popular approaches. A thorough understanding of these approaches is essential for students to apply probability to solve real-life problems.

> The greatest challenge in solving a probability problem is that there are usually many ways to solve it but no obvious way to verify the results. Suppose that a student is presented with a probability problem. The first thing to decide is which of the approaches is the most appropriate one to solve it. There are several set theoretic laws, rules, permutation and combination, urn models, principle of inclusion and exclusion, and so on that are used to solve probability problems. Fundamental laws of set theory give rise to analogous laws of probability. Majority of these approaches are classical, as exemplified in the following discussion. The answer obtained in a problem can be verified only in some particular cases where the numbers involved are small.

5.2 PROBABILITY

Definition 5.2 Probability is a quantitative measure of uncertainty or chance associated with future events or random experiments.

In gambling or games of uncertain outcomes, it is referred to as "the odds." For example, it may be mentioned that the odds are three to two that a horse will win a race. In estimation theory, it is called the "likelihood." The reliability in engineering and plausibility in management refers to probability.

Probability is always associated with one or more future events, a happening, an unknown process, or a working condition. Probability is also associated with random samples, random variables, and uncertain outcomes. The "likelihood" that is mentioned earlier associates a probability to a random sample drawn from a population with a known functional form. Probability associated with random variates is mathematical expressions that return a real number in [0,1] range for each possible value of the variate. These can be too low for some x values when the range of the variate is infinite or for particular parameter values. In numerical probability problems that are discussed in the following sections, it represents the chance of a specified event as a real number. This probability is the same irrespective of the method used to arrive at it. To simplify our discussion, we assume the chance mechanism as logically fine grained. The chance mechanism may be a fine-grained event (likelihood of error-free transmission of a data packet, chance of winning a game, likelihood that two political contestants will address the same location or share the same podium, etc.), a random phenomenon (chance that an electronic component will fail in a computer), or an experiment (probability of survival after a surgery, probability that a new drug will be more effective than the existing ones). Note that in each of them there exist many levels of uncertainties. For example, transmission of a data packet depends on network bandwidth, transmission media, network congestion, and the proper functioning of other hardware or software components. Thus, there are multiple interacting simple events involved in the main event see Table 5.1 for a set of symbols). In all probability problems, we will unambiguously identify the events at the root level. Probability is a ratio-measure. A "probability of zero" indicates an impossibility. A 'probability of one' indicates a complete certainty (in common parlance "in all probability" denotes a very likely or certain event). These occur quite often in theoretical problems but are a rarity in practice.

Probabilities encountered in some fields are extremely small. Consider manufactured products from a company that has implemented six-sigma. As all processes are streamlined and quality control techniques ensure stringent restrictions, chances of defects in newly manufactured items are extremely small. Other examples are survival chances in some terminal diseases, chances of product returns in newly introduced items (like new models of cell phones), chances of natural calamities in some locations, and so on. Each of these events has a "complementary event" (defined below) for which the corresponding probability is quite high (close to 1). For instance, probability that an electronics component will work without failure is high. This shows that the magnitude of probability depends on how we define events.

TABLE 5.1   Some Common Symbols in Probability

Symbol	Description	Probabilistic Interpretation
Ω	Set of all outcomes	Sample Space
ω	A member of set	An outcome
A	Subset of Ω	An event (an outcome in A occurs)
$\overline{A}$	Complement of A	No outcomes in A occurs
$A \cup B$	Union of sets	An outcome in either A or B occurs
$A \cap B$	Intersection of sets	Both A and B occurs
$A - B$	Difference of sets	Event in A but not in B occurs

Set theory and probability theory use the same operator symbols like $\cup, \cap, -$ and complements. However, set theory symbols combine subsets to produce other sets, whereas probability symbols combine numbers to produce probabilities.

5.3   DIFFERENT WAYS TO EXPRESS PROBABILITY

It was mentioned in Section 5.2 (p. 112) that probability is a real number between 0 and 1. The information content in probability statements can be expressed in multiple ways. Popular ways to express a probability are (i) fractional form, (ii) decimal form, (iii) scientific form, (iv) percentage form, (v) literal form, (vi) pictorial form, and (vii) as tail areas under empirical curves or functions [2].

The fractional form represents a probability as a fraction p/q where p and q are assumed to be without common factors (called proper form of a fraction). The decimal form represents a probability in the form 0.dddd where "d" denotes a decimal digit that may or may not repeat. In case of repeating fractions, the digits repeat either individually or as a group. For example $1/3 = 0.33\underline{3}$ is a single digit repeating fraction (here the digits that are underlined denotes the repeating part). Consider $5/11 = 0.45\underline{45}$. This is a double-digit repeating fraction (the repeating part 45 is underlined). Such repeating fractions are encountered in several applications. The fractional form has the advantages that it is easy to remember and compact for permanent computer storage. Fortunately, the decimal form of probability can be converted into its fractional equivalent by some simple algorithms described in the following. For this, we consider three cases depending on whether any of the trailing digits cyclically repeat or not.

5.3.1   Converting Nonrepeating Decimals to Fractions

Suppose we have a non-repeating decimal number. How do we convert it into the equivalent fractional form p/q? As the trailing decimal digits do not repeat, multiply the decimal number by an appropriate power of 10 (say $m = 10^k$) to remove all decimal places. Let the number after multiplication be n. Find the greatest common divisor (GCD) of m and n (say $p = \text{GCD}(m, n)$). If $p \neq 1$ and $p \neq n$, divide both m and n by p to get the answer. This method will work only when the number (n) is divisible by 2, 5, or their multiples (such as 4, 10, and so on.). We can only give approximate result when the trailing decimal digits cyclically repeat over a wide interval. We summarize it as an algorithm for positive fraction in the following. Extension to negative fractions is straightforward. Line 6 in the listing means that the result is returned in the form p/q.

■   **EXAMPLE 5.1   Decimal to fractional form example-1**

Express the following probabilities in fractional form p/q:
(i) 0.18, (ii) 0.0015, (iii) 0.125, (iv) 0.29, (v) 0.032

Solution 5.1 We need to multiply 0.18 by 100 to discard all decimal digits. Thus, $n = 18, m = 100$. The GCD(18,100) is 2. Dividing both 18 and 100 by 2 gives the answer as 9/50. (ii) In this case, we have $n = 15$ and $m = 10,000$. The

GCD(15,10,000) = 5. Dividing both 15 and 10,000 by 5 gives the resulting fractional form as 3/2000. (iii) Here $n = 125$, $m = 1000$, and GCD(125,1000) = 125. Dividing by 125 gives the resulting fractional form as 1/8. (iv) Here $n = 29$, $m = 100$, and GCD(29,100) = 1 (as 29 is a prime). Hence, the resulting fractional form is 29/100. (v) In this case, we have $n = 32$ and $m = 10^3 = 1000$, GCD(32, 1000) = 8 giving the result 4/125.   ■

Algorithm 5.1 Convert a non-repeating decimal number into fractional form

---

1: Input the decimal number into X

{∗ ignore trailing zeros if any ∗}

Ensure: $(-1 < X < +1)$

2: Count the total number of decimal places k in X

3: Multiply X by 10^k to make it an integer (say Y)

4: Form the fraction p = $Y/10^k$

5: Find the GCD of Y and 10^k as m = GCD(Y, 10^k)

6: **if** (m == 1) **then**

7:      **return** "Y / 10^k "

8: **else**

9:      Divide both Y and 10^k by m

10: **return** Irreducible fraction "(Y/m) / (10^k/m)"

11: **end if**

5.3.2   Converting Repeating Decimals to Fractions

This is more challenging than the nonrepeating case. Here we consider two cases. In the first case, the repeating block starts as the very first digit. If a set of trailing digits repeat cyclically within a reasonable size, we identify the decimal number as $p = 0.dd$ where d is the cyclically repeating part. As done earlier, we multiply it by $m = 10^k$ to move the decimal point to the right position of the last digit of the first repeating block (k is the size of the repeating block). Let the resulting value be $Y = 10^k * p$. Compute $Z = Y - p$, which is devoid of fractions. Now find $r = $ GCD$(Z, m - 1)$. Divide both Z and $m - 1$ by r to get the desired fractional representation.

◨   **EXAMPLE 5.2   Decimal to fractional form example-2**

Express the following probabilities in fractional form p/q:
(i) 0.6̲6̲6̲, (ii) 0.18̲1̲8̲, (iii) 0.315 3̲1̲5̲

Solution 5.2 (i) Let $p = 0.6̲6̲6̲$. Here the first digit itself repeats indefinitely. Hence $d = 6$ (repeating block), $k = 1$ (its size). Multiply p by $m = 10$ to get

$Y = 6.6\underline{6}$. Subtract p from Y to get $Z = 6$. As $m - 1 = 9, r = $ GCD$(Z, m - 1) = $ GCD$(6, 9) = 3$. Divide numerator and denominator by 3 to get the fractional equivalent as $(6/3)/(9/3) = 2/3$. This is of the form p/q without common factors. (ii) Let $p = 0.18\underline{18}$. Here $d = 18, k = 2$, so that $m = 100$ (as there are two digits that cyclically repeats) and $Y = 18.18\underline{18}$. Compute $Z = Y - p = 18$, and $r = $ GCD$(Z, m - 1) = $ GCD$(18, 99) = 9$. Divide both Z and $m - 1$ by r to get $p = 2/11$. (iii) Here $d = 315$ repeats indefinitely. Hence, we need to multiply by $m = 10^3 = 1000$ to move the decimal place. This gives $Z = 315, m = 1000$, $r = $ GCD$(315, 999) = 9$. The answer is $(315/9)/(999/9) = 35/111$. We give below an algorithm for this purpose. ∎

Algorithm 5.2 Convert a Repeating decimal number into fractional form

---

1: Input the decimal number into X

{∗ Assumption: blocks of digits repeat starting with the first digit ∗}

Ensure $(-1 < X < +1)$

2: Count the total number of decimal places K in X

3: Find the repeating cycle length k in X

4: Multiply X by 10^k to make it an integer followed by a fraction (say Y)

5: Subtract X from Y to get an integer Z

6: Form the fraction p = $Z/(10^k - 1)$

7: Find the GCD of Z and $(10^k - 1)$ as m = GCD$(Z, 10^k - 1)$

8: **if** (m == 1)**then**

9:      **return** "Z / $(10^k - 1)$"

10: **else**

11:     Divide both Z and $10^k - 1$ by m

12:     **return** Irreducible fraction "(Z/m) / $[(10^k - 1)/m]$"

13: **end if**

5.3.3   Converting Tail-Repeating Decimals to Fractions

This is a variant of the aforementioned in which the trailing digits repeat cyclically, after a nonrepeating block of digits. This is the hardest case to consider. We identify the decimal number as $p = 0. d_1 d\underline{d}$ where d_1 is the non-repeating part and d is the cyclically repeating part. Note that d_1 can be a single digit or zero too (as in $0.63\underline{3}, 0.01\underline{51}5$). As done earlier, we multiply p by $m = 10^n$ (where n is the number of digits in d_1) to move the decimal point to the right position of the last digit of d_1. Let $Y = p * 10^n$. Next multiply Y by 10^k to move the decimal point to the right position of the first block of repeating digits and store it in Z. Then $Q = Z - Y$ is devoid of fractions. Next find $r = $ GCD$(Q, 10^n(10^k - 1))$. Divide both Q and $10^n(10^k - 1)$ by r to get the desired fractional representation. These are explained in the following sections.

Algorithm 5.3 Convert Tail Repeating decimal into a fraction

1: Input the decimal number into X

{∗ Assumption: Blocks of digits repeat after a non-repeating block ∗}

Ensure $(-1 < X < +1)$

2: Find the repeating cycle length k in X, and non-repeating block length n

3: Multiply X by 10^n to make it an integer followed by a repeating fraction (say Y)

4: Multiply Y by 10^k to make it an integer followed by a fraction (say Z)

5: Subtract Y from Z to get an integer Q

6: Find GCD of Q and $[10^n(10^k - 1)]$ as m = GCD(Q, $[10^n(10^k - 1)]$)

7: **if** (m == 1) **then**

8:      **return** "Q / $10^n(10^k - 1)$"

9: **else**

10:     Divide both Q and $10^n(10^k - 1)$ by m

11:     **return** Irreducible fraction "(Q/m) / $(10^n(10^k - 1)/m)$"

12: **end if**

■ **EXAMPLE 5.3   Decimal to fractional form example-3**

Convert the following probabilities (i) 0.6333 (ii) 0.21515, (iii) 0.0571428 571428 into the form p/q.

Solution 5.3 Let $X = 0.6333$. As the nonrepeating block is of size 1, first multiply X by 10 to get $Y = 6.333$, then multiply Y by 10 to get $Z = 63.33$. Subtract Y from Z to get $Q = 63 - 6 = 57$. Find $r =$ GCD(57, 10 ∗ (10 − 1)) = GCD(57, 90) = 3. Divide both 57 and 90 by 3 to get $p = (57/3)/(90/3) = 19/30$. In part (ii) $p = 0.21515$. Here repeating cycle length is $k = 2$ digits, and nonrepeating block size is $n = 1$ so that $10^k = 100, 10^{k+n} = 1000$, and $[10^n(10^k - 1)] = 990$. This gives $Q = 215 - 2 = 213$. Form the fraction $p = 213/990$. Find the GCD as $m = $ GCD(213, 990) = 3. Divide both the numerator and denominator of p by 3 to get the required answer $p = 71/330$. In Case (iii), we have $K = 6$ and $n = 1$, so that $10^k = 1000000$, and $[10^n(10^k - 1)] = 9999900$. This gives $Q = 571428$, and $p = 571428/9999900$. Next, we need to find the GCD(571428,9999900). We write $571428 = 2^2 ∗ 3^3 ∗ 11 ∗ 13 ∗ 37$ and $9999900 = 2^2 ∗ 3^2 ∗ 5^2 ∗ 37$, from which the GCD is 2857140. Divide both the numerator and denominator of p by 2857140 to get the required answer $p = 2/35$.   ■

■ **EXAMPLE 5.4   Repeating decimals to fractional form**

Convert the probabilities (i) 0.01515, (ii) 0.006363 into fractional form.

Solution 5.4 Let $X = 0.01\underline{51}$. Here the repeating cycle length is $k=2$ digits, and $n=1$ so that $10^k = 100$ and $10^n(10^k - 1) = 990$. Multiply X by 10^n to make it an integer (in this case 0) followed by a fraction as $Y=0.1\underline{51}5$. Next multiply Y by 10^k to make it an integer (in this case 15) followed by a fraction as $Z = 10^{k+n} * X = 0.1\underline{51}5 * 100 = 15.1\underline{51}5$. Now subtract Y from Z to get an integer $Q = Z-Y = 15.1\underline{51}5 - 0.1\underline{51}5 = 15$. Form the fraction $p = Q/[10^n(10^k - 1)]$ $= 15/990$. Find the GCD of Z and $10^n(10^k - 1)$ as $m = $ GCD$(Z, 10^n(10^k - 1))$ $= $ GCD$(15,990) = 15$. Finally, divide both the numerator and denominator of p by 15 to get the required answer $p = 1/66$. (ii) Here also $k=2$ digits, so that $10^k = 100$. Proceed as earlier and find the GCD of 63 and 99 as GCD$(63,99)$ $= 9$. Divide both 63 and 99 by 9 to get the answer 7/11. ∎

We could improve upon our GCD in some particular cases. For example, if Q is an odd number, GCD$(Q, [10^n(10^k - 1)])$ is the same as GCD$(Q, [(10^k - 1)])$. If the nonzero digits in the nonrepeating block is an exact divisor of the repeating block, we could reduce it to the above-mentioned form. Consider $X = 0.00021\ 4\underline{2}42$, in which the nonrepeating block has a 21, which divides the repeating block 42 $\underline{42}$. This reduces to $X = 0.0000102\underline{0}20$ in which the nonrepeating block has nonzero digit as a single 1, and the repeating block is "02" of length 2 (or nonrepeating block "10" followed by repeating digits "20"). When there are several leading zeros in the nonrepeating block as in this example, we could consider the nonrepeating block as the nonzero digits (by simply sliding the decimal place over all zeros) and make a final adjustment to the result. This is described in the following algorithm.

Algorithm 5.4 Tail Repeating decimal with many leading zeros into p/q form

---

1: Input the decimal number into X

{* Assumption: Blocks of digits repeat after a non-repeating block, the first few of which are all 0's *}

Ensure $(-1 < X < +1)$

2: Find the repeating cycle length k in X, and block size of leading zeros of length m, and non-repeating nonzero-digit block of length n

3: Multiply X by 10^m to move the decimal place over the zeros (say Y)

4: Multiply Y by 10^n to make it an integer followed by a repeating fraction (say Z)

5: Multiply Z by 10^k to make it an integer followed by a fraction (say T)

6: Subtract Z from T to get an integer Q

7: Find GCD of Q and $[10^{n+m}(10^k - 1)]$ as m = GCD$(Q, [10^{n+m}(10^k - 1)])$

8: **if** (m == 1) **then**

9:     **return** "Q / $[10^{n+m}(10^k - 1)]$"

10: **else**

11:    Divide both Q and $[10^{n+m}(10^k - 1)]$ by m

12:    **return** Irreducible fraction "(Q/m) / $([10^{n+m}(10^k - 1)]/m)$"

13: **end if**

▣ EXAMPLE 5.5 Tail repeating decimal to fractional form

Convert the decimal 0.00022 4545 to fractional form.

Solution 5.5 Here repeating block is of length $k = 2$, nonrepeating block is of length $n = 2$ (5–"3 zeros"), and m is 3 (as there are three leading zeros). First multiply X by 10^m. We have $10^k = 100$, $10^{k+n} = 10,000$, and $10^n(10^k - 1) = 9900$. This gives $Q = 2245 - 22 = 2223$, and $p = 2223/9900000$. Next, find $m = \text{GCD}(2223, 9900000) = 9$. Dividing both the numerator and denominator of p by 9 gives $p = 247/1100000$, which is the required answer. ∎

The repeating block may be too long for some fractions, especially involving ratios of primes. Consider $2/17 = 0.1176470588235294$ $\underline{117}$.., which repeats itself after 16 decimal places. Similarly, there are many fractions for which the cycle of digits repeats well beyond the calculator display. Consider $7/29 = 0.2413\ 7931\ 0344$ $8275\ 8620\ 6896\ 5517\ 241$, and so on, which repeats after 28 decimal places! (they are in general of the form $k/(k*n+ 1)$ with cycle block size $k*n$). We could either approximate such decimals or employ other algorithms.

If we truncate it at the wrong decimal place (say 8th or 16th place), the resulting fraction will not come even close to the true value (7/29 in the above-mentioned example). For example, truncating at second decimal place gives 6/25 and truncating at 12th decimal place gives 30172413793/125000000000. An astute reader will notice that our original digits are repeated after a nonrepeating block of length 4. This property can be used to approximate the fractional value using the second algorithm given earlier. This means that we may sometimes be able to approximate a nonrepeating decimal number (or a repeating decimal with a large cycle length) by dividing it by a small number. In the above-mentioned case, we get the approximation as $3017/12500 = 0.24136$. This is correct to the fourth decimal place.

Assuming that all our decimal numbers are positive, we could store any decimal number in just two memory locations (one for storing p and the other for storing q). Signed decimals need an extra 1 bit to store the sign as 0 for positive and 1 for negative. As an unsigned int type can store numbers between 0 and 65,535 in just 2 bytes of memory, we could represent a great majority of fractions that we encounter in practice using this method, provided that both the numerator and denominator are less than 65,536. We could use the unsigned long int data type (4 bytes of memory) when larger numbers are involved, as it can store up to 4,294,967,295.

Percentage form of probability is obtained by multiplying the decimal form by 100. These are usually used in conversations and correspondences. Scientific form is preferred when a probability is too small with several leading zeros. The pictorial form is used in geometric probability problems.

5.4 SAMPLE SPACE

Random experiments are at the core of experimental probability. Here, the word "experiment" has a different meaning in statistics than its literal meaning. A simple measurement of a physical or other characteristics of an object, a count of objects that satisfy one or more conditions, an observation of the duration of a phenomena (like the lifetime of a device) can all be considered as an experiment in statistics.

Definition 5.3 Random experiments are those that are repeated under identical conditions every time and always produce one among several outcomes.

Here, the clause "under identical conditions" needs some scrutiny. It only means that the conditions are replicable and statistically insignificant. For example, consider the measurement of the storage of water in a reservoir. If measurements are taken over a period of time (say on successive days or weekends), the conditions may not be exactly identical in the strict scientific sense. Owing to the pull exerted by celestial bodies on the surface of the earth, the reservoir levels could go up when the moon has just passed overhead. This gravitational pull is more in the equatorial region when the moon and the Sun are both oriented in more or less the same direction over the place of observation (this is why very high tides occur on some days), which is maximum during the closest approach of the moon to the Earth. Similarly, the amount of water evaporated depends on the day-time temperature, wind speed, humidity, and reservoir area among other things. It is our tacit assumption that random experiments are conducted in rapid succession or in short duration of time. Extraneous factors, if any, that could affect the measurements should be accurately maintained in highly sensitive and time-dependent scientific experiments. These are often negligible when the sample is collected over a short duration. The purpose of an experiment could also be the identification of such differences (as in agricultural experiments). The qualifier "random" indicates that the outcomes are unpredictable until the results are observed. In other words, the results will vary from trial to trial even when the conditions of the experiment are the same.

Definition 5.4 The set of all possible outcomes of a random experiment is called its sample space.

The sample space itself is an event because it always occurs. By convention, it is indicated by the Greek symbol Ω (pronounced capital omega). Its complement is denoted by $\Omega^c = \phi$ (the null set, pronounced small phi). The complement of an event X is denoted as X', $\overline{X}$, or X^c. As $\overline{X}$ is used in subsequent chapters to denote the arithmetic mean, we will use X^c for complement. The very first step in solving any probability problem is to identify the sample space. These are quite often easy to find. We illustrate it with various examples.

■ EXAMPLE 5.6   Sample space for simple experiment

What is the sample space of an experiment of throwing two fair coins?

Solution 5.6 Denote a Head turning up by H and a tail turning up by T. Then, the sample space is {HH, HT, TH, TT} where HH denotes that both throws resulted in Heads, and so on. Here "H" and "T" are simply labels. We could assign any label we wish (because the English letter H is a silent syllable in Spanish (words) and is pronounced differently in Greek, Russian, etc.). For example, if Head is denoted by a "1" and the tail by a "0," our sample space becomes {11, 10, 01, 00}.   ■

EXAMPLE 5.7   Circuits in series in a device

There are two circuits in series in a device, both of which can be open or closed. Identify the sample space when the device is turned on.

Solution 5.7 Denote the open circuit by a 0 and closed circuit by a 1. Then, the possibilities are {00, 01, 10, 11} where 00 indicates that both circuits are open and 11 indicates that both are closed.   ■

EXAMPLE 5.8   Balls in urns

Find the sample space for (i) drawing two balls from an urn containing three red and two blue balls that are indistinguishable except for the color (ii) two throws of a dice that result in a sum of 10.

Solution 5.8 (i) Denote the red ball by R and blue ball by B. The possible outcomes are {R, R}, {R, B}, {B, R}, {B, B}, (ii) Denote the numbers on the die by {1, 2, 3, 4, 5, 6}. Then the possible 36 values in the sample space are {1, 1}, {1, 2}, {1, 3}, {1, 4}, {1, 5},{1, 6}, {2, 1}, {2, 2}, {2, 3}, {2, 4}, {2, 5}, {2, 6}, ... , {5, 6}, and {6, 6}. Here {1, 6} and {6, 1} are considered to be different, even if the two dice are thrown simultaneously. For part (ii), the favorable cases are {(5,5), (4,6), (6,4)}.   ■

The sample space obviously depends on the defined event. If an event U is defined as the sum of the numbers that show up when two dice are thrown, the sample space of U becomes {2, 3, 4, 5, 6, 7, 8, 9, 0, 11, 12}. If another event V is defined as the absolute value of the difference between the numbers that show up, the sample space of V becomes {0, 1, 2, 3, 4, 5}. This shows that multiple sample spaces can be obtained on the same random experiment.

5.5   MATHEMATICAL BACKGROUND

Probability problems are unlike the problems in other sciences. Beginning students sometimes find it difficult to solve probability problems because there are either several ways or no obvious way to solve it. Different problems may require a different approach, concept, or tool. There are many such tools and techniques needed to solve every problem in probability. Examples are Venn diagrams, permutations and combinations, principle of inclusion and exclusion, urn models, recurrence relations, divide and conquer or decrease and conquer principles, sampling with and without replacement, bipartite graphs, and De'Morgan's laws to name a few. In addition, independence of events, conditional events, and other event algebra discussed in the following sections may be needed individually or in combination in some problems. There are still other problems that can be solved easily by geometric reasoning, properties of probability distributions, and so on. A thorough understanding of these tools and techniques are essential to solve all probability problems with ease. The following section first describes the essential tools and then applies it to individual problems.

5.5.1 Sets and Mappings

A set is a collection of distinguishable elements logically considered as a group. The elements may be homogeneous or heterogeneous. For example, consider fruits and vegetables as two separate sets. The fruit set can comprise of apples, oranges, berries, bananas, and the like, whereas the vegetable set may consist of potatoes, tomatoes, carrots, and so on. Total number of distinct elements in a set S is called the size of the set or its cardinality. It is denoted as $|S|$. This is always an integer ≥ 1. To extend the set theory to various situations involving intersect and complement operations, we will denote an empty set (without any elements in it) by the Greek symbol ϕ (pronounced "small phi"). The size of the empty set by convention is zero (i.e., $|\phi| = 0$). The totality of all elements under consideration in a set is called the universal set, super-set, or set space. It is symbolically denoted by Ω. Any element of Ω is called a member or point of the set and is denoted by ω. Multiple elements can be combined to get subsets of the set Ω. In probability theory, our main interest is in counting proper subsets of Ω.

Definition 5.5 The collection of all subsets of a set S (including the null set ϕ and the set itself) is called the power-set (it is denoted by 2^S and has $2^{|S|}$ elements).

◼ EXAMPLE 5.9   Cardinality of Power-set

Use induction to prove that the power set $P(S)$ of a finite set S has cardinality $2^{|S|}$.

Solution 5.9 Consider a singleton set S (with just one element, say b). Its power-set is $\{\phi, b\}$ of cardinality 2. Next consider a set with two elements $S = \{a, b\}$. Its power-set is $\{\phi, \{a\}, \{b\}, \{a, b\}\}$ of cardinality 4. Thus, the assumption is true for $n = 1, 2$ where n is the number of elements in the set. Assume that it is true for an arbitrary set S of size $k > 2$. Obviously, cardinality of S is $2^{|S|} = 2^{|k|} = 2^k$. Label all elements currently in S by a group symbol σ. Now add a single new element x to S to make it $S' = \{\sigma, x\}$ of cardinality $|S'| = k + 1$. The power-set of S' comprises the power-set of S, plus new subsets formed by adding x to each of them. As sets are unordered collections, adding x to each subset of σ produces at most $2^{|S|}$ new subsets. Thus, the total number of subsets in S' is $2^{|S|} + 2^{|S|} = 2 * 2^{|S|} = 2^{|S|+1} = 2^{k+1} = 2^{|S'|}$. This shows that if the assumption is true for $n = k$, it is true for $n = k + 1$. By induction, it is true for all positive integers $n \geq 1$.												∎

EXAMPLE 5.10   Powerset example-2

Find the power-set of the set $S = \{a, b, c\}$

Solution 5.10 We will tackle the problem by the divide-and-conquer approach. First consider all one-element subsets. There are three singleton subsets as $\{a\}, \{b\}, \{c\}$. Next consider two-element subsets. There are $\binom{3}{2} = 3$ two-element subsets as $\{a, b\}, \{a, c\}, \{b, c\}$ (see Section 5.9.5 in page 145). To this add the null set ϕ (with no elements), and the set S itself to get the power-set

$$2^S = \{\phi, \{a\}, \{b\}, \{c\}, \{a, b\}, \{a, c\}, \{b, c\}, \{a, b, c\}\} \tag{5.1}$$

having $2^{|S|} = 2^3 = 8$ elements. This is pictorially shown in Figure 5.1.			∎

A special decomposition of a finite set S is of importance in probability theory. This is called the *partition* of S or *set partition*.

Definition 5.6 A partition of a finite set S with at least two distinct elements is a collection of mutually exclusive and collectively exhaustive subsets $S_1, S_2, \ldots, S_m$ such that $S = S_1 \cup S_2 \cup \cdots , \cup S_m = \cup_{i=1}^m S_i$, and $S_i \cap S_j = \phi$ for all $i \neq j$. Note that ϕ and S are not counted in a *set partition*.

Each element of a *set partition* can be mapped to a real number p_i. If this mapped number has the property that they add up to 1 ($\sum_i p_i = 1$), it is called a distribution defined over S.

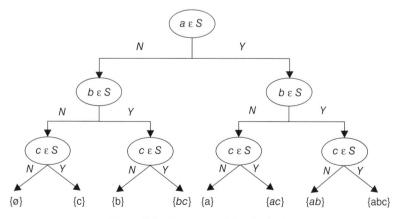

Figure 5.1   Power-set of $S = \{a, b, c\}$.

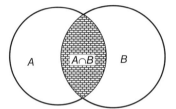

Figure 5.2   Venn diagram for $A \cap B$.

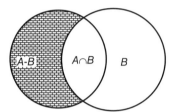

Figure 5.3   Venn diagram for AB^c.

5.5.2   Venn Diagrams

The British mathematician and cleric John Venn (1834–1923) introduced Venn diagrams in 1881 for representing sets and operations on them. They became instantly popular because there are just two symbols used in its graphical representation—a rectangle denotes the universal set U, and one or more labeled circles or ellipses drawn wholly within the rectangle denote subsets of U. Event interactions are represented by intersecting labeled circles (see Figures 5.2 and 5.3). The area that is common to intersecting circles can map the actual amount of interaction of the events. In most of the problems, this need not be so fine-grained because the Venn diagram is not used to compute the probabilities directly; rather it is a visual device simply to ascertain if events interact or not. Sets without common elements are drawn as nonintersecting circles. This is useful when the number of events is small (say 2–6). The importance of Venn diagrams in probability arises because events (both discrete and continuous) that underlie probabilities are easily represented by sets. They are valuable tools in breaking complex probability problems involving multiple intersecting events into simpler subproblems that are easy to solve. Venn diagrams have been extended by many researchers to suit problems in engineering, geology, chemistry, and other sciences. Examples are Karnaugh maps, Euler diagrams, Johnston diagrams, Edwards' Venn Diagrams, and Peirce diagrams. Euler diagrams are an extension of Venn diagrams to represent more than one sample space (see Reference 110). Venn diagrams may not be easy to comprehend when there are too many intersecting events. In such a case, we could form a hierarchical Venn diagram by labeling events with a common denomination to the top of the hierarchy.

🖥 **EXAMPLE 5.11   Union of events**

Sixty percentage of the people in an office read newspaper "*A*," and 50% read newspaper "*B*." If 10% of the people read neither "*A*" nor "*B*," what percentage of the people read both newspapers?

Solution 5.11 This problem is easily solved using a Venn diagram. Let "*A*" denote the event that people read newspaper "*A*" and "*B*" denote the event that they read newspaper "*B*." Then $A \cup B$ denotes the event that people read either of the newspapers and $A \cap B$ denotes the event that people read both newspapers. As this problem involves count or percentage, the event and count can be considered as synonymous. As the number of people who reads either of them is given as $A \cup B$, the number of people who read neither is U-$A \cup B = 10\%$ (given). From this, we get $10 = 100 - [(60 + 50) - A \cap B]$ or $A \cap B = 20\%$. Hence, 20% of the people read both newspapers.   ∎

5.5.3   Tree Diagrams

Several probability problems involve mutually exclusive subcases or subevents. These are best represented as rooted trees or forests (a collection of disjoint trees is called a forest). A tree in computer science is a nonlinear data structure with a distinguished node called the root. A pictorial representation of trees makes it much easy to comprehend. For this purpose, the root is always drawn either at the top or at the left. A tree is a special case of a graph. Although a graph can be directed or undirected, a tree is almost always undirected. The branches (straight lines) drawn from a node represent subproblems, subsets, or subcases. This representation can sometimes decompose a complex probability problem into two or more simple ones or as a hierarchy of subproblems. Each such subcase can be further broken down into smaller trees. This subdivision usually uses a categorical variable such as sex and religion or outcomes of an experiment. Quantitative variables could also be used to subdivide a node into smaller subtrees if (i) the number of cases are small or (ii) discretization is used to categorize the continuous variable.

🖥 **EXAMPLE 5.12   Tree-diagram for coin toss experiment**

A fair coin is tossed three times. Draw the tree diagram and find the sample space.

Solution 5.12 As there are just two possible outcomes in each throw, we denote it by two branches from the nodes. Consider the first throw. It could result in either an H or a T. The second and subsequent throws are denoted as further branching as in the figure. The sample space is obtained as the union of labels at the leaf nodes as {HHH, HHT, HTH, HTT, THH, THT, TTH, TTT}. This is pictorially shown in Figure 5.4.   ∎

In some problems, there exist more than one way to draw a tree. Sometimes, the tree is formed by the occurrence of a related happening as in sports tournaments in which the winning team encounters other players or opponents.

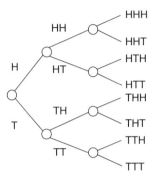

Figure 5.4   Coin-tossing sample space.

5.5.4   Bipartite Graphs

A bipartite graph consists of two sets of nodes (say V and W) such that each node in V is connected to some node in W and vice versa. This means that nodes in V are not connected to other nodes in V and similarly nodes in W are not connected to other nodes in W. The elements of V and W are nominal or ordinal type in most of the applications. The bipartite graph is extremely useful in simplifying some of the probability problems.

▉ **EXAMPLE 5.13   Jobs and applicants**

A company has five vacancies that require different skill sets for which 10 applications are received. Describe how bipartite graphs can be used to model the matching of applicants and jobs.

Solution 5.13 Here, both sets are nominal type. Represent 10 applicants by 10 labeled nodes on the left and five jobs by five nodes on the right. Make a link among the two sets of nodes if the skill set of ith applicant fits the job j. For each job j, if there is only a single applicant, remove it from the bipartite graph. If the remaining graph is a forest, we could identify groups of applicants that are clustered in groups of jobs (or a single job). Otherwise, the bipartite graph will show the choices for hiring applicants.   ▪

5.5.5   Bipartite Forests

These are special cases of bipartite graphs in which the entire graph can be decomposed into two or more disjoint bipartite graphs. The smallest possible bipartite graph is one in which one node on the left is connected to just one node on the right. In the jobs and applicants example, if one applicant is connected to one job and there are no other links between these two nodes, this can be removed as there are no other choices. Hence, bipartite forests decompose a larger problem into smaller subproblems that can be independently solved.

5.6 EVENTS

There are many types of events encountered in probability and related fields. A good understanding of them can greatly simplify some of the probability problems. In addition, they provide a parallel between the axioms of set theory and probability. We assume in the subsequent discussions that $P(\phi) = 0$ (probability associated with the null set is zero), and $P(\Omega) = 1$.

Definition 5.7 An event is a well-defined outcome of an experiment or a subset of a sample space.

The literal meaning of an event is *a thing that happens*. In probability, an event is a well-defined outcome associated with a random experiment or a trial. Set theoretically, events are subsets of the set of all possible outcomes.

Objective probability has two basic building blocks. First, there should be a random experiment that generates uncertain outcomes. These can be discrete or continuous. A unique label is assigned to each outcome of the experiment to distinguish among themselves. Secondly, we must have events that are either single outcomes or a collection of outcomes that satisfy a user-specified condition or criterion. The set of all possible outcomes of a random experiment is called sample space. The sample space is specific to each random experiment. It may or may not depend on time. Most of the problems that are encountered below are time independent sample spaces.

Each discrete event is uniquely identified by a label, a symbol, a number, or other identifying mark. There are no *hard and fast rules* to name events. Event labels are usually denoted by capital letters of an alphabet (English, Greek, etc.). We can combine events and denote them by other labels or abbreviated letters. One experimenter may label the events resulting in a single toss of a coin as {H, T} while another may label it as {1, 0}. These labels are meant to distinguish the events among themselves. If you conduct such a trial or experiment, every event will eventually occur. Some events that have high chance of occurring materializes more. Thus, the probability associated with each event can tell us which are more likely to materialize than others.

■ EXAMPLE 5.14   Event identification

Consider contaminants in drinking water sources collected from different parts of a city. Describe what are some possible events and how to combine them.

Solution 5.14 Assume that there are a dozen possible contaminants in drinking water. Label each contaminant by a unique letter. Then a simple event can be defined as "Presence of contaminant in drinking water above the prescribed limit." For instance, let "A" denote the presence of Arsenic above its permitted limit and let "C" denote the presence of chromium, and so on. If a sample contains both arsenic and chromium (and not the others), it can be labeled as "AC" or "CA." This type of concatenated labeling is inconvenient when there are a large number of possibilities. A solution is to give separate codes of fixed size or numbers to each combination event. ∎

5.6.1 Deterministic and Probabilistic Events

Events may be deterministic or probabilistic. Outcomes of deterministic events are always predictable using mathematical equations or various laws of physics, chemistry, or logic. For instance, chances of getting six spades in a hand of 10 cards is exactly predictable as the number of spades in any deck of cards is 13. The reservoir capacity example given in page 5–18 is predictable using the position of the Sun and path of the moon. Probabilistic events can be predicted by past analysis of outcomes. An aircraft's engine failure can be predicted well in advance if past data on engine failures are available for specific engines of certain age and type.

5.6.2 Discrete Versus Continuous Events

Events can be discrete or continuous. In probability problems (especially textbook examples), we seldom encounter continuous events. However, continuous events are encountered in engineering applications.

■ **EXAMPLE 5.15 Continuous events**

Give examples of continuous events.

Solution 5.15 Consider a microchip that has just been manufactured. Its lifetime (in hours) is an event that can take any positive value. The sample space in this case is $S = \{t | t \geq 0\}$ and the variable involved is time. Let F denote the event that it does not fail during the first $c = 1000$ hours of operation. Then $P(F) = P(t > c)$. This probability becomes smaller as c is increased. Consider a swimming pool with an optimal capacity. If the water inflow and outflow are ignored, the amount of water in the pool (say in cubic feet or inches) can be considered as a continuous sample space. The complementary event in this case is the optimal level less the current capacity. Both of them are continuous. In general, if the sample space is continuous, the complement of an event defined on the sample space is also continuous. ∎

Probabilistic events are associated with random experiments, random variations in some processes (like manufacturing), or unknown variations in some variables (probability that an air-bag in a car will fail to inflate upon a collision depends on the speed at impact and various circuitry characteristics. Probability of a rain or snow (at an appropriate location) tomorrow depends on hundreds of interacting atmospheric variables). One need not understand the variables that drive a phenomenon to predict the probability. Past data collected over a period of time can be used in such situations using the frequency approach. In this chapter, we are more interested in probability mechanisms involved in random experiments.

Each event of a random experiment is mapped to a probability between 0 and 1. This is denoted by $P(A)$ where A is the event label, name, or token; and $P()$ is the notation for probability. An event is said to occur if the outcome of a random experiment results in that event.

◼ EXAMPLE 5.16   Combination event examples

Give examples of events defined on each of the following: (i) software-controlled machine failures and (ii) viscous flow through a pipe.

Solution 5.16 Machines may fail due to (i) mechanical faults, (ii) electrical faults, (iii) software faults, (iv) wrong handling or wrong initial settings, and (vi) other reasons. Let these be denoted by events M, E, S, H and O. Combination events can then be represented as described earlier. In the case of viscous flow, we could define events using flow-rate or average amount of liquid transported since last overhaul work. This could vary slightly depending on the pressure applied, viscosity of liquid, surface corrosion, and outside temperature variations.   ◼

Almost certain events have probability one. As examples, the probability that an email message with a correctly specified address will be delivered to an existing person within a reasonable time interval is 1 (unless the recipient's mailbox is full or the server is down, the probability of which are small), probability that a payment for an online transaction processed and approved through a payment gateway will be credited to the merchant's account is 1. Uncertain or unlikely events have probability near 0. For instance, the probability that an ATM machine will eject an amount larger than requested by a customer is zero. A probability of 0.5 implies a "fifty–fifty" chance for the occurrence or nonoccurrence of an event. Experimentalists, managers, and practitioners are more interested in probabilities that deviate much away from 0.5 owing to this simple reason.

5.6.3   Event Categories

There are many types of events encountered in probability problems.

1. Simple and compound events
   A simple event (also called elementary event) cannot be decomposed into simpler events. All compound events are built using either simple events or other compound events using event operations. In most cases, a compound event can be sliced up into several fine-grained simple events. Consider the working condition of a device. It may be defective (D) or nondefective (N). These are simple events. An event can comprise of a set of items. Consider the tossing of two dice, each with six faces marked 1–6. The possible events are $\{(1,1), (1,2), ..., (6,6)\}$.
2. Mutually exclusive events
   Events are mutually exclusive if they are disjoint. Symbolically, two events A and B are mutually exclusive or disjoint if $A \cap B = \phi$ or equivalently

$P(A \cap B) = 0$. In the coin-tossing experiment, the mutually exclusive events are Head and Tail. This definition can be extended to any number of events. Let $A_1, A_2, \ldots, A_n$ be n events. If $n > 2$, they are totally mutually exclusive if $P(A_1 \cap A_2 \cap \cdots \cap A_n) = \phi$. If they are totally mutually exclusive, they need not be pairwise mutually exclusive. For instance, events A and B can have a common portion and B,C can have a common portion, but A and C can be disjoint. This implies that $P(A \cap C) = \phi$. As $\phi \cap X$ is ϕ where X is any other event, we can simply add events to totally mutually exclusive events. Suppose $B = A$. Incorporating B to $P(A \cap C) = \phi$ we get $P(A \cap B \cap C) = \phi$ although $A \cap B = A$. A set of events is minimally mutually exclusive if there are events $A_1, A_2, \ldots, A_k$ such that $P(A_1 \cap A_2 \cap \cdots \cap A_k) = \phi$, but this relationship does not hold for any $m < k$. The mutually exclusivity property is extremely useful in decomposing some of the complex probability problems into simpler ones.

3. Equally likely events (ELEs)

   If every event of an experiment has an equal chance of occurring, they are called equally likely events (ELEs). Probability problems are greatly simplified in such situations. Examples are dice and coin tosses. Consider the outcomes of an unbiased coin (with two possible events H and T), tosses of a six-faced die, a regular prism with four faces, or a regular pyramid with five faces. The probability of any event occurring in ELE is one divided by the total number of events in the sample space. Consider a regular five-faced pyramid with faces numbered from 1 to 5 that is tossed to a hard surface. As there is a unique face (at the bottom) on which it will rest, we can define an event as "face number that is hidden at the bottom." Then, each of the faces is equally likely with probability 1/5.

   Two or more independent experiments with ELE may be combined. Let V denote an event defined on the sample space Ω of n equally likely outcomes. Then $P(V) = |V|/n$ where $|V|$ denotes the cardinality of V (number of favorable elementary events in V). Consider two tosses of our pyramid. Define an event that V="the sum of the numbers at the bottom is even." There are $5 \times 5 = 25$ total possibilities. The 13 favorable cases are (1,1), (1,3), (1,5), (2,2), (2,4), (3,1), (3,3), (3,5), (4,2), (4,4), (5,1), (5,3), and (5,5). Hence $P(V) = 13/25$. Next consider the toss of a fair die. Let V be the event that the face that shows up is a prime number. The favorable cases are $V = \{1, 2, 3, 5\}$ with $|V| = 4$. This gives $P(V) = 4/6 = 2/3$. These results are summarized in the following theorem.

   Equally Likely Principle (ELP): If the sample space Ω of a random experiment consists of a finite number of equally likely outcomes, then any non-null event E defined on Ω has probability of occurrence $|E|/|\Omega|$.

   These probabilities can easily be calculated directly using the count-and-conquer techniques or indirectly calculated using one of the do-little principles (Section 5.6.4 (p. 131), Section 5.15.4(p. 159)).

4. Complementary events

   Complementary events are those that do not include the outcomes of another event. This complement operation is taken with respect to the entire sample space Ω. This means that complement of an event belongs to the sample space. Some probability problems can be substantially simplified by the complementary event principle. One common example is those problems that contain *at least one* outcome. If the number of outcomes is large, we could take the complement of the event to get a simple favorable set. A related operator is event difference (denoted by $-$), which is taken with respect to another event. Thus, X-Y represents an event that contains events in X but not in Y. Consider the event V defined on the toss of a fair die. The complementary event comprises $V^c = \{4, 6\}$.

5. Dependent and independent events
Events may be dependent on one another. This dependence can be due to a shared property or some underlying commonalities. Two or more events are independent if the occurrence of one in no way affects the occurrence of others. Consider the following examples:
(i) Let X denote the event that an e-commerce customer at a video store has "Red hair" and Y denote the event that the order is for "Adventure movie." Then, X and Y are independent.
(ii) Consider a school kid with six shirts, five pants, and three ties. Wearing any of the combinations are independent events. However, the decision to choose a shirt that matches the color of the pants or a tie that matches the shirt color may preclude some possibilities. Thus, the events may be considered as dependent. See 5.15.7 in page 163 for further discussion.

6. Conditional events
Events may depend conditionally on other events. These are called conditional events. In other words, you have knowledge that some other event has occurred. This filters out a subset of the sample space, thereby reducing the computational burden.
Consider a simple experiment of throwing a fair coin until you get the first Head. As the coin is fair, probability of getting a Head in the first throw is 1/2. The event of Head occurring in the second throw is conditional on the first throw resulting in a Tail. Similarly for subsequent events. These are the sequence of events considered in geometric distribution.
Consider an electronic board with a parallel circuit in each of which there are three components. If any component in one circuit fails, the device will continue operating. However if at least one component in both circuits fails, the device will stop working. Thus, the nonworking condition of a device is conditionally dependent on both circuits in the board.

5.6.4 Do-Little Principle for Events

Complementary events are sometimes easy to find when the sample space consists of a large number of discrete events as in the above-mentioned example. These are especially true in "at least k" and "at most k" type problems (or a combination of both) that can be considerably simplified by taking the complement or opposite event. It is also called complement-and-conquer principle. It has two versions—a count version to count the complementary events and a probability version to obtain the probability of complementary events. Both of them are related.

5.7   EVENT ALGEBRA

Every random experiment involves two or more events. Events are usually combined using the logical operators {AND, OR, NOT}. Combining events using OR operator often increases the probability, while combining events using AND operator often decreases the probability. If A and B are two events, $P(A \text{ OR } B) = P(A)$ only when B is a proper subset of A. Similarly $P(A \text{ AND } B) = P(A)$ only when A and B are exactly identical. The NOT operator may or may not increase the probability—this depends on how big a chunk of the sample space is spanned by the defined event. The events resulting from applying a NOT operator to another event is called its complement. For example, consider the throw of a six-faced die numbered 1–6. If event X is defined as "an even number shows up," then it has probability 0.5 as the possible outcomes are $X = \{2, 4, 6\}$. Its complement event also has probability 0.5, as the NOT operator returns $X^c = \{1, 3, 5\}$. Next consider the event "the number that shows up is a prime." This has associated probability $4/6 = 2/3$ as the possible outcomes are $\{1, 2, 3, 5\}$. Its complement event consists of $\{4, 6\}$ with probability $2/6 = 1/3$. Here, the complementation has reduced the probability. Unless otherwise stated, the complement is always taken with respect to the entire sample space Ω. In other words, the NOT operator is to be interpreted as anything that remains in the sample space other than those in the considered event or subspace. Sometimes, we seek the probability of occurrence of event combinations. Events can be combined using set theoretic operations union ($\cup$), intersections ($\cap$), complements, and differences. Any of these can be combined to produce compound events.

5.7.1   Laws of Events

There are several laws of events that are direct descendants of corresponding laws of set theory. These laws are helpful to solve some of the discrete probability problems. More importantly, they form the theoretical foundations on which classical probability theory is built up. Most of the axioms of probability are direct generalizations of corresponding event axioms. Union of events represents the occurrence of either or both of them, whereas intersection of two events represents their joint occurrence (both occur).

◼ EXAMPLE 5.17   Flight delay

Consider a commercial flight that is scheduled for departure at a fixed time. A delay in departure can happen due to many reasons:–(i) technical problems with the aircraft, (ii) delay in one or more connecting flights that have passengers for current flight, (iii) delay of flight crew in reporting for duty (iv) delay in security checking, (v) delay due to runway problems or congestion, and (vi) other reasons.

Assume that each of these is independent occurrences. Form meaningful events using union and intersection of events.

Solution 5.17 Let T, V, W, X, Y, and Z denote each of the six events. As each of them are independent, a union represents an "either or" situation. For example, $T \cup V$ represents the event that the flight is delayed either due to technical problems with the aircraft or connecting flights are delayed. The event $T \cup V \cup W \cup X \cup Y$ denotes the event that there is a likely flight delay due to the occurrence of individual events or a combination of events mentioned. The event $T \cap V$ denotes that there is technical problem with the aircraft AND connecting flights are delayed. If all events are represented by single character labels, we could drop the $\cap$ operator and represent multiple event occurrences by a concatenated label. For instance, $W \cap X \cap Y$ can be represented as WXY. This is just a new name or label given to a combination of events.   ∎

5.7.1.1 *Law of Total Probability*

Let X and Y be two nonempty sets with common elements. Then, we can partition the set X into two parts as $X = X \cap Y + X \cap \overline{Y}$, where $X \cap Y$ contains members of X with both traits and $X \cap \overline{Y}$ contains members of X without the trait of Y. In terms of probability, this is written as $P(X) = P(X \cap Y) + P(X \cap \overline{Y})$. Similarly, $Y = Y \cap X + Y \cap \overline{X}$ gives $P(Y) = P(Y \cap X) + P(Y \cap \overline{X})$. This result is used in the derivation of Bayes theorem.

5.7.1.2 *Commutative Laws*

The literal meaning of "commutative" is "unchanged in result by a reordering of operands." These are meaningful for binary operators that take two operands. These laws are formed in event algebra and set theory using $\cup$ and $\cap$ set-theoretic operators. Simply put, these laws state that the events on either side of these operators can be swapped. Let X and Y denote two events. Then, the commutative law states that $X \cup Y = Y \cup X$ and $X \cap Y = Y \cap X$. In Example 2.17, $T \cup V$ and $V \cup T$ represent the same event. Similarly, $W \cap X$ and $X \cap W$ represent the same thing. Humans can easily conceive the meaning of these expressions by perception. This is especially easy when the events are disjoint. However, the law holds even when two events overlap. They are more useful when more than two events are involved.

▮ EXAMPLE 5.18   Weight-loss program

Consider a weight-loss clinic for over fat people that offers three programs:– (i) restricted diet (RD) program that can decrease the weight on the average by 10 pounds in 4 weeks, (ii) a fat-burning exercise regime (ER) with a thermal belt that sheds on the average 8.5 pounds in 4 weeks, (iii) a bariatric surgery (BS) program that sheds on the average 16 pounds in 4 weeks. A patient can opt for either individual programs or for a combination. Does the commutative law make sense in this example? Describe the following event combinations in plain English:– (i) ER∩BS and (ii) RD∪BS.

Solution 5.18 As events represent various programs offered by the clinic, the intersection of events indicate the programs for which a patient has opted. Thus, ER∩BS indicates that a patient is registered in both ER and BS programs. However, union of events in this problem does not make sense. If the events are defined in terms of counts (total number of people registered for the program), then union of events make sense. For instance, ER∪BS indicates the total number of people registered for either of the programs or both.   ■

5.7.1.3 Associative Laws Let X, Y, and Z denote three distinct events. Then, the associative law states that $X \cup (Y \cup Z) = (X \cup Y) \cup Z$ and $X \cap (Y \cap Z) = (X \cap Y) \cap Z$. The meaning of each of them is that the flight is delayed due to a delay in security checking OR a delay due to runway problems or congestion OR due to other reasons. Here, the operator that is commuted is the same. As in the case of commutative law, humans can easily conceive the meaning when the events are disjoint.

5.7.1.4 Distributive Laws The name distributive comes from the fact that two non-identical event combinations are simplified by distributing one of the operators. Let X, Y, and Z denote three events. Then, the distributive law states that

$$X \cup (Y \cap Z) = (X \cup Y) \cap (X \cup Z) \quad \text{and} \quad X \cap (Y \cup Z) = (X \cap Y) \cup (X \cap Z), \quad (5.2)$$

where the ∪ operator outside the parenthesis in the first expression is distributed and the ∩ operator outside the parenthesis in the second expression is distributed. This law is more meaningful when some events have a "combined effect." Set-theoretically this means that the intersection of some of the events is non-null. These rules are extremely useful in reducing the favorable sample space of compound events.

In the case of associative law, we had the same operator (either ∪ or ∩). If the operators are different, we get the distributive law given below discussion.

EXAMPLE 5.19 Distributive Laws

Consider the pre-requisite courses for enrolling in a statistics course. A student who has finished College Algebra (X) is eligible, as also those who have finished both of Computer Science 100 (Y) and Maths 104 (Z). Express this using distributive law.

Solution 5.19 Label the events as X, Y, and Z. Then, the condition can be expressed as $X \cup (Y \cap Z)$.   ■

EXAMPLE 5.20 Event Combinations

Consider example 5.18 given earlier. Describe the following event combinations in plain English:– (i) RD ∩ (BS ∪ ER) and (ii) (ER ∩ RD) ∪ BS

Solution 5.20 As the operator inside the bracket in (i) is ∪, the meaning is to select a weight-loss program with RD AND either BS OR ER. Similarly, the meaning of (ii) is to opt for a program with either BS alone or both ER and RD.   ■

5.7.2 De'Morgan's Laws

These laws relate the complement of compound events in terms of individual complements. In the following section, we use overline to denote complements.

Rule 1 Complement of an intersection is the union of their complements.

Let A and B be two arbitrary events. Then $\overline{A \cap B} = \overline{A} \cup \overline{B}$. Consider the newspaper readership problem. If there are just two newspapers and the percentage of people who read both of them are known, the percentage of people who read neither of them can be found using the above-mentioned law.

Rule 2 Complement of a Union is the intersection of their complements.

Symbolically, $\overline{A \cup B} = \overline{A} \cap \overline{B}$. These rules can be extended to any number of events as follows: $(\overline{\cup_{i=1}^{n} A_i}) = \cap_{i=1}^{n} \overline{A_i}$, and $(\overline{\cap_{i=1}^{n} A_i}) = \cup_{i=1}^{n} \overline{A_i}$. These are proved using induction and Venn diagrams.

5.8 BASIC COUNTING PRINCIPLES

A great majority of probability problems can be solved by a good mastery of a few counting principles. These are more applicable to discrete sample spaces in 1D and 2D than for others. They are intended to count the number of objects, events, possibilities, occurrences, or arrangements that satisfy zero or more properties or constraints. There are a myriad of constraints possible. These may be related to adjacency, occupancy, linear or circular arrangement, observation of some events, and so on.

5.8.1 Rule of Sums (ROS)

This is also known as the principle of disjunctive counting. Consider a set S of objects that has been divided into disjoint subsets $S_1, S_2, \ldots, S_m$ so that $S = S_1 \cup S_2 \cup \cdots \cup S_m$. If there are n_1 favorable cases for an event in S_1, n_2 favorable cases for the same event in S_2, and so on, n_m favorable cases for the same event in S_m, then the total number of favorable cases for the event in S is $n_1 + n_2 + \cdots + n_m$. Symbolically, this can be written as $|S| = |S_1 \cup S_2 \cup \cdots \cup S_m| = |S_1| \cup |S_2| \cup \cdots \cup |S_m|$. Another way to state it is as follows:– There are m cases or events with no common options (i.e., they are mutually exclusive). If ith case or event can occur in n_i ways, then the total number of options or ways in which *one* of them can occur is $n_1 + n_2 + \cdots + n_m$. The principle of inclusion and exclusion (p. 158) is an extension when at least two of the subsets have common elements.

■ EXAMPLE 5.21 Breakfast choices

The McDonalds restaurant offers eight varieties of breakfast, whereas Burger King offers six varieties. Joe has a choice of going either to McDonalds or to Burger King on any day (but not both) for breakfast. How many choices of breakfast are possible?

Solution 5.21 This problem can be cast using rule of sums (ROS), where $S_1 =$ the choices available at McDonald's and $S_2 =$ the choices available at Burger King. Total possible choices are $|S_1| + |S_2| = 8 + 6 = 14$. ■

5.8.2   Principle of Counting (POC)

This is also called multiplication law of counting (MLOC) or multiplication principle. It has direct applications in counting the number of occurrences of outcomes of experiments such that the first experiment can result in n_1 possible outcomes, and for *each* outcome, there exist another independent experiment with n_2 possible outcomes, and so on. It is also useful in classical approach to probability in which we need to count the favorable number of cases of an experiment.

Lemma 1 If one thing (or activity) can be done in "m" ways, and another in succession in "n" ways, the two together can be done in $m * n$ different ways.

■ **EXAMPLE 5.22   Computer file types**

A software allows an image to be saved in three different file types (as .JPEG, .GIF, or .TIFF) in four different resolutions. How many possible ways are there to save the image?

Solution 5.22 As the file types and resolutions are independent, there are $3 * 4 = 12$ different ways to save the image.   ■

■ **EXAMPLE 5.23   Car colors**

A car manufacturer offers eight exterior colors and four interior designs. How many varieties of cars can be produced if (i) each of them can be manufactured in luxury and sedan models? (ii) if each of them can be made as petrol, diesel and hybrid (electric) versions?

Solution 5.23 We define three events as follows:– (i) E = {Exterior color}, (ii) I = {Interior design}, and (iii) M = {Model}. As the number of possibilities for E is 8, I is 4, and M is 2, by the principle of counting there exist $8 \times 4 \times 2 = 64$ possible choices. For Case (ii), there are three types (petrol, diesel, and hybrid) so that the number of ways is $8 \times 4 \times 3 = 96$.   ■

In some problems, we may have to combine both ROS and POC multiple times to reach a final result. This is illustrated in the following example.

■ **EXAMPLE 5.24   Multiple choice exam**

A multiple choice exam has 15 questions, each with 4 answer choices (say A, B, C, and D). How many possible ways are there to answer the questions assuming that multiple markings are not allowed, and (i) all questions are to be answered and (ii) questions can be skipped (kept unanswered).

Solution 5.24 As the questions are independently answered, any of the questions can be marked in four ways. Hence, the total possible answer combination in (i) is 4^{15}. For instance, if there are just 2 questions, the 16 answer choices are *(A,A), (A,B), (A,C), (A,D), (B,A), (B,B), (B,C), (B,D), (C,A), (C,B), (C,C), (C,D), (D,A), (D,B), (D,C),* and *(D,D),* where the first letter denotes the answer for question-1 and second letter is the answer for question-2. In Case (ii), suppose that k questions are answered and (15-k) are skipped. As any of the questions can be answered in four ways, there are 4^k answer combinations for k answered questions. However, the k questions can be any among the 15 questions. A student could select arbitrary k questions in $\binom{15}{k}$ different ways (see 5.9.5 in p. 145). By the multiplication principle, the total number of possible combinations is $\binom{15}{k} * 4^k$. By summing this expression over the possible range of k gives the answer as $\sum_{k=0}^{15} \binom{15}{k} * 4^k$. Here, $k = 0$ means that none of the questions are answered. This can be done in just one way. Similarly, $k = 15$ means that all questions are answered (in one way only). This is of the form $\sum_{k=0}^{n} \binom{n}{k} * x^k$, which is the binomial expansion of $(x + 1)^n$. Thus, the above-mentioned sum is $(4 + 1)^{15} = 5^{15}$. If there are just two questions, we have nine new combinations in addition to the 16 listed earlier as *(*,A), (*,B), (*,C), (*,D), (A,*), (B,*), (C,*), (D,*),* and *(*,*),* where "*" indicates an unanswered question and *(*,*)* means that both questions are skipped. This gives a total of $16 + 9 = 25 = 5^2$ combinations.   ∎

EXAMPLE 5.25   Cloth washing

A schoolchild has 10 colorless and 6 colored dresses to be washed on a weekend. Colored dresses are of two types—Red and Blue. Both of them cannot be loaded into the same washing cycle due to color dissolving. The color-less dresses can be washed in any of the three settings: (i) hot, (ii) lukewarm, and (iii) cold and rinsed after the wash cycle in two settings (lukewarm-rinse or cold-rinse); whereas the colored dresses of same color can all be washed and rinsed in a cold or lukewarm wash only. How many ways are there to wash all the clothes?

Solution 5.25 This problem is most easily solved using a tree. There are two branches at the top for colored and colorless. The colored branch is further broken down as Red and Blue. First consider colorless dresses. They cannot be mixed with colored ones due to color staining. Thus, there exist three ways to wash them and two ways to rinse them. By the POC, there exist $3 \times 2 = 6$ ways to wash them. Next consider colored clothes. How many of the colored ones are Red or Blue is not known. Let c of them be Red and 6-c Blue. The c Red ones can be washed in four ways {C-C, C-L, L-C, L-L}, where C indicates a Cold and L indicates a Lukewarm wash or rinse in the first and second place. Similarly 6-c Blue clothes have four washing choices. Thus, there are eight choices for the colored clothes. Adding both cases together, we get the answer as $6 + 8 = 14$ choices.   ∎

In some experiments, each of the outcomes has an equal number of occurrences (see Table 5.2). In other words, the probabilities are equally likely. These are much easier to solve as illustrated in the following.

EXAMPLE 5.26   Car license plates

A car license plate comprises of two English capital letters followed by four digits. How many license plates are possible if (i) each of the letters and digits can be repeated and (ii) only digits can be repeated.

Solution 5.26 There are 26 capital English letters and 10 digits (between 0 and 9). As repetitions are allowed for (i) there are $(26 \times 26) \times (10 \times 10 \times 10 \times 10) = 676 \times 10^4$ possible ways. Our assumption is that lower case letters are not used on license plates (which need not be true in some countries). As letter repetitions are not allowed for (ii), there are $(26 \times 25) \times (10 \times 10 \times 10 \times 10) = 650 \times 10^4$ possible ways. ∎

EXAMPLE 5.27   Cylindrical number lock

Consider a cylindrical number lock on a briefcase with three wheels or rings. Assume that each of the wheels is marked with the digits 0–6 (total seven digits or rollings possible). Using a lever, a user can set the lock to any desired number (formed by the three digits chosen in succession from the wheels, say from left to right). What is the total number of possible lock codes?

Solution 5.27 As there are three independent wheels, each with seven possibilities, the total number of combinations is $7 \times 7 \times 7 = 343$. Thus, the briefcase can be locked in 343 possible ways. ∎

5.8.3   Complete Enumeration

As the name implies, this method enumerates (count one by one) all possibilities. This is more relevant in discrete probability problems involving throws of a coin or dice, arrangement of digits, alphabets, assignments of elements in two finite sets, spin of a numbered wheel, and so on. Sometimes, we need to enumerate only a small subset by eliminating commonalities as in the following problem.

TABLE 5.2   Some Equally Likely Experiments and Their Probabilities

Experiment	Sample Space	Probability
Fair coin toss	{H, T}	$1/2 = 0.5$
Toss of a fair die	{1, 2, 3, 4, 5, 6}	$1/6 = 0.166\overline{6}$
Playing cards	{spade, heart, diamond, club}	$1/13 = 0.076923\overline{076923}$

The labels given to events are fixed in the case of playing cards but are arbitrary for others.

■ EXAMPLE 5.28   Leap-year

How many ways are there for a leap year (with 366 days) to have (i) 53 Sundays?, (ii) 53 Saturdays and 53 Sundays?, (iii) exactly 52 Saturdays and 52 Sundays?, (iv) exactly 53 Fridays or 53 Sundays?, and (v) exactly 52 Tuesdays and 52 Thursdays?

Solution 5.28 As $52 \times 7 = 364$, every year will have 52 weekdays each for sure. There is an extra day (strictly speaking 1.24219879 days) for nonleap years, and there are two extra days in leap years (the 0.24 days add up to approximately 1 day in 4 years and is counted as February 29 to get a leap-year). As these two extra days are consecutive, we can do a complete enumeration of these days as (Sunday, Monday), (Monday, Tuesday), (Tuesday, Wednesday), (Wednesday, Thursday), (Thursday, Friday), (Friday, Saturday), and (Saturday, Sunday). These are the only seven possible combinations for the extra 2 days. (i) As Sunday occurs in two of the seven combinations, the number of ways a leap year will have 53 Sundays is 2. (ii) As (Saturday, Sunday) occurs once, the desired number of favorable cases is 1. (iii) Neither Saturday nor Sunday occurs in four out of the seven possible pairs. (iv) There exist four pairs containing either a Friday or a Sunday. (v) There are three favorable cases, namely, (Sunday, Monday), (Friday, Saturday), and (Saturday, Sunday) using the complement rule. ■

■ EXAMPLE 5.29   Roots of quadratic equation

Consider a quadratic equation $px^2 + qx + r = 0$, whose nonzero coefficients (p, q, r) are determined by the number that turns up when a die with six faces numbered 1–6 is thrown. Find the number of ways in which (i) the equation will have real roots, (ii) equal roots, (iii) imaginary roots, (iv) both integer roots, and (vi) exactly one integer root.

Solution 5.29 As each of the coefficients p, q, r is determined by the number that shows up in the throw of a die, we need three throws to decide them (say choose p first, then q, and finally r). We do a complete enumeration as follows. As each of them can be in $\{1, 2, 3, 4, 5, 6\}$, there exist a total number of $6 \times 6 \times 6 = 6^3 = 216$ equations by the POC. (i) We know that the condition for real roots is $q^2 - 4pr \geq 0$. As repetitions are allowed, the least value of $4pr$ is $4 \times 1 \times 1 = 4$. However, q^2 is greater than 4 when $q = 3, 4, 5, 6$. This means that there exist four favorable cases when $p = r = 1$ (and five cases if $q = 2$ is also counted in which case we have equal roots). Next consider ($p = 2$ and $r = 1$) or ($p = 1$ and $r = 2$). In both cases $4pr = 8$, and q^2 is greater than 8 when $q = (3, 4, 5, 6)$. Proceed similarly with ($p = 3$ and $r = 1$), ($p = 1$ and $r = 3$), or ($p = 2$ and $r = 2$). In the first two cases $4pr = 12$ and q^2 is greater than 12 for $q = (4, 5, 6)$. For ($p = 2$ and $r = 2$), ($p = 1$ and $r = 4$), or ($p = 4$ and $r = 1$) $4pr = 16$ and $q = (5,6)$. For ($p = 1$ and $r = 5$) or ($p = 5$ and $r = 1$) $4pr = 20$ and $q = (5,6)$. Similarly for ($p = 1$ and $r = 6$), ($p = 6$ and $r = 1$), ($p = 2$ and $r = 3$) and ($p = 3$ and $r = 2$), $4pr = 24$ and $q = (5,6)$. Finally

TABLE 5.3   Roots of Quadratic Equation

No.	p	r	$4pr$	q	No.	p	r	$4pr$	q
1	1	1	4	3, 4, 5, 6	9	1	5	20	5, 6
2	1	2	8	3, 4, 5, 6	10	5	1	20	5, 6
3	2	1	8	3, 4, 5, 6	11	1	6	24	5, 6
4	3	1	12	4, 5, 6	12	6	1	24	5, 6
5	1	3	12	4, 5, 6	13	2	3	24	5, 6
6	1	4	16	5, 6	14	3	2	24	5, 6
7	4	1	16	5, 6	15	2	4	32	6
8	2	2	16	5, 6	16	4	2	32	6

for ($p = 2$ and $r = 4$) or ($p = 4$ and $r = 2$), $4pr = 32$ and $q = (6)$. For ($p = 3$ and $r = 3$), real roots are not possible as $4pr = 36$. Summing the counts, we get the total number of cases as 38. In addition, there are five cases (given the following discussion) for the roots to be equal.   ∎

There exists $(38 + 5) = 43$ ways (see Table 5.3). In Case (ii), the five favorable cases are (1,2,1), (1,4,4), (2,4,2), (3,6,3), and (4,4,1). A quadratic equation can have either real roots or equal roots or imaginary roots only. Hence, the favorable cases for (iii) can be directly obtained using complement rule as $216 - 43 = 173$. Consider Case (iv). Both roots are integers in two cases: (a) both q and $q^2 - 4pr$ are odd and (b) both are even. The 10 favorable cases are (1,2,1),(1,3,2),(1,4,3),(1,4,4),(1,5,4),(1,5,6),(1,6,5),(2,4,2), (2,6,4),(3,6,3). (v) The eight favorable cases are (2,3,1), (2,5,2), (2,5,3), (3,4,1), (3,5,2), (4,5,1), (4,6,2), and (5,6,1).

5.9   PERMUTATIONS AND COMBINATIONS

The literal meaning of permutation is *an ordering or arrangement*. Mathematically, a permutation of a set S is a one-to-one mapping of S onto itself. In other words, it is the total number of arrangements of a set of elements. The elements being arranged are all uniquely distinguishable to the human eye. This arrangement can be linear (in 1D) or circular (in 2D space). An ordered subset of a larger set is also called a permutation. As a great majority of probability problems are valid in 1D or 2D only, we will not discuss higher dimensional permutations. Consider a set of three students {Amy, John, and Mary}. Denoting each of them by their first letter, there are six ways to arrange them linearly as {A,J,M}, {A,M,J}, {J,A,M}, {J,M,A}, {M,A,J}, and {M,J,A}. These are the only possible linear permutations.

Definition 5.8 A permutation is an arrangement of the whole or part (with at least two elements) of a finite set of distinguishable elements without repetition, where the order is considered as important.

There is no universally accepted notation for permutations. The four most widely used notations are ^nP_r, $P_{r,n}$, $P(r,n)$ and P_r^n, where both n and r are integers such that $r \leq n$. Permutation can also be interpreted as selection of elements from a group without replacement.

Theorem 5.1 Prove that the total number of permutations of r objects from among n distinguishable objects is ^nP_r where $r \leq n$.

Proof: As there are n elements initially, there exist n ways to choose the first element. Now there are $(n - 1)$ elements remaining, as one element is already removed from the set. Thus, there are $(n - 1)$ ways to select the second object. Continuing in like fashion r times, we see that there are

$$^nP_r = n * (n - 1) * \ \dots \ (n - r + 1) = n!/(n - r)! = (n)_r \qquad (5.3)$$

ways to choose r objects from n objects ($(n)_r$ is the Pochhammer notation for falling factorial). This is the same as the number of samples of size r without replacement from n distinguishable objects (see Table 5.4).   ∎

Lemma 2 A set of n distinguishable objects can be linearly arranged among themselves in $n!$ ways.

Proof: Mark the positions of n objects. The first position can be filled by any of them (in n possible ways). Once this position is filled, there are $(n - 1)$ objects remaining and $(n-1)$ positions to put them into. Next we fix second of the $(n-1)$ possible positions. There are $(n-1)$ ways to choose an object to this position. Thus, the first two positions can be filled in $n*(n-1)$ ways. Continuing this way we find that for filling the last position, we have only one choice. Hence, the total number of ways to fill all the positions is $^nP_n = n * (n - 1) * (n - 2) * \cdots * 2 * 1 = n!/(n - n)! = n!$ ways (as $0! = 1$ by convention).

EXAMPLE 5.30   National flags

A political summit is attended by delegates from five countries. All five national flags are to be arranged in a row at the entrance. In how many ways can this be done?

TABLE 5.4   Some Permutation Formulas

Objects	Type	Number of Ways
n distinguishable	Linear	$n!$
k among n distinguishable	Linear	$P(n, k) = n(n - 1) \dots (n - k + 1)$
n items with duplicates	Linear	$\binom{n}{k_1, k_2, \dots, k_j} = n!/k_1!k_2! \cdots k_j!$
n distinguishable	Circular	$(n-1)!$

Solution 5.30 As $n = 5$, there are $n! = 5! = 120$ possible ways. ∎

5.9.1 Permutations with Restrictions

In most of the practical applications, we have restrictions on the elements. The most common restriction is duplicates (property restriction) discussed in the following. Other types of restrictions include adjacency restrictions, locational restrictions (such as fixed positions), and end point (extreme position) restrictions.

■ **EXAMPLE 5.31   Book arrangement**

A schoolchild has five books, one each on Mathematics, Gaming, English, Physics, and Biology. How many ways are there to arrange the books on a rectangular rack if (i) no order is maintained among them?, (ii) the leftmost book must be Gaming book?, (iii) the left-most and right-most places are occupied by Mathematics and Physics books?, (iv) Physics and Mathematics books are always adjacent, and (v) English and Biology books cannot be next to each other?

Solution 5.31 (i) If no order is maintained among them, the possible number of ways is $5! = 120$. In Case (ii), the Gaming book occupies a fixed position. There are four books to be arranged. This can be done in $4! = 24$ ways. In Case (iii), two places are preoccupied. The remaining three books can be arranged in $3! = 6$ ways. In Case (iv), we consider Physics and Mathematics as a single logical bundle. Then, it reduces to arranging four books among themselves. This can be done in $4! = 24$ ways. The easiest way to tackle (v) is using the do-little principle (complement-and-conquer). We consider the complement event that English and Biology books are together. As it is similar to Case (iv), there are 24 possibilities. The required answer is then found using the complement as $5! - 4! = 120 - 24 = 96$. ∎

5.9.2 Permutation of Alike Objects

If there are n things of which n_1 of them are of one kind, n_2 of them are of another kind, $\dots$, n_k of them are of kth kind, then there are

$$^nP_{n_1,n_2,\dots,n_k} = n!/[n_1! * n_2! * \dots * n_k!] \tag{5.4}$$

different permutations, where $n = n_1 + n_2 + \cdots + n_k$. This is called the multinomial coefficient. It can alternately be written as

$$^nP_{n_1,n_2,\dots,n_k} = \binom{n}{n_1}\binom{n-n_1}{n_2}\binom{n-n_1-n_2}{n_3}\cdots\binom{n-\sum_{j=1}^{k-1}n_j}{n_k}, \tag{5.5}$$

where the missing operator is a "*." Some of the n_i in this theorem can be one. When all of the n_i's are 1's, the denominator reduces to 1 and we get the number of permutations as $n!$. This result can be stated alternately as follows. The number of possible divisions of n distinct objects into r groups of respective sizes $n_1, n_2, \ldots, n_r$ is $\binom{n}{n_1, n_2, \ldots n_r} = n!/[n_1! * n_2! * \cdots * n_r!]$. Multinomial coefficients are further discussed in Chapter 6.

▣ EXAMPLE 5.32   Shelving of books

A library has received a shipment of 12 books of which 2 are duplicate copies of a Statistics book, 4 are duplicate copies of a Database book, and 3 each are duplicate copies of C programming book and Java programming book. These need to be kept on a reserve shelf. In how many ways, can this be arranged?

Solution 5.32 As several items are exactly alike, we use the above-mentioned formula with $n_1 = 2, n_2 = 4$, and $n_3 = n_4 = 3$, to get $12!/(2!*4!*3!*3!)$. One of the 3! cancels with 6 and 2!*3! cancels with 12 leaving $= 5*7*8*9*10*11$ in the numerator. This simplifies to 277,200 possible ways.   ■

▣ EXAMPLE 5.33   Train coaches

A train has five ordinary coaches and three AC coaches in addition to an engine. How many ways are there to connect the coaches if the engine is always at the front?

Solution 5.33 This problem is most easily solved using permutations of alike objects. The eight coaches can be considered as objects of which five are of one kind and three are of another kind. Hence using the above-mentioned theorem, the total number of possibilities is $8!/(5!*3!) = 336/6 = 56$ ways.   ■

5.9.3   Cyclic Permutations

If the permutations of distinguishable objects occur along a "logical circle," it is called cyclic or circular permutation. Here, logical circle means that the objects can be thought of as forming an imaginary circle (although physically it can be any closed shape including a triangle, square, rectangle, or pentagon). The circle can be rotated by fixing the objects in place. Hence, it is not the circle, but the order of occurrence of objects that is more important. As an example, if several people sit around a rectangular table, we could consider it as a logical circle as long as the sides of the table are not distinguished or considered with respect to the persons.

Lemma 3 The number of permutations of n distinguishable objects along a circle is $(n - 1)!$.

Proof: Keep any one of the objects as fixed. There are $(n - 1)$ others remaining. They can be arranged among themselves in $(n - 1)!$ ways.   ■

◨ EXAMPLE 5.34   Roundtable seating

How many ways are there to seat four people W, X, Y, Z around a circular table?

Solution 5.34 According to the above-mentioned lemma, there are $(4-1)!$ $= 6$ different ways. In clockwise order, they are $\{W, X, Y, Z\}, \{W, Y, X, Z\},$ $\{W, Z, X, Y\}, \{W, X, Z, Y\}, \{W, Y, Z, X\},$ and $\{W, Z, Y, X\}$.   ■

◨ EXAMPLE 5.35   Circular arrangement

Suppose that n boys and n girls are to be seated around a circular table. (i) How many ways can this be done if no two of the same sex are seated next to each other, (ii) there are no restrictions on males and females?, and (iii) three or more pairs of boys cannot be together, but at most two pairs of boys are allowed?

Solution 5.35 First fix the n boys around the circle with an empty chair between them. This can be done in $(n-1)!$ ways, as they can be rearranged among themselves using circular permutation formula given earlier. As there are n empty chairs, the n girls can be circularly arranged in $(n-1)!$ ways. This gives a total of $(n-1)!^2$ possible ways. (ii) If there are no restrictions, we need to arrange $2n$ persons along a circle. This can be done in $(2n-1)!$ ways. Case (iii) is most easily solved using the do-little principle. The complement of the problem is to find the number of ways in which any three males are together. Mark the group of three males by M. Then, there are $n-3$ remaining males (plus one M). They can be arranged among themselves in $(n-3+1-1)! = (n-3)!$ ways. As the three males can be fixed in $\binom{n}{3}$ ways, the total number of ways is $\binom{n}{3} * (n-3)!$ ways. Take complement from $(n-1)!^2$ ways to get the desired answer.   ■

5.9.4   Cyclic Permutations of Subsets

Consider n distinct objects. If r is an integer between 1 and n, there are $(r-1)! \binom{n}{r}$ different ways to circularly permute the r objects.

◨ EXAMPLE 5.36   Train coaches

A train has five ordinary coaches and three AC coaches in addition to the engine. How many ways are there to connect the coaches if (i) the coach immediately behind the engine and the rear-end coach are both ordinary?, (ii) if all AC coaches cannot be together?, (iii) at most three ordinary coaches can be together?

Solution 5.36 Denote ordinary coach by O and AC coach by C. This problem has restrictions. In Case (i), two of the five "O coaches" are fixed. This leaves three O and three C coaches remaining to be connected. Using permutations of alike objects, the answer is $6!/(3!*3!) = 20$ ways. (ii) Fix the five O coaches with an empty space in-between them. There are two extra empty spaces at the beginning and end (extremes behind the engine). This gives a total of six empty spaces where we could place three C coaches together. Total number of ways in which five O coaches and three C coaches can be connected together is $8!/(5!*3!) = 56$ ways. By subtracting the number of ways in which all the AC coaches are together, we get the answer to part (ii) as $56 - 6 = 50$. (iii) The complementary event of "at most three ordinary coaches can be together" is either four coaches are together or all five coaches are together. These are more easier to count. Number of ways in which four coaches are together is found as follows: fix the three C coaches with a space in between them (including beginning and end). We can place four O coaches in four ways. The remaining one O coach can be placed in three ways. This gives a total of $4 \times 3 = 12$ ways. Next consider all five coaches together. These can be placed in four different ways. By the ROS principle, total number of ways for the complementary event is $12 + 4 = 16$ ways. Hence, the desired number of ways is $8!/(5!*3!)-16 = 56-16 = 40$ ways. ∎

5.9.5 Combinations

Permutation is an arrangement technique in which the order of elements matters, but order of elements does not matter in combinations. This means that if X and Y are two elements, XY and YX are considered the same in combination but not in permutations. The combination of n things taken r at a time was introduced in Chapter 1. It is denoted by $\binom{n}{r}$, ${}^nC_r$, C^n_r, or $C(n, r)$. Symbolically, it is expressed as

$$\binom{n}{r} = n!/(r! * (n-r)!) = (n)_r/r! = n!/((n-r)! * r!) = \binom{n}{n-r}. \qquad (5.6)$$

This denotes the number of ways in which r objects can be selected from n distinguishable objects without regard to order and without replacement. For a fixed r, there exist $r!$ permutations that give the same combination. Hence,

$$ {}^nP_r/r! = n!/[r!(n-r)!] = {}^nC_r = \binom{n}{r}. \qquad (5.7) $$

By writing $r! * (n-r)!$ as $(n-r)! * r!$, it follows that $\binom{n}{r} = \binom{n}{n-r}$ ($i.e., {}^nC_r = {}^nC_{n-r}$). Particular cases are $\binom{n}{n} = 1$, $\binom{n}{0} = 1$, and $\binom{n}{1} = \binom{n}{n-1} = n$.

EXAMPLE 5.37   Pilot choices

A flight has to be scheduled using a pilot. There are 12 persons in the pool for the pilot among whom 8 are males, 5 of the 8 speak English and Spanish, and the rest 3 speak English only. Two of the four females speak English and Spanish, and the rest of them speak English only. How many ways are there to select a pilot and a copilot such that (i) there is one male and one female, both speak Spanish? (ii) two males, at least one of whom speak Spanish?

Solution 5.37 Define events A, B as follows:–A: = Event that the pilot candidate is Male and B: = Event that the candidate speaks Spanish. We seek the number of possibilities of the event $A \cap B$. There are five out of eight males who speak Spanish and English. The number of ways to choose a bilingual male is $\binom{5}{1} = 5$. Similarly, number of ways to choose a bilingual female is 2. By the product rule, total number of ways is $5*2 = 10$. (ii) As there are five males who speak Spanish, we consider the two cases: (i) both chosen persons speak Spanish. (ii) only one of them speak Spanish. The favorable cases for (i) is $\binom{5}{2} = 10$ and for (ii) is $\binom{5}{1} * \binom{3}{1} = 5 * 3 = 15$. By the ROS principle, the total favorable cases are $10 + 15 = 25$. ∎

☐ EXAMPLE 5.38   Poker game

Find the number of ways of obtaining a hand of cards in a poker game.

Solution 5.38 This problem is easy to solve using combination law. As a hand contains five cards in a poker game, the number of ways is $\binom{52}{5} = 2,598,960$. ∎

☐ EXAMPLE 5.39   Irrigation plot

An irrigation plot is divided into 6×6 blocks of equal size (with 36 subplots). A sample of four subplots is to be selected at random. What is the number of ways in which the four subplots will (i) lie along any row or column, (ii) lie along the main diagonal or parallel to the main diagonal (from top left to bottom right), (iii) they stick together as a 2×2 subplot anywhere, and (iv) if nine subplots are selected, find the number of ways they stick together as 3×3 subplots.

Solution 5.39 Total number of ways to select four subplots from 36 plots is $\binom{36}{4}$. In case (i), there are two possibilities to consider (1) they lie along the rows and (2) they lie along the columns. In the first case, there are $\binom{6}{4} = 15$ ways for all four to lie along any fixed row. As there are six rows, the total number of ways is $6 \times 15 = 90$ ways. Owing to symmetry, there are 90 ways for the columns too. This gives 180 total possibilities. In case (ii), the main diagonal (with six slots), its immediate above and below diagonals with five slots and those at distance 2 from it (with four slots) are the only favorable positions. There exist $\binom{6}{4} = 15$ ways for the main diagonal, $\binom{5}{4} = 5$ ways for its immediate above and below diagonals, and $\binom{4}{4} = 1$ way each for distance 2 diagonals. By the ROS principle, total number of ways is $15 + 2*5 + 2*1 = 27$ ways. For case (iii), we fix the 2×2 subplot as a square and use cell (2,2) (second column in second row) as an "anchor" for alignment. This anchor can be aligned in a 5×5 subplots giving 25 possible ways. Similarly for case (iv), consider cell (3,3) as the anchor. This can be anchored along a 4×4 matrix of subplots giving a total of 16 possible ways. ∎

EXAMPLE 5.40   Plant operators

A small production plant needs eight operators, two shipping and handling persons, two clerks, and one supervisor for a day. If there are 10 operators, four shipping and handling persons, three clerks, and two supervisors available for work, how many ways are there to staff the plant?

Solution 5.40 As each of the jobs are disjoint, we could apply the above-mentioned principle and get the answer as $\binom{10}{8} * \binom{4}{2} * \binom{3}{2} * \binom{2}{1} = 45 * 6 * 3 * 2 = 1620$. ∎

EXAMPLE 5.41   Chess players

A college has 10 chess players of which 6 are males and 4 are females. Two students are to be sent for an inter-collegiate festival. How many ways are there to send a team of 2 if: – (i) the gender is not considered, (ii) exactly one is a male, (iii) at least one must be female, and (iv) both are females?

Solution 5.41 For case (i), the total number of possibilities is $\binom{10}{2} = 10 * 9/2 = 45$. In case (ii), the total favorable cases is $\binom{6}{1} * \binom{4}{1} = 24$. In case (iii), there are two possibilities {FM and FF}. The possible ways for FM is found above as 24. Possible ways for FF is $\binom{4}{2} = 6$. Adding these two gives the answer as $24 + 6 = 30$. In case (iv), answer to this is found above as $\binom{4}{2} = 6$. ∎

5.10   PRINCIPLE OF INCLUSION AND EXCLUSION (PIE)

This is one of the most widely used principles when events or sets interact (have subsets or subevents in common). It has two interpretations:– in terms of counts and in terms of probabilities. Both are analogous at the conceptual level. We discuss the count version below. The extension to probability is given in a later section. The count version provides an answer to the query "How many elements or objects are there in the union of a finite number of sets, some of which have elements or properties in common?.""

Theorem 5.2 If A_1, A_2, ... , A_n are finite sets, some of which have common elements, then

$$|A_1 \cup A_2 \cup \cdots A_i \cup \cdots A_n| = \sum_i |A_i| - \sum_{i<j} |A_i \cap A_j| + \sum_{i<j<k} |A_i \cap A_j \cap A_k|$$

$$- \cdots + (-1)^{n-1}|A_1 \cap A_2 \cap \cdots A_n|, \qquad (5.8)$$

where vertical bars denote the cardinalities (number of elements) of respective sets, and $i < j$, and so on, on the summation sign denotes that the sum is carried out only for those values of indices satisfying respective conditions.

Proof: Consider the special case with just two sets say X and Y. The above-mentioned theorem takes the form $|X \cup Y| = |X| + |Y| - |X \cap Y|$. If X and Y do not overlap, then $X \cap Y = \phi$ so that $|X \cap Y| = 0$, and the results follow. If $X = Y$, then $X \cap Y = X = Y$, so that the negative term cancels out with one of the X or Y giving the result. Next, suppose that X and Y overlap (with c common elements where $c \geq 1$) and $X \neq Y$. In counting $|X| + |Y|$, the c common elements are counted twice. Hence, we need to subtract one of the c counts to get the number of elements in $X \cup Y$. This gives $|X \cup Y| = |X| + |Y| - |X \cap Y|$. Assume that the theorem is true for an arbitrary m. Consider

$$|A_1 \cup A_2 \cup \cdots A_i \cup \cdots A_m \cup A_{m+1}|. \qquad (5.9)$$

Write $A = A_1 \cup A_2 \cup \cdots A_i \cup \cdots A_m$. Then (5.9) becomes $|A \cup A_{m+1}|$. Expand it using the special case to obtain $|A \cup A_{m+1}| = |A| + |A_{m+1}| - |A \cap A_{m+1}|$. Substitute for $A = A_1 \cup A_2 \cup \cdots A_i \cup \cdots A_m$ and use the fact that intersection distributes over union operator to get the RHS as

$$= \sum_{i=1}^{m+1} |A_i| - \sum_{i<j=1}^{m+1} |A_i \cap A_j| + \sum_{i<j<k=1}^{m+1} |A_i \cap A_j \cap A_k|$$

$$\cdots + (-1)^m |A_1 \cap A_2 \cap \cdots A_m \cap A_{m+1}|. \qquad (5.10)$$

∎

This shows that if the theorem is true for m, it is also true for $m + 1$. As it is true for $m = 2$, it is also true for $m = 3, 4, \dots$.

Corollary 1 If $A_1, A_2, \dots A_n$ are finite sets, some of which have common elements, then $|A_1 \cap A_2 \cap \cdots A_i \cap \cdots A_n| = |U| - \sum_{i=1}^{n} |A_i| + \sum_{1 \leq i \leq j}^{} |A_i \cap A_j| - \sum_{1 \leq i \leq j \leq k}^{} |A_i \cap A_j \cap A_k| + .. + (-1)^{n-1} |A_1 \cap A_2 \cap \dots \cap A_{n-1}|$.

▣ EXAMPLE 5.42　Divisible integers

How many integers between 1 and 100 are divisible by 3, 5 or 7?

Solution 5.42　Let S denote the set of 100 integers $S = \{1, 2, 3, \ldots, 100\}$. Define three events as follows:– (i) E_1 = count of all integers in S that are divisible by 3, (ii) E_2 = count of all integers in S that are divisible by 5, and (iii) E_3 = count of all integers in S that are divisible by 7. Then, $E_1 \cap E_2$ is the count of all integers in S that are divisible by both 3 and 5, and so on. Using the PIE principle $E_1 \cup E_2 \cup E_3$ is the set of integers divisible by 3, 5, or 7. The results needed to compute this are given in Table 5.5. Using (5.11),

$$|E_1 \cup E_2 \cup E_3| = |E_1| + |E_2| + |E_3| - |E_1 \cap E_2| - |E_1 \cap E_3| - |E_2 \cap E_3|$$
$$+ |E_1 \cap E_2 \cap E_3|. \tag{5.11}$$

Substitute the values to get $|E_1 \cup E_2 \cup E_3| = 33 + 20 + 14 - 6 - 4 - 2 + 0 = 67 - 12 = 55$. ∎

5.11　RECURRENCE RELATIONS

A recurrence relation is a recursive relationship that relates the nth term of a sequence or task in terms of lower order terms. If the nth term is related to the $(n-1)$th term, it is called first-order recurrence. Most of the recurrence relations encountered in this book are first order recurrences. If the nth term is related to two prior terms, it is called second-order recurrence. A special case is the recurrence relation connecting successive probabilities of discrete distributions. These are formed by reducing one of the integer parameters. These are explained in subsequent sections.

5.11.1　Derangements and Matching Problems

Matching problems comprise two sets of objects (such as husband and wife, person and hat, person and overcoat, and letter and envelope) that have a one-to-one correspondence among themselves. These types of problems seem to have fascinated mathematicians for centuries. The first "person and hat" problem was documented by de Montfort in Reference 111. They arise in many situations. For example, consider 13 cards numbered 1–13 without duplicates that are kept face down on a table.

TABLE 5.5　Divisibility of Integers by 3, 5, or 7

3	5	7	(3,5)	(3,7)	(5,7)	(3,5,7)
$\left\lfloor \dfrac{100}{3} \right\rfloor$	$\left\lfloor \dfrac{100}{5} \right\rfloor$	$\left\lfloor \dfrac{100}{7} \right\rfloor$	$\left\lfloor \dfrac{100}{(3*5)} \right\rfloor$	$\left\lfloor \dfrac{100}{(3*7)} \right\rfloor$	$\left\lfloor \dfrac{100}{(5*7)} \right\rfloor$	$\left\lfloor \dfrac{100}{(3*5*7)} \right\rfloor$

A person utters a number between 1 and 13 and then picks up a card and notes the number. This is repeated 13 times, such that each time a different number is uttered. Obviously, after 12 tries, the last number can easily be guessed. In how many ways, can the person get $k(< 13)$ correct matches? As another example, suppose that there are n books kept on a book rack in some specific order (say in alphabetical order of first author name, increasing order of accession numbers, or using call numbers). During the "library-hour," kids take out all the books and return it arbitrarily back to the rack. What is the chance that exactly k of the books are returned back to their original position? What is the probability that none of the books are in their proper position? These problems can easily be modeled by the bipartite graph described in p. 126. In such a mapping, the first pair of the n objects are represented as n nodes on the left (say S), and the second pair is represented by n nodes on the right (say T). Each node in S can be connected to at most 1 node in T. An undirected arc from node i in S to node j in T denotes a new order or assignment. A perfect match (original order is maintained) is indicated by a forest in which each node in S is connected to the matching node in T. These are easy to solve when n is small. In the following discussion, it is assumed that n is fairly large. A few of the situations where such problems arise are listed in the following:

1. Consider n married couples (H_1, W_1), (H_2, W_2), .. , (H_n, W_n) at a party. Assume that the men and women are randomly paired for a dance. A complete match occurs if each couple happens to be paired together.
2. Suppose that n letters are to be sent to n different people in n envelopes. The addresses are already printed on the envelope, and the letters are shuffled. An absent-minded clerk randomly puts the letters, one each, into the n envelopes. A complete match occurs if each letter is put in its correct envelope.
3. Suppose n people with overcoats go for a party. They give the coat to the waiter for safe keeping. While leaving the party, the waiter randomly grabs a coat and gives it to the people. A complete match occurs if each person gets his or her own coat.
4. A defective electronic device has n exactly looking parts. A repair person removes each of them without labeling them, tests it individually, and returns them back to the original positions arbitrarily. A complete match occurs if each part ends up in its correct slot.

Each of these problems is mathematically equivalent. If none of them match, it is called a derangement. They can be modeled by different techniques such as recurrence relations and using inclusion–exclusion principle.

Theorem 5.3 Total number of derangements of n elements is

$$D_n = n! \left[1 - \frac{1}{1!} + \frac{1}{2!} - \frac{1}{3!} + \ldots + (-1)^n \frac{1}{n!} \right]. \tag{5.12}$$

Proof: This can be proved in many ways. We give the following two simple proofs. The first one uses the PIE principle. Let A_i denote the event that ith object is correctly paired with matching pair (ith letter is put in its correct envelope, etc.). Let $\bar{A}_i$ denote the complementary event. Then $\bar{A}_1\bar{A}_2$ denotes that the first two objects are not paired with their matches. The event $\bar{A}_1\bar{A}_2 \ldots \bar{A}_n$ denotes that none of the objects are matched with their peers. This is what is meant by a derangement. By generalized DeMorgan's law, we have $\bar{A}_1 \cap \bar{A}_2 \cap \ldots \cap \bar{A}_n = [A_1 \cup A_2 \cup \cdots A_i \cup \cdots A_n]^c$, where the superscript denotes the complement. Using the "do-little" principle (Section 5.6.4 (p. 131)), the complementary event $A_1 \cup A_2 \cup \ldots \cup A_n$ on the RHS is much easier to evaluate. Using the PIE principle, this can be expanded as

$$|A_1 \cup A_2 \cup \cdots A_i \cup \cdots A_n| = \sum_{i=1}^{n} |A_i| - \sum_{i<j=1}^{n} |A_i \cap A_j| + \sum_{i<j<k=1}^{n} |A_i \cap A_j \cap A_k|$$

$$- \cdots + (-1)^{n-1} |A_1 \cap A_2 \cap \cdots A_n|. \qquad (5.13)$$

There are n ways in which ith object can be paired with its match. For any two arbitrary pairs (i,j), there are $n(n-1)$ ways to pair them using the multiplication law. Owing to the restriction on i and j, there are $\binom{n}{2}$ such pairs, and so on. In general, the number of ways in which k items are paired with their peers is $(n-k)!$. Substitute in equation (5.13) to get

$$|A_1 \cup A_2 \cup \cdots A_i \cup \cdots A_n|$$

$$= \binom{n}{1}(n-1)! - \binom{n}{2}(n-2)! + \ldots + (-1)^n \binom{n}{n}. \qquad (5.14)$$

∎

Expand $\binom{n}{i} = n!/(i!(n-i)!)$, take $n!$ as common factor, subtract from $n!$ and simplify to get $n! \sum_{i=0}^{n} (-1)^i/i!$. If n is large, $n!/e$ is a good approximation to equation (5.14) because $e^{-1} = \sum_{k=0}^{\infty} (-1)^k/k!$.

The derangement problem is easy to solve using recurrence relations. Let u_n denote the number of derangements of n objects and S_n denote the corresponding set. Fix any two objects say P and Q. There are four possibilities:– (i) both P and Q are paired with their own match, (ii) either P or Q is paired with own partner, (iii) P is paired with Q's match and Q is paired with P's, and (iv) only one of P or Q is paired with the other's match. Obviously, options 1 and 2 do not belong to S_n because they violate the derangement condition. Only favorable cases are options 3 and 4. Consider option 3 first. As they are not matched to their peers, the remaining $(n-2)$ objects can be deranged in u_{n-2} ways. Obviously, S_{n-2} is a subset of S_n. Object P can be matched to $(n-1)$ other objects j (excluding its peer say (i). This automatically determines one such match for Q as i. Thus, there are $(n-1)$ ways in which pairs of objects that occupy each others' place can be formed. If one among the fixed objects occupy another place but not vice versa, there exist u_{n-1} ways in which others can go wrong. This gives the recurrence $u_n = (n-1)*(u_{n-1} + u_{n-2})$. Write the RHS as

$$n * u_{n-1} - u_{n-1} + (n-1) * u_{n-2}, \qquad (5.15)$$

and take $n * u_{n-1}$ to the LHS to get

$$u_n - n * u_{n-1} = (-1) * (u_{n-1} - (n-1) * u_{n-2}). \tag{5.16}$$

Repeated application of equation (5.16) results in $u_n - n * u_{n-1} = (-1)^{n-2}(u_2 - 2u_1)$. Substitute $u_1 = 0$ and $u_2 = 1$ to get $u_n - n * u_{n-1} = (-1)^n$. Divide throughout by $n!$ and cancel out n from the second term to get

$$u_n/n! - u_{n-1}/(n-1)! = (-1)^n/n!. \tag{5.17}$$

Replace n successively by $(n-1), (n-2), ..., 2$ and add them together to get $(u_n/n!) = \{1/2! - 1/3! + \cdots (-1)^n/n!\}$. Add and subtract 1 in the RHS and write it as $1-1/1!$ to get the final result. For two objects (A, B), there is only one derangement (namely (B, A)) with $D_2 = 1$. For three objects say (A, B, C), there exist two derangements (B, C, A) and (C, A, B) so that $D_3 = 2$. Similarly $D_4 = 6$ and so on.

■ EXAMPLE 5.43   Hat-Check problem

There are n customers at a club, each of whom wears a cap. Each member puts his cap in a basket while entering the club in the evening. While going out, each one picks a cap randomly and walks out. What is the possible number of ways that (i) all of them picks their own cap, (ii) no one picks their own cap, and (iii) exactly half of them gets back their own hats (where n is even)?

Solution 5.43 Number the caps from 1 to n. The total number of possible ways to arrange the n caps is $n!$. Out of this, there is only one way in which everyone can get their own caps, so that the answer to (i) is 1 out of $n!$, (ii) number of ways in which no one picks their own hat is the derangement D_n, and (iii) if exactly half of them gets back their own hats, the other half do not get their hat. This can happen in $D_{n/2}$ ways. As there are $\binom{n}{n/2}$ ways to fix the half (for n even), there are $\binom{n}{n/2} D_{n/2}$ total ways.   ■

5.12   URN MODELS

An urn model is a conceptual framework for representing a set of problems that satisfy the following conditions:–

- problem involves a collection of (preferably three or more) items, where each item belongs to a group or has a type, and elements of the same group or type are indistinguishable;
- all items are put together or assigned as a whole such that any subset of them can be selected at random, without looking at their type;

- Each of the groups are distinguishable, but their order or arrangement is unimportant.

There are a large number of problems that can be cast as urn model. These are also called occupancy problems.

▣ EXAMPLE 5.44   Unique ID numbers

A university wishes to assign a unique 4 digit ID number to each of the enrolled students with the following restrictions that the student number cannot start with digit "0." What is the maximum number of IDs that can be generated if (i) the digits can be repeated any number of times and (ii) digits cannot be repeated?.

Solution 5.44 Consider the four positions as four numbered urns arranged along a line. We can fill these urns from left to right. As the student number cannot start with digit "0," the first urn can be filled in nine ways (with digits one through nine). In case (i), we are allowed to repeat the already used digit. This means that the second, third, and fourth urns can be filled with any of the 10 digits. This gives $9 * 10^3 = 9000$ possible numbers. In case (ii), the first urn can be filled in nine ways as before. The second urn can be filled in nine ways as digits cannot be repeated. Similarly, the third and fourth urns can be filled in eight and seven ways, respectively. By the multiplication rule, we get the answer as $9*9*8*7 = 4536$ ways. Thus, up to 4536 student, ID numbers can be generated if digits are not repeated.   ■

Theorem 5.4 Total number of ways in which n indistinguishable balls can be put in k distinguishable urns (see Table 5.6) where none of the urns can be empty, and maximum capacity of each urn is n is k^n.

Proof: As the urns are distinguishable, arrange them in a linear order. Start with the leftmost urn. We can put any of the balls there. Thus, there are n ways. This is true for each of the k urns. By the multiplication law, the total number of ways is k^n.   ■

▣ EXAMPLE 5.45   Common birthday

Suppose that there are $n - 1$ (<365) other persons along with you in a room, none of whom are twins. How many ways are there for each of the following events to realize assuming that leap-years are not accounted for? (i) None of the people shares a common birthday, (ii) at least one other person in the room shares a birthday with you, (iii) at most three people share a common birthday, and (iv) find the value of n such that the probability for at least two persons to share a common birthday is 0.6?

Solution 5.45 This can be cast in the urn-model framework by assuming days of the year consecutively numbered as urns and people as balls (Table 5.6). Then

TABLE 5.6   Urns and Balls Without Restrictions

	k Urns	
n Balls	Distinguishable	Indistinguishable
Distinguishable	k^n	$\sum_{i=1}^{n} \left\{ {k \atop i} \right\}$
Indistinguishable	$\binom{n+k-1}{k}$	$\pi(n+k, n)$

The literal meaning of these rules can be worded simply as follows:

1. The number of ways in which k ordered items can be sampled from n items with replacement is n^k and without replacement is $(n)_k$.

case (i): "None have a common birthday" means that all birthdays are different. This is the same as the number of ways to choose n different days from 365 days, which is $(365)_n$. (ii) The complementary event of "at least one other person" is that discussed in case (i). As the total number of ways for the birthdays of n persons is 365^n, the required answer is $365^n - (365)_n$. (iii) At most three people will have common birthdays if either two or three people have the same birthday. These are, respectively, $\binom{n}{2} 365 * (364)_{n-2}$ and $\binom{n}{3} 365 * (363)_{n-3s}$. (iv) The value of n is found by solving $(365)_n/365^n \simeq 0.6$.   ∎

5.13   PARTITIONS

We saw in Section 5.9.5 (p. 5–63) that the number of partitions of n things into two groups of sizes r and $n - r$ is $\binom{n}{r} = n!/[r!(n - r)!]$. In some problems, we need to divide a finite set S of size n elements into *all possible* subsets. This is not to be confused with set partitions defined in Section 5.5.1 in page 122. The trivial subsets are S itself and one-element subsets. For simplicity, consider a set with three elements $S = \{1, 2, 3\}$. Then, the seven possible partitions are $\{\{1\}, \{2\}, \{3\}, \{1,2\}, \{1,3\}, \{2,3\}\}$, and $\{1,2,3\}$. When S has four elements, there are $m = 15$ partitions. In general, when there are n elements, there exist $S(n, k)$ partitions, where $S(n, k)$ is called the Stirling number of second kind. These numbers satisfy the recurrence $S(n, k) = S(n - 1, k - 1) + k * S(n - 1, k)$, where $S(n, 1) = S(n, n) = 1$, $S(n, 2) = 2^{n-1} - 1$.

5.14   AXIOMATIC APPROACH

With the solid mathematical footing given earlier, we are ready to define the axiomatic approach to probability. Consider a finite set of mutually exclusive and collectively exhaustive set of events A_i such that $\cup_{i=1}^{n} A_i = \Omega$. The literal meaning of an axiom is "a statement that is always true or obviously true." Events lie at the core of axiomatic approach. We break the sample space of a random experiment into events that do not occur together. Then, we define probability as a real valued function that obeys certain conditions.

5.14.1 Probability Measure

A probability measure has two fundamental ingredients. A sample space Ω of outcomes of a random experiment and a function that maps each elementary outcome A_i to a real number between 0 and 1 such that they add up to 1. These are stated as three axioms:–

1. $0 \leq P(A_i) \leq 1$,
2. $P(\Omega) = 1$, $P(\phi) = 0$,
3. if A_i is a sequence of disjoint events, then $P(\cup_{i=1}^{n} A_i) = \sum_{i=1}^{n} P(A_i)$.

The third axiom can be extended to countably infinite collective mutually exclusive events. Such a function is called a probability measure.

The postulation $P(\phi) = 0$ follows because $\Omega \cup \phi = \Omega$ and $P(\Omega) = 1$. A direct consequence of these axioms is the following set of properties that are stated in set theoretic symbols and operators.

Theorem 5.5 If X and Y two arbitrary events defined on a sample space, then (i) $0 \leq P(X) \leq 1$, (ii) $0 \leq P(Y) \leq 1$, (iii) if $X \subset Y \rightarrow P(X) \leq P(Y)$, and (iv) $P(X \cup Y) = P(X) + P(Y) - P(X \cap Y)$.

Proof: The first two results follows from the above-mentioned theorem. To prove the third result, we write $X + (Y - X) = Y$. Then apply the third axiom to get $P(X) + P(Y - X) = P(Y)$. As $P(Y - X) \geq 0$, it follows that $P(X) \leq P(Y)$. ∎

To prove (iv), write $X \cup Y$ as the disjoint unions as $X \cup Y = (X \cap Y^c) \cup (Y \cap X^c) \cup (X \cap Y)$. As the subsets on the RHS are disjoint, axiom 3 can be applied to get

$$P(X \cup Y) = P(X \cap Y^c) + P(Y \cap X^c) + P(X \cap Y). \tag{5.18}$$

Add and subtract $P(X \cap Y)$ on the RHS and combine $P(X \cap Y^c) + P(X \cap Y)$ as $P(X)$. Similarly, write $P(X \cap Y)$ as $P(Y \cap X)$ and combine $P(Y \cap X^c) + P(Y \cap X)$ as $P(Y)$. Substitute the values on the RHS to get $P(X) + P(Y) - P(X \cap Y)$. This is known as the addition rule of probability.

5.14.2 Probability Space

A probability space is a triplet $\{\Omega, \mathbb{A}, \mathbb{P}\}$, where Ω is the sample space, $\mathbb{A}$ the set of events defined on Ω, and $\mathbb{P}$ the probability measure that maps events in $\mathbb{A} \rightarrow [0, 1]$ such that $P(\Omega) = 1, P(A) \in [0, 1] \forall A \in \mathbb{A}$, which is countably additive. This forms the foundation for several theoretical studies in probability. Note that all three are related, but second and third components are more related than others. This is because the third component is a mapping from elementary events of Ω to the real line $[0,1]$. In other words, $\mathbb{P}$ has domain $\mathbb{A}$ and range $[0,1]$ (it is assumed here that the probabilities are represented as decimals and not as percentages). It is used to mathematically represent a random phenomenon or an unknown experiment.

A probability model is a triplet $\mathbb{P} = (\Omega, S, p(x))$, where Ω is the sample space, S a set of events associated with an experiment, and $p(x)$ the probability associated with each event in S such that $\sum_i p(x_i) = 1$.

5.15   THE CLASSICAL APPROACH

The sample space Ω is well-defined and often enumerable in the classical approach. In addition, there are no conditional events involved. Assume that there are n equally likely, mutually exclusive, and collectively exhaustive outcomes of a random experiment. If m of them are favorable to an event E, the classical approach states that the desired probability is m upon n (i.e., $p = m/n$). Symbolically, this can be written *as*

$p(E) = $ number of outcomes favorable to E/ total number of outcomes in Ω.

This definition holds only when the sample space is finite.

5.15.1   Counting Techniques in Classical Probability

Several counting techniques were discussed in Section 5.8 (starting p. 135). These form the foundation of the classical approach. Some of the counting techniques developed there have direct analogs in probability. Consider for example, the principle of inclusion and exclusion discussed in page 147. It was mentioned there that the PIE has two variants in terms of counts and probabilities. The "probability version" given below has direct application in finding the probability of a union of events, at least some of which have common elements.

5.15.2   Assigning Probabilities to Events

It is fairly straightforward to find probabilities of events by the classical approach. First find the total number of possible events say n. Then find the number of favorable events say m. Then divide m by n to get the probability.

Lemma 4 Classical probability $=$ number of favorable cases/total number of cases $(p = m/n$, where $m = $ # favorable cases, $n = $ total number of cases).

Numerator can be either enumerated, estimated by other means, or evaluated recursively in most of the problems. These are exemplified in the following.

■ **EXAMPLE 5.46   Cards in a box**

A box contains cards marked with numbers 1–10. (A) What is the probability that a number drawn at random is (i) prime number and (ii) divisible by 3. (B) What is the probability that the sum of two numbers drawn at random without replacement is (i) even and (ii) odd integer greater than or equal to 15.

Solution 5.46 The sample space is well defined. There are no conditional probabilities involved. Total favorable cases are easy to enumerate. Hence, we could easily find the probability by dividing the total favorable cases by the number of points in the sample space. For Case (A), we need to enumerate all prime numbers. There are 5 of them as $\{1,2,3,5,7\}$ are all primes. Hence, required probability by lemma 8 is $5/10 = 1/2$. In case (ii), the favorable cases are $\{3,6,9\}$. From this, the required probability follows easily as $3/10$. In Part (B), we are drawing the cards without replacement. Total number of ways to draw two numbers from 10 is $\binom{10}{2}$. As the sum of two numbers is even when both are even or both are odd, we can easily enumerate the 20 favorable cases in the sample space as $S = \{(1,3),(1,5),(1,7),(1,9),(3,5),(3,7),(3,9),(5,7),(5,9),(7,9),(2,4),(2,6),(2,8),(2,10), (4,6),(4,8),(4,10),(6,8),(6,10),(8,10)\}$. The required probability is then $20/\binom{10}{2} = 20/45 = 4/9$. In case (ii), the favorable cases in the sample space are $(5,10),(7,10),(9,10),(6,9),(8,9),(7,8),(9,8)$. Hence, the required probability by Lemma 8 is $p = $ total favorable cases/number of points in the sample space $= 7/\binom{10}{2} = 7/45$. ∎

5.15.3 Rules of Probability

This section refreshes some of the rules that are necessary for laying a foundation for subsequent discussions.

Rule 3 Probability is always between 0 and 1 ($0 \leq P(A) \leq 1$).

In Section 5.2, we have seen various ways to express probability. All of the methods described there (except the percentage method) map the probability into the interval $[0,1]$.

Rule 4 Probability of the entire sample space is 1. That is $P(\Omega) = 1$.

The proof follows trivially because the probability of all the events occurring is certainty.

Rule 5 Probability of occurrence of either of two *disjoint* events is the sum of their individual probabilities (i.e., $P(A \cup B) = P(A) + P(B)$).

◼ EXAMPLE 5.47 Playing card problem

What is the probability that a card selected from a deck of playing cards will be either an Ace or a Queen?

Solution 5.47 Let "A" denote the event that it is an Ace and "B" denote the event that it is a Queen. These two are disjoint events. Hence, the required probability by Rule 5 is $P(A) + P(B)$. But $P(A) = 4/52 = 1/13 = P(B)$. The answer follows as $2/13$. This rule can be extended to any number of disjoint events. Let $A_1, A_2, \ldots, A_n$ be disjoint events. Then $(P(A_1 \cup A_2 \cdots \cup A_n) = \sum_{i=1}^{n} P(A_i))$. ∎

Rule 6 Product Rule

If A and B are two independent events, the probability of occurrence of both of these events is the product of their individual probabilities:– $P(A \cap B) = P(A)P(B)$.

Proof: As the events are independent, the occurrence of A has nothing to do with the occurrence of B. The probability of occurrences of A and B is the product of their individual probabilities. This rule can be generalized to any number of independent events as $P(A_1 A_2 \cdots A_n) = P(A_1)P(A_2)...P(A_n)$. ∎

▮ EXAMPLE 5.48 Furniture making

A furniture is made through three processes:–(i) cutting process, (ii) drilling process, and (iii) assembly and finishing process. The respective probabilities of a defect in each of the stages are 1/60, 1/20, and 1/80. Find the probability that a finished furniture is (i) defective and (ii) has no cutting or drilling defect.

Solution 5.48 Assume that the processes are independent. The probability that it is defective is $1/60*1/20*1/80 = 1/96,000$. (ii) Probability that it has no cutting or drilling defect is $(1 - 1/60) * (1 - 1/20) = 0.93416$. ∎

Rule 7 Sum rule

The probability of occurrence of either of two events (not necessarily independent) is $P(A \cup B) = P(A) + P(B) - P(A \cap B)$.

Proof follows trivially using the principle of inclusion and exclusion. Let "X" denote the common intersection of events A and B($X = A \cap B$). Then $P(A) + P(B)$ will contain the "X" portion twice. Therefore, we need to subtract it once to get $P(A \cup B)$. Another proof appears in Theorem 5.5 (p. 155).

The probability of non-occurrence of an event is the complement of the probability of occurrence. Symbolically, $P(\overline{A}) = 1 - P(A)$.

The complement of an event comprises all events in the sample space Ω except the event. As the probability of the sample space is 1, it follows that the probability of the event union the probability of its complement is 1. Symbolically $P(A) + P(\overline{A}) = 1$, from which the result follows.

Theorem 5.6 If $A_1, A_2, \ldots, A_n$ are events defined on a sample space, at least some of which have common elements, then

$$P(A_1 \cup A_2 \cup \cdots A_i \cup \cdots A_n) = \sum_i P(A_i) - \sum_{i<j} P(A_i \cap A_j)$$

$$+ \sum_{i<j<k} P(A_i \cap A_j \cap A_k) \cdots$$

$$+ (-1)^{n-1} P(A_1 \cap A_2 \cap \cdots A_n). \tag{5.19}$$

Proof: Consider a special case with just two events say X and Y. We know that

$$|X| + |Y| - |X \cap Y| = (|X - Y| + |X \cap Y|) + (|Y - X| + |X \cap Y|) - |X \cap Y|$$
$$= |X - Y| + |X \cap Y| + |Y - X|$$
$$= |X \cup Y|.$$

Divide both sides by the total number of points in the sample space and swap the LHS and RHS. Then, the above-mentioned expression takes the form $P(X \cup Y) = P(X) + P(Y) - P(X \cap Y)$. If X and Y are disjoint events, then $X \cap Y = \phi$ so that $P(X \cap Y) = 0$, and the results follow. Next assume that the result is true for an arbitrary $m > 2$. Then $P(\cup_i A_{i=1}^m) = \sum_{i=1}^m P(A_i) - \sum_{i<j} P(A_i \cap A_j) + \sum_{i<j<k} P(A_i \cap A_j \cap A_k) \cdots + (-1)^{n-1} P(A_1 \cap A_2 \cap \cdots A_n)$. Write

$$\cup_{i=1}^{m+1} A_i = \cup_{i=1}^m A_i + A_{m+1} - [(\cup_{i=1}^m A_i) \cap A_{m+1}]$$
$$= \cup_{i=1}^m A_i + (A_{m+1} - \cup_{i=1}^m (A_i \cap A_{m+1})). \qquad (5.20)$$

As done earlier, divide by $|\Omega|$ to get the probabilities as $P(\cup_{i=1}^{m+1} A_i) = P(\cup_{i=1}^m A_i) + P(A_{m+1}) - P(\cup_{i=1}^m (A_i \cap A_{m+1}))$. Now apply the above-mentioned equation to get the RHS in desired form. Thus, the result follows by induction.   ∎

Corollary 2 If $A_1, A_2, \ldots A_n$ are finite sets, some of which have common elements,   then $|A_1 \cap A_2 \cap \cdots A_i \cap \cdots A_n| = |U| - \sum_{i=1}^n |A_i| - \sum_{1 \le i \le j} |A_i \cap A_j| + \sum_{1 \le i \le j \le k}^n |A_i \cap A_j \cap A_k| + \ldots + (-1)^{n-1} |A_1 \cap A_2 \cap \ldots \cap A_n|$.

5.15.4   Do-Little Principle of Probability

It was mentioned in Section 5.6.4 (p. 131) that complementary events are sometimes easy to find when the sample space consists of a large number of discrete events as in the above-mentioned example. These are especially true in "at least k" and "at most k" type problems. These are called the do-little (or complement-and-conquer) principle of probability. See page 162 and 166 for numerical examples.

EXAMPLE 5.49   Multiple choice exam

A multiple choice exam has 15 questions, each with 4 answer choices (say A,B,C,D). If a student guesses the answer to every question, what is the probability of getting at least two questions correct?

Solution 5.49 Here, the keyword is "at least 2." The complement event is 0 or 1 correct answers that are much easier to find. As there are four choices, probability of guessing the answer is 1/4 (so that the probability of incorrect answer is 3/4). We can consider the 15 questions as independent. This gives the probability of 0 correct answers as $(3/4)^{15} = 0.0133635$. Now consider getting at least one correct answer. This correct answer may correspond to any of the questions 1 through 15 (in other words, there exist 15 possibilities). Hence, this has probability $\binom{15}{1}(1/4)(3/4)^{14} = 0.0668173$. Subtracting the sum of these probabilities from 1 gives the required answer of getting at least two questions correct as $1 - 0.0668173 - 0.0133635 = 1 - 0.080181 = 0.919819$.   ∎

EXAMPLE 5.50   At least type problem

There are 10 students in a class. What is the probability that at least two of them have a common birthday if none were born in a leap-year?

Solution 5.50 Here, the keyword is again "at least 2." The complementary event is "none of them have a common birthday." This means that each student has a different birthday. Arrange the students in an arbitrary order. The first student has 365 choices. Having fixed the birthday of first student, there are 364 choices for the second student and so on. Thus, by the multiplication principle, total number of ways (favorable cases) in which all birthdays are different is $m = 365*364*363*356$. Total number of ways in which the birthdays can be distributed (including those counted above) is $n = 365^{10}$. The required probability (in which all birthdays are different) is obtained by dividing m by n. Note that one of the 365's cancel out from the numerator and denominator giving the answer as $(364)_9/365^9 = 0.88305$. Subtract this from 1 to get the probability that at least two of them have a common birthday. In general, if there are m students, the probability for all birthdays to be different is $(364)_{m-1}/365^{m-1}$.   ∎

EXAMPLE 5.51   Shipping container

A shipping container is loaded with 50 food cartons. The probability that any of the cartons will get damaged during transshipment is $0.003 = 3/1000$. What is the probability of finding at least one defective carton when the container reaches its destination?

Solution 5.51 As the probability that it will get damaged during shipment is 0.003, the probability that it will not be damaged is 0.997. Hence, the probability that at least one of them gets damaged = 1-probability that none of them is damaged = $1 - (0.997)^{50} = 1 - 0.8605 = 0.1395$.   ∎

EXAMPLE 5.52   Defective circuits

An electronic board has three parallel circuits, each of which contains three, eight, and five components. The probability for each component to malfunction is 0.0015. The board will stop working when at least one of the parallel circuits has a defect. What is the probability that the board does not work?

Solution 5.52 Probability that at least one of the circuits does not work = 1-probability that none of them are defective. Probability that the first circuit is not defective = 0.9985^3. Similarly, the corresponding probabilities for second and third circuits can be found. ∎

5.15.5 Permutation and Combination in Classical Approach

Permutation is useful to solve a variety of probability problems involving place-ment of objects (such as books, people, and electronic components). Probabilities can be assigned to the events that make up a random experiment using the axiomatic approach. This is easily done in the case of equally likely experiments using the classical approach. When the favorable cases for an event involve counting several arrangements, we can use the techniques developed in the permutation and combina-tion theorems.

▨ EXAMPLE 5.53 Leap year

Consider Example 5.28 in page 139. What is the probability for a leap year with 366 days to have (i) 53 Sundays?, (ii) 53 Saturdays and 53 Sundays?, (iii) exactly 52 Saturdays and 52 Sundays?, (iv) exactly 53 Fridays or 53 Sundays?, and (v) exactly 52 Tuesdays and 52 Thursdays?

Solution 5.53 As mentioned before, there are only seven possible combinations for the extra 2 days. (i) As Sunday occurs in two of the seven combinations, the probability that a leap year will have 53 Sundays is 2/7. (ii) As (Saturday, Sunday) occurs once, the desired probability is 1/7. (iii) As neither Saturday nor Sunday occurs in four out of the seven possible pairs, the desired probability is 4/7. (iv) There exist four pairs containing either a Friday or a Sunday required probability is 4/7. (v) There are three favorable cases, namely, (Sunday, Monday), (Friday, Saturday), and (Saturday, Sunday) using the complement rule. Hence, the answer is 3/7. ∎

▨ EXAMPLE 5.54 Roots of quadratic equation

Consider the quadratic equation $px^2 + qx + r = 0$ considered in Example 5.29 (p. 139). Find the probability that (i) the equation will have real roots, (ii) equal roots, (iii) imaginary roots, (iv) both integer roots, and (v) exactly one integer root.

Solution 5.54 As each of the coefficients p, q, r is determined by the number that shows up in the throw of a die, we need three throws to decide them (say choose p first, then q, and finally r). In Example 5.29, we found that there exists $(38 + 5) = 43$ ways for the equation to have real roots. Hence, the answer to (i) is 43/216. (ii) As the five favorable cases are $(1,2,1), (1,4,4), (2,4,2), (3,6,3)$, and $(4,4,1)$, answer to (ii) is 5/216. A quadratic equation can have real roots, equal roots, or imaginary roots only. Hence, the favorable cases for case (iii) can be directly obtained using complement rule as $216 - 43 = 173$. This gives the probability for case (iii) as $173/216$. Consider case (iv). Both roots are integers in 10 favorable cases, so that the required probability is $10/216 = 5/108$. (v) There are eight favorable cases so that the probability is $8/216 = 1/27$.   ∎

EXAMPLE 5.55   Equal number of Heads and Tails in coin toss

An unbiased coin is tossed $2n$ times where $n \geq 1$. What is the probability of observing an equal number of heads and tails?

Solution 5.55 As we are interested in "an equal number of heads and tails," this can be considered as an arrangement of n Heads and n Tails in $2n$ trials. There are $\binom{2n}{n}$ ways in which n Heads and n Tails can occur. Each of them has the associated probability $p^n q^n$ where $q = 1 - p$. By the ROS principle, the answer is $\binom{2n}{n} p^n q^n$. As we are given that the coin is unbiased, $p = q = 0.5$. Substitute in the above-mentioned equation to get the answer as $\binom{2n}{n}$ $(1/2)^{2n}$.   ∎

EXAMPLE 5.56   Common birthday

Suppose that there are n (<365) passengers in a plane. What is the probability that at least two people have a common birthday? What is the minimum value of n such that the probability that (i) none will have a common birthday is 0.4313? (ii) Two or more people will share a common birthday is at least 0.9?

Solution 5.56 Assume that the birthdays are randomly distributed, and none were born on February 29 of a leap year. Then we could consider the 365 days as the equivalent of numbered urns. A person whose birthday is January 10 is assigned to 10th urn, and one whose birthday is December 26 is assigned to urn 360. The desired probability is found by enumerating the number of ways in which these urns can be filled by people such that at least two people are assigned to an urn. This sample space is not easy to enumerate. Next apply the "complement-and-conquer" principle. Consider the complement event. As a common birthday occurs with at least two people, the complement event is that none of the passengers have a common birthday. This is equivalent to counting the number of ways in which people can be assigned to urns such that each urn is either empty or has at most one assigned person. This event is greatly simplified. The total number of ways in which birthdays of n passengers may fall among the 365 days is 365^n. Order the persons arbitrarily from 1 to n. There are 365 possibilities for the first person's birthday. As that day is taken, the second person's birthday can fall in 364 days, and so on. As our assumption is that $n < 365$, the last person's birthday can be chosen in $365*364*363*...*(365 - n + 1)$ ways. This can be denoted using factorials as $365!/(365 - n)!$ or using Pochhammer notation as $(365)_n$. Thus, the number of ways in which each person's birthday is different is $365!/(365 - n)!$ because we do not care which person's birthday is on a particular date. From this, the probability that each of the birthdays is different is obtained as $p = 365!/((365 - n)!365^n) = (365)_n/365^n$. Hence, the probability that at least two people have a common birthday is $1 - p = 1 - (365)_n/365^n$. For part (i), we need to find n such that $(365)_n/365^n \simeq 0.4313$. Take log of both sides and try successive values to get $n = 25$. In part (ii), we have to find that value of n for which $1 - p = 1 - (365)_n/365^n \leq 0.1$. For $n = 40$, the probability that all birthdays are different is 0.108768, and for $n = 41$, it is 0.0968. Hence $n = 41$. ∎

EXAMPLE 5.57   No Common birthday

Consider the above-mentioned example where $n > 365$. What is the probability that (i) none have a birthday on Sundays?. (ii) Exactly k persons have a common birthday on the X'mas day?.

Solution 5.57 (i) A year has either 52 or 53 Sundays (if 1 January is a Sunday, then that year will have 53 Sundays, as 31 December is also Sunday). In the former case, there are $365 - 52 = 313$ days that are not Sundays. Hence, the total number of possibilities is 313^n. In the later case, there are 312 days that are not Sundays with 312^n possibilities. Thus, the probability is either $(312/365)^n$ or $(313/365)^n$ depending on whether January 1st is a Sunday or not. (ii) As k persons birthday fall on an X'mas day, there are $n - k$ persons whose birthday falls on other 364 days. There are $\binom{n}{k}$ ways for k persons to have birthday on X'mas day and $(364)^{n-k}$ ways for other birthdays. Hence, the required probability is $\binom{n}{k}(364)^{n-k}/(365)^n$. ∎

5.15.6   Sequentially Dependent Events

Events cyclically repeat in some applications. Consider a working traffic light. In each cycle, the signal changes color from Green (G) to Yellow (Y) to Red (R) and then to Green. Hence, the events are {G, Y, R}. These are not equally likely because the duration of these signals are preset based on the traffic density in different directions. Assume that Green signal is shown for 50 seconds, Yellow for 5 seconds, and Red for 35 seconds in one direction. Then $P(\text{Green}) = 50/90 = 5/9$, $P(\text{Yellow}) = 5/90 = 1/18$, and $P(\text{Red}) = 35/90 = 7/18$. This may differ in other directions.

5.15.7   Independence of Events

Independence of events is an important condition to check, as this can considerably simplify probability calculations. This is most often intuitively clear to humans but not to machines. Independence of events is often assumed in random experiments.

Definition 5.9 Two events "*A*" and "*B*" are independent if the occurrence of either of them is not influenced by prior knowledge about the occurrence of the other event.

Symbolically, we denote it as $P(A \text{ and } B) = P(A) * P(B)$. Note that independence is a "logical relation" among events, but it is mathematically cast using the probability notation. This can also be expressed as $P(A) = P(A|B)$ (or $P(B) = P(B|A)$), where the vertical bar denotes conditioning (it is read as "Probability of *A* equals probability of *A* given *B*," etc.). The first notation is easier than others to generalize the concept to *n* events. Symbolically events $E_1, E_2, \dots, E_n$ are independent if $P(E_1 \cap E_2 \cdots \cap E_n) = P(E_1) * P(E_2) * \cdots * P(E_n)$.

◼ EXAMPLE 5.58   Student selection

One class has 5 girls and 10 boys. Another class has 8 girls and 7 boys. If one student each is selected from both classes, what is the probability that (i) both are boys, (ii) both are girls, and (iii) one boy and one girl?

Solution 5.58 Let "A_i" denote the event of selecting a boy and "B_i" denote the event of selecting a girl from *i*th class. $P(A_1) = 10/15 = 2/3$ and $P(A_2) = 7/15$. The probability that both are boys $= 2/3 * 7/15 = 14/45$, (ii) probability that both are girls $= 1/3 * 8/15 = 8/45$, and (iii) one boy and one girl can come in two ways (boy from first class or second class). Thus, the probability that one is a boy and other is a girl is $(2/3) * (8/15) + (1/3) * (7/15) = 23/45$ as the events "*A*" and "*B*" are independent. ◼

◼ EXAMPLE 5.59   Restaurant menu

A restaurant offers 6 varieties of soup; of which 4 are vegetarian and 2 are nonveg soups; 10 varieties of the main course meal; of which 8 are nonveg and the rest 2 are vegetarian meals. If 80% of the customers take vegetarian soup, and among those 90% orders nonveg main meal, what is the probability that a randomly chosen customer will order a veg soup followed by a vegetarian main meal? If 95% of the people who orders nonveg soup also orders nonveg meals, what is the probability that a randomly chosen customer will eat vegetarian meal?

Solution 5.59 As there are interacting events, this problem is easy to crack using a table. Probability that a randomly chosen customer will order a nonveg soup is 80%. Probability that this is followed by vegetarian main meal is $0.80 * 0.10 = 0.08$ or 8%. From Table 5.7, we see that 20% of the customer's order nonveg soup, among which 95% (or 19 customers) order nonveg main meal. This means that only 1% of the customers who order nonveg soup also orders veg meal. Hence, the probability that a randomly chosen customer will eat vegetarian meal (irrespective of soup type) is $8 + 1 = 9\%$. ◼

TABLE 5.7   Soup and Meal Combination

Soup	Main meal		
	Veg	Nonveg	
Veg	8	72	80
Nonveg	1	19	20

90% of 80 is 72, and 95% of 20 is 19. Other entries are found by subtraction.

5.15.8   Independent Random Variables

Independence of events discussed earlier can be expressed in terms of conditional probabilities as $P(X|Y = y) = P(X)$. This immediately leads to independence of random variables. Let X and Y be discrete or continuous random variables. We define the independence in terms of probability of joint occurrence and individual occurrences as follows:

Definition 5.10 Two random variables X and Y are independent if $P(XY) = P(X) * P(Y)$.

This definition can be extended to any number of random variables.

As random variables have probability and distribution functions, we have several choices to define independence. Two random variables X and Y are independent if any of the following conditions is satisfied:– (i) $f(X|Y = y) = f(X)$, (ii) $f(Y|X = x) = f(Y)$, (iii) $F(X, Y) = F(X) * F(Y)$, (iv) $F(X|Y = y) = F(X)$, and (v) $F(Y|X = x) = F(Y)$. In addition, generating functions can also be used.

An "empirical" probability is estimated after an experimental trial using known or observed frequencies of outcomes. Here, the assumption is that the trials are independent. It may also be estimated using a computer simulation. Experimental probability is derived numerically through the use of existing or simulated data. In the coin-tossing example, if we toss the coin 100 times and observe the number of Heads that turn up, we could find the experimental probability of observing a Head. Objective probability is a ratio measure that expresses the likelihood of an event occurring in many repeated and identical trials of a random experiment.

■ EXAMPLE 5.60   Birthday sharing

A class has 60 students, of which 20 are males. Find (i) Probability that the birthday of at least one student falls on a Sunday. (ii) Probability that at least three female students will share the same birthday on Wednesday. (iii) Probability that at most two male students will have their birthday on a weekend.

Solution 5.60 We assume that there are 52 weeks in a year ($52 \times 7 = 364$ days). As the extra day in a year can be a Sunday (for nonleap years) with probability 1/7 and other days with probability 6/7, we get the exact probability as follows. The probability that the birthday of an arbitrary student falls on a Sunday is $p = 1/7$ and the probability that it does not fall on a Sunday is $q = 6/7$. To answer (i), we use the complement-and-conquer principle (p. 131), which is the probability that *none* of the student birthdays fall on a Sunday. Hence, the required probability is $1 - q^{60} = 1 - (6/7)^{60}$. This answer is not exact. For non-leap years, the extra day can be a Sunday with probability $1/7$, we get the exact result as $(6/7)[1 - (6/7)^{60}] + (1/7)[1 - (53/365)^{60}]$. For leap years, the multipliers are $5/7$ and $2/7$. (ii) Using do-little principle, the answer is one-probability that less than two female students share a birthday on Wednesday. As there are 40 female students, this is $1 - \sum_{i=0}^{2} \binom{40}{i} q^{40-i}$. As we have used the word "shares," the other possibilities (only one student's birthday is on Wednesday or none have their birthday on Wednesday) are irrelevant. (iii) The answer can be broken into three groups: (a) none have their birthday on a weekend, (b) only one male student has birthday on a weekend, and (c) exactly two male students have birthday on a weekend. These are, respectively, $(5/7)^{20}$, $\binom{20}{1} * (2/7) * (5/7)^{19}$, and $\binom{20}{2} * (2/7)^2 * (5/7)^{18}$. ∎

◼ EXAMPLE 5.61   Chessboard squares

If two squares are chosen at random on a chessboard, what is the probability that they will form a rectangle?

Solution 5.61 There are 64 squares in total on the chessboard. The chosen squares will form a rectangle when they are adjacent and either horizontally or vertically aligned (but not diagonally). Let these be denoted by events X and Y. The total favorable cases for X to materialize on any row are seven (as this could happen in $(1,2),(2,3), ..., (7,8)$) squares. As there are eight rows, the total number of favorable cases for X is $8 \times 7 = 56$. Similarly, there are 56 cases for vertical alignment along any of the columns. Thus, the total number of favorable cases is $56 + 56 = 112$. Total number of ways to choose two squares on a chessboard of 64 squares is $\binom{64}{2}$. Hence, the required probability = total favorable cases/number of points in the sample space = $112/\binom{64}{2} = 112/[32 * 63] = 7/[2 * 63] = 7/126 = 1/18$. ∎

5.16   FREQUENCY APPROACH

Consider a random experiment that is conducted n times under identical conditions. If an event X occurs r times out of the n equally likely outcomes, the ratio r/n can be considered as the probability of occurrence of the event X. This probability may fluctuate for small values of n but will stabilize for large values. Symbolically, $P(X) = r/n = $ number of favorable cases/total number of trials.

TABLE 5.8   Frequency Distribution of BMI Values

BMI Range	Frequency	BMI Range	Frequency
"12–15"	1	"15–18"	6
"18–21"	24	"21–24"	56
"24–27"	67	"27–30"	33
"30–34"	11	"> 34"	2

Definition 5.11 Probabilities computed using frequency distributions or random trials that are repeated under identical conditions are called empirical probability. This approach can be used when the sample space is fuzzy, uncountable, or even unknown.

▣ EXAMPLE 5.62   BMI values

The BMI values of 200 patients are given in Table 5.8. Find the probability that a new patient will have a BMI in the range (i) "24–27," (ii) between 21 and 27, and (iii) at least 30?

Solution 5.62 From the table, the relative frequency of patients with BMI in the range "24–27" is $p = 67/200$, which is the required answer. For part (ii) by the frequency approach, we get $p = (56 + 67)/200 = 123/200$; for part (iii) by the frequency approach, we get $p = (11 + 2)/200 = 0.065$. The probabilities obtained are only estimates of the towards true value. If the sample size is increased from 200 to 2000, some of these probabilities may improve slightly toward true value.   ■

▣ EXAMPLE 5.63   Newspaper readership

Consider the Example 11 in page 5–26. What is the probability that a randomly chosen person reads either of the newspapers?

Solution 5.63 Define the events A and B as before. The required probability is $P(A \cup B) = P(A) + P(B) - P(A \cap B)$. As we are given that $P(\overline{A} \cap \overline{B}) = 0.10$, we could directly obtain $P(A \cup B)$ as $1 - P(\overline{A} \cap \overline{B}) = 1 - 0.10 = 0.90$.   ■

▣ EXAMPLE 5.64   Human blood groups

The human blood is categorized into four groups called "A," "B," "O," and "AB" using the presence of an antigen on the cell marker. Suppose that the percentage of people with these blood groups is 40, 12, 43, and 5, respectively. Find the probability that (i) two persons getting married are of blood group "A," (ii) two persons getting married are of the same blood group, and (iii) a child will be born with blood group "O."

Solution 5.64 Using the frequency approach, we expect the probability of any person with blood group "A" as 0.40. Denote this as $P(A) = 0.40$. Thus, the probability that both couples are of type "A" is $0.4*0.4 = 0.16$ by the product rule. (ii) We need to add the probabilities for each couple to be of the same type. This gives $p = 0.4 * 0.4 + 0.12 * 0.12 + 0.43 * 0.43 + 0.05 * 0.05 = 0.3618$. (iii) Assuming that all possible blood types are present among the parents, there are 16 possibilities. From the Table 5.9 page 173, we see that an "O" occurs in nine cases. Hence, the required probability is 9/16. ∎

5.16.1 Entropy Versus Probability

Entropy is a term that originated in data communication. It is a measure of the uncertainty in a system. Small entropy values indicate the presence of structure and large entropy values indicate randomness. Probability and entropy are inversely related. This means that the probability of certainty is 1 while entropy of certainty is 0. While the probability quantifies the *degree of belief*, the entropy quantifies the *lack of pattern or organization*. It is used in data communications, decision tree induction, and many other fields [2].

5.17 BAYES THEOREM

This theorem was invented by the English mathematician and cleric Thomas Bayes (1702–1761) but was published posthumously in 1763. The basic ingredient of Bayes theorem is conditional probability. Here, the word "conditional" implies that an event depends on one or more conditions being fulfilled. Usually, the condition is the occurrence of another event. Conditional probability concept is always based on two or more events (in the same sample space) or random variables.

Definition 5.12 Conditional probability is the probability of occurrence of an event with prior knowledge or assumption about another event defined on the same sample space.

■ **EXAMPLE 5.65 Soxes and colors**

Suppose that a drawer contains n pairs of soxes. All soxes are exactly alike except for the color. In utter darkness, a boy wishes to grab just enough number of soxes so that at least two of them are of the same color (he need not have to go and grab another one). What is the minimum number of soxes to grab if (i) there are only two possible colors ((black and white), (ii) there are three possible colors?, (iii) what is the probability of obtaining two whites in a grab of size 3?, and (iv) a kid grabs four soxes. One of them is found to be a Black. What is the conditional probability that the other three together will make two matching pairs?.

Solution 5.65 Let the two colors be Black (B) and White (W). Minimum grabs cannot be 2 as they could be of opposite color. Let it be three. The possible cases are {B,B,B}, {B,B,W}, {B,W,B}, {W,B,B}, {B,W,W}, {W,B,W}, {W,W,B}, and {W,W,W}. Because the order is unimportant, some of these are exactly identical. (i) As every combination should include either two Blacks or two Whites, the minimum number of soxes to grab is 3. For case (ii), let the three colors be Black (B), Red (R), and White (W). If the minimum number of soxes grabbed is 3, there is only one case {B,R,W} (or its permutations) where a match cannot occur. However, if the minimum number of soxes grabbed is 4, a match will always occur. (iii) The answer is easily seen to be 0.5 from above. (iv) As one of them is Black, the other three should contain two Whites and one Black to make two matching pairs or all three Blacks. The favorable cases are {B,W,W}, {W,B,W},{W,W,B}, and {B,B,B}. The required probability is 4/8 = 1/2. ∎

Bayes theorem is a convenient way to compute the conditional probability of a hypothesis H given that an observation (evidence) E has occurred using the probability of an observation, given that a hypothesis has occurred.

Lemma 5 Probability of hypothesis given evidence is the ratio of joint occurrence of hypothesis and evidence over probability of evidence. $P(H|E) = P(H \cap E)/P(E)$.

Corollary 3 The unconditional probability of hypothesis is the sum of the products of the probabilities of hypothesis given evidence and probability of evidence; and probability of hypothesis given no evidence and probability of no-evidence. $P(H) = P(H|E).P(E) + P(H|\overline{E}).P(\overline{E})$.

Here, $P(H|E)$ is the posterior probability. These can be obtained from each other with the help of prior probabilities and likelihood as given by Bayes theorem.

5.17.1 Bayes Theorem for Conditional Probability

This theorem is also known as the law of inverse probability. Bayes theorem is used to calculate posterior probability in terms of priors. In other words, Bayes theorem analyzes the root causes and associated risks of alternatives using empirical data to come up with the best plausible aposteriori probability or probability of occurrence of hypothetical causes. Conceptually, posterior = likelihood × prior/evidence where likelihood is estimated from sample data or found by other means. It expresses *aposteriori* probability in terms of *apriori* probabilities using newly acquired information.

Let X and Y be two arbitrary events. Suppose that Y has already occurred. If X and Y have *some outcomes in common*, X will occur iff $X \cap Y$ occurs. This is symbolically denoted as $P(X|Y) = P(X \cap Y)/P(Y)$. Cross-multiply to get $P(X \cap Y) = P(Y) * P(X|Y)$. As X and Y are arbitrary, this can also be expressed as $P(X \cap Y) = P(X) * P(Y|X)$. This is the multiplicative law of probability discussed earlier. In $P(X|Y) = P(X \cap Y)/P(Y)$, replace the numerator $P(X \cap Y)$ by $P(X).P(Y|X)$. Substitute in the aforementioned to get $P(X|Y) = P(X).P(Y|X)/P(Y)$. Using the law of total probability (page 133), we have $P(Y) = P(Y \cap X) + P(Y \cap \overline{X})$. Reorder the events to get $P(Y) = P(X \cap Y) + P(\overline{X} \cap Y)$. Now write $P(X \cap Y) = P(X) * P(Y|X)$ and $P(\overline{X} \cap Y) = P(\overline{X}) * P(Y|\overline{X})$.

Replace the denominator $P(Y)$ by $P(X) * P(Y|X) + P(\overline{X}) * P(Y|\overline{X})$ to get

$$P(X|Y) = P(X).P(Y|X)/[P(X) * P(Y|X) + P(\overline{X}) * P(Y|\overline{X})]. \quad (5.21)$$

This is the simplest form of Bayes theorem. Using the hypothesis and evidence notation used earlier, if the nonoccurrence of the hypothesis is denoted by $P(\overline{H})$, we get

$$P(H|E) = P(H) * P(E|H))/[P(H) * P(E|H) + P(\overline{H}) * P(E|\overline{H})].$$

Next consider n events $A_1, A_2, \ldots, A_n$. Let B be an event that spans at least two of the $A_i's$. If the apriori probabilities of occurrence of $P(B|A_i)$ are known, we could utilize the information in obtaining an estimate of the aposteriori probability using Bayes theorem. Symbolically, it can be written as $P(A_i|B) =$

$$[P(A_i).P(B|A_i)]/[P(A_1).P(B|A_1) + P(A_2).P(B|A_2) + \cdots + P(A_n).P(B|A_n)]. \quad (5.22)$$

Proof: Let A_i denote possible explanations for a given set of data B. As the data size increases, the probability $P(B|A_i)P(A_i)$ increases. If A is decomposed as $A = A_1 \cup A_2 \cup \cdots A_n$, then B can be represented as $B = BA_1 \cup BA_2 \cup \cdots \cup BA_n$. Thus $P(B) = \sum_i P(BA_i) = \sum_i P(A_i).P(B|A_i)$. As $P(A_i|B) = P(A_i).P(B|A_i)/P(B)$, the proof follows by substituting the value for $P(B)$. This proves the theorem for the general case.   ∎

In multiple hypotheses situations, Bayes theorem provides a "best" estimate for the probability of evidence under the assumption that each hypothesis is true. A generalization to the three event case easily follows as $P(AB|C) = P(A|BC) * P(B|C) = P(B|AC) * P(A|C)$.

▣ EXAMPLE 5.66   ATM Cash Withdrawal

Consider cash withdrawals at an ATM booth. From analysis of prior fraudulent transaction, a Bank has found that the probability of any transaction to be fraudulent is one in thousand ($P(\text{Fraud}) = 0.001$), 90% of fraudulent transactions are for amounts above 2000 (i.e., $P(\text{Amount} > 2000|\text{Fraud}) = 0.90$), and 99% of cash withdrawals for amounts > 2000 are genuine. Using this information, what is the probability that a transaction is fraudulent, given that the withdrawal amount is 4000?

Solution 5.66 We have $P(\text{Fraud}) = 0.001$, $P(\text{Amount} > 2000|\text{Fraud}) = 0.90$, $P(\text{Amount} > 2000|\text{Not Fraud}) = 0.99$ By Bayes theorem, $P(\text{Fraud}|\text{Amount}>2000) = P(\text{Fraud}) * P(\text{Amount} > 2000|\text{Fraud})/[P(\text{Fraud}) * P(\text{Amount} > 2000|\text{Fraud}) + P(\text{Not Fraud}) * P(\text{Amount} > 2000| \text{Not Fraud})] = 0.001 * 0.90 / [0.001 * 0.90 + 0.999 * 0.99] = 0.0009/(0.0009 + 0.98901) = 0.90917$ E-3 $= 0.000909$.   ∎

5.17.1.1 Odds-Likelihood Ratio Form of Bayes Theorem In some applications, we are interested in finding the ratio of the likelihoods

$$\frac{P(\text{Hypothesis}_1 \,|\, \text{Evidence})}{P(\text{Hypothesis}_2 \,|\, \text{Evidence})} = \frac{P(\text{Hypothesis}_1)P(\text{Evidence/Hypothesis}_1)}{P(\text{Hypothesis}_2)P(\text{Evidence/Hypothesis}_2)}. \quad (5.23)$$

EXAMPLE 5.67 Blood type of parents

Table 5.10 gives the break-down of the actual count of patients who visited a clinic, with the combination blood type of parents, where columns denote father's and rows denote mother's blood type[1]. A newly admitted patient only knows that her father was "O" blood type. Find the probabilities that (i) her mother had blood type AB and (ii) mother was also "O" blood type. (iii) If another patient knows only that mother's blood type is AB, what is the probability that the father's blood type is A or O?

Solution 5.67 Let X denote the event that Father was "O" blood type. Let Y_i denote the event that mother's blood type is as given on the ith row of Table 5.10. From the Table 5.9 below, we find that if father is of type "O" and mother is of type "AB," there are two possibilities for the child to have blood types A or B. ■

From the above-mentioned table, we get $P(Y_1) = 183/456, P(Y_2) = 52/456$, $P(Y_3) = 200/456$, and $P(Y_4) = 21/456$. Similarly $P(X|Y_1) = 68/183, P(X|Y_2) = 23/52, P(X|Y_3) = 72/200, P(X|Y_4) = 9/21$. For question (i), we need to find $P(Y_i|X)$ for $i = 4$ and 3. Using Bayes theorem $P(Y_i|X)$=

$$[P(Y_i).P(X|Y_i)]/[P(Y_1).P(X|Y_1) + P(Y_2).P(X|Y_2) + \cdots + P(Y_4).P(X|Y_4)].$$

The denominator is $183/456 * 68/183 + 52/456 * 23/52 + 200/456 * 72/200 + 21/456 * 9/21 = 68/456 + 23/456 + 72/456 + 9/456 = 172/456$. The numerator

TABLE 5.9 Child's Blood-type from Those of Parents'

Entries Are		Father's Blood Type			
Child's Type		A	B	O	AB
Mother's	A	A, O	A,B,O,AB	A,O	A,B,AB
Blood	B	A,B,O,AB	B, O	B, O	A,B,AB
Type	O	A,O	B,O	O	A,B
	AB	A,B,AB	A,B,AB	A,B	A,B,AB

[1]The blood types of people differ in various countries and among different ethnic groups. See www.wikipedia.org/wiki/Blood_type_distribution_by_country for a break up. Data in Table 5.10 can be further broken down using Rh-factor +ve or −ve.

TABLE 5.10 Blood Type Frequency Data

Entries Are		Father's Blood Type				
Patient Counts		A	B	O	AB	Total
Mother's	A	82	24	68	9	183
Blood	B	20	5	23	4	52
Type	O	90	28	72	10	200
	AB	8	2	9	2	21
Total		200	59	172	25	456

is $21/456 * 9/21 = 9/456$. Substitute the values to get the answer to $P(Y_4|X) = 9/456/[172/456] = 9/172$. For part (ii), the numerator is $P(Y_3).P(X|Y_3) = 200/456 * 72/200 = 72/456$, so that the required probability is $72/172$. As the GCD$(72,172) = 4$, divide both numerator and denominator by the GCD to get the answer as $18/43$.

(ii) Let Y denote the event that mother's blood type is AB. Let X_i denote the event that father's blood type is as given on the ith column of Table 5.10. As done earlier, we get $P(X_1) = 200/456, P(X_2) = 59/456, P(X_3) = 172/456$, and $P(X_4) = 25/456$. Similarly $P(Y|X_1) = 8/200, P(Y|X_2) = 2/59, P(Y|X_3) = 9/172$, and $P(Y|X_4) = 2/25$. For part (iii), we need to find $P(X_i|Y)$ for $i = 1$ and 3 (blood group "A" or "O"). Using Bayes theorem, this is $P(X_i|Y)=$

$$[P(X_i).P(Y|X_i)]/[P(X_1).P(Y|X_1) + P(X_2).P(Y|X_2) + \cdots + P(X_4).P(Y|X_4)].$$

The denominator is $200/456 * 8/200 + 59/456 * 2/59 + 172/456 * 9/172 + 25/456 * 2/25 = 21/456$. The numerator is $200/456 * 8/200 = 8/456$. Substitute the above-mentioned values to get the answer for blood group of father = "A" as $(8/456)/(21/456) = 8/21$. Answer for father's blood group "O" differs only in the numerator. As the numerator is $172/456 * 9/172 = 9/456$, we get the answer for subpart as $(9/456)/(21/456) = 9/21$. Add these two probabilities to get the answer that father is of type A or O as $8/21 + 9/21 = 17/21 = 0.809523\ \underline{809523}$.

5.17.1.2 Product Rule for Conditional Probability $P(AB|C) = P(A|C) \cdot P(B|AC) = P(B|C).P(A|BC)$, where AB denotes $A \cap B$, and so on.

5.17.2 Bayes Classification Rule

Consider two propositions A and B whose apriori probabilities are known. Let $U(A)$ and $U(B)$ denote the utilities of propositions A and B, respectively. Then, A is preferred over B if $U(A) > U(B)$. In data mining applications, Bayes' rule is used to specify how the learning system updates its beliefs as new data instances arrive. This is the basic principle of statistical decision theory.

5.17.2.1   *Rule of Expected Utility*   Assuming A as the action and B as the consequence, this rule gives the utility of A as

$$U(A) = P(B|A)U(A \cap B) + P(\overline{B}|A)U(A \cap \overline{B}). \tag{5.24}$$

A disadvantage of this approach is that they depend on prior probabilities of propositions explicitly. If these are unknown, they need to be estimated (using point estimation, EM algorithm [2], stochastic sampling, or parametric approximations) before starting the decision process.

Now consider the problem of classifying a dichotomous attribute using data instances. If the two attribute values are "Yes" and "No," we could obtain a measure of entropy using the probabilities of Yes and No responses as

$$E(S) = -p_{\text{yes}}\log{}_2(p_{\text{yes}}) - p_{\text{no}}\log{}_2(p_{\text{no}}).$$

As the logarithm of a number in the range $(0,1)$ is negative, it combines with the minus sign to return a positive number.

5.18   SUMMARY

This chapter introduced the concepts, tools, and techniques of probability in an intuitive way. Several examples drawn from different fields help the readers in honing the problem-solving skills, and applying it with confidence to practical problems. Several self-understanding and concrete examples make the book accessible to even average students. See Reference 112 for a historical review, References 113 and 114 for theoretical aspects, Reference 115 for urn models, and References 116–120 for further examples.

Data uncertainty prevail in most experiments. With scientific rules and regulations of probability, the exactitude and their remedies in the data could be understood and interpreted. Two or more outcomes in an application might be dependent. As explained in this chapter, the level of their dependence could be calculated and utilized. When two outcomes are dependent, the prediction of one outcome becomes more precise based on the occurrence of a connected outcome using conditional probability. This type of prediction is the basic foundation of decision making in engineering and applied sciences.

EXERCISES

5.1 Mark as True or False

   a) A probability of 0.5 is realized only for discrete event,

---

[2]Expectation Maximization(EM) algorithm is used for maximum likelihood estimation of parameters from missing or incomplete data

b) Addition and subtraction are the only operations on probabilities,

c) A Venn diagram can quantify the probability of an event,

d) If $P(A|B) = P(A)$ then $P(B|A) = P(B)$ always,

e) Combining events using OR operator often increases the probability,

f) Venn diagrams can represent only discrete events,

g) $P(A \cap B)$ is always $\leq P(A \cup B)$.

5.2 The mechanism that generates uncertain outcomes is called (A) event (B) random experiment (C) sample space (D) combination

5.3 Permutations of objects in which nothing is in its original position is (A) Power-set (B) circular permutation (C) Combination (D) Derangement.

5.4 The identifiable outcomes of a random experiment is called (A) event (B) probability (C) sample space (D) power-set

5.5 Decimal form of probability always take values in the range (A) -1 to $+1$ (B) 0 to 1 (C) -0.5 to $+0.5$ (D) any positive value

5.6 A pharmacist has six medicine packs, three of which are of one kind, two of another kind and a last one of a single kind. How many ways are there to arrange them on a shelf?

5.7 A video store has nine cassettes, of which four are of one kind, three are of a second kind, and two are of a third kind to be arranged on a rack. In how many ways, can this be done?

5.8 Evaluate the multinomial coefficients (i) $\binom{7}{3,3,1}$. (ii) $\binom{10}{4,3,2,1}$, (iii) $\binom{12}{4,4,2,2}$

5.9 Convert to fractional form p/q (i) 0.27$\underline{27}$ (ii) 0.428571 $\underline{428571}$, (iii) 0.285714 $\underline{285714}$, (iv) 0.809523 $\underline{809523}$

5.10 An event E1 can happen in m ways and a mutually exclusive event E2 can happen in n ways. In how many ways, can $E1$ and $E2$ happen?

5.11 An elevator starts with six girls from the ground floor to all other floors of a four storeyed building. If all disembarks, define appropriate events.

5.12 Sets of outcomes of a sample space meeting some specifications is (A) subspace (B) partition (C) cardinality (D) Event

5.13 If 4 squares are chosen at random on a chessboard, what is the probability that they will form a bigger square?

5.14 A man has five pairs of soxes of different styles and colors. One night when the power was off, he selects two soxes at random. Find the probability that they form a matching pair.

5.15 Consider a set of ordered events of a random experiment. Express following events using probability notation:
(a) at least one has occurred.
(b) at most one has occurred.

5.16 Prove or disprove the following: (i) if $A \subseteq B$ and $B \subseteq A$ then $A = B$.

5.17 Prove $(n)_k / n^k = \prod_{i=0}^{k-1}(1 - i/n)$

5.18 There are 60 students in a class. What is the probability that (i) two or more students share the same birthday? (ii) there are exactly k days in which no one's birthday falls?

5.19 How many people are there in a room to have the probability that two or more people will have the same birthday is (i) greater than 0.5? and (ii) less than 0.75?

5.20 A diabetes medicine comes as a tablet, capsule, nasal spray, or injection. If each of them is available in regular and generic varieties, in how many ways, can it be prescribed?

5.21 Let X and Y be two finite sets. Define $X \oplus Y$ as the set of all elements in X or Y but not in both. Verify whether $\oplus$ is commutative and associative.

5.22 If X and Y are nondisjoint events, prove (i) $P(X \cup Y) + P(X \cap Y) = P(X) + P(Y)$, (ii) $P[(X \cap \overline{Y}) \cup (Y \cap \overline{X})] = P(X) + P(Y) - 2 * P(X \cap Y)$

5.23 If X and Y are nondisjoint events, arrange the following probabilities in increasing order of magnitude: $P(X), P(X \cap Y), P(X \cup Y), P(X) + P(Y)$.

5.24 A pizza can be ordered in 3 varieties of crust (thin crust, medium, and thick crust), 4 varieties of cheese, and 12 varieties of toppings. How many different varieties of pizzas can be ordered?. If crust is fixed as thick how many choices are left?

5.25 A chocolate bar is in the form of 6×8 pieces of equal size. The bar can be broken only along a straight line horizontally or vertically but not diagonally. If the bars are broken one at a time, what is the probability of obtaining eight 2×3 pieces in seven tries?.

5.26 A school kid has 10 varieties of shirt, 7 varieties of pants, and 5 varieties of tie. In how many ways, can the kid dress up?

5.27 Consider n tosses of a die with faces numbered 1–6. What is the probability that the top face number is greater than the bottom face number?.

5.28 An electronic circuit has n^2 components that look identical. A technician has time to inspect all except n of the components in any trip. If the components are chosen, one after another find the number of ways to choose the components in any trip. Using Stirling's formula for factorials obtain a simplified expression for it.

5.29 Describe a suitable sample space Ω for the following experiments:–(i) absolute value of the difference between the numbers at the top and bottom when a die is rolled; (ii) number of Red balls in x draws of a ball from an urn containing m White and n Red balls; (iii) two dice are thrown and an event is specified as "the number at top of second die is greater than the one on the first die."

5.30 A committee consists of five members of whom the three males are (X, Y, Z) and the females are (U, V). If a meeting is attended by only three members, what is the probability that (i) at least one female is present; (ii) If V was present in the meeting, what is the probability that Y and Z were also present? How many ways are there to form a subcommittee of size 3 comprising (i) at least two males, (ii) at most one female?, and (iii) exactly two males?

5.31 A furniture shop makes a variety of furnitures. Each piece goes through three processes:–(i) cutting process, (ii) drilling process, and (iii) assembly and finishing process. A quality inspector inspects each furniture before it is shipped. The respective probability of a defect in each of the stages is 1/60, 1/20 and 1/80. If a finished furniture is found to be defective, what is the probability that (i) it is a cutting defect and (ii) it is due to either drilling or assembly process?

5.32 Is the event $X \cup Y$ defined when either of them is discrete and the other is continuous? Is the concept of independence defined for continuous events?

5.33 There are n different tasks to be assigned to m employees where $n > m$. How many ways are there if every employee is assigned at least one task?

5.34 Two persons P and Q play a game with respective initial amounts of 70 and 30. Probability of P winning is p, and for Q it is $1 - p$.

Each winning person gets an amount of 1 from the loser. Find the expected value of the amount owned by the winner at nth game. What is the probability of (i) P winning the game?, (ii) the game being over in 120 plays?, and (iii) both have 50:50 in n trials.

5.35 There are 160 customers who buy electronics, and 120 customers who buy other items. Among the 160 customers, 30 also buy other items. If nine customers are randomly chosen, what is the probability that (i) a customer who bought only other items will get selected? And (ii) a customer who bought both items will get selected?

5.36 A safe locker has two locks. An intruder has gained access to "n" keys. If two keys are chosen at random each time, (i) what is the probability that the intruder will be able to open the locker in first try and (ii) the locker in third try?

5.37 If $X \cup Y$ is translated into words as "X or Y occurs," $X \cap Y$ as "X and Y occurs," $X - Y$ as "X but not Y occurs," $\overline{X}$ as "X does not occur," translate the following expressions into words:

(i) $\overline{X \cap Y}$ (ii) $X - X \cap Y$, and (iii) $X \cup Y - X \cap Y$.

5.38 A train has 3 general (unreserved) compartments, 12 reserved coaches, and a pantry car in addition to the engine. Out of 12 reserved coaches, 3 are AC coaches and the rest are ordinary coaches. How many ways are there to connect the coaches if (i) the pantry car can never be the first or last and (ii) the coach immediately behind the engine and the rear-end coach are both ordinary?. (iii) If all AC coaches cannot be together?

5.39 A university wishes to assign a unique 6 digit number to each of the enrolled students with the following restrictions that :– (i) the student number cannot start with digit "0," (ii) digits cannot be repeated in the first three places, but repetition is allowed in subsequent digits. How many student numbers can be generated?

5.40 A fruit merchant has five baskets of apples (all of one kind), three baskets of oranges (all of one kind), and two baskets of bananas to be displayed in front of the shop. (i) How many ways can this be put if all of them are placed in one straight line (ii) if they are arranged as a circle?

5.41 A restaurant offers 5 varieties of soup, 10 varieties of the main course meal, and 5 varieties of ice cream or cake after meal. How many choices are possible for a person who will take any of the choices? How many choices are possible if a person does not take 2 of the 10 varieties of the main course and 3 varieties of ice-cream choices?

5.42 The passenger area of a jumbo-jet can be divided into an executive section (XS) and an economy section (ES). There are three different ways in which XS can be arranged and five different ways in which ES can be arranged. Total

how many ways are there for seating arrangement in the aircraft?

5.43 There are 22 people in a hospital including 2 twins who were born on the same day. What is the mathematical expression to find the probability that at least three persons have the same birthday? At most two people have the same birthday? no-one except the twins have a common birthday?

5.44 A customer needs change for a 10 dollar bill in 5 dollar, 2 dollar, and 1 dollar bills. How many ways are there to make the change?

5.45 Use the PIE principle to find how many integer solutions exist for the equation $x_1 + x_2 + x_3 = 11$ where $x_1 \le 3, x_2 \le 4$, and $x_3 \le 6$.

5.46 A street has 10 houses on one side and 12 houses on the other side. Each of the houses should be numbered sequentially by starting from either end of the road with three digits. How many ways can this be done if only the digits $\{0,1,2,3,4,5\}$ are used? If all houses on one side get even house numbers and all houses on the other side get odd numbers? If 0 cannot be used as a first digit for numbering.

5.47 A rocket can fail independently due to navigation error(NE), software error(SE), or hardware fault(HF). The probability of NE is twice as large as that of SE, and the probability of SE is three times as large as HF. Assume that it failed. (i) If there were no navigation errors, what is the probability that it was due to one of the other faults? (ii) What is the probability that it was due to hardware faults given that there was no software fault?

5.48 There are four blood group types {A, B, AB, and O} and two types of Rh-factors {+, −}. Assume that all of them are equally likely. Among a group of 50 students, what is the probability that (i) there are at least 5 students with O blood group.(ii) Probability for at least 10 students with +ve Rh-factor and blood group A or B.

5.49 Consider a quadratic equation $px^2 + qx + r = 0$, whose nonzero coefficients are determined by the number that turns up when a die is thrown. Find the probability that (i) the discriminant $b^2 − 4 * a * c$ is an integer, (ii) the roots are integers, (iii) at least one integer root, and (iv) there are no real roots.

5.50 A family has n friends. They invite m $(1 < m < n/7)$ friends randomly on each day from Sunday to Saturday to their house where some of the invited guests may overlap on different days, but the group as a whole are different on each day (no identical groups invited twice). For instance, if $m = 2$ and $\{X,Y\}$ are invited on Sunday, $\{X,Z\}$ or $\{Y,Z\}$ may be invited on another day. What is the minimum and maximum number of friends who visit the house in a week?

5.51 Consider a die with six faces. They are not numbered from 1 to 6, but it is known that two of the numbers repeat once (resulting in four numbers). A quadratic equation $px^2 − qx + r = 0$ is formed, whose nonzero coefficients (p, q, r) are determined by the number that turns up when this die is thrown. What is the probability that the roots are real if the numbers that repeat are 1 and 2 and nonrepeating are 3 and 4?. What is the probability that the roots are equal?

5.52 A class comprises 35 males and 25 females. If five students each have to give a seminar randomly each day, find the probability that (i) all five of them on a day are males and (ii) three are males and rest females. How many ways are there if at least two boys are to give the seminar each day?

5.53 A trailer truck has 10 identical looking wheels. A mechanic removes the brake pedals for cleaning and returns them back to the wheels after some time. What is the probability that (i) all brake pedals are returned to their correct wheels? And (ii) none of the brake pedals match their corresponding wheels?

5.54 An online examination has n questions, which are taken together by m (≥ 2) students sitting in a computer laboratory. To avoid copying, the instructor sets it up in such a way that the questions *for each student* are generated using unique random numbers between 1 and n (so that the same question is not displayed twice, and adjacent students may get different question orders). Find the probability that (i) all questions are generated in exactly the same order to two students sitting next to each other?. (ii) Exactly k of the questions are generated the same order to two students sitting next to each other?. (iii) All questions are generated differently for two or more students?

5.55 Assume that n pairs of husbands and wives enter a club, each one wearing a hat. The hats are handed over to a waitress for keeping. After a short while, all n pairs of people assemble for a dance each one wearing a hat. If the waiter distributes the hats at random, and the dancing pairs are formed at random, find the probability that (i) no couple are properly paired, and nobody gets their own hat; (ii) exactly k of the couple are matched, but no one get their matching hat; (iii) exactly k of the hats are matched, but no couple are matched; (iv) exactly k of the couple and m of the hats are matched; and (v) all husbands and wives are matched but none of the hats are matched.

5.56 A biased coin has probability p of heads showing up. If it is tossed 12 times, find the conditional probability that for each of the following if it is known that a total of six heads have been obtained: (i) the first four outcomes are HTHT and (ii) they are TTTH.

5.57 A telephone number has eight digits. What is the probability for each of the following, if starting (leftmost) digits cannot be zeros? (i) four or more digits are repeated?, (ii) at most six digits are repeated?, and (iii) none of the digits are repeated?

5.58 Two exactly identical deck of cards is shuffled. Then, two cards each are drawn from the pool and kept face down. A player is allowed to take one pair of face down cards at a time until k identical pairs are obtained. What is the probability of obtaining k matching pairs?

5.59 An urn contains m Blue and n Red balls. A second urn contains a Blue and b Red balls. Two balls are drawn at random from the first urn and put into the second urn. Then, a ball is drawn from the second urn. Find the probability that (i) it is Blue and (ii) it is Red.

5.60 There are n pairs of shoes in a box (total $2n$ shoes). If $m(<n)$ shoes are chosen at random from the box without looking at the shoe, find the probability that (i) none of the m shoes have a matching pair, (ii) at most two of them have a matching pair, and (iii) in how many ways can you choose m pairs $(<n)$ such that at least one matching pair is obtained.

5.61 A lottery selects the winner by drawing 5 numbers between 1 and 39 randomly. Find probability of the following events if (i) none of the numbers can repeat (all numbers are unique or it is like sampling without replacement) and (ii) numbers can repeat: (a) all 5 numbers are odd, (b) all numbers are below 25, (c) at least 2 numbers are above 30, and (d) none of the numbers are primes.

5.62 A device is manufactured in m independent successive steps. Probability of making an error at step k is p_k. Find the probability that out of n manufactured items, (i) at least one item is defective, (ii) at most two are defectives, (iii) a lot contains between two and five defectives, and (iv) none are defective.

5.63 Consider a quadratic equation $px^2 - qx + r = 0$, whose nonzero coefficients (p, q, r) are determined by the number that turns up when a regular pyramid with five faces numbered 1–5 is thrown. Find the number of ways in which (i) the equation will have real roots, (ii) equal roots, (iii) imaginary roots, (iv) both integer roots, and (vi) exactly one integer root.

5.64 A software company has 8 VB experts, 5 C++ experts, 10 Java experts, and 4 C# expert programmers. A new project that requires 3 VB, 2 Java, and 3 C++ experts is to be initiated. In how many ways can the team be formed? If another project requires two each of VB and C# experts, three each of C++ and Java experts, how many ways are there to form the team?

5.65 The simple matching coefficient (SMC) used in cluster analysis is a similarity coefficient defined on binary strings as $SMC(x, y) = \frac{1}{d}$(Number of positions in which x and y match), where $d =$ total number of bits or the size of the data. If x and y are d bits long, what is the probability that (i) SMC takes the value 1, (ii) SMC takes the value $\geq 1/d$, and (iii) SMC is 1/2 (d even).

5.66 How many numbers between 1 and 200 are divisible by (i) 3, 5,

and 7? And (ii) at least by 3 and 7 but not by 5?

5.67 Verify whether (i) $P(A|B) * P(B|A) = P(A \cap B)$ or (ii) $P(A^c) * P(B^c) = P(A^c \cap B^c)$ when events A and B are independent.

5.68 A faulty electronics appliance has eight exactly looking components. Bob samples four components arbitrarily, tests each of them individually, writes his initial "B" on each of them and puts them back. After he is finished, Peter comes and samples three components arbitrarily and does the same testing, writes his initial "P" on each of the 3 and puts them back. What is the probability that (i) none of the components have both marks, (ii)exactly two of the components will have marks "P" and "B"? and (iii) at least two of them have both the marks?.

5.69 A group of 12 school kids are on a sightseeing trip. The instructor wants to stock enough drinks of each kind. There are seven students who drink coffee, four students who drink tea, nine students who drink fruit juice, three students who drink coffee and tea, four who drinks coffee and juice, and two who drink tea and juice. (i) How many students drink all three beverages? (ii) A student is selected at random and is found to drink fruit juice. What is the probability that student does not drink coffee.

5.70 There are three classes X,Y,Z with respective male and female strengths (30, 10), (27,15), and (32,12). An aptitude test is given in all the three classes. If a girl scored the highest marks followed by two boys overall, what is the probability that (i) all came from Y?. (ii) Female topper came from Z and males came from X and Y?

5.71 The underground water supply system in a city in northern latitude has 5 major pipes and 50 minor pipes with respective probabilities of cracks in a year as POIS (1/1000) and POIS (1/200). What is the probability of one major and two minor cracks in a year? If a crack has indeed occurred, what is the probability that both of them have cracked?

5.72 A tourist has to visit n tourist-spots in a city. In how many possible order can this be visited if (i) all of them are visited on a single day?, (ii) $n/2$ each in two days (for n even)?, and (iii) m of them on first day and rest on second day.

5.73 If the probability of Head showing up is $p = 0.4$ for a coin, find the probability that (i) the second Head is obtained in an odd numbered trial and (ii) the third head is obtained in at least 10 and at most 15 trials.

5.74 A satellite that failed to reach orbit is falling down to the Earth. Some internal parts made of steel, titanium, and beryllium that have a high melting point are likely to make through the descent without burning up. The exact location where it will hit the surface is unknown owing to its eccentric path. Assume that 70% of Earth surface is covered with water. The Atlantic ocean is 16.67% of the total ocean area. The probable hit point is 85% in ocean. What is the probability that it will (i) fall in the Atlantic ocean and (ii) it will hit land mass or Atlantic?

5.75 Two medical tests are being developed for a new virus infection. The first test T1 has probability of identifying the presence of the virus in 99% of the cases but is expensive. The second test T2 is cheap, but it can detect the disease in 96% of the cases. The first test has a false positive rate of 0.05, whereas the second test has a false positive rate of 0.06. If a person is tested positive using T2, what is the probability that the disease is truly present? If both tests show positive what is the probability that the person truly has the disease.

5.76 There are nine rings, all of which have exactly identical look. Three of them are gold and the rest are brass. If you are given a balance, what is the probability of identifying the three gold rings in (i) two weightings?, (ii) three weightings?, and (iii) four weightings?

5.77 A container has 12 machinery parts, all looking alike. Seven of them are known to be good, three of them have mild defect, and the rest have severe defect. Two parts are selected at random. Find probability that (i) both of them are good and (ii) one is good and other has mild defect?

5.78 A course on probability theory is attended by 28 students with statistics major or computer science major. If there are 18 statistics majors and 14 computer science majors, how many are double majors?

5.79 Suppose you are in a room with n other people. What is the probability that no one else shares the

birthday with you?. If m among the n people are males, find the probability that at least one male shares the birthday with you.

5.80 Using Table 5.9 (p. 5.9), compute the conditional probability that parent blood types are A or B given that child's type is O.

5.81 If X and Y are independent, which of the following are also independent? (a) X and $\overline{Y}$, (b) $\overline{X}$ and Y, and (c) $\overline{X}$ and $\overline{Y}$.

5.82 The customer breakdown to a store is given in Table 5.11. Find the probability that (i) a randomly chosen customer to the store is a female and (ii) conditional probability that a customer will visit the store on Friday, given that the customer is male.

TABLE 5.11   Customers to a Store

Day	Males	Females	Total
Monday	11	3	14
Tuesday	6	8	14
Wednesday	7	6	13
Thursday	5	6	11
Friday	6	7	13
Total	35	30	65

TABLE 5.12   Cancer Incidence among Smokers and Nonsmokers

Entries Are	Smoker type		
Patient Counts	Direct	Exposed	Total
Malignant tumour	186	14	200
Benign tumour	124	26	150
Total	310	40	350

5.83 Cancer incidence among first-hand smokers and second-hand smokers (who are exposed to smokers inside enclosed areas) are given in Table 5.12. A new patient is found to be a nonsmoker. What is the probability that he has benign cancer? A retiree is having malignant tumor. Find the probability that he is a smoker.

5.84 If $A_1, A_2, \ldots A_n$ are finite events, some of which have overlaps, prove that

$$P(A_1 \cup A_2 \cup \cdots A_i \cup \cdots A_n)$$
$$= \sum_i P(A_i) - \sum_{i<j} P(A_i \cap A_j) + \sum_{i<j<k} P(A_i \cap A_j \cap A_k) - \cdots$$
$$+ (-1)^{n-1} P(A_1 \cap A_2 \cap \cdots A_n), \tag{5.25}$$

where summations are carried out using conditions specified.

TABLE 5.13   Impurity in Minerals

Entries Are	Mining process		
Impurity in Percentage	Process P1	Process P2	Total
Impurity A	2.5	0.81	3.31
Impurity B	5.0	1.69	6.69
Total	7.50	2.50	10

5.85 A mineral is extracted using two processes $P1$ and $P2$. Two types of impurities in the mineral are examined by a quality inspector. Data appear in Table 5.13. A lot produced by process P1 is randomly selected. Find the probability that it contains Impurity B. If a lot is known to contain Impurity A, what is the chance that it was produced by process $P2$?

6

DISCRETE DISTRIBUTIONS

After finishing the chapter, students will be able to

- Understand binomial theorem and its forms
- Explain Bernoulli trials and Bernoulli distribution
- Describe binomial distribution and its properties
- Apply Poisson distribution in practical situations
- Understand geometric, hypergeometric distribution, and its properties
- Describe negative binomial distribution and its properties
- Describe logarithmic and multinomial distribution and its properties
- Apply the Power method to find the MD of discrete distributions

---

6.1 DISCRETE RANDOM VARIABLES

A real-valued function defined on the sample space of a random experiment is called a random variable. We denote the random variables by capital English letters (X, Y, etc.) and particular values by lowercase letters (x, y, etc.). Random variables can be discrete or continuous. A random variable that can take a countable number of possible values in a finite or infinite interval is called a discrete random variable. In most of the applications, the values assumed are positive ($x \geq 1$) or nonnegative ($x \geq 0$) integers that are equispaced. Theoretically, this is not a restriction. Consider, for example,

Statistics for Scientists and Engineers, First Edition. Ramalingam Shanmugam and Rajan Chattamvelli.
© 2015 John Wiley & Sons, Inc. Published 2015 by John Wiley & Sons, Inc.

the portion of a fruit (say apples) taken by a person at a dining table per day. If it is cut evenly and eaten by different family members, the random variable of interest takes values $0, \frac{1}{2}, 1, 1\frac{1}{2}, 2$, and so on. Similarly, if an employer allows an employee to take either a half-day leave or a full-day leave only, the variable of interest takes integer or half-integer values. However, the majority of discrete distributions discussed in the following are defined on "counts" or "occurrences" that can take nonnegative integer values $(0, 1, 2, \ldots)$. Displaced distributions are those obtained by a change of origin transformation $(Y = X \pm c)$. The constant c is assumed to be a nonzero integer for discrete distributions and a real number for continuous distributions. Left-truncated distributions are exceptions in which the starting value is offset by a positive integer.

This chapter discusses popular discrete distributions. The x values are assumed to be equispaced integers, unless otherwise noted. The domain of X can be finite (as in binomial, discrete uniform, and hypergeometric distributions (HGDs)) or infinite (as in Poisson, geometric, and negative binomial distributions). In the case of infinite range, we naturally expect the probabilities to tail-off to zero beyond a cutoff. Discrete distributions with finite range are more popular in practical applications, whereas those with infinite range are more important theoretically. This is because some statistical distributions with finite range asymptotically converge to discrete distributions with infinite range as shown below. The cumulative distribution functions (CDFs) of discrete random variables are step functions. The CDF of binomial, negative binomial, and Poisson distributions can be expressed as continuous functions such as the incomplete beta and gamma functions as shown below. It may be noted that there exist many more statistical distributions than those mentioned below (see References 121–123. Here, we discuss only those that are widely used in the applications of probability and statistics in everyday life.

6.2 BINOMIAL THEOREM

The binomial theorem with positive and negative exponents has many applications in statistical distribution theory. This section provides an overview of this theorem, which will be used in the sequel. We first consider an expansion for integer powers of a sum or difference of two quantities. More specifically, if n is a positive integer, and x and y are nonzero real numbers, the power $(x + y)^n$ can be expressed as a sum of $n + 1$ quantities in either of two ways as follows:

$$(x + y)^n = \sum_{k=0}^{n} \binom{n}{k} x^k y^{n-k} = \sum_{k=0}^{n} \binom{n}{k} y^k x^{n-k}, \tag{6.1}$$

where $\binom{n}{k}$ denotes $n!/(k!(n - k)!)$. This is most easily proved by induction on n (see exercise). Here, the indexvar k is used as an exponent and a function (in $\binom{n}{k}$). The numbers $\binom{n}{k}$ (also denoted as nC_k, see page 1–16) are called *binomial coefficients*, which are always integers when n and k are integers. The special case $\binom{n}{0}$ is defined to be 1 by convention. In the particular case when $x = y = 1$, the aforementioned becomes $2^n = \sum_{k=0}^{n} \binom{n}{k}$. As $\binom{n}{k} = \binom{n}{n-k}$, the coefficients in the above-mentioned expansion are symmetric (hence $2^n = \sum_{k=0}^{n} \binom{n}{n-k}$, which follows by summing in reverse). If n is odd, there are $(n + 1)$ terms with $(n + 1)/2$ coefficients symmetrically placed. For instance, if $n = 5$, there are three coefficients $\binom{5}{0} = \binom{5}{5} = 1, \binom{5}{1} = \binom{5}{4} = 5, \binom{5}{2} = \binom{5}{3} = 10$. If n is even, there are $n/2$ symmetric coefficients with a unique middle coefficient $\binom{n}{n/2}$.

If y is negative, we write $x - y$ as $x + (-y)$ and the above-mentioned expansion gives

$$(x - y)^n = \sum_{k=0}^{n} \binom{n}{k} x^k (-y)^{n-k} = \sum_{k=0}^{n} \binom{n}{k} (-y)^k x^{n-k} = \sum_{k=0}^{n} \binom{n}{k} (-1)^k y^k x^{n-k}. \quad (6.2)$$

When the index n in the above-mentioned expansion is negative, we get an infinite series as given below:

$$(x + y)^{-n} = \sum_{k=0}^{\infty} \binom{n+k-1}{k} (-x)^k y^{n-k} = \sum_{k=0}^{\infty} \binom{-n}{k} x^k y^{n-k}. \quad (6.3)$$

In the particular case when $y = 1$, we get

$$(1 + x)^{-n} = \binom{-n}{0} + \binom{-n}{1} x + \binom{-n}{2} x^2 + \cdots = \sum_{k=0}^{\infty} \binom{n+k-1}{k} (-x)^k. \quad (6.4)$$

By differentiation, it is easy to prove for $k = 0, 1, 2, \ldots, n$ that [124]

$$\frac{\partial^n}{\partial x^k \partial y^{n-k}} (x + y)^n = n!, \quad \frac{\partial^n}{\partial x^k \partial y^{n-k}} (x - y)^n = (-1)^{n-k} n!. \quad (6.5)$$

We have not placed any restrictions on x and y values in the above-mentioned expansions, other than that they are nonzero real numbers. As the total probability of statistical distributions must sum to unity, we make the restriction that $x + y = 1$. These are usually denoted by p and q (or θ and $1 - \theta$) instead of x and y in statistical applications. This implies that $q = 1 - p$, so that both p and q lie in the interval $[0, 1]$. As shown in the following, p and q are the probabilities associated with the occurrence or nonoccurrence of well-defined events in distribution theory.

6.2.1 Recurrence Relation for Binomial Coefficients

Binomial coefficients satisfy many recurrences. These are most often proved by "combinatorial arguments" because $\binom{n}{r}$ denotes the number of ways of choosing r objects from among n objects without replacement. We give only the simplest and most popular recurrences in the following:

1. $\binom{n}{r} = \binom{n-1}{r} + \binom{n-1}{r-1}$.

This is known as Pascal's identity. As the only arithmetic operation involved is addition, this always returns an integer result.

2. $\binom{n}{r} = \frac{n}{r}\binom{n-1}{r-1}$

This recurrence simultaneously decrements both the arguments and is useful in computing the coefficients for small r values. It is used in Chapter 8, Example 8.35 (p. 8–11). As shown in the following, this could result in approximations owing to truncation error resulting from (n/r). A remedy is suggested below.

3. $\binom{n}{r} = \frac{n-r+1}{r} * \binom{n}{r-1}$

This form is useful when n is large and r is small.

4. $\binom{n}{r} = \frac{n}{n-r} * \binom{n-1}{r}$.

This form is useful when n is very large and r is close to n, so that the decrementing of n is continued until n becomes r. It is used to simplify the MD of binomial distribution (p. 6–23).

5. $\binom{n}{r} = (-1)^r \binom{r+n-1}{r}$

This is used in negative binomial distribution.

6. $\binom{n}{r}\binom{r}{m} = \binom{n}{m}\binom{n-m}{r-m}$

This form is useful when n and m are large and close-by.

7. $\sum_{r=m}^{n} \binom{r}{m} = \binom{n+1}{m+1}$

This combines multiple summations of combinations into a single combination. It is used in finding factorial moments (see page 6–45).

8. $\sum_{k \leq n} \binom{r}{k}\binom{s}{n-k} = \binom{r+s}{n}$

This is called Vandermonde convolution. It is used in deriving factorial moments of some discrete distributions (see page 6–82).

Binomial coefficients evaluated by a computer can sometimes result in approximations. For instance, $\binom{5}{3}$ when evaluated by $\binom{n}{r} = \frac{n}{r} * \binom{n-1}{r-1}$ gives ((5/3) $*$ (4/2) $* \binom{3}{1}$) = 1.6666666 $* 2 * 3$ = 9.999999999999) owing to truncation error. This is because the expression inside the bracket is forcibly evaluated. If the order of evaluation is modified as $\binom{n}{r} = n * \binom{n-1}{r-1} / r$ without parenthesis, we will get the correct integer result. Alternatively, use Pascal's identity (1). It always returns an integer as it involves only additions (see Reference 22).

6.2.2 Distributions Obtainable from Binomial Theorem

There are many statistical distributions that are derived from the above-mentioned form of the binomial theorem. Taking $n = 1, x = p$, and $y = q = 1 - p$, we get the Bernoulli distribution (Section 6.4). Setting $n > 1$ to be an integer, $x = p$ and $y = q$

$= 1 - p$, results in the Binomial distribution (Section 6.5). Putting $n = -1, p = -P, q = Q$, we get $f(x) = 1/(Q - P)$, which is a special case of discrete uniform distribution. Setting $n = -m, x = p$, and $y = q = 1 - p$ results in the negative binomial distribution (Section 6.8). Writing $(x + y)^n$ as $x^n(1 + y/x)^n$, putting $y/x = -Q$, $1/x = P$, and $n = -1$ we get the geometric distribution (which has infinite range). Put $n = -1$ and write $(x - y)^{-1}$ as $1/x\,(1 - y/x)^{-1}$. Setting $y/x = \theta$, taking logarithm and expanding using $-\log(1 - x) = x + x^2/2 + x^3/3 + \cdots$ results in logarithmic series distribution. Write $(1 + y)^n = (1 + y)^{n_1} * (1 + y)^{n_2}$ where $n_1 + n_2 = n$. Expand each one using binomial theorem, equate identical coefficients on both sides and divide RHS by LHS constant to get the HGD.

6.3 MEAN DEVIATION OF DISCRETE DISTRIBUTIONS

Finding the MD of discrete distributions is a laborious task, as it requires a lot of arithmetical work. It is also called the mean absolute deviation or L_1-norm. The MD is closely associated with the Lorenz curve used in econometrics, Gini index and Pietra ratio used in economics and finance, and in reliability engineering. In 1730, the French mathematician Abraham De Moivre (1667–1754) gave a surprisingly simple and computationally appealing closed-form expression for the MD of a binomial distribution (which is given in p. 201). This is perhaps the very first published work on MD. This was followed by several interesting investigations, which are given in the summary section (p. 201). Johnson [125] surmised that the MD of some discrete distributions can be put in the form $2\mu_2 f_m$, where $\mu_2 = \sigma^2$ and f_m is the probability mass evaluated at the integer part of the mean $m = \lfloor \mu \rfloor$. This holds good for Poisson, binomial, negative binomial, and geometric distributions. Kamat [126] generalized Johnson's result to several discrete distributions.

The following theorem greatly simplifies the work and is very helpful to find the MD of a variety of discrete distributions. It can easily be extended to the multivariate case and for other types of mean deviations such as mean deviation from the median and medoid.

Theorem 6.1 The MD of any discrete distribution that tails off to the left is expressed in terms of the CDF as

$$MD = 2 \sum_{x=ll}^{\mu-1} F(x), \tag{6.6}$$

where ll is the lower limit of the distribution, μ the arithmetic mean, and $F(x)$ the CDF.

Proof: By definition

$$E|X - \mu| = \sum_{x=ll}^{ul} |x - \mu| p(x), \tag{6.7}$$

where ll is the lower and ul the upper limit of the distribution.   ∎

Split the range of summation from ll to $\mu - 1$ and μ to ul and note that $|X - \mu| = \mu - X$ for $x < \mu$. This gives

$$E|X - \mu| = \sum_{x=ll}^{\mu-1} (\mu - x)p(x) + \sum_{x=\mu}^{ul} (x - \mu)p(x). \qquad (6.8)$$

As $E(X) = \mu$, we can write $E(X - \mu) = 0$, where $E()$ is the expectation operator. Expanding $E(X - \mu)$ as

$$E(X - \mu) = \sum_{x=ll}^{ul} (x - \mu)p(x) = 0. \qquad (6.9)$$

As done earlier, split the range of summation from ll to $\mu - 1$ and μ to ul to get

$$E(X - \mu) = \sum_{x=ll}^{\mu-1} (x - \mu)p(x) + \sum_{x=\mu}^{ul} (x - \mu)p(x) = 0. \qquad (6.10)$$

Substitute $\sum_{x=\mu}^{ul} (x - \mu)p(x) = -\sum_{x=ll}^{\mu-1} (x - \mu)p(x)$ in (6.8) to get

$$E|X - \mu| = \sum_{x=ll}^{\mu-1} (\mu - x)p(x) - \sum_{x=ll}^{\mu-1} (x - \mu)p(x) = 2 \sum_{x=ll}^{\mu-1} (\mu - x)p(x).$$

Split this into two sums to get

$$E|X - \mu| = 2 \left(\mu F(\mu - 1) - \sum_{x=ll}^{\mu-1} xp(x) \right). \qquad (6.11)$$

As the MD is always positive, the first term in (6.11) is greater than the second for positive random variables.

Expand the summation inside the bracket in reverse order of indexvar as

$$\sum_{x=ll}^{\mu-1} xp(x) = (\mu - 1) * p(\mu - 1) + (\mu - 2) * p(\mu - 2) + \cdots + ll * p(ll). \qquad (6.12)$$

Collect the first term from each expression on the RHS to get

$$\sum_{x=ll}^{\mu-1} xp(x) = \mu * F(\mu - 1) - \sum_{k=ll}^{\mu-1} (\mu - k) * p(k), \qquad (6.13)$$

where $F(\mu - 1) = p(\mu - 1) + p(\mu - 2) + \cdots + p(ll)$ so that both partial expectations are bounded, for finite μ. Now substitute in (6.11). The $\mu F(\mu - 1)$ term cancels out, leaving behind

$$E|X - \mu| = 2 \left(\sum_{k=0}^{\mu-1} (\mu - k) * p(k) \right), \qquad (6.14)$$

which is same expression obtained above.

Write (6.14) as two summations

$$E|X - \mu| = 2 \left(\sum_{x=ll}^{\mu-1} \sum_{i=ll}^{x} p(i) \right) \tag{6.15}$$

and substitute $\sum_{i=ll}^{x} p(i) = F(x)$ to get the final result as

$$\text{MD} = 2 \sum_{x=ll}^{\mu-1} F(x). \tag{6.16}$$

If the mean μ is a half-integer, a correction term $F(\lfloor \mu \rfloor)$ must be added to get the correct MD. If the distribution of X is symmetric, we can write the aforementioned as

$$\text{MD} = \sum_{x=ll}^{\mu-1} F(x) + \sum_{x=\mu}^{ul} S(x), \tag{6.17}$$

where $S(x)$ is the survival function. If the distribution tails off to the right extreme, the aforementioned is evaluated as

$$\text{MD} = 2 \sum_{x=\mu}^{ul} S(x). \tag{6.18}$$

If the mean μ is neither an integer nor a half-integer, the summation is carried out to the nearest integer. In this case, the results are only approximate (see Example 6.40 in p. 6–78). Nevertheless, the above-mentioned theorem is of enormous use, as it can be easily extended to find the MD of bivariate and multivariate discrete distributions. There are two other novel methods to find the mean deviation. The first one uses generating functions (Chapter 9, Section 9.4, p. 9–11) to fetch a single coefficient of $t^{\mu-1}$ in the power series expansion of $(1 - t)^{-2} P_x(t)$, where $P_x(t)$ is the probability generating function. This works best for discrete distributions. The second method is using the inverse of distribution functions (Chapter 10, Section 10.10, p. 10–9), the discrete analog of which is obtained by replacing integration by summation.

▮ EXAMPLE 6.1   Variance of discrete distribution as tail probability

Prove that the variance of discrete distributions can be expressed in terms of tail probabilities when the mean is an integer or a half-integer.

Solution 6.1 We know that the MD is an L_1-norm and σ^2 is an L_2-norm. We found earlier that MD $= 2 \sum_{x=ll}^{\mu-1} F(x) = 2 \sum_{x=\mu}^{ul} S(x)$. Equating Johnson's result that MD $= 2\mu_2 f_m$, where $\mu_2 = \sigma^2$ and f_m is the probability mass evaluated at the integer part of the mean $m = \lfloor \mu \rfloor$ we get $\mu_2 * f_m = \sum_{x=ll}^{\mu-1} F(x)$. Divide both sides by f_m to get

$$\sigma^2 = (1/f_m) \sum_{x=ll}^{\mu-1} F(x) = (1/f_m) \sum_{x=\mu}^{ul} S(x). \tag{6.19}$$

TABLE 6.1   Mean Deviation of Binomial Distribution Using Our Power Method (equation 6.8) for np a Half-integer

n, p (np)	0	1	2	3	4	eq (6.66)	Final
15, 0.3 (4.5)	0.0047	0.0353	0.1268	0.2969	0.5155	0.92742	1.442913
18, 0.25 (4.5)	0.0056	0.0395	0.1353	0.3057	0.5187	1.39742	1.91171
8.8125 (6.5)	0.0009	0.0078	0.0455	0.1762	0.4594	0.46098	0.920416

First column gives n, p (np) values of binomial. Second column onward are the values computed using (6.84). Seventh column gives the uncorrected MD using equation (6.6). Correction term $F(\lfloor np \rfloor)$ is added to get the correct MD in the last column.

When $m = \lfloor \mu \rfloor$ is a half-integer, the correction term mentioned earlier must be applied (Table 6.1).   ∎

6.3.1   Recurrence Relation for Mean Deviation

Mean deviation of some distributions involves complicated terms. In the above-mentioned theorem, we have obtained an expression for MD in terms of CDF. It is possible to develop recurrences for MD in those situations where the CDF has closed-form expressions in terms of incomplete beta or gamma functions, normal distribution, confluent hypergeometric functions, or orthogonal polynomials. This argument applies to both discrete and continuous distributions. As examples, the CDF of binomial and negative binomial distributions are expressed in terms of incomplete beta functions. However, the beta function satisfies several recurrences like

$$a I_x(a + 1, b) = (a + b) I_x(a, b) - b I_x(a, b + 1). \tag{6.20}$$

Equation (6.20) allows one to successively reduce the first argument of beta function, which in turn results in a recurrence for the MD. Similarly

$$I_x(a, b) = x I_x(a - 1, b) + (1 - x) I_x(a, b - 1), \tag{6.21}$$

and

$$I_x(a + 1, b - 1) = [1 + bx/(a(1 - x))] I_x(a, b) - bx/[a(1 - x)] I_x(a - 1, b + 1). \tag{6.22}$$

Equation (6.21) allows one to successively reduce both parameters, which results in another recurrence for MD.

6.4   BERNOULLI DISTRIBUTION

The Bernoulli distribution results from a random experiment in which each outcome is either a success (denoted by 1) with probability p or a failure (denoted by 0) with a probability q so that $p + q = 1$. This means that fixing the value of p automatically fixes the value of q. A question that naturally arises is what should be chosen as p?. This is not an issue because p and q are simply place holders for probabilities. It depends more on the research hypothesis. Here, the meaning of the word success and failure should not be taken literally – it simply means two dichotomous outcomes of an experiment. In engineering, it can denote faulty or nonfaulty, working or defunct, closed or open (as in electrical circuits), and detected or undetected (radioactivity, smoke, abnormality, etc.). If we wish to check if something is faulty, we choose p as the probability of a fault. In medical sciences, p is chosen as the probability of the presence of a symptom or condition. Thus, this distribution finds applications in a variety of fields. It is named after the Swiss mathematician Jacques Bernoulli (1654–1705). Such an experiment is known as a *Bernoulli trial*. The probability density function (PDF) of a Bernoulli random variable is given by $f(x;p) = p^x q^{(1-x)}$, $x = 0$ or 1, and $0 \leq q = 1 - p \leq 1$. We will denote the Bernoulli distribution by BER(p). This could also be expressed in the following convenient form:

$$f(x;p) = \begin{cases} p^x q^{1-x} & \text{for } x = 0, 1; \\ 0 & \text{otherwise.} \end{cases}$$

The mean and variance of a Bernoulli distribution are $\mu = p, \sigma^2 = pq$. As the only values of x are 0 and 1, we get the mean as $E(X) = 0^* q + 1^* p = p$. Similarly, $E(X^2) = 0^2 * q + 1^2 * p = p$, so that the variance becomes $E(X^2) - E(X)^2 = p - p^2 = p(1 - p) = pq$. $\beta_1 = (1 - 2p)/\sqrt{(pq)}, \beta_2 = 3 + (1 - 6p)/pq$.

Bernoulli distribution has only one unknown parameter p. This unknown probability is usually estimated either from past experiments or from empirical studies. If we observe k successes in n Bernoulli trials, an estimate of p is obtained as $p = k/n$. The probability generating function is easily obtained as $P_X(t) = (q + pt)$, and characteristic function is $\phi(t) = q + pe^{it}$. Hence, all moments about zero are p.

There are many other probability distributions based on Bernoulli trials. For example, the binomial, negative binomial, and geometric distributions mentioned earlier; success-run distributions are all defined in terms of independent Bernoulli trials.

■ EXAMPLE 6.2 CDF of Bernoulli distribution

Suppose p denotes the probability of a trait in a group of persons. Define a random variable X that takes the value 1 if trait is present and is 0 otherwise. Find the PDF and CDF of X.

Solution 6.2 Assign a random variable X to the two possible outcomes as $P(\text{trait}) = p$ and $P(\text{trait not present}) = q = 1 - p$. The PDF is expressed as

$$f(x) = \begin{cases} q = 1 - p & \text{if } x = 0 \text{ (trait not present)} \\ p & \text{if } x = 1 \text{ (trait).} \end{cases}$$

TABLE 6.2 Properties of Bernoulli Distribution

Property	Expression	Comments
Range of X	$x = 0, 1$	Discrete, finite
Mean	$\mu = p$	
Variance	$\sigma^2 = pq$	$\Rightarrow \mu > \sigma^2$
Skewness	$\gamma_1 = (1 - 2p)/\sqrt{pq}$	$= (q - p)/\sqrt{pq}$
Kurtosis	$\beta_2 = 3 + (1 - 6pq)/pq$	
Mean deviation	$2pq$	
Median Moments	$\mu'_r = p$	
MGF	$(q + pe^t) = 1 + p(e^t - 1)$	$= 0.5(1 + e^t)$ if $p = q$
PGF	$(q + pt) = 1 + p(t - 1)$	$= 0.5(1 + t)$ if $p = q$
Additivity	$\sum_{i=1}^{n} \text{BER}(p) = \text{BINO}(n, p)$	Independent

[a]Bernoulli distribution is the building block of binomial, geometric, negative binomial, and success-run distributions.

As there are only two possible values, the CDF is obtained as

$$F(x) = \begin{cases} q = 1 - p & \text{if } x = 0 \quad (\text{trait not present}) \\ 1 & \text{if } x = 1 \quad (\text{trait}). \end{cases}$$

The $p = 0$ or $p = 1$ cases are called degenerate cases, as there is no randomness involved. See Table 6.2 for summary of properties. ∎

◩ EXAMPLE 6.3 Product of two Bernoulli random variables

If X and Y are IID BER(p), find the distribution of $U = X * Y$

Solution 6.3 X and Y both takes the values 1 with probability p, and 0 with probability $q = 1 - p$. Hence, XY takes the value 0 when either or both of X and Y take the value 0 with probability $q^2 + qp + pq$. Here, q^2 is the probability that both of them takes the value 0, and qp and pq are the probabilities that either of them takes value 0 and other takes value 1. Write $q^2 + qp = q[q + p]$ and use $q + p = 1$ to get q. Next combine q with pq to get $q + pq = q(1 + p)$. Write q as $(1 - p)$ to get $(1 - p) * (1 + p) = 1 - p^2$. Probability that XY takes the value 1 is p^2. Hence, XY is Bernoulli with probability of success p^2. This can be extended to the case $\prod_{i=1}^{n}(1 + kX_i)$, where k is a constant and X_i are iid BER(p) (see Exercise 6.13). ∎

6.5 BINOMIAL DISTRIBUTION

Binomial distribution is a natural extension of Bernoulli distribution for two or more independent trials ($n > 1$). It was first derived in its present form by the Swiss mathematician Jacques Bernoulli (1654–1705), which was published posthumously in 1713 [127], although the binomial expansion (for arbitrary n) was studied by Blaise Pascal [128]. It can be interpreted in terms of random trials or in terms of random variables. Consider n independent Bernoulli trials. We assume that the trials have already occurred. We are interested in knowing how many successes have taken place among the n trials. This number is any integer from 0 to n inclusive. Assuming that there are x successes, this can happen at any of the n positions in $\binom{n}{x}$ ways. Since the probability of success remains the same from trial to trial, the probability of x successes and $n - x$ failures is given by

$$
f(x; n, p) = \begin{cases} \binom{n}{x} p^x q^{n-x} & 0 \le q = 1 - p \le 1 \text{ if } x = 0, 1, 2, \ldots, n \\ 0 & \text{elsewhere.} \end{cases}
$$

It is called binomial distribution because it is the xth term in the binomial expansion of $(p + q)^n$. It belongs to the exponential family.

The random variable interpretation of binomial distribution is based on independent Bernoulli trials. Let $X_1, X_2, \ldots, X_n$ be a sequence of IID Bernoulli random variables with the same parameter p. Then, the sum $X = X_1 + X_2 + \cdots + X_n$ has a binomial distribution with parameters n and p. We denote this as BINO(n, p).

6.5.1 Properties of Binomial Distribution

There are two parameters for this distribution (see Figure 6.1 and Table 6.3), namely the number of trials ($n > 1$; an integer), and the probability of success in each trial

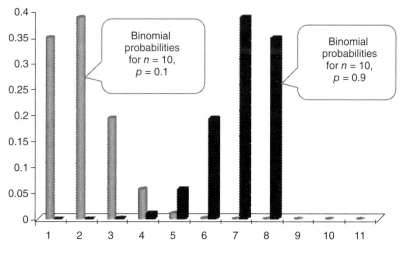

Figure 6.1 BINO(10,0.1) and BINO(10,0.9).

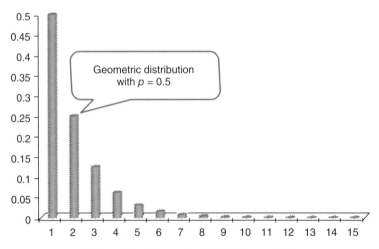

Figure 6.2   Geometric distribution $p = 0.5$.

(p), which is a real number between 0 and 1. This probability remains the same from trial to trial, which are independent. This distribution is encountered in sampling with replacement from large populations. If p denotes the probability of observing some characteristic (there are x individuals that have the characteristic in the population, so that $p = x/N$ where N is the population size), the number of individuals in a sample of size n from that population has the characteristic is given by a binomial distribution BINO(N, p).

6.5.1.1   Moments   As the trials are independent, and $X = X_1 + X_2 + \ldots + X_n$, the mean is $E(X) = p + p + .. + p = np$. Similarly, the variance of X is $V(X) = V(X_1) + V(X_2) + \ldots + V(X_n) = pq + pq + \ldots + pq = npq$.   Hence $\mu_1 = E(X) = np$, $\mathrm{Var}(X) = npq = \mu_1 * q$. Note that when $p \to 0$ from above, $q \to 1$ from below and the variance $\to \mu_1$. This results in a distribution with the same mean and variance (Table 6.3). If $np = \mu$ is a constant, we could reparametrize the binomial distribution by putting $\mu = np$ to get

$$f(x; n, \mu) = \binom{n}{x} (\mu/n)^x((n - \mu)/n)^{(n-x)} \tag{6.23}$$

TABLE 6.3   Binomial Probabilities Example

	0	1	2	3	4	5	sum
$p\backslash$	q^n	$\binom{n}{1}pq^{n-1}$	$\binom{n}{2}p^2q^{n-2}$	$\binom{n}{3}p^3q^{n-3}$	$\binom{n}{4}p^4q^{n-4}$	$\binom{n}{5}p^5q^0$	
0.1	0.59049	0.32805	0.0729	0.0081	0.00045	0.00001	1.0
0.5	0.03125	0.15625	0.3125	0.3125	0.15625	0.03125	1.0
0.9	1E-05	0.00045	0.0081	0.0729	0.32805	0.59049	1.0

with mean μ and variance $\mu(n - \mu)/n$. The symmetry of variance for fixed n indicates that the variance of BINO(n, p) and BINO(n, q) are the same.

◼ EXAMPLE 6.4   Maximum variance of a binomial distribution

Prove that the maximum variance of a binomial distribution BINO(n, p) as a function of p is $n/4$.

Solution 6.4 We know that the variance is given by $V(n, p) = npq = np - np^2$. Differentiating with respect to p, we get $\frac{\partial}{\partial p}V(n, p) = n - 2np$. Equating to zero and solving for p gives $p = 1/2$. As the second derivative is $-2n$, this indeed gives the maxima. Substitute in the above to get $V(n, p) = n*(1/2)*(1/2) = n/4$. This can be increased without limit by letting $n \to \infty$ (see discussion in page 6–54). ◼

The ratio of two probabilities of a discrete distribution at distinct ordinal (x) values provides the relative likelihood of the random variable taking a value at one level versus the other. This is useful to calculate the probabilities recursively, locate the mode, and develop moment recurrences [129].

◼ EXAMPLE 6.5   Mode of binomial distribution

Find the mode of the binomial distribution.

Solution 6.5 Consider the ratio

$$f(x; n, p)/f(x - 1; n, p) = (n - x + 1)p/(xq). \qquad (6.24)$$

Add and subtract xq in the numerator and combine $-xp - xq$ to get $-x$. Then the ratio becomes $1 + [(n + 1)p - x]/(xq)$. The bracketed expression in the numerator is positive when $(n + 1)p > x$ and negative otherwise. If $(n + 1)p = x$, the bracketed expression vanishes giving two modes at $x - 1$ and x. Otherwise, the mode is $\lfloor(n + 1)p\rfloor$. This shows that the binomial distribution is unimodal for all values of p and n, except when $(n + 1)p$ is an integer. ◼

6.5.2   Moment Recurrences

Low order moments could be obtained using the density recurrences $f_x(n, p)/f_x(n - 1, p) = (n/(n - x))q$, which upon cross multiplication and rearrangement becomes

$$xf_x(n, p) = n[f_x(n, p) - qf_x(n - 1, p)]. \qquad (6.25)$$

Multiply both sides of (6.25) by x^k, denote the moments by $\mu_k(n, p)$, and sum over the range of x to get

$$\mu_{k+1}(n, p) = n[\mu_k(n, p) - q * \mu_k(n - 1, p)] \text{ with } \mu_0(n, p) = 1. \qquad (6.26)$$

Put $k = 1$, to get $\mu_2(n, p) = n[\mu_1(n, p) - q\mu_1(n - 1, p)]$. Substituting $\mu_1(n, p) = np, \mu_1(n - 1, p) = (n - 1)p$, the RHS becomes $n[np - q(n - 1)p] = np[np + q]$. Higher order moments are obtained similarly.

■ EXAMPLE 6.6   Binomial ordinary moment recurrence

Prove that the ordinary moments of a binomial distribution satisfy the recurrence

$$\mu'_{k+1} = p\left(q\,\frac{\partial}{\partial p}\mu'_k + n\mu'_k\right). \tag{6.27}$$

Solution 6.6 Write

$$\mu'_k = \sum_{x=0}^{n} x^k\binom{n}{x}p^x(1 - p)^{n-x}. \tag{6.28}$$

Differentiate (6.28) by p to get

$$\frac{\partial}{\partial p}\mu'_k = \sum_{x=0}^{n} x^k\binom{n}{x}[(1 - p)^{n-x}xp^{x-1} - p^x(n - x)(1 - p)^{n-x-1}]. \tag{6.29}$$

Write $q = 1 - p$ and consider the terms $q^{n-x}xp^{x-1} + xp^xq^{n-x-1} = xp^{x-1}q^{n-x-1}$ $[q + p] = xp^{x-1}q^{n-x-1}$. Substitute in equation (6.29) and simplify to get

$$\frac{\partial}{\partial p}\mu'_k = \sum_{x=0}^{n} x^{k+1}\binom{n}{x}[q^{n-x-1}p^{x-1} - n\sum_{x=0}^{n} x^k\binom{n}{x}p^xq^{n-x-1}]$$

$$= \mu'_{k+1}/(pq) - (n/q)\mu'_k. \tag{6.30}$$

Multiply throughout by pq and rearrange to get $\mu'_{k+1} = p\left(n\mu'_k + q\,\frac{\partial}{\partial p}\mu'_k\right)$ ■

■ EXAMPLE 6.7   Binomial central moment recurrence

Prove that the central moments of a binomial distribution satisfy the recurrence

$$\mu_{k+1} = pq\left(\frac{\partial}{\partial p}\mu_k + nk\mu_{k-1}\right). \tag{6.31}$$

Solution 6.7 Consider

$$\mu_k = \sum_{x=0}^{n} (x - np)^k\binom{n}{x}p^xq^{n-x}. \tag{6.32}$$

Note that there are three functions of p on the RHS (as $q = 1 - p$). Differentiate both sides with respect to p using chain rule to get

$$\frac{\partial}{\partial p}\mu_k = -nk\sum_{x=0}^{n}(x-np)^{k-1}\binom{n}{x}p^x q^{n-x} + \sum_{x=0}^{n}(x-np)^k\binom{n}{x}xp^{x-1}q^{n-x}$$

$$-\sum_{x=0}^{n}(x-np)^k\binom{n}{x}(n-x)p^x q^{n-x-1}$$

$$= -nk\mu_{k-1} + \sum_{x=0}^{n}(x-np)^k\binom{n}{x}p^{x-1}q^{n-x-1}(xq-(n-x)p). \tag{6.33}$$

Write $xq - (n-x)p$ as $x(1-p) - (n-x)p = x - np$ and combine with $(x-np)^k$; multiply both numerator and denominator by pq to get $-nk\mu_{k-1} + \mu_{k+1}/pq$. Rearrange the expression to get the result. ∎

EXAMPLE 6.8   Binomial central moment recurrence

Prove that the central moments of a binomial distribution satisfy the recurrence

$$\mu_{k+1}(n,p) = nq[\mu_k(n,p) - \sum_{j=0}^{k}\binom{k}{j}(-p)^{k-j}\mu_j(n-1,p)]. \tag{6.34}$$

Solution 6.8 Write

$$\mu_k = \sum_{x=0}^{n}(x-np)^k\binom{n}{x}p^x(1-p)^{n-x}. \tag{6.35}$$
∎

Consider equation (6.25) as $(n-x)f_x(n,p) = nqf_x(n-1,p)$. Write $(n-x)$ as $n(p+q) - x = nq - (x - np)$ on the LHS. Multiply throughout by $(x-np)^k$, write $(x-np) = [(x-(n-1)p)-p]$ on the RHS, expand using binomial theorem and sum over the proper range to get

$$nq\,\mu_k(n,p) - \mu_{k+1}(n,p) = nq\sum_{x}\sum_{j=0}^{k}\binom{k}{j}(x-(n-1)p)^j(-p)^{k-j}f_x(n-1,p)$$

$$= nq\sum_{j=0}^{k}\binom{k}{j}(-p)^{k-j}\mu_j(n-1,p). \tag{6.36}$$

Rearrange (6.36) to get

$$\mu_{k+1}(n,p) = nq[\mu_k(n,p) - \sum_{j=0}^{k}\binom{k}{j}(-p)^{k-j}\mu_j(n-1,p)] \tag{6.37}$$

with $\mu_0(n-1,p) = 1$.

Theorem 6.2 The factorial moment $\mu_{(r)}$ is given by $\mu_{(r)} = n^{(r)}p^r$.

Proof: Consider the expression for factorial moment as

$$\mu_{(r)} = E[x_{(r)}] = E[x(x-1)..(x-r+1)] = \sum_x [x(x-1)..(x-r+1)]f(x) \quad (6.38)$$

Substitute the PDF and sum over the proper range of x to get the RHS as

$$\sum_x x(x-1)..(x-r+1)n!/(x!(n-x)!)p^x q^{(n-x)}. \quad (6.39)$$

Write $n!$ in the numerator as $n(n-1)..(n-r+1) * (n-r)!$, this becomes

$$n(n-1)..(n-r+1)p^r \sum_x (n-r)!/((x-r)!(n-x)!)p^{x-r}q^{(n-x)} = n^{(r)}p^r. \quad (6.40)$$

Write

$$x(x-1)..(x-r+1)/(n-x)! = \sum_{i=0}^{r} s(r,i)x^i \quad (6.41)$$

where $s(r,i)$ is the Stirling number of first kind. The factorial moments are found using Stirling numbers as

$$\mu_{(r)} = \sum_{i=0}^{r} s(r,i)\mu_i', \quad (6.42)$$

where μ_i' denotes the ith ordinary moment. The reverse relationship is

$$x^r = \sum_{i=0}^{r} S(r,i)x!/(x-i)!, \quad (6.43)$$

where $S(r,i)$ is the Stirling number of second kind. This allows us to write

$$\mu_r' = \sum_{k=0}^{r} S(r,k)\mu_{(k)}'. \quad (6.44)$$

∎

Theorem 6.3 Prove that the mean deviation from the mean of a binomial distribution is

$$2npq \binom{n-1}{\mu-1} p^{\mu-1}q^{n-\mu}, \quad (6.45)$$

where μ is the largest integer less than np (symbolically $\mu = \lfloor np \rfloor$).

Proof: By definition,

$$MD = \sum_{x=0}^{n} |x-np| \binom{n}{x} p^x q^{n-x}. \quad (6.46)$$

Split the RHS into two sums for $x \le np$ and $x > np$, respectively:

$$\sum_{x=0}^{np}(np-x)\binom{n}{x}p^x q^{n-x} + \sum_{x=np+1}^{n}(x-np)\binom{n}{x}p^x q^{n-x}. \quad (6.47)$$

Using Theorem 2.4 in page 2–6, we have $\sum_{x=0}^{n}(np-x)\binom{n}{x}p^xq^{n-x}=0$. Substitute in the aforementioned to get

$$2\sum_{x=0}^{np}(np-x)\binom{n}{x}p^xq^{n-x}. \tag{6.48}$$

Write $(np-x)=(np-x(p+q))=p(n-x)-xq$ and split the aforementioned sum as

$$2\left(p\sum_{x=0}^{np}(n-x)\binom{n}{x}p^xq^{n-x}-q\sum_{x=0}^{np}x\binom{n}{x}p^xq^{n-x}\right). \tag{6.49}$$

Expand $\binom{n}{x}$ and cancel out common terms to get

$$MD=2\left(npq\sum_{x=0}^{np}\binom{n-1}{x}p^xq^{n-x-1}-npq\sum_{x=1}^{np}\binom{n-1}{x-1}p^{x-1}q^{n-x}\right). \tag{6.50}$$

Taking npq as common factor, we notice that the alternate terms in the left and right sums of 6.50 cancel out giving

$$MD=2npq(\binom{n-1}{np}p^{np}q^{n-np-1}) \tag{6.51}$$

when np is an integer. If np is noninteger, we carry out the summation up to $v=\lfloor np\rfloor$ so that the last term of the LHS remains (all others are canceled out). In this case, we get

$$MD=2npq(\binom{n-1}{v}p^vq^{n-1-v}), \tag{6.52}$$

where $v=\lfloor np\rfloor$ is the greatest integer $\leq np$. This is the form obtained originally by De Moivre and subsequently by Bertrand [131]. Johnson [125], Diaconis and Zabell [132] among others have discussed other equivalent forms and approximations. In Section 6.5.6, page 208, we provide a new method to find the MD using the Power method introduced in Section 6.3, page 189. ∎

6.5.2.1 *Generating Functions*   Generating functions are extensively discussed in Chapter 9. Here we give the main results, which are proved in that chapter.

Theorem 6.4 The probability generating function is $P_X(t)=(q+pt)^n$, and the moment generating function is $M_X(t)=(q+pe^t)^n$.

Proof: The pgf is

$$E(t^x)=\sum_{x=0}^{n}t^x\binom{n}{x}p^xq^{n-x}=\sum_{x=0}^{n}\binom{n}{x}(pt)^xq^{n-x}=(q+pt)^n, \tag{6.53}$$

where $E()$ is the expectation operator. The mgf is found similarly by replacing t^x by e^{tx}. See also Chapter 9, page 379. ∎

TABLE 6.4   Properties of Binomial Distribution

Property	Expression	Comments
Range of X	$x = 0, 1, .., n$	discrete, finite
Mean	$\mu = np$	need not be integer
Variance	$\sigma^2 = npq = \mu q$	$\Rightarrow \mu > \sigma^2$
CV	$q/(np)^{1/2}$	
Mode	$(x-1), x$ if $(n+1) * p$ is not integer	$x = (n+1) * p$ else
Skewness	$\gamma_1 = (1-2p)/\sqrt{npq}$	$= (q-p)/\sqrt{npq}$
Kurtosis	$\beta_2 = 3 + (1-6pq)/npq$	
Mean deviation	$2npq(\binom{n-1}{\lfloor np \rfloor} p^{\lfloor np \rfloor} q^{n-1-\lfloor np \rfloor})$	$2 \sum_{x=0}^{\lfloor np-1 \rfloor} I_{1-p}(n-x, x+1)$
$E[X(X-1)\cdots$ $(X-k+1)]$	$n^{(r)} p^r$	
MGF	$(q + pe^t)^n$	$= p^n(1+e^t)^n$ if $p = q$
PGF	$(q + pt)^n$	$= p^n(1+t)^n$ if $p = q$
Additivity	$\sum_{i=1}^{m} B(n_i, p) = B(\sum_{i=1}^{m} n_i, p)$	independent
Recurrence	$f(x; n, p)/f(x-1; n, p) =$ $(n-x+1)p/[xq]$	$1 + ((n+1)p-x)/(xq)$
Tail probability	$\sum_{x=k}^{n} \binom{n}{x} p^x q^{n-x} = I_p(k, n-k+1)$	$I =$ Incomplete beta

Symmetric when $p = q = 1/2$.

6.5.3   Additivity Property

If $X_1 \sim$ BINO(n_1, p) and $X_2 \sim$ BINO(n_2, p) are independent binomial random variables with the same probability of success p, the sum $X = X_1 + X_2$ is distributed as BINO$(n_1 + n_2, p)$ (Table 6.4).

Proof: The easiest way to prove the above-mentioned result is using the MGF. As X_1 and X_2 are independent, $M_{X_1+X_2}(t) = M_{X_1}(t) * M_{X_2}(t)$. Substituting $M_X(t) = (q + pe^t)^n$, the RHS becomes $(q + pe^t)^{n_1} * (q + pe^t)^{n_2} = (q + pe^t)^{n_1+n_2}$, which is the mgf of BINO$(n_1 + n_2, p)$. An interpretation of this result in terms of Bernoulli trials is the following – "if there are n_1 independent Bernoulli trials with the same probability of success p and another n_2 independent Bernoulli trials with the same probability of success, they can be combined in any desired order to produce a binomial distribution of size $n_1 + n_2$."   ∎

Another way of stating the above-mentioned theorem is that if $X + Y$ is distributed as BINO$(n_1 + n_2, p)$, and either of X or of Y is distributed as BINO(n_1, p), the other random variable must be BINO(n_2, p) (or BER(p) is $n_2 = 1$). This result can be extended to any number of independent binomial distributions with the same probability of success. Symbolically, if $X_i \sim$ BINO(n_i, p), then $\sum_i X_i \sim$ BINO$(\sum_i n_i, p)$.

▌ **EXAMPLE 6.9   Distribution of $Y = n - X$**

If X has a binomial distribution with parameters n and p, derive the distribution of $Y = n - X$ and obtain its PGF and MGF. Obtain the mean and variance. What is the additive property for Y?

Solution 6.9 As X takes the values $0, 1, \ldots, n$; Y also takes the same values in reverse. Thus, the range of X and Y are the same.

$$P[Y = y] = P[X = n - y] = \binom{n}{n-y} p^{n-y} q^{n-(n-y)} = \binom{n}{n-y} p^{n-y} q^y. \qquad (6.54)$$

Using $\binom{n}{n-y} = \binom{n}{y}$, this becomes $\binom{n}{y} q^y p^{n-y}$. This is the PDF of a binomial distribution with p and q reversed. Hence $Y = n - X \sim$ BINO(n, q), so that all properties are obtained by swapping the roles of p and q in the corresponding property of BINO(n,p). The PGF is

$$P_Y(t) = E(t^y) = \sum_{y=0}^{n} t^y \binom{n}{y} q^y p^{n-y} = \sum_{y=0}^{n} \binom{n}{y} (qt)^y p^{n-y} = (qt + p)^n = (p + qt)^n.$$

Similarly, the MGF is $M_Y(t) = E(e^{ty}) = (p + qe^t)^n$. From this, the cumulant generating function follows as $n * \ln(q + pe^t)$. The mean and variance are nq and npq, respectively. This shows that X and Y have the same variance, but the mean is $nq = n(1 - p) = n - np$. As it is binomial distributed, the additive property remains the same. This means that if Y and Z are BINO(n_1, q) and BINO(n_2, q) random variates, then $Y + Z$ is distributed as BINO$(n_1 + n_2, q)$, provided q (or equivalently $p = 1 - q$) is the same.   ■

6.5.4   Distribution of the Difference of Successes and Failures

The number of successes and failures in a binomial distribution is related through n. If there are x successes, there exist $n - x$ failures and vice versa. In other words, they must add up to the total number of trials. The following example derives the distribution of the absolute difference of them.

▌ **EXAMPLE 6.10   Distribution of $U = |X - Y|/2$.**

Let X denotes the number of successes (or Heads) and Y denotes the number of failures (or Tails) in n independent Bernoulli trials with the same probability of success p. Find the distribution of $U = |X - Y|/2$ for n even.

Solution 6.10 Obviously, U takes the values $(0, 1, \ldots, n/2)$. As n is even, it can take the value 0 in just one way – when both X and Y are $n/2$. The probability of this case is $\binom{n}{\frac{n}{2}} p^{n/2} q^{n-n/2} = \binom{n}{\frac{n}{2}} (pq)^{n/2}$. There exist two ways in which all other values are materialized. First consider the number of successes exceeding the number of failures by x. Let t be the number of failures (so that the number of successes is $t + x$). Then $t = (n - x)/2$, and $t + x = (n + x)/2$. Probability of this happening is

$$f_u(x; n, p) = \binom{n}{\frac{n+x}{2}} p^{\frac{n+x}{2}} q^{\frac{n-x}{2}} = (pq)^{n/2} \binom{n}{\frac{n+x}{2}} (p/q)^{x/2}, \qquad (6.55)$$

for $x = 2, 4, \ldots, n$. Next consider the number of failures exceeding the number of successes by x. Let t be the number of failures. Then $t = (n + x)/2$ and $t - x = (n - x)/2$. Probability of this happening is

$$f_u(x; n, p) = \binom{n}{\frac{n-x}{2}} p^{\frac{n-x}{2}} q^{\frac{n+x}{2}}. \qquad (6.56)$$ ∎

Using $\binom{n}{x} = \binom{n}{n-x}$, this becomes

$$f_u(x; n, p) = \binom{n}{\frac{n+x}{2}} p^{\frac{n-x}{2}} q^{\frac{n+x}{2}} = (pq)^{n/2} \binom{n}{\frac{n+x}{2}} (q/p)^{x/2}, \qquad (6.57)$$

for $x = 2, 4, \ldots, n$. Adding (6.55) and (6.57) gives the probability of U assuming the value u as

$$f_U(u; n, p) = \begin{cases} \binom{n}{n/2+u} (pq)^{\frac{n}{2}} [(p/q)^u + (q/p)^u] & \text{for } u = (1, \ldots, n/2); \\ \binom{n}{\frac{n}{2}} (pq)^{n/2} & \text{for } u = 0; \\ 0 & \text{otherwise.} \end{cases}$$

Putting $t = n/2 + u$, this can also be written as

$$f_T(t; n, p) = \begin{cases} \binom{n}{t} [q^n(p/q)^t + p^n(q/p)^t] & \text{for } t = (n/2 + 1, \ldots, n); \\ \binom{n}{\frac{n}{2}} (pq)^{n/2} & \text{for } t = n/2; \\ 0 & \text{otherwise.} \end{cases}$$

Take q^n as a common factor and write $(q/p)^t = (p/q)^{-t}$ to get the alternate form

$$f_T(t; n, p) = q^n \binom{n}{t} [(p/q)^t + (p/q)^{n-t}]. \qquad (6.58)$$

TABLE 6.5 Properties of Discrete Uniform Distribution

Property	Expression	Comments
Range of X	$x = 1, .., N$	Discrete, finite
Mean	$\mu = (N+1)/2$	Need not be integer
Variance	$\sigma^2 = (N^2-1)/12 = \mu * (\mu-1)/3$	$\Rightarrow \mu > \sigma^2$ for $N < 7$
Mode	any x	
Skewness	$\gamma_1 = 0$	Special symmetry
Kurtosis	$\beta_2 = \dfrac{3}{5}(3 - \dfrac{4}{N^2-1})$	
CV	$\{(N-1)/[3(N+1)]\}^{1/2}$	
MD	$(N-1)(N+1)/4N$	$(N^2-1)/4N$
Moments	$\mu'_r = (1^r + 2^r + \cdots + N^r)/N$	Bernoulli numbers
MGF	$e^t(1-e^{Nt})/[N(1-e^t)]$	
$\phi_x(t)$	$(1-e^{itN})/[N(e^{-it}-1)]$	
PGF	$[t(1-t^N)]/[N(1-t)]$	
Recurrence	$f(x; N)/f(x-1; N) = 1$	$f(x) = f(x-1)$
Tail probability	$\displaystyle\sum_{x=k}^{N} 1/N = (N-k+1)/N$	

Truncation results in the same distribution with higher probability for each x value.

TABLE 6.6 Distribution of $U = |X - Y|/2$ for n Even

$x\backslash(n,p)$	(6,0.2)	(10,0.6)	(20,0.3)	(20,0.9)	(20,0.5)
0	0.0819	0.2007	0.0308	0.0000	0.17620
1	0.2611	0.3623	0.0774	0.0001	0.32040
2	0.3948	0.2575	0.1183	0.0004	0.24030
3	0.2622	0.1315	0.1653	0.0020	0.14790
4		0.0419	0.1919	0.0089	0.07390
5		0.0062	0.1789	0.0319	0.02960
6			0.1304	0.0898	0.00920
7			0.0716	0.1901	0.00220
8			0.0278	0.2852	0.00040
9			0.0068	0.2702	0.000038
10			0.0008	0.1216	0.000002
SUM	1.0000	1.0000	1.0000	1.0000	1.0000

[a]Second column onward gives values of n and p. As u varies between 0 and $n/2$, there are $n/2 + 1$ values in each column.

For $p = q$, this simply becomes $\binom{n}{t}/2^{n-1}$. A similar result could be derived when n is odd. See Table 6.5 for some sample values, and Exercise 6.11 (p. 247).

There is another way to derive the above-mentioned distribution using a result in Chapter 10 (p. 401). Let there be x successes and $(n - x)$ failures in n trials (Table 6.6). Then $S - F = x - (n - x) = 2x - n = y$(say), where S denotes the successes and F

denotes the failures. Clearly, y takes the values $-n, -n+2, \ldots, 0, \ldots, n-2, n$.

$$P(Y = y) = P(2x - n = y) = P(x = (n+y)/2) = \binom{n}{(n+y)/2} p^{(n+y)/2} q^{(n-y)/2}.$$

$$(6.59)$$

The distribution of $|Y|$ is given in Section 10.4.1 as $f(y) + f(-y)$. Put $y = -y$ in equation (6.59) and add to get

$$f(y) = \binom{n}{(n+y)/2} p^{(n+y)/2} q^{(n-y)/2} + \binom{n}{(n-y)/2} p^{(n-y)/2} q^{(n+y)/2}.$$

$$(6.60)$$

Use $\binom{n}{x} = \binom{n}{n-x}$ and take common factors outside to get the above-mentioned form (6.57).

6.5.5 Algorithm for Binomial Distribution

Successive probabilities of the binomial distribution are found using the recurrence relationship $f(x)/f(x-1) = ((n-x+1)/x)p/q$ with starting value $f(0) = \binom{n}{0} q^n = q^n$. This could also be written as

$$f(x)/f(x-1) = [(n+1)/x - 1]p/q \text{ or as } 1 + [(n+1)p - x]/(qx), \qquad (6.61)$$

where the last expression is obtained by adding and subtracting $1/q$, writing $-1/q$ as $-x/qx$ and using $(-p/q + 1/q) = 1$. When n is very large and $x > n/2$, we could reduce the number of iterations by starting with $f(n) = \binom{n}{n} p^n = p^n$ and recurring backward using the relationship $f(x-1) = \frac{q}{p} \frac{x}{(n-x+1)} f(x)$.

■ EXAMPLE 6.11   Winning group

A class has b boys and g girls, both ≥ 2. A competition is conducted between the boys (who form group G1) and the girls (who form group G2), where each competition is independent of others and it is between the groups. If there are n prizes to be distributed to the winning groups, find (i) probability that girls bag more prizes than boys, (ii) number of prizes bagged by boys is odd, (iii) number of prizes bagged by girls is even number, and (iv) boys get no prizes.

Solution 6.11 As there are b boys and g girls, the proportion of boys is $b/(b+g)$ and that of girls is $g/(b+g)$. As there are n prizes, the distribution of the prizes in favor of the boys is a BINO$(n, b/(b+g))$, where we have assumed that a "success" corresponds to the boys winning a prize. We assume that this probability remains the same because the prizes are distributed to the groups independently.
   (i) Probability that girls bag more prizes than boys = Probability that boys get less prizes than girls = $\Pr[x < n-x]$ = $\Pr[2x < n]$ = $\Pr[x < n/2]$ = $\sum_{i=0}^{\lfloor n/2 \rfloor} \binom{n}{i} p^i q^{n-i}$, where $p = b/(b+g)$ and the summation is from 0 to

$(n-1)/2$ if n is odd and to $n/2$ if n is even. (ii) The number of prizes bagged by boys is odd $= \Pr[x = 1,3,\dots] = \sum\limits_{x \text{ odd}} \binom{n}{x} p^x q^{n-x}$. To evaluate this sum, consider the expression $(p+q)^n - (p-q)^n$. Expanding using binomial theorem and canceling out all even terms, we get

$$(p+q)^n - (p-q)^n = 2\left[\sum_{x \text{ odd}} \binom{n}{x} p^x q^{n-x}\right]. \tag{6.62}$$

Hence, the required probability is $\frac{1}{2}[(p+q)^n - (p-q)^n]$. However, $p+q = 1$ and $p - q = b/(b+g) - g/(b+g) = (b-g)/(b+g)$. Substitute in the afore-mentioned to get the required probability as $\frac{1}{2}[1 - [(b-g)/(b+g)]^n]$. When the number of boys is less than that of girls, the second term can be negative for odd n. (iii) Number of prizes bagged by girls is even $= \left[\sum\limits_{x \text{ even}} \binom{n}{x} q^x p^{n-x}\right]$, where we have swapped the roles of p and q. To evaluate this sum, consider the expression $(p+q)^n + (q-p)^n$. Expanding using binomial theorem, all odd terms cancel out giving $2\left[\sum\limits_{x \text{ even}} \binom{n}{x} q^x p^{n-x}\right]$. Hence, the required probability is $\frac{1}{2}[1 + [(g-b)/(b+g)]^n]$. (iv) Probability that boys get no prizes $= q^n = [g/(b+g)]^n$. ∎

EXAMPLE 6.12   Rolling a die

Consider rolling a die 20 times. What is the probability of getting at least 10 sixes?

Solution 6.12 The probability p of getting a six on any roll is $1/6$, and the count X of sixes has a $B(20, 1/6)$ distribution. Hence, the required probability is obtained by summing the individual probabilities as $\sum_{x=10}^{20} \binom{20}{x} (1/6)^x (5/6)^{20-x}$. ∎

6.5.6   Tail Probabilities

The CDF of a binomial distribution $\text{BINO}(n, p)$ is $F_x(n, p) = \sum_{k=0}^{x} \binom{n}{k} p^k q^{(n-k)}$. We could compute this by the straightforward method of adding the successive probabilities. However, for large n and k, this method is very inefficient. A better approach is to use the relationship between the binomial distribution and the incomplete beta function as follows.

$$F_x(k; n, p) = P[X \le k] = \sum_{x=0}^{k} \binom{n}{x} p^x q^{n-x} = I_{1-p}(n-k, k+1). \tag{6.63}$$

(see Chapter 7, Section 7.6, p. 7–36). The LHS of equation (6.63) is a discrete sum, whereas the RHS is a continuous function of p. When $k > np/2$, this is computed as

$F_x(k) = 1 - I_p(k + 1, n - k)$. Alternatively, the tail probabilities (SF) can be expressed as

$$\sum_{x=k}^{n} \binom{n}{x} p^x q^{n-x} = \frac{n\,!}{(k-1)!(n-k)!} \int_{y=0}^{p} y^{k-1}(1-y)^{n-k} dy. \qquad (6.64)$$

Replacing the factorials by gamma functions, this is seen to be equivalent to

$$\sum_{x=k}^{n} \binom{n}{x} p^x q^{n-x} = I_p(k, n - k + 1). \qquad (6.65)$$

As the incomplete beta function is widely tabulated, it is far easier to evaluate the RHS. This is especially useful when n is large and k is not near n.

■ EXAMPLE 6.13

Find the MD of binomial distribution using the Power method in Section 6.1 (p. 189).

Solution 6.13 We know that the mean of binomial distribution is np. The lower limit ll for the BINO(n, p) is $x = 0$, so that $xF(x) = 0$. Hence using Theorem 6.1, the MD is given by

$$MD = 2 \sum_{x=ll}^{\mu-1} F(x) = 2 \sum_{x=0}^{\lfloor np-1 \rfloor} I_{1-p}(n - x, x + 1). \qquad (6.66)$$

The results obtained by equation (6.66) and Theorem 6.1 are given in the Table 6.7 (see also Table 6.1). Both results totally tally when np is an integer or half-integer (with the correction term). Otherwise, the results are only approximate when np is small, but the accuracy increases for large np values.

TABLE 6.7   Mean Deviation of Binomial Distribution Using Equation (6.6)

n, p	0	1	2	3	4	5	6	7	Equation (6.66)
10, 0.10	0.349								0.6974
15, 0.40	0.000	0.005	0.027	0.091	0.217	0.4032			1.4875
20, 0.50	0.000	0.001	0.006	0.021	0.058	0.132	0.252	0.412	1.762
30, 0.10	0.042	0.184	0.4114						1.2749
40, 0.05	0.129	0.399							1.0552
50, 0.14	0.001	0.005	0.022	0.067	0.153	0.281	0.438		1.934
8, 0.875	0.000	0.000	0.000	0.001	0.011	0.067	0.264		0.6872
50, 0.20	0.001	0.006	0.018	0.048	0.103	0.190	0.307	0.444	2.2371
80, 0.1125	0.001	0.004	0.016	0.046	0.102	0.191	0.310	0.448	2.2339
25, 0.36	0.000	0.002	0.007	0.025	0.068	0.148	0.271	0.425	1.8937

First column gives n, p values of binomial. Second column onward are the values computed using equation (6.66). Last column finds the MD using equation (6.6), which is the same as that found using equation (6.66) when np is an integer. When np is not an integer, results are only approximate. Values have been left shifted by two places for (20, 0.50), (50, 0.20) rows and by one place to the left for few other rows, as the first few entries are zeros.

If n is large and p is small, the first few $F(x)$ terms in equation (6.66) could be nearly zeros. A solution is to start the iterations at $x = \mu - 1$ and recur backward or start the iterations at a higher integer from lower limit ll and recur forward. ∎

6.5.7 Approximations

As there are two parameters for this distribution, the binomial tables are lengthy and cumbersome. When the probability of success p is very close to 0.5, the distribution is nearly symmetric (see Figure 6.6, p. 6–61). From Figure 6.6, it is evident that the normal approximation is not good for x values away from the modal value (10) when n is small due to the disparity in the variance. If we reduce the variance of the approximating normal, the peak probabilities will increase. When n is large, the central limit theorem can be used to approximate binomial by a normal curve. The accuracy depends both on n and whether the value of p is close to 0.5. Not only the probabilities but also the cumulative probabilities can also be approximated using normal tail areas. This approximation is quite good when p is near 0.5 rather than near 0 or 1 (use the normal approximation when $np > 10$ or $np(1 - p) > 10$ or both np and $n(1 - p)$ are > 5). Symbolically, $P[x \le k] = Z(\frac{k-np}{\sqrt{npq}})$, where $Z()$ is the standard normal distribution. As this is an approximation of a discrete distribution by a continuous one, a continuity correction could improve the precision for small n. This gives us

$$P[x \le k] = Z((k - np + 0.5)/\sqrt{npq}) \text{ and } P[x \ge k] = 1 - Z((k - np - 0.5)/\sqrt{npq}).$$
(6.67)

See Reference 133 for normal approximations, References 123, 133–135 for further discussions.

6.5.8 Limiting Form of Binomial Distribution

The binomial distribution tends to the Poisson law (p. 6–67) when $n \to \infty, p \to 0$ such that np remains a constant. This result was known to S.D. Poisson (1837), which is why the limiting distribution is called *Poisson distribution*. This is easily derived from the PDF as follows (see Figures 6.3 and 6.7). Write the PDF as

$$f_X(k; n, p) = \binom{n}{k} p^k q^{n-k} = \frac{n(n - 1)(n - 2)..(n - k + 1)}{k!} p^k q^{n-k}.$$
(6.68)

Multiply the numerator and denominator by n^k, combine it in the numerator with p^k and write q^{n-k} as $(1 - p)^n * (1 - p)^{-k}$ to obtain:

$$f_X(k; n, p) = \frac{n}{n} \cdot \frac{n - 1}{n} \cdot \frac{n - 2}{n} \cdots \frac{n - k + 1}{n} \frac{(np)^k}{k!} (1 - p)^{-k}(1 - p)^n.$$
(6.69)

According to our assumption, np is a constant (say λ) so that $p = \lambda/n$. Substitute in the aforementioned and let $n \to \infty$

$$\underset{n \to \infty}{Lt} f_X(k; n, p) = \underset{n \to \infty}{Lt} \frac{n}{n} \cdot \frac{n - 1}{n} \cdot \frac{n - 2}{n} \cdots \frac{n - k + 1}{n} \frac{\lambda^k}{k!} (1 - \lambda/n)^{-k}(1 - \lambda/n)^n.$$
(6.70)

If k is finite, the multipliers all tend to 1 and $(1 - \lambda/n)^{-k}$ also tends to 1. The last term tends to $e^{-\lambda}$ using the result that $\underset{n \to \infty}{Lt}(1 - x/n)^n = e^{-x}$. Hence in the limit, the RHS tends to the Poisson distribution $e^{-\lambda}\lambda^k/k!$. We could write this in more meaningful form as

$$f_X(x; n, p) = \binom{n}{x}p^x q^{n-x} \to e^{-np}(np)^x/x! + O(np^2/2) \text{ as } n \to \infty, \qquad (6.71)$$

where $O(np^2/2)$ is the asymptotic notation [8, 22]. An interpretation of this result is that the binomial distribution tends to the Poisson law when $p \to 0$ faster than $n \to \infty$. In other words, the convergence rate is *quadratic* in p and *linear* in n. This allows us to approximate the binomial probabilities by the Poisson probabilities even for very small n values (say $n < 10$), provided that p is comparatively small.

The above-mentioned result can also be proved using the PGF. We know that $P_x(t; n, p) = (q + pt)^n$. Write $q = 1 - p$ and take logarithm of both sides to get $\log(P_x(t; n, p)) = n \log(1 - p(1 - t))$. Write $n = -(-n)$ on the RHS and expand as an infinite series using $-\log(1 - x) = x + x^2/2 + x^3/3 + \cdots$ to get

$$\log(P_x(t; n, p)) = -n[p(1 - t) + p^2(1 - t)^2/2 + p^3(1 - t)^3/3 + \ldots]. \qquad (6.72)$$

Write $np = \lambda$ and take negative sign inside the bracket. Then, the RHS becomes $\log(P_x(t; n, p)) = \lambda(t - 1) - np^2(t - 1)^2/2 + \cdots$. When exponentiated, the first term becomes the PGF of a Poisson distribution (p. 6–69). The rest of the terms contain higher order powers of the form np^r/r for $r \geq 2$.

We have assumed that $p \to 0$ in the above-mentioned proof. This limiting behavior of p is used only to fix the Poisson parameter. This has the implication that we could approximate both the left-tail and right-tail areas, as well as individual probabilities using the above-mentioned approximation. If p is not quite small, we use the random variable $Y = n - X$, which was shown to have a binomial distribution (see example in p. 6–27) with probability of success $q = 1 - p$, so that we could still approximate probabilities in both tails when p is very close to 1.

When p is near 0.5, a normal approximation is better than the Poisson approximation due to the symmetry. However, a correction to the Poisson probabilities could improve the precision. For large values of n, the distributions of the count X and the sample proportion are approximately normal. This result follows from the Central Limit Theorem. The mean and variance for the approximately normal distribution of X are np and $np(1 - p)$, identical to the mean and variance of the binomial(n, p) distribution. Similarly, the mean and variance for the approximately normal distribution of the sample proportion are p and $(p(1 - p)/n)$.

◼ EXAMPLE 6.14   Political parties

Consider a group of n individuals who support one of two political parties say P1 and P2. Assuming that none of the votes are invalid, what is the probability that a candidate of party P1 wins over the other candidate?

Solution 6.14 If the voting decision of an individual is not influenced by the decision of another (for example, husband's and wife's decision or decision among friends), the proportion of individuals who support one of two political parties can be regarded as a binomial distributed random variable with probability p. To find winning chances, we need to consider whether n is odd or even. If n is odd, P1 will win if the number of votes received is $\geq \frac{n+1}{2}$. Thus, the required probability is $\sum_{x=\frac{n+1}{2}}^{n} \binom{n}{x} p^x q^{n-x}$, where p = probability that the vote is in favor of the candidate of P1. If n is even, the summation needs to be carried out from $(n/2) + 1$ to n. ∎

⬛ EXAMPLE 6.15 Malfunctioning electronic device

Consider an electronic device containing n transistors from the same manufacturer. The probability of each transistor malfunctioning is known from previous observations over a long period of time to be p. Find the probability that (i) at most three transistors malfunction and (ii) none of the transistors malfunction.

Solution 6.15 We assume that the transistors malfunction independent of each other. Then, the number of transistors that malfunction has a binomial distribution. Hence, the required probability is $P[X \leq 3] = \sum_{x=0}^{3} \binom{n}{x} p^x q^{n-x}$. Probability that none of them malfunction is $\binom{n}{0} p^0 q^{n-0} = q^n$. ∎

See Reference 136 for dependent Bernoulli trials and References 135, 137, and 138 for further examples.

6.6 DISCRETE UNIFORM DISTRIBUTION

A random variable that takes equal probability for each of the outcomes has a discrete uniform distribution (DUNI[N]). The PDF is given by

$$f(x) = \Pr[X = k] = 1/N, \quad \text{for } k = 1, 2, \ldots, N \text{ where } N > 1. \tag{6.73}$$

For $N = 2$, we get the Bernoulli distribution with $p = 1/2$. As each of the probabilities is equal, $f(x)/f(x+k) = 1$ for all k in the range. It is also called discrete

TABLE 6.8 Variance of Discrete Distributions

Distribution	Mean	Variance	σ^2 as μ	Ordering
Binomial	np	npq	μq	$\sigma^2 < \mu$
Poisson	λ	λ	μ	$\sigma^2 = \mu$
Geometric	q/p	q/p^2	μ/p	$\sigma^2 > \mu$
Negative binomial	kq/p	kq/p^2	μ/p	$\sigma^2 > \mu$
Hypergeometric	$\dfrac{nk}{N} = t$	$\dfrac{t}{N-1}$ $(1-k/N)(N-n)$	$\dfrac{\mu}{N-1}$ $(1-k/N)(N-n)$	$\sigma^2 < \mu$
Discrete uniform	$(N+1)/2$	$(N^2-1)/12$	$\mu(\mu-1)/3$	

For discrete uniform and HGDs, the inequality depends on parameter values. For discrete uniform distribution $\sigma^2 < \mu$ when $N < 7$ and $\sigma^2 > \mu$ for $N > 7$. They are equal when $N = 7$.

rectangular distribution. A displaced discrete uniform distribution DUNI[a, b] (where $b > a$) can be defined as $f(x) = \Pr[X = a + k] = 1/N$, for $k = 0, 1, 2, \ldots, b-a$ (or $f(x) = \Pr[X = k] = 1/N$ for $k = a, a+1, a+2, \ldots, b-1$). Choosing $a = 0, b = N$ gives another form of the distribution as $f(x) = 1/(N+1)$, for $x = 0, 1, 2, \ldots, N$. In general, we could shift the origin by c (positive or negative) to get the generalized DUNI(N) as $f(x) = \Pr[X = k] = 1/N$, for $k = c, c+1, \ldots, c+N-1$.

EXAMPLE 6.16 CDF of DUNI(N)

Find the CDF of DUNI[N] and obtain the mean using $E(X) = \sum_k P(X \geq k)$.

Solution 6.16 Assume that the PDF is $f(x) = \Pr[X = k] = 1/N$. The CDF $F(x) = P[X \leq x] = \sum_{k=1}^{x} 1/N = x/N$. From this, we get $P(X > x) = 1 - [x/N]$ and $P(X \geq x) = 1 - [(x-1)/N]$. Now $E(X) = \sum_k P(X \geq k) = \sum_{k=1}^{N}(1 - [(k-1)/N]) = \sum_{k=1}^{N} 1 - 1/N \sum_{k=1}^{N}(k-1) = N - (1/N)[1 + 2 + 3 + .. + (N-1)] = N - (1/N)(N-1)N/2$. This simplifies to $E(X) = \mu = (N+1)/2$. ■

EXAMPLE 6.17

Find the MD of DUNI[N] distribution using the Power method in Section 6.1 (p. 6–7) when N is odd.

Solution 6.17 We know that the mean of DUNI[N] distribution is $c = (N+1)/2$. Using Theorem 6.1, the MD is given by

$$\text{MD} = 2\sum_{x=ll}^{\mu-1} F(x) = 2\sum_{x=1}^{c} x/N, \quad \text{where } c = ((N+1)/2) - 1 = (N-1)/2. \tag{6.74}$$

Take $(1/N)$ outside the summation and evaluate $\sum_{x=1}^{c} x = c(c+1)/2$. This gives

$$\text{MD} = (2/N) * c(c+1)/2 = c(c+1)/N. \tag{6.75}$$

Now put $c = (N-1)/2$ to get MD $= (N-1)(N+1)/4 = (N^2 - 1)/4$. ■

6.6.1 Properties of Discrete Uniform Distribution

This distribution has a single parameter. The MGF is easy to find as

$$E(e^{tx}) = \sum_{x=1}^{N} e^{tx}/N = \frac{1}{N}[e^t + e^{2t} + \cdots + e^{Nt}] = \frac{e^t}{N}\frac{1 - e^{Nt}}{1 - e^t} \tag{6.76}$$

for $t \neq 0$ and $= 1$ for $t = 0$. The PGF is obtained by replacing e^t in the above by t as $P_x(t) = \frac{t}{N}\frac{1-t^N}{1-t}$. The characteristic function is written as $\phi_x(t) = (1 - e^{itN})/[N(e^{-it} - 1)]$, where we have divided both numerator and denominator by e^t. The mean and variance are easily seen to be $E(X) = (N+1)/2$ and $V(X) = (N^2 - 1)/12$. The coefficient of kurtosis is $\beta_2 = \frac{3}{5}(3 - \frac{4}{n^2-1})$. This shows that it is always platykurtic. Truncated discrete uniform distributions are of the same type with the probabilities simply enlarged (because dividing by the truncated sum of probabilities simply enlarges each individual probability. See Table 6.5).

EXAMPLE 6.18   Variance as a function of μ for the DUNI(N)

Express the variance of DUNI(N) as a function of μ alone

Solution 6.18 We know that $\mu = (N+1)/2$ and $\sigma^2 = (N^2 - 1)/12$. From $\mu = (N+1)/2$, we get $N = 2 * \mu - 1$. Write $\sigma^2 = (N^2 - 1)/12 = (N-1)(N+1)/12$. Substitute for $(N+1)/2 = \mu$ to get $\sigma^2 = [(2 * \mu - 1) - 1] * \mu/6 = (\mu - 1) * \mu/3$. If the variance of a DUNI(N) distribution is estimated from data, we can obtain an estimate of the mean as follows. Write the afore-mentioned as a quadratic equation $x^2 - x - 3k = 0$, where $x = \mu$ and $k = \sigma^2$. This has positive root $x = (1 + \sqrt{1 + 12 * k})/2$. Put the values for x and k to get $\mu = (1 + \sqrt{1 + 12 * \sigma^2})/2$. Alternatively, we could first estimate N from variance as $N = \sqrt{1 + 12 * \sigma^2}$ and obtain the mean as $(N+1)/2$ (see Table 6.8).

EXAMPLE 6.19   Factorial moments of DUNI(N)

Find the factorial moments of DUNI[N], and obtain the mean.

Solution 6.19 By definition

$$\mu_{(k)} = E[X(X-1)\cdots(X-k+1)] = \sum_{x=1}^{N} x(x-1)\cdots(x-k+1)(1/N). \quad (6.77)$$

As $(1/N)$ is a constant while summing with respect to X, take it outside the summation and adjust the indexvar to vary from k to N. This gives $\mu_{(k)} = (1/N) * \sum_{x=k}^{N} x(x-1)\cdots(x-k+1)$. Multiply and divide by $1, 2, \ldots (x-k)$ and write this as $\mu_{(k)} = (1/N) * \sum_{x=k}^{N} x!/(x-k)!$. Next multiply and divide by $k!$ and write $x!/[k!(x-k)!]$ as $\binom{x}{k}$. The LHS becomes $(k!/N) * \sum_{x=k}^{N} \binom{x}{k}$. Now use $\sum_{x=k}^{N} \binom{x}{k} = \binom{N+1}{k+1} = (N+1)!/[(k+1)! * (N-k)!]$ (identity 7 in p. 6–5). Write $(k+1)!$ in the denominator as $(k+1) * k!$ and cancel out the $k!$ to get

$$\mu_{(k)} = (1/N(k+1)) * (N+1)!/(N-k)! \quad (6.78)$$

EXAMPLE 6.20   Distribution of U = X+Y.

If X and Y are IID DUNI(N) with the same range, find the distribution of $U = X + Y$.

Solution 6.20 Without loss of generality, we assume that X and Y take values $1, 2, \ldots, N$. Then $\Pr[X + Y = k] = \Pr[X = t \cap Y = k - t] = \Pr[X = t] \cap \Pr[Y = k - t]$ due to independence. As t is arbitrary, this becomes $\sum_{t=1}^{k} 1/N^2 = k/N^2$. Hence $f_u(u) = u/N^2$, for $u = 2, 3, \ldots, 2N$. $\blacksquare$

6.6.2 An Application

The DUNI(N) is used in lotteries and random sampling. Let there be m prizes in a lottery. If N tickets are sold, the chance that an arbitrary ticket will win a prize is m/N. If each ticket is printed with the same number of digits (say the width is 6), and each of the digits (0, 1, ..., 9) is equally likely, the PDF of kth digit is DUNI(10). Similarly, in random sampling with replacement, if the population has N elements, the probability distribution of the kth item in the sample is $1/N$.

6.7 GEOMETRIC DISTRIBUTION

Consider a sequence of independent Bernoulli trials with the same probability of success p. We observe the outcome of each trial, and either continues it if it is not a success or stop it if it is a success. This means that if the first trial results in a success with probability p, we stop further trials. If not, we continue observing failures until the first success is observed. Let X denotes the number of trials needed to get the first success. Naturally X is a random variable that can theoretically take any value from 0 to ∞. In summary, practical experiments that result in a geometric distribution can be characterized by the following properties:

1. The experiment consists of a series of IID Bernoulli trials
2. The trials can be repeated independently without limit (as many times as necessary) under identical conditions. The outcome of one trial has no effect on the outcome of any other, including next trial
3. The probability of success, p, remains the same from trial to trial until the experiment is over.
4. The random variable X denotes the number of trials needed to obtain the first success.

If the probability of success is reasonably high, we expect the number of trials to get the first success to be a small number. This means that if $p = 0.9$, the number of trials needed is much less than if $p = 0.5$ in general. Let X_k denote the random variable for observing the first success. If a success is obtained after getting x failures, the probability is $q^x p$ by the independence of the trials. This is called the geometric distribution (see Figure 6.3). It gets its name from the fact that the PDF is a geometric progression with first term p and common difference q with closed form $(p/(1 - q))$. In other words, the individual probabilities (divided by the first probability p) form a geometric progression. Some authors define the PDF of a geometric distribution as $f(x; p) = q^{x-1}p$, where $x = 1, 2, \ldots, \infty$. We could combine the above-mentioned two cases and write the PDF as

$$f(x; p) = \begin{cases} q^{x-1}p & \text{if } x \text{ ranges from } 1, 2, 3, \ldots, \infty \\ q^x p & \text{if } x \text{ ranges from } 0, 1, 2, \ldots, \infty \\ 0 & \text{elsewhere.} \end{cases}$$

The second form follows easily by a change of origin transformation $Y = X - 1$ in the first form. The mean, mode, and other location measures are simply displaced in this case (see the following discussion). The variance remains the same because $V(X - 1) = V(X)$. The Polya distribution

$$f(x; \lambda) = [\lambda/(1 + \lambda)]^x / \lambda \tag{6.79}$$

is obtained by setting the mean $q/p = \lambda$ so that $1/p = (1 + \lambda)$ or $p = 1/(1 + \lambda)$. Now substitute in $q^{x-1}p$, multiply numerator and denominator by λ to get the above-mentioned form (Figure 6.4).

This can be considered as the distribution of waiting time until the occurrence of first success. Consider a sequence of customers in a service queue. Assume that either a new customer joins the queue (with probability p) or none arrives (with probability $q = 1 - p$) in a short-enough time interval. The time T until the next arrival is distributed as GEO(p). It has a single parameter p, the probability of success in each trial.

6.7.0.1 Relationship with Other Distributions It is a special case of the negative binomial distribution when $r = 1$. If $X_1, X_2, \ldots, X_n$ are IID geometric variates with parameter p, then $Y = X_1 + X_2 + \ldots + X_n$ has a negative binomial distribution with

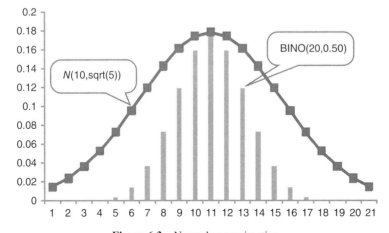

Figure 6.3 Normal approximation.

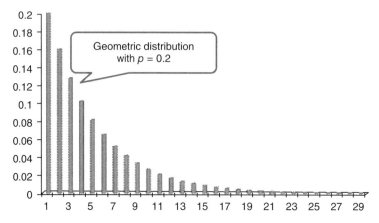

Figure 6.4   Geometric distribution $p = 0.2$.

parameters n, p. This is easily proved using the MGF. It can also be considered as the discrete analog of the exponential distribution.

6.7.0.2   Moments and Generating Functions   The mean and variance are $\mu = 1/p, \sigma^2 = q/p^2$ if $f(x; p) = q^{x-1}p$, with support $x = 1, 2, \ldots$; and $q/p, q/p^2$ if $f(y; p) = q^y p$ with support $y = 0, 1, 2, \ldots$. The mean is easily obtained

$$\mu = \sum_{x=0}^{\infty} xq^x p = p[q + 2q^2 + 3q^3 + \cdots] = pq[1 - q]^{-2} = pq/p^2 = q/p. \quad (6.80)$$

If the PDF is taken as $f(x; p) = q^{x-1}p$, the mean is $1/p$. The ordinary moments of this distribution are easy to find using PGF or MGF. We find higher order moments using the MGF technique.

$$M_x(t) = E(e^{tx}) = \sum_{x=0}^{\infty} e^{tx} q^x p = p \sum_{x=0}^{\infty} (qe^t)^x = p/(1 - qe^t). \quad (6.81)$$

(the MGF for range 1 to ∞ is $pe^t/(1 - qe^t)$). The characteristic function is obtained from the MGF as $\Phi_x(t) = p/(1 - qe^{it})$.

◼ **EXAMPLE 6.21   Moments of geometric distribution**

Obtain the moments of GEO(p) distribution using MGF

Solution 6.21   Take logarithm to the base e of the MGF and differentiate once to obtain

$$M_x'(t)/M_x(t) = qe^t/[1 - qe^t] = q/[e^{-t} - q]. \quad (6.82)$$

Putting $t = 0$, we get $M'_x(0)/M_x(0) = q/(1 - q) = q/p$. Hence $\mu = q/p$, as expected. Differentiating again, we get

$$[M_x(t)M''_x(t) - M'_x(t)^2]/M_x(t)^2 = q[(1 - qe^t)e^t + e^t qe^t]/(1 - qe^t)^2, \qquad (6.83)$$

from which $M''_x(0) = q/p^2 + q^2/p^2$, so that the variance is q/p^2. The CDF is obtained as $F(x) = \sum_{k=0}^{x} q^k p = p[1 + q + q^2 + \cdots + q^x]$. As q is a probability, each power of q is between 0 and 1. Hence, the above-mentioned series converges for all values of q, giving the summed value $p[1 - q^{x+1}]/[1 - q]$. As $(1 - q) = p$, the p in the numerator and denominator cancels out giving the CDF as $F(x) = [1 - q^{x+1}]$.   ∎

EXAMPLE 6.22   Mean deviation of geometric distribution

Find the MD of geometric distribution using Theorem 6.1 (p. 189).

Solution 6.22 We know that the mean of GEO(p) distribution is $\mu = q/p$. Using Theorem 6.1, the MD is given by

$$\text{MD} = 2 \sum_{x=\lfloor l \rfloor}^{\mu-1} F(x) = 2 \sum_{x=0}^{c} [1 - q^{x+1}], \quad \text{where } c = \lfloor q/p \rfloor - 1. \qquad (6.84)$$

Split this into two sums. The first one becomes $2(c + 1) = 2\lfloor q/p \rfloor$. The sum

$$\sum_{x=0}^{c} q^{x+1} = q[1 + q + q^2 + \cdots + q^c] = q[1 - q^{c+1}]/p = (q/p)[1 - q^{\lfloor q/p \rfloor}]. \qquad (6.85)$$

Combine with the first term to get

$$\text{MD} = 2(\lfloor q/p \rfloor) - 2(q/p)[1 - q^{\lfloor q/p \rfloor}]. \qquad (6.86)$$

Write $q = 1 - p$ and cancel out $+2$ and -2. This simplifies to

$$2\lfloor 1/p \rfloor (q^{\lfloor q/p \rfloor}). \qquad (6.87)$$

Write $q = 1 - p$ in the exponent to get an alternate expression

$$\text{MD} = (2/q)\lfloor 1/p \rfloor (q^{\lfloor 1/p \rfloor}). \qquad (6.88)$$

∎

EXAMPLE 6.23 $Y = \lfloor X \rfloor$ **of an Exponential Distribution**

If X has an exponential distribution, find the distribution of $Y = \lfloor X \rfloor$.

Solution 6.23 As X is continuous, $\Pr[Y = y] = \Pr[y \leq X < y + 1]$. Now consider

$$\Pr[y \leq X < y + 1] = \int_y^{y+1} \lambda \exp(-\lambda x)dx = -\exp(-\lambda x)|_y^{y+1}$$

$$= \exp(-\lambda y) - \exp(-\lambda(y + 1)) = \exp(-\lambda y)[1 - \exp(-\lambda)] \tag{6.89}$$

Write $\exp(-\lambda y)$ as $[\exp(-\lambda)]^y$. Then, equation (6.89) is of the form $q^y p = (1 - q)q^y$, where $q = \exp(-\lambda)$. This is the PDF of a geometric distribution with probability of success $p = 1 - q = [1 - \exp(-\lambda)]$. Hence, $Y = \lfloor X \rfloor$ is GEO($[1 - \exp(-\lambda)]$). ∎

EXAMPLE 6.24   **Moments of geometric distribution** $q^{x/2}p$

Find the mean of a distribution defined as

$$f(x; p) = \begin{cases} q^{x/2}p & \text{if } x \text{ ranges from } 0, 2, 4, 6, \ldots, \infty \\ 0 & \text{elsewhere.} \end{cases}$$

Solution 6.24 By definition $E(X) = \sum_x x q^{x/2}p = p[2q + 4q^2 + 6q^3 + \cdots]$. Take $2q$ as common factor and simplify using

$$(1 - x)^{-2} = 1 + 2x + 3x^2 + 4x^3 + \cdots \tag{6.90}$$

to get $2pq(1 - q)^{-2} = 2pq/p^2 = 2q/p$ (see also p. 8–31). ∎

EXAMPLE 6.25   **Geometric probability exceeding** $1/p$

If $X \sim$ GEO(p) find the probability that X takes values larger than the mean.

Solution 6.25 Let $\lfloor 1/p \rfloor$ denote the integer part. Then, the required probability is

$$p \sum_{x=\lfloor 1/p \rfloor}^{\infty} q^x = pq^{\lfloor 1/p \rfloor}(1 + q + q^2 + \cdots) = pq^{\lfloor 1/p \rfloor}(1 - q)^{-1} = q^{\lfloor 1/p \rfloor}. \tag{6.91}$$
∎

EXAMPLE 6.26   **Factorial moments of geometric distribution**

Obtain the factorial moments of GEO(p) distribution.

Solution 6.26 Differentiate the identify $\sum_{x=0}^{\infty} q^x = 1/(1-q)$ with respect to q multiple times to obtain the factorial moments. Differentiating it once, we get

$$\sum_{x=0}^{\infty} xq^{x-1} = 1/(1-q)^2 = 1/p^2. \tag{6.92}$$

Multiply both sides by pq. Then, the LHS becomes $\sum_{x=0}^{\infty} xq^x p = E(X)$. The RHS is $pq/p^2 = q/p$. Differentiating it again results in

$$\sum_{x=1}^{\infty} x(x-1)q^{x-2} = 2/(1-q)^3 = 2/p^3. \tag{6.93}$$

Multiply both sides by $q^2 p$ and simplify to get $E[X(X-1)] = 2q^2/p^2$. Differentiating k times gives

$$\sum_{x=k}^{\infty} x(x-1)\cdots(x-k+1)q^{x-k} = 1, 2, 3, \ldots k/(1-q)^{k+1}. \tag{6.94}$$

Multiply both sides by $q^k p$ to get

$$E[X(X-1)\cdots(X-k+1)] = k!/(1-q)^{k+1}q^k p = k!q^k/p^k = k!(q/p)^k. \tag{6.95}$$

We could reparameterize the geometric distribution by putting $\mu = q/p$ to get

$$f(x;\mu) = (\mu/(1+\mu))^x \, 1/(1+\mu), \tag{6.96}$$

with mean μ and variance $\mu(1+\mu)$. Left truncated geometric distribution is obtained by truncating at a positive integer K. The resulting PDF is $f(x;p) = q^{x+K}p/[1 - \sum_{y=0}^{K-1} q^y p]$. ∎

6.7.1 Properties of Geometric Distribution

Both the above-mentioned densities in equation 6.7 are related through a change of origin transformation $Y = X-1$. This simply displaces the distribution to the left or right. Using $E(Y) = E(X) - 1$, we get $E(Y) = (1/p - 1) = (1-p)/p = q/p$. Variance remains the same because $V(Y) = V(X)$. As $\sigma^2 = \mu/p > \mu$, the distribution is over-dispersed. Similarly, $Z = \min(X_1, X_2, \ldots, X_n)$ has the same geometric distribution. A geometric distribution of order k is an extension:–in a sequence of independent Bernoulli trials, we look for the first consecutive block of k successes (either in the beginning itself or surrounded by failures). For example, in SSFFSFSSF SSSF, an SSS occurs at position 10.

Coefficient of skewness is $\beta_1 = (2-p)/\sqrt{q}$. As the numerator never vanishes for valid values of $0 \leq p \leq 1$, the geometric distribution is never symmetric (in fact, it is always positively queued). The kurtosis is $\beta_2 = (p^2 + 6q)/q = 6 + p^2/q$. As p^2/q can never be negative, the distribution is always leptokurtic. Probability generating function is $p/(1-qs)$ and the characteristic function is $p/(1-qe^{it})$ (see Table 6.9).

In Chapter 3, it was mentioned that the sample variance is a measure of the spread of observations around the sample mean. As the expected value of a constant multiple of the sample variance is the population variance $(E(s^2) = ((n-1)/n)\sigma^2)$, we categorize statistical distributions using the boundedness property of population variance.

Several statistical distributions have a single unknown parameter. Examples are Bernoulli, geometric, Poisson, χ^2, exponential, Rayleigh, and T distributions. The population variance is a linear function of this parameter for Poisson $(= \lambda)$, $\chi^2 (= 2n)$, and other distributions. It is a nonlinear function for exponential $(1/\lambda^2)$, $T (n/(n-2))$ and geometric distributions $((1-p)/p^2)$. For normal distributions, it is σ^2. It is the square of a parameter for double exponential and logistic distributions, constant multiple of the square for Rayleigh and extreme value distributions, and square of the difference of the parameters for uniform distribution. It is a quadratic for binomial $(np(1-p))$ distribution. It is a linear combination of parameters for noncentral $\chi^2 (2(n+2\lambda))$. This discussion shows that the variance can be increased without limit by increasing the respective parameter(s) in the numerator or decreasing the parameters in the denominator for some distributions. However, there are some statistical distributions *with strictly bounded parameter values*, whose variance is either a ratio of parameters or a function of two or more unknown parameters (e.g., as transcendental functions), and *cannot be increased* without limit. Examples are the BETA-I(a, b) with variance $ab/[(a+b)^2(a+b+1)]$, Student's $T (n/(n-2)$ for $n > 2)$, and the distribution of the correlation coefficient. This has interesting implications in the asymptotics of statistical distributions with respect to a subset of the parameter space. For example, the variance of both geometric and negative binomial distributions can be increased without limit by letting $p \to 0$. We could reparametrize these distributions appropriately to have this asymptotic behavior at the extreme right end of the parameter space. For instance, let $\mu = q/p$ for the geometric distribution, so that $p = 1/(1 + \mu)$. Then $p \to 0$ is equivalent to $\mu \to \infty$. These distributions have a characteristic property that the variance is greater than the mean. Similar results could be obtained for higher order moments and cumulants. Naturally, we expect this property to hold in samples drawn from such populations. Such samples are called over-dispersion samples $(s^2 \geq \bar{x})$.

Let $X_1, X_2, \ldots, X_n$ be IID geometric random variates with common parameter p. Then, $Y = X_1 + X_2 + \ldots + X_n$ has a negative binomial distribution with parameters n,p. This is easily proved using the characteristic function. This property can be used to generate random numbers from negative binomial distribution using a random number generator for geometric distribution. Similarly, $Z = \min(X_1, X_2, \ldots, X_n)$ has the same geometric distribution. See Reference 139 for characterizations and Reference 140 for applications.

■ **EXAMPLE 6.27   Variance of geometric distribution**

Prove that the ratio of variance to the mean of a geometric distribution is $1/p$. Express the variance as a function of μ and discuss the asymptotic behavior.

Solution 6.27 We know that the variance is $q/p^2 = (1 − p)/p^2$ and mean is q/p. As $p \to 0$, numerator of variance $\to 1$ and the denominator $\to 0$. The ratio $\sigma^2/\mu = (q/p^2)/(q/p) = 1/p$, which is obviously > 1 as $0 < p < 1$. Thus, the ratio tends to ∞. This has the interpretation that as $p \to 0$, the number of trials needed to get the first success increases without limit. The variance is expressed as a function of the mean as $\sigma^2 = \mu(1 + \mu)$. ∎

■ **EXAMPLE 6.28 Conditional distribution of geometric laws**

If X and Y are IID GEO(p), find the conditional distribution of $(X|X + Y = n)$.

Solution 6.28 As X and Y are independent,

$$\Pr(X|X + Y = n) = \Pr(X = x) * \Pr[Y = n − x]/\Pr[X + Y = n]. \qquad (6.97)$$

We will evaluate the denominator expression first. $X + Y$ takes the value n when $x = k$ and $y = n − k$. Hence, $\Pr[X + Y = n] = \sum_{k=0}^{n} P[X = k]P[Y = n − k]$ (here we have terminated the upper limit at n because Y is positive) $= \sum_{k=0}^{n} q^k p q^{n−k} p = (n + 1)p^2 q^n$. Thus

$$\Pr(X|X + Y = n) = q^x p q^{n−x} p/[(n + 1)p^2 q^n] = 1/(n + 1), \qquad (6.98)$$

which is the PDF of a discrete uniform distribution DUNI($(n+1)$). ■

■ **EXAMPLE 6.29 Geometric probabilities**

If $X \sim$ GEO(p), find the following probabilities:– (i) X takes even values and (ii) X takes odd values.

Solution 6.29 As the geometric distribution takes $x = 0, 1, 2, \dots \infty$ values, both the above-mentioned probabilities are evaluated as infinite sums. (i) $P[X$ is even$] = q^0 p + q^2 p + \cdots = p[1 + q^2 + q^4 + \cdots] = p/(1 − q^2) = 1/(1 + q)$. (ii) $P[X$ is odd$] = q^1 p + q^3 p + \cdots = qp[1 + q^2 + q^4 + \cdots] = qp/(1 − q^2) = q/(1+q)$, which could also be obtained from (i) because $P[X$ is even$] = 1−P[X$ is odd$] = 1−[1/(1+q)] = q/(1+q)$. ■

6.7.2 Memory-less Property

The geometric density function possesses an interesting property called memory-less property.

Theorem 6.5 If m and n are natural numbers, and $X \sim$ GEO(p), then $\Pr(X > m+n \mid X > m) = \Pr(X > n)$.

Proof: We know that $P(A|B) = P(A \cap B)/P(B)$. Applying this to the LHS we get $\Pr(X > m + n|X > m) = \Pr(X > m + n \cap X > m)/P(X > m)$. However, the numerator is simply $\Pr(X > m + n)$, so that the ratio becomes $\Pr(X > m + n)/P(X > m)$.

TABLE 6.9   Properties of Geometric Distribution

Property	Expression	Comments
Range of X	$x = 0, 1, \ldots, \infty$	Discrete, infinite
Mean	$\mu = q/p$	Need not be integer
Variance	$q/p^2 = \mu/p = \mu(\mu + 1)$	$\Rightarrow \mu < \sigma^2$
Mode	0	
Skewness	$\gamma_1 = (1 + q)/\sqrt{q}$	$= (2 - p)/\sqrt{q}$
Kurtosis	$\beta_2 = 9 + p^2/q$	$= 7 + (q + 1/q)$
CV	$1/\sqrt{q}$	
Mean deviation	$(2/q)\lfloor 1/p \rfloor (q^{\lfloor 1/p \rfloor})$	
$E[X(X - 1) \cdots (X - k + 1)]$	$k!(q/p)^k$	Diverges if $p \to 0, k \to \infty$
CDF	$[1 - q^{x+1}]$	
MGF	$p/(1 - qe^t)$	
PGF	$p/(1 - qt)$	
Recurrence	$f(x; n, p)/f(x - 1; n, p) = q$	
Tail probability	q^{x+1}	

Never symmetric, always leptokurtic.

Substituting the PDF, this becomes $\sum_{x=m+n+1}^{\infty} q^x p / \sum_{x=m+1}^{\infty} q^x p = q^n$ (see the following), which is $\Pr(X > n)$. The above-mentioned result holds even if the $>$ operator is replaced by $\geq$.   ∎

6.7.3   Tail Probabilities

The survival probabilities from $x = c$ is

$$\sum_{x=c}^{\infty} q^x p = p[q^c + q^{c+1} + \cdots] = pq^c[1 + q + q^2 + \cdots] = pq^c/(1 - q) = pq^c/p = q^c.$$

As $q < 1$, this goes down to zero for large c. The left-tail probabilities can be found from complementation as $\Pr[0 \leq x \leq c] = 1 - \Pr[x > c] = 1 - q^c$.

6.7.4   Random Samples

Random samples from this distribution can be generated using a uniform random number u in $(0,1)$ by first finding a c such that $1 - q^{c-1} < u < 1 - q^c$. Subtract 1 from each term and change the sign to get $q^{c+1} < 1 - u < q^{c-1}$. Now consider $q^c < 1 - u$. As $1 - U$ and U have the same distribution, taking log we get $c * \log(q) < \log(u)$ from which $c < \log(u)/\log(q)$. Similarly, taking log of both sides of $1 - u < q^{c-1}$, we get $(c - 1)\log(q) > \log(1 - u)$ or equivalently $c > 1 + \log(u)/\log(q)$. Combine both the conditions to get $c = \lfloor 1 + \log(u)/\log(q) \rfloor$. This value being an integer is returned as the random variate from the geometric distribution.

6.8   NEGATIVE BINOMIAL DISTRIBUTION

This distribution gets its name from the fact that the successive probabilities are obtained from the infinite series expansion of the expression $p^k(1-q)^{-k}$, where $q = 1 - p$ and $p > 0$ (see also Exercise 6.20, p. 6–97).

Consider a sequence of independent Bernoulli trials. Instead of counting the number of trials needed to get the first success, we count the number of trials needed to get the kth success, where k is a fixed constant integer greater than 1 known in advance (we are actually counting the number of failures, as the number of successes is fixed at k). Hence, in $x + k - 1$ trials, we have observed $k - 1$ successes, and the $(x + k)$th trial must result in the kth success. The probability of occurrence is thus

$$f(x; k, p) = \binom{x+k-1}{k-1} p^{k-1} q^x \times p = \binom{x+k-1}{k-1} p^k q^x. \tag{6.99}$$

Using $\binom{n}{x} = \binom{n}{n-x}$, the PDF becomes

$$f(x; k, p) = \binom{x+k-1}{x} p^k q^x = \Gamma(x+k)/[\Gamma(k)\, x!]\, p^k q^x, \tag{6.100}$$

for $x = 0, 1, 2, \ldots$ and $k = 1, 2, \ldots$. For $k = 1$, this reduces to the geometric distribution because $\binom{x}{0} = 1$. The second form in equation (6.100) is more general, as k is not restricted to be an integer. Put $y = x + k$ in equation (6.99) to get an alternate form

$$f(y; k, p) = \binom{y-1}{k-1} p^k q^{y+k} \quad \text{for } y = k, k+1, \ldots, \tag{6.101}$$

6.8.1   Properties of Negative Binomial Distribution

By using $\binom{-n}{x} = (-1)^x \binom{n+x-1}{x}$, the PDF can be written alternatively as

$$f(x; k, p) = \binom{-k}{x} p^k (-q)^x. \tag{6.102}$$

Putting $p = k/(u+k)$ and $q = u/(u+k)$, this could also be written in alternate form as

$$\binom{x+k-1}{x} (k/(u+k))^k (u/(u+k))^x = \Gamma(x+k)/[\Gamma(k)\, x!]\, (k/(u+k))^k (u/(u+k))^x.$$

The MGF of NBINO(k, p) is

$$M_x(t) = E[e^{tx}] = \sum_{x=0}^{\infty} e^{tx} \binom{-k}{x} p^k (-q)^x = \sum_{x=0}^{\infty} \binom{-k}{x} p^k (-qe^t)^x = [p/(1 - qe^t)]^k.$$

Differentiating with respect to t and putting $t = 0$ gives $E(X) = kq/p$.

■ **EXAMPLE 6.30   Candidate interviews**

A company requires k candidates with a rare skill set, See Figure 6.6 for p = 0.2,
0.8. As there is a scarcity of local candidates perfectly matching the required skill
set, the company decides to conduct a walk-in interview until all k candidates
have been found. If the probability of a candidate who matches perfectly is p, find
the expected number of candidates interviewed, assuming that several candidates
whose skill set is not completely matching also walks-in.

Solution 6.30 We are given that the probability of perfect match is p. Each
interviewed candidate is either rejected if the skill set is not 100% match or
hired. As the company needs k such candidates, the distribution of finding
all k candidates is negative binomial with parameters (k,p). The expected
number of candidates is kq/p. Owing to the rarity of the sought skill set, p is
small so that q/p is large. For instance, if $p = 0.1, q/p = 9$ and if $p = 0.005$,
$q/p = 199$.   ■

6.8.1.1   Factorial Moments   The falling factorial moments are easier to find than
ordinary moments. Let $\mu_{(r)}$ denote the rth factorial moment.

Theorem 6.6 The factorial moment $\mu_{(r)}$ is given by $\mu_{(r)} = k^{(r)} (q/p)^r$.

Proof: Consider $\mu_{(r)} = E[x_{(r)}] = E[x(x-1)..(x-r+1)]$. Substitute the PDF and sum
over the proper range of x to get the RHS as $\sum_x x(x-1) \ldots (x-r+1) \binom{k+x-1}{x} p^k q^x$.

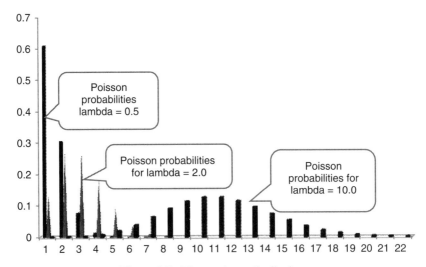

Figure 6.5   Three poisson distributions.

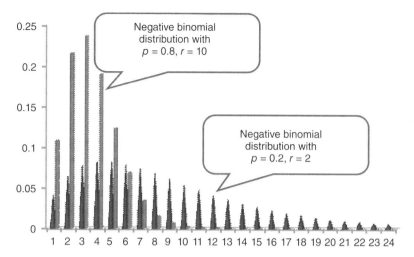

Figure 6.6 Two negative binomials.

Write $x!$ in the denominator as $x(x-1) \ldots (x-r+1) * (x-r)!$, multiply both numerator and denominator by $k(k+1)(k+2)\cdots(k+r-1)$ this becomes

$$k(k+1)(k+2)\cdots(k+r-1)\, p^k q^r \sum_{x=r}^{\infty} \binom{k+x-1}{x-r} q^{x-r}. \qquad (6.103)$$

Put $y = x - r$ in equation (6.103) and rearrange the indexvar. This gives

$$k(k+1)(k+2)\cdots(k+r-1)\, p^k q^r \sum_{y=0}^{\infty} \binom{k+r+y-1}{y} q^y. \qquad (6.104)$$

The infinite sum in equation (6.104) is easily seen to be $(1-q)^{-(k+r)}$. As $(1-q) = p$, the p^k cancels out giving $\mu_{(r)} = E[x_{(r)}] = k^{(r)}(q/p)^r$. This can be written in terms of gamma function as $\mu_{(r)} = [\Gamma(k+r)/\Gamma(k)]\,(q/p)^r$. ∎

6.8.1.2 Relationship with Other Distributions Tail areas of binomial and negative binomial distributions are related. Consider equation (6.101). Then $\Pr[Y \geq n - c] = \Pr[X \leq c]$, where X is distributed as BINO(n, p) and Y as NBINO(c, p). As $k \to \infty$ and $p \to 1$ such that $k(1-p)$ is a constant, the negative binomial distribution approaches a Poisson law with parameter $\lambda = k(1-p)$. Similarly, NBINO(k, p/k) as $k \to \infty$ tends to the Poisson law $\exp(-\lambda)\lambda^x/x!$ [121] with complexity $O(kq^2/2)$. This means that "the negative binomial distribution tends to the Poisson law when $p \to 1$ faster than $k \to \infty$." If k is an integer or a half-integer, the SF can be written as $\Pr[Y > y] = I_{1-p}(y,k)$, where $I(a,b)$ denotes the incomplete beta function. This can also be written in terms of an F distribution as $\Pr[Y > y] = F_t(2k, 2y)$ where $t = p * y/(q * k)$.

Proposition 1 The negative binomial distribution can be regarded as a sum of k independent geometric distributions with the same parameter p. We know from Section 6.7.0.2 in p. 216 that the MGF of a GEO(p) distribution is given by $M_x(t) = p/(1 - qe^t)$. Hence, the MGF of k IID GEO(p) is given by $M_Y(t) = [p/(1 - qe^t)]^k$.

■ EXAMPLE 6.31   Gamma mixture of the Poisson parameter

Prove that a gamma(m, p) mixture of the Poisson parameter (λ) gives rise to a NBINO(p, $m/(m + 1)$) distribution.

Solution 6.31 The PDF of Poisson and Gamma variates are, respectively,

$$f(x; \lambda) = e^{-\lambda} \lambda^x / x!, \quad \text{and} \quad g(\lambda; m, p) = \frac{m^p}{\Gamma(p)} e^{-m\lambda} \lambda^{p-1}. \tag{6.105}$$

The unconditional distribution is obtained as $f(x) =$

$$\int_{\lambda=0}^{\infty} e^{-\lambda} \lambda^x / x! \, \frac{m^p}{\Gamma(p)} e^{-m\lambda} \lambda^{p-1} d\lambda \tag{6.106}$$

$$= \frac{m^p}{x! \Gamma(p)} \int_{\lambda=0}^{\infty} e^{-\lambda(1+m)} \lambda^{x+p-1} d\lambda = \frac{m^p}{x! \Gamma(p)} \frac{\Gamma(p + x)}{(m + 1)^{p+x}}. \tag{6.107}$$

This upon rearrangement becomes

$$f(x) = \frac{\Gamma(p + x)}{x! \Gamma(p)} \left(\frac{m}{m + 1} \right)^p \left(\frac{1}{m + 1} \right)^x, \quad \text{for } x = 0, 1, 2, \ldots, \tag{6.108}$$

■ EXAMPLE 6.32   Variance of negative binomial

Prove that the ratio of variance to the mean of a negative binomial random variable is $1/p$.

Solution 6.32 We know that the variance is given by $V(r, p) = kq/p^2$. Obviously, this can be increased by increasing the parameter k without limit. As the mean is kq/p, the ratio of variance to the mean is $(kq/p^2)/(kq/p) = 1/p$, which is obviously greater than 1 (as $0 < p < 1$). This gives $p * \sigma^2 = \mu$.   ■

■ EXAMPLE 6.33

Find the MD of negative binomial distribution using Theorem 6.1 (p. 189).

Solution 6.33 We know that the mean of negative binomial distribution is $\mu = kq/p$. Using Theorem 6.1, the MD is given by

$$\text{MD} = 2 \sum_{x=\mathrm{ll}}^{\mu-1} F(x) = 2 \sum_{x=0}^{c} I_p(k, x+1), \quad \text{where } c = \lfloor kq/p \rfloor - 1, \qquad (6.109)$$

where $I(a, b)$ is the incomplete beta function. This simplifies to $2c \binom{k+c-1}{c}$ $q^c p^{k-1} = 2\mu_2 * f_c$, where f_c is the probability mass evaluated at the integer part of the mean. ∎

6.8.2 Moment Recurrence

The central moments satisfy the recurrence

$$\mu_{r+1} = q((kr/p^2)\mu_{r-1} - \partial \mu_r/\partial p), \qquad (6.110)$$

where $\mu_r = E(x - kq/p)^r$. Consider

$$\mu_r = \sum_{x=0}^{\infty} (x - kq/p)^r \binom{k+x-1}{x} p^k q^x. \qquad (6.111)$$

As $\binom{k+x-1}{x}$ is independent of p, and $q = 1 - p$, write the aforementioned as

$$\mu_r = \sum_{x=0}^{\infty} \binom{k+x-1}{x} \{(x + k - k/p)^r p^k (1-p)^x\}. \qquad (6.112)$$

Differentiate the expression within the curly brackets with respect to p using the function of a function rule to get

$$\partial \mu_r/\partial p = \sum_{x=0}^{\infty} \binom{k+x-1}{x} \{r(x + k - k/p)^{r-1} p^k (1-p)^x (+k/p^2)$$

$$+ (x + k - k/p)^r k p^{k-1} (1-p)^x - (x + k - k/p)^r p^k x (1-p)^{x-1}\}. \qquad (6.113)$$

Combine the last two terms as $[k(1-p) - px] = -p(x - k(1-p)/p) = -p(x + k - k/p)$ to get $-p (x + k - k/p)^{r+1} p^{k-1} (1-p)^{x-1}$. Multiply and divide by pq and combine the terms as $-(1/q)(x + k - k/p)^{r+1} p^k (1-p)^x$. This gives

$$\frac{\partial}{\partial p} \mu_r = -\mu_{r+1}/q + rk/p^2 \mu_{r-1}. \qquad (6.114)$$

Cross-multiply and rearrange the expressions to get the result.

Theorem 6.7 Additivity theorem: If $X_1 \sim \text{NB}(n_1, p)$ and $X_2 \sim \text{NB}(n_2, p)$ are independent NB random variables, then $X_1 + X_2 \sim \text{NB}(n_1 + n_2, p)$

TABLE 6.10   Properties of Negative Binomial Distribution

Property	Expression	comments
Range of X	$x = 0, 1, .., \infty$	Discrete, infinite
Mean	$\mu = kq/p$	Need not be integer
Variance	$\sigma^2 = kq/p^2 = \mu/p$	$\mu < \sigma^2$
Mode	$(x-1), x$	$x = [(q/p)(k-1)]$ is int.
Skewness	$\gamma_1 = (1+q)/\sqrt{kq}$	$= (2-p)/\sqrt{kq}$
Kurtosis	$\beta_2 = 3 + 6/k + p^2/(kq)$	Always leptokurtic
CV	$1/\sqrt{kq}$	
CDF	$F_c(k,p) = I_p(k, c+1)$	
Mean deviation	$2 \sum\limits_{x=0}^{\lfloor kq/p \rfloor - 1} I_p(k, x+1)$	$2\mu_2 * f_m$
Factorial mom	$k^{(r)}(q/p)^r = [\Gamma(k+r)/\Gamma(k)] \ (q/p)^r$	
MGF	$p^k/(1 - qe^t)^k$	$[p/(1-qe^t)]^k$
PGF	$p^k/(1 - qt)^k$	$p^k(1 - t + pt)^{-k}$
FMGF	$(1 - qt/p)^{-k}$	
Additivity	$\sum\limits_{i=1}^{m} NB(k_i, p) = NB(\sum\limits_{i=1}^{m} k_i, p)$	Independent
Recurrence	$f(x; k, p)/f(x-1; k, p) = q(k + x - 1)/x$	
Tail probability	$\sum\limits_{x>c} \binom{x+k-1}{x} p^k q^x = I_q(c+1, k)$	I = Incomplete beta

Proof: This is most easily proved by the MGF method. We have seen in equation 6.103 that MGF is $[p/(1 - qe^t)]^k$. As p is the same, replace k by n_1 and n_2 and take the product to get the result. This result can be extended to any number of NBIN(r_i, p) as follows: If $X_i \sim$ NBIN(r_i, p), then $\sum_i X_i \sim$ NBIN($\sum_i r_i, p$). The pgf is obtained by replacing e^t by t.   ∎

6.8.3   Tail Probabilities

As the random variate extends to ∞, the right-tail probabilities are more challenging to evaluate (Table 6.10 and Figure 6.6). The left-tail probabilities of NBIN(r, p) are related to the right-tail probabilities of binomial distribution as $F_k(r, p) = P(X \le k) = P(Y \ge r) = 1$-BINO($k + r, p$). The upper tail probabilities of an NB distribution can be expressed in terms of the incomplete beta function as

$$\sum_{x>c} \binom{x+k-1}{x} p^k q^x = I_q(c+1, k). \qquad (6.115)$$

The lower tail probabilities can be found from the complement rule as

$$\sum_{x=0}^{c} \binom{x+k-1}{x} p^k q^x = I_p(k, c+1).$$ (6.116)

This can also be expressed as tail areas of an F distribution (A. Meyer [141]; see also Guenther [142].

🖥 **EXAMPLE 6.34   Negative binomial probabilities**

If $X \sim$ NBIN (r, p), find the following probabilities:– (i) X takes even values and (ii) X takes odd values

Solution 6.34 Let $P_x(t)$ denote the PGF of NBIN (r, p). (i) $P[X$ is even] has PGF given by

$$[P_x(t) + P_x(-t)]/2 = (p^r/2)[1/(1-qt)^r + 1/(1+qt)^r].$$ (6.117)

This can be simplified and expanded into an even polynomial in t with the corresponding coefficients giving the desired sum. (ii) The PGF for X taking odd values is $\frac{1}{2}[P_x(t) - P_x(-t)]$. Substitute for $P_x(t)$ to get

$$[P_x(t) + P_x(-t)]/2 = (p^r/2)[1/(1+qt)^r - 1/(1-qt)^r].$$ (6.118)

Proceed as above and expand as an odd polynomial in t whose coefficients give desired probabilities. ■

See References 143 and 144 for a generalizations and Reference 145 for MLE.

6.9   POISSON DISTRIBUTION

The Poisson distribution was invented by S.D. Poisson (1781–1840) in 1838 as counts (arrivals) of random discrete occurrences in a fixed time interval. It can be used to model temporal, spatial or spatiotemporal rare events that are open-ended. For example, it is used to predict the number of telephone calls received in a small time interval, number of accidents in a time period, number of automobiles coming at a gas station, number of natural disasters (like earthquakes) in a year, and so on. These are all temporal models with different time intervals. Examples of spatial frame of reference include predicting defects in newly manufactured items such as clothing sheets, paper rolls or newsprints, cables and wires, and micro-chips. Spatiotemporal applications include predicting earthquakes and tsunamis in a particular region over a time period, outbreak of epidemics in a geographical region over a time period, and so on. It is also used in many engineering fields. The unit of the time period in these cases is implicitly assumed by the modeler. A wrong choice of the time period may lead to convoluted Poisson models.

The PDF is given by

$$p_x(\lambda) = e^{-\lambda}\lambda^x/x!, x = 0, 1, 2, \ldots, \tag{6.119}$$

where e is the natural logarithm. Obviously, summing over the range of x values gives $\sum_{x=0}^{\infty} e^{-\lambda}\lambda^x/x! = e^{-\lambda}(1 + \lambda + \lambda^2/2! + \cdots) = e^{-\lambda}e^{\lambda} = 1$, where the indexvar is varied as an exponent and a function. It belongs to the exponential family.

It can be considered as the limiting case of a binomial distribution as shown in Section 6.5.8 in page 6–37. Most of the textbooks give this limiting behavior as follows: "When n, the number of trials is large, and p, the probability of success is small, such that np remains a constant λ, then BINO(n, p) $\rightarrow$ POIS(λ)." Johnson et al. [123, 306] mentions in page 152 that "It is the largeness of n and smallness of p that are important." The product $\lambda = np$ can remain a constant in two limiting cases: (i) $n \rightarrow \infty$ faster than $p \rightarrow 0$ and (ii) $p \rightarrow 0$ faster than $n \rightarrow \infty$. As shown in Section 6.5.8, this limiting property is valid only when np remains finite, and $np^2/2$ and higher order terms are negligible. We give the revised rule that "the binomial distribution tends to the Poisson law when p tends to zero faster than n tends to infinity." Thus, the Poisson approximation is valid even for low values of n, provided that p is comparatively very small. In most practical applications, the value of n is at the hands of a researcher, and the value of p is observed from the data. When p is near 0.5, the above-mentioned condition may not hold. In such cases, a correction term is needed to get higher accuracy for the approximation. Consider equation (6.72) in page 6–38, which is reproduced in the following:

$$\log(P_x(t; n, p)) = -n[p(1 - t) + p^2(1 - t)^2/2 + p^3(1 - t)^3/3 + \ldots]. \tag{6.120}$$

Keeping first term intact, and collecting constant terms from the rest, we get the RHS as

$$np(t - 1) - [np^2/2 + np^3/3 + \cdots] = np(t - 1) - n[-\log(1 - p) - p]$$
$$= np(t - 1) + n[\log q + p] = \lambda(t - 1) + \lambda + n\log(q). \tag{6.121}$$

Exponentiating LHS and RHS, we see that the first term becomes the PGF of the Poisson distribution.

6.9.1 Properties of Poisson Distribution

This distribution has a single parameter λ, which is both the mean and variance of the distribution. It is easy to compute for small λ values. It is an excellent choice for forming mixture distributions (like noncentral χ^2 distribution).

The difference of two independent Poisson random variables has the Skellam distribution with PDF (see Figure 6.5)

$$f(x, \lambda_1, \lambda_2) = e^{-(\lambda_1 + \lambda_2)}(\lambda_1/\lambda_2)^{x/2}I_x(2\sqrt{\lambda_1\lambda_2}), \tag{6.122}$$

where $I_x()$ is the modified Bessel function of the first kind.

6.9.1.1 Moments and MGF   The first moment is readily obtained as

$$\mu = E(X) = \sum_{x=0}^{\infty} xe^{-\lambda}\lambda^x/x! = \lambda e^{-\lambda} \sum_{x=1}^{\infty} \lambda^{x-1}/(x-1)! = \lambda e^{-\lambda}e^{+\lambda}. \qquad (6.123)$$

Using $e^m * e^n = e^{m+n}$, the aforementioned reduces to λ. To find the second moment $E(X^2)$, write x^2 as $x * (x-1) + x$ to get

$$E(X^2) = \sum_{x=0}^{\infty} x^2 * e^{-\lambda}\lambda^x/x! = \sum_{x=0}^{\infty} [x(x-1)+x] * e^{-\lambda}\lambda^x/x! = \lambda^2 + \lambda. \qquad (6.124)$$

From this, the variance is found as $V(X) = E(X^2) - E(X)^2 = \lambda$. Factorial moments of a Poisson distribution are easier to find because of the presence of $x!$ in the denominator of the PDF. The rth factorial moment is

$$\mu_{(r)} = E[x_{(r)}] = E[x(x-1)..(x-r+1)] = \sum_{x=0}^{\infty} x(x-1)..(x-r+1)e^{-\lambda}\lambda^x/x!$$

$$= \lambda^r e^{-\lambda} \sum_{x=r}^{\infty} \lambda^{x-r}/(x-r)! = \lambda^r e^{-\lambda}e^{+\lambda} = \lambda^r. \qquad (6.125)$$

Higher order moments can be obtained from this as $\mu_2' = \lambda^2 + \lambda$, $\mu_3' = \lambda^3 + 3\lambda^2 + \lambda$.

6.9.1.2 Moment Generating Function   The moment generating function is

$$M_x(t) = E[e^{tx}] = \sum_{x=0}^{\infty} e^{tx}e^{-\lambda}\lambda^x/x! = e^{-\lambda} \sum_{x=0}^{\infty} (\lambda e^t)^x/x! = e^{-\lambda}e^{\lambda e^t} = e^{\lambda(e^t-1)}.$$

From this, the PGF is obtained by replacing e^t by t as $P_x(t) = e^{\lambda(t-1)}$.

▪ EXAMPLE 6.35   Mode of Poisson distribution

Prove that the mode of the Poisson distribution is $\lfloor \lambda \rfloor$ if λ is noninteger and is bimodal with the modes located at $[\lambda - 1, \lambda]$ otherwise.

Solution 6.35   Consider the ratio $f_x(k, \lambda)/f_x(k-1, \lambda) = \lambda/k$. If $k \leq \lambda$, the LHS is strictly increasing. Otherwise, it is strictly decreasing. If λ is integer, λ/k will assume the last integer value at $k = \lambda$ (if λ is a prime number, this occurs only once, but if it is composite, the ratio could be integer for more than one value of k). Thus, if λ is an integer, the RHS becomes 1 when $k = \lambda$ giving $f_x(\lambda, \lambda) = f_x(\lambda - 1, \lambda)$ (we have simply substituted $k = \lambda$). Thus, the maximum occurs at $k = \lambda - 1$ and λ. Otherwise, there is a single mode at $[\lambda]$, the integer part.   ▪

TABLE 6.11   Properties of Poisson Distribution

Property	Expression	Comments
Range of X	$x = 0, 1, .., \infty$	Discrete, infinite
Mean	$\mu = \lambda$	Real number
Variance	$\sigma^2 = \lambda$	$\Rightarrow \mu = \sigma^2$
Mode	$[\lambda - 1, \lambda]$ if λ is integer	$\lfloor \lambda \rfloor$ if not integer
Skewness	$\gamma_1 = 1/\sqrt{\lambda}$	
Kurtosis	$\beta_2 = 3 + 1/\lambda$	Leptokurtic
Even sum	$\frac{1}{2}(1 + e^{-2\lambda})$	
SF	$P[x > r] = \dfrac{1}{\Gamma(r+1)} \displaystyle\int_0^\lambda e^{-x} x^r dx$	
CV	$\sqrt{\lambda}$	
Mean deviation	$2 \displaystyle\sum_{x=0}^{\lfloor \lambda \rfloor - 1} \gamma(x+1, \lambda)/\Gamma(x+1)$	$2 * \exp(-\lambda)\lambda^{\lfloor \lambda \rfloor + 1}/\lfloor \lambda \rfloor!$
Moments	$\mu_r = \lambda \displaystyle\sum_{i=0}^{r-2} \binom{r-1}{i} \mu_i$	$r > 1, \mu_0 = 1$
rth cumulant	λ	
Factorial moments	λ^r	
FMGF $= E(1+t)^x$	$\exp(t\lambda)$	
MGF	$e^{\lambda(e^t - 1)}$	
PGF	$e^{\lambda(t-1)}$	
Additivity	$\displaystyle\sum_{i=1}^{m} P(\lambda_i) = P(\sum_{i=1}^{m} \lambda_i)$	independent
Recurrence	$f(x; n, p)/f(x-1; n, p) = \lambda/x$	
Tail probability	$\displaystyle\sum_{x=0}^{m} \dfrac{e^{-\lambda}\lambda^x}{x!} = \dfrac{\gamma(m+1, \lambda)}{\Gamma(m+1)}$	$\gamma(m, \lambda) = \int_\lambda^\infty e^{-y} y^{m-1} dy$ Incomplete gamma

Approaches normality when $\lambda \to \infty$.

■ **EXAMPLE 6.36   Defectives in shipment**

Consider a collection of items such as light bulbs and transistors, of which some are known to be defective with probability $p = 0.001$. Let the number of defectives in a shipment follow a Poisson law with parameter λ (Tables 6.11 and 6.12). How is p and λ related? What is the probability of finding (i) no defectives and (ii) at least two defective items in a shipment containing 20 items?

Solution 6.36 If n is the number of items in the shipment, p and λ are related as $np = \lambda$. To find the probability of at least two defectives, we use the complement-and-conquer rule. The complement event is that of finding

TABLE 6.12 Mean Deviation of Poisson Distribution Using Our Power Method (equation 6.8) for λ Integer

λ	Direct	1	2	3	4	5	6	7	8	9
5	1.755	0.0135	0.0943	0.3436	0.8737	1.755				
7	2.086	0.0018	0.0164	0.0757	0.2392	0.5852	1.186	2.086		
8	2.233	0.0007	0.0067	0.0342	0.1190	0.3182	0.7007	1.328	2.233	
10	2.502	0.0012	0.0066	0.0273	0.0858	0.220	0.480	0.921	1.586	2.502

First column gives λ values of Poisson distribution. Second column onward is the values accumulated using equation (6.134) in page 6–78. Row for $\lambda = 10$ has been left shifted by one column.

either 0 or 1 defective. The corresponding probabilities are $e^{-\lambda}$ and $\lambda e^{-\lambda}$. As $n = 20, n * p = 20 * 0.001 = 0.02$. (i) The probability of finding no defectives $= e^{-0.02} = 0.98019867$ and (ii) substitute for λ to get $e^{-0.02} + 0.02 * e^{-0.02} = 0.9801986 + 0.0196039 = 0.9998$ as the complement probability. From this, the required answer follows as $1 - 0.9998 = 0.0002$ ∎

6.9.1.3 *Additivity Property* If $X_1 \sim \text{POIS}(\lambda_1)$ and $X_2 \sim \text{POIS}(\lambda_2)$ are independent, then $X_1 + X_2 \sim \text{POIS}(\lambda_1 + \lambda_2)$.

This is most easily proved using the MGF. Using $M_{X_1+X_2}(t) = M_{X_1}(t) * M_{X_2}(t)$, we get $M_{X_1+X_2}(t) = e^{(\lambda_1+\lambda_2)(e^t-1)}$. This result can be extended to an arbitrary number of random variables (see Table 6.11).

■ EXAMPLE 6.37 Distribution of $X_1|(X_1 + X_2 = n)$

If $X_1 \sim \text{POIS}(\lambda_1)$ and $X_2 \sim \text{POIS}(\lambda_2)$ are independent, then the distribution of $X_1|(X_1 + X_2 = n)$ is $\text{BINO}(n, \lambda_1/(\lambda_1 + \lambda_2))$.

Solution 6.37 Consider the conditional probability $P[X_1|(X_1 + X_2 = n)] = P[X_1 = x_1] \cap P[X_2 = n - x_1]/P(X_1 + X_2 = n)$. Substitute the density to get $e^{-\lambda_1}\lambda_1^{x_1}/x_1! * e^{-\lambda_2}\lambda_2^{n-x_1}/(n - x_1)!/e^{-(\lambda_1+\lambda_2)}(\lambda_1 + \lambda_2)^n/n!$. Canceling out common terms from the numerator and denominator and writing $(\lambda_1 + \lambda_2)^n$ in the denominator as $(\lambda_1 + \lambda_2)^{x_1} * (\lambda_1 + \lambda_2)^{n-x_1}$ this becomes

$$n!/[x_1!(n - x_1)!](\lambda_1/(\lambda_1 + \lambda_2))^{x_1}(\lambda_2/(\lambda_1 + \lambda_2))^{n-x_1}$$

which is the binomial PDF with probability of success $p = \lambda_1/(\lambda_1 + \lambda_2)$. ∎

■ EXAMPLE 6.38 Poisson probabilities

If $X \sim \text{POIS}(\lambda)$, find (i) $P[X$ is even$]$, (ii) $P[X$ is odd$]$.

Solution 6.38 (i) $P[X$ is even$] = e^{-\lambda}[\lambda^0/0! + \lambda^2/2! + \cdots]$. To evaluate this sum, consider the expansion of $\cosh(x) = [1 + x^2/2! + x^4/4! + \cdots]$. The above-mentioned sum in the square bracket is then $\cosh(\lambda) = \frac{1}{2}(e^\lambda + e^{-\lambda})$. From this, we get the required probability as $e^{-\lambda} * \frac{1}{2}(e^\lambda + e^{-\lambda}) = \frac{1}{2}(1 + e^{-2\lambda})$ (ii) $P[X$ is odd$] = 1 - P[X$ is even$] = 1 - \frac{1}{2}(1 + e^{-2\lambda}) = \frac{1}{2}(1 - e^{-2\lambda})$ (see Example 9.4 in Chapter 8, p. 9–7). ∎

▣ EXAMPLE 6.39   Conditional distribution

If $X \sim$ POIS(λ), find the conditional distribution of the random variable (i) $X|X$ is even and (ii) $X|X$ is Odd.

Solution 6.39 Let Y denote the random variable obtained by conditioning X to even values and Z denote the random variable obtained by conditioning X to odd values. As the Poisson variate takes values $x = 0, 1, 2, \ldots$ the variate Y takes the values $Y = 0, 2, 4, 6, \ldots \infty$, and Z takes the values $Y = 1, 3, 5, \ldots \infty$. Using above-mentioned example,

$$\sum_{i=0,2,4\ldots}^{\infty} f(y) - \sum_{i=1,3,5,\ldots}^{\infty} g(z) = \sum_{i\ even} e^{-\lambda}\lambda^i/i! - \sum_{i\ odd} e^{-\lambda}\lambda^i/i!$$

$$= \sum_{k=0}^{\infty} (-1)^k e^{-\lambda}\lambda^k/k! = e^{-2\lambda}. \tag{6.126}$$

From conditional probability, $P[X = k|X$ is even$] = P[X = k \cap X$ is even$]/P[X$ is even$] = f(y) = 2e^{-\lambda}\lambda^y/[y!(1 + e^{-2\lambda})]$. This gives

$$f(y; \lambda) = \begin{cases} 2e^{-\lambda}\lambda^y/[y!(1 + e^{-2\lambda})] \\ 0 \qquad\qquad\qquad\qquad \text{otherwise.} \end{cases}$$

Proceed exactly as above to get the PDF of z as $P[X = k|X$ is odd$] = P[X = k \cap X$ is odd$]/P[X$ is odd$]$ as

$$g(z; \lambda) = \begin{cases} 2e^{-\lambda}\lambda^z/[z!(1 - e^{-2\lambda})] \\ 0 \qquad\qquad\qquad\qquad \text{otherwise.} \end{cases} \qquad ∎$$

6.9.1.4 Relationship with Other Distributions   The tail probabilities of a Poisson distribution is related to the incomplete gamma function as follows:

Theorem 6.8 Prove that the survival function of POIS(λ) is related to incomplete gamma function as

$$F(r) = P[x > r] = \sum_{x=r+1}^{\infty} e^{-\lambda}\lambda^x/x! = \frac{1}{\Gamma(r+1)} \int_0^\lambda e^{-x}x^r dx. \tag{6.127}$$

Proof: Consider $\int_\lambda^\infty e^{-x} x^r dx$. Put $y = x - \lambda$, so that the range becomes 0 to ∞, and we get

$$\int_\lambda^\infty e^{-x} x^r dx = \int_0^\infty e^{-(y+\lambda)}(y+\lambda)^r dy = e^{-\lambda} \int_0^\infty e^{-y} \sum_{j=0}^r \binom{r}{j} y^j \lambda^{r-j} dy. \quad (6.128)$$

Take constants independent of y outside the integral to get $e^{-\lambda} \sum_{j=0}^r \binom{r}{j} \lambda^{r-j} \int_0^\infty e^{-y}$ $y^j dy$. Put $\int_0^\infty e^{-y} y^j dy = \Gamma(j+1) = j!$ in equation (6.128) and expand $\binom{r}{j} = r!/[j!(r-j)!]$. The $j!$ cancels out giving

$$\int_\lambda^\infty e^{-x} x^r dx = e^{-\lambda} \sum_{j=0}^r \lambda^{r-j} r!/(r-j)!. \quad (6.129)$$

Divide both sides by $r!$ and write $r!$ as $\Gamma(r+1)$.

$$\frac{1}{\Gamma(r+1)} \int_\lambda^\infty e^{-x} x^r dx = \sum_{j=0}^r e^{-\lambda} \lambda^{r-j}/(r-j)!. \quad (6.130)$$

Put $r - j = k$ on the RHS. When $j = 0, k = r$ and when $j = r, k = 0$. Thus, the sum is equivalent to $\sum_{k=0}^r e^{-\lambda} \lambda^k/k!$. Subtract both sides from 1. The LHS is then $\frac{1}{\Gamma(r+1)} \int_0^\lambda e^{-x} x^r dx$. The RHS is $\sum_{k=r+1}^\infty e^{-\lambda} \lambda^k/k!$. This shows that the left-tail area of the gamma function is the survival probability of Poisson distribution. This proves our result. ∎

Theorem 6.9 Prove that the survival function of a central chi-square distribution with even degrees of freedom is a Poisson sum (Fisher [146]) as

$$1 - F_n(c) = \int_c^\infty e^{-x/2} x^{\frac{n}{2}-1}/2^{n/2}\Gamma(n/2) dx = \sum_{x=0}^{n/2-1} \frac{e^{-\lambda}\lambda^x}{x!}, \text{ where } \lambda = \frac{c}{2}.$$

Proof: The proof follows easily because the χ^2 and gamma distributions are related as $\chi_n^2 \equiv \text{GAMMA}(n/2, 1/2)$.

$$F_n(x) = 1 - e^{-x/2} \sum_{i=0}^{(n-2)/2} (x/2)^i/i!, \quad (6.131)$$

Putting $n = 2m$, we find that the CDF of central χ^2 with even df can be expressed as a sum of Poisson probabilities.

$$F_{2m}(x) = \sum_{j=m}^\infty P_j(x/2), \text{ and } \overline{F}_{2m}(x) = \sum_{j=0}^{m-1} P_j(x/2), \quad (m = 1, 2, 3, \ldots), \quad (6.132)$$

where $P_j(u) = e^{-u} u^j/j!$. These are discussed in Chapter 7 (Figure 6.7). ∎

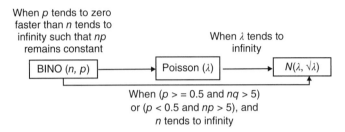

Figure 6.7   Limiting behavior of binomial distributions.

6.9.2   Algorithms for Poisson Distribution

Individual probabilities can be calculated using the forward recurrence

$$f_x(k+1; \lambda) = [\lambda/(k+1)]f_x(k; \lambda), \text{ with } f_x(0; \lambda) = e^{-\lambda}. \tag{6.133}$$

When λ is large, $e^{-\lambda}$ is too small. This may result in loss of precision or even underflow (in computer memory). As subsequent terms are calculated using the first term, error may propagate throughout the subsequent computation steps. A solution is to use the log-recursive algorithm suggested in Reference 4. Another possibility is an algorithm that starts with the mode of the Poisson distribution, which then iteratively calculates subsequent values leftward (reverse) and rightward (forward). This may be combined with the log-recursive algorithm to provide a reliable and robust algorithm for Poisson distributions, and other mixture distributions that use Poisson weighting [4, 5]. The left-tail probabilities (CDF) $F_c(\lambda) = \sum_{j=0}^{c} P(j)$ converge rapidly for small λ values. CDF can be evaluated efficiently using $F_c(\lambda) = \gamma(c+1, \lambda)/\Gamma(c+1)$, where $\gamma(c+1, \lambda) = \int_\lambda^\infty e^{-t}t^c dt$ is the incomplete gamma integral.

◼ EXAMPLE 6.40

Find the MD of Poisson distribution using the Power method Section 6.1 (p. 189). See Figure 6.5.

Solution 6.40   We know that the mean of Poisson distribution is λ. The lower limit ll is $x = 0$, so that $xF(x) = 0$. Hence using Theorem 6.1, the MD is given by

$$MD = 2\sum_{x=ll}^{\mu-1} F(x) = 2\sum_{x=0}^{c} \gamma(x+1, \lambda)/\Gamma(x+1), \tag{6.134}$$

where $c = \lfloor \lambda - 1 \rfloor$. See Table 6.12.   ◼

6.9.2.1   Approximations   The Poisson distribution provides a good approximation to the Binomial distribution $B(n, p)$ when p is small, provided $\lambda = np > 10$, and n is large enough. The accuracy of this approximation increases as p tends to zero. As mentioned earlier, this limiting behavior is more dependent on the rate at which $p \to 0$ faster than $n \to \infty$ (see Figure 6.7).

As the variance of the distribution is λ, normal approximations are not applicable for small λ values. However when $\lambda \to \infty$, the variate $Y = (X - \lambda)/\sqrt{\lambda}$ is approximately normal. The continuity correction can be applied as before to get $P[(X; \lambda) \lessgtr k] = \Pr[Z \lessgtr \frac{k - \lambda \pm 0.5}{\sqrt{\lambda}}]$. The square root transformation is a variance stabilizing transformation for this distribution. Many approximations have appeared in the literature based on this observation. For example, the Anscombe [147] approximation uses $2\sqrt{X + 3/8} \sim \text{NORM}(0,1)$. An improvement to this is the $\sqrt{X} + \sqrt{X + 1}$ transformation to normality suggested by Freeman and Tukey [148] . As the Poisson left-tail areas are related to the χ^2 right-tail areas, the individual Poisson probabilities can be approximated using the χ^2 probabilities.

6.9.2.2 Applications The Poisson distribution has been applied to various problems involving high uncertainty (low probability of occurrence). Examples are the number of false fire alarms in a building, number of flaws in a sheet roll of newly manufactured fabric, number of phone calls received by a telephone operator in a fixed time interval, number of natural calamities such as earthquakes and tsunamis in a fixed time interval (say 1 month), number of epidemics in a locality, number of deaths due to a rare disease, and so on.

◧ EXAMPLE 6.41   Structural damage

A dam is built to withstand water pressure and mild tremors. Let X denote the number of damages resulting from a major quake. If X is distributed as POIS(0.008), find the following probabilities: (i) probability of no damage, (ii) probability of at least two damages, and (iii) probability of at most four damages.

Solution 6.41 The PDF is $f(x, \lambda) = e^{-0.008}(0.008)^x/x!$, for $x = 0, 1, 2 \dots$. Answer to (i) is $p_0 = e^{-0.008}(0.008)^0/0! = e^{-0.008} = 0.9920$. Answer to (ii) is $1 - P(0) - P(1) = 1 - 0.992032 - 0.007936 = 1 - 0.99996817 = 3.18298E - 05$. (iii) Probability of at most four damages $= \sum_{x=0}^{4} e^{-0.008}(0.008)^x/x!$.   ■

6.9.3   Truncated Poisson Distribution

A useful distribution in epidemiological studies is a Poisson distribution truncated at 0. It is also used in search engine optimization. Assume that a user query returns a large number of matches that are displayed by a search engine in discrete screenfuls of say 10 matches each. Then, the number of pages viewed by a surfer can be modeled as a zero-truncated Poisson law or a zipf law [2]. The PDF is given by

$$p_x(\lambda) = e^{-\lambda}\lambda^x/[(1 - e^{-\lambda})x!] = \lambda^x/[(e^\lambda - 1)x!], x = 1, 2, \dots, \qquad (6.135)$$

where the second expression is obtained from the first by multiplying the numerator and denominator by e^λ. The mean and variance are $\lambda/(1 - e^{-\lambda})$. See Shanmugam [149]–[151] for incidence rate restricted Poisson distribution, Reference 152 for spinned Poisson distribution, and Reference 46 for a discussion on Poisson dispersion.

6.10 HYPERGEOMETRIC DISTRIBUTION

Consider a "lot" containing N items of which k are of one kind, and the rest $(N - k)$ are of another kind. We assume that the two kinds are indistinguishable. Suppose we sample n items *without replacement* from the lot. The number of items x of first kind is then given by

$$p(x) = \binom{k}{x}\binom{N-k}{n-x} / \binom{N}{n}, \quad \text{where } x = 0, 1, 2, ..., \min(n, k). \tag{6.136}$$

This is called the HGD, which has three parameters k, N and n. This can be derived using the following argument. As there are k items of one kind, we can choose x items from it in $\binom{k}{x}$ ways. To make the count to n, we need to select further $n - x$ items. However, these can be selected from $(N - k)$ items of second kind in $\binom{N-k}{n-x}$ ways. Using the product rule for selection (Chapter 5), the total number of ways is the expression in the numerator of equation (6.136). To make it a PDF, we need to divide it by the total number of ways to select n items, namely $\binom{N}{n}$. We have not made any assumptions on the items being sampled. In practical applications of this distribution, it could be defective and nondefective items, marked and unmarked items, successes and failures in independent Bernoulli trials, and so on.

As the expression involves binomial coefficients, there is a natural symmetry involved in the above-mentioned PDF. Instead of sampling x items from the first kind, we could take x items from the second kind and $(n - x)$ items from the first kind. This gives us the alternate PDF:

$$p(x) = \binom{N-k}{x}\binom{k}{n-x} / \binom{N}{n}. \tag{6.137}$$

To impose the range for both these forms, we modify the range of x values as $0, 1, 2, ..., \min(m, N - m, n)$. As all combination terms must exist, x varies between $\max(0, n + m - N)$ and $\min(m, N)$.

6.10.1 Properties of Hypergeometric Distribution

This distribution has three parameters, all of which are integers. The recurrence relation for the PDF is

$$h(x + 1; k, n, N) = h(x; k, n, N) * \frac{(n - x)(k - x)}{(x + 1)(n - k + x + 1)}. \tag{6.138}$$

As the parameters are all related, the following symmetries follow easily (i) $h(x; k, n, N) = h(x; n, k, N)$,   (ii) $h(x; k, n, N) = h(k - x; k, N - n, N)$,   and   (iii) $h(x; k, n, N) = h(n - x; N - k, n, N)$. Replace x by $x - 1$ in equation (6.138) to get

$$h(x; k, n, N) = h(x - 1; n, k, N) * \frac{(n - x + 1)(k - x + 1)}{x(N - k - n + x)}. \tag{6.139}$$

6.10.2 Moments of Hypergeometric Distribution

Factorial moments are easier to find due to the $x!$ in the denominator (of both the forms (6.136) and (6.137)). The rth falling factorial moment

$$\mu_{(r)} = E[x_{(r)}] = E[x(x - 1)..(x - r + 1)]$$

$$= \sum_x x(x - 1)..(x - r + 1) \binom{k}{x} \binom{N - k}{n - x} / \binom{N}{n}$$

$$= [1/\binom{N}{n}] \sum_{x=r}^{n} x(x - 1)..(x - r + 1)(k)_x/x! \binom{N - k}{n - x}. \tag{6.140}$$

Cancel out $x(x - 1) \ldots (x - r + 1)$ from $x!$ in the denominator and write $(k)_x = k(k - 1) \cdots (k - r + 1)(k)_{x-r}$ and take it outside the summation. This gives $\mu_{(r)} = (k)_r / \binom{N}{n} \sum_{x=r}^{n} \binom{k-r}{x-r} \binom{N-k}{n-x}$. Change the indexvar using $u = x - r$ to get

$$\mu_{(r)} = (k)_r / \binom{N}{n} \sum_{y=0}^{n-r} \binom{k - r}{y} \binom{N - k}{n - y - r}. \tag{6.141}$$

Using Vandermonde's identity (p. 6–6), this becomes

$$\mu_{(r)} = (k)_r \binom{N - r}{n - r} / \binom{N}{n}. \tag{6.142}$$

The mean is easily obtained from the above by putting $r = 1$ as nk/N. The variance is $(nk/N)(1 - k/N)(N - n)/(N - 1)$. Replace nk/N on the RHS by μ and write the multiplier as $(1 - m/N) * [(N - n)/(N - 1)]$. This shows that $\sigma^2 < \mu$ as both $(1 - m/N)$ and $(N - n)/(N - 1)$ are fractions. The mode of the distribution is $\lfloor (k + 1)(n + 1)/(N + 2) \rfloor$, which is greater than the mean. The MGF does not have simple form but is expressed in terms of hypergeometric functions as

$$M_x(t) = \binom{N - k}{n} / \binom{N}{n} \, {}_2F_1(-n, -k; N - k - n + 1; e^t). \tag{6.143}$$

Covariance is given by $\text{Cov}(x_i, x_j) = np_ip_j\frac{N-n}{N-1}$. The coefficient of skewness is as follows:

$$\beta_1 = \frac{(N - 2k)(N - 2n)\sqrt{N - 1}}{(N - 2)\sqrt{nk(N - k)(N - n)}}. \tag{6.144}$$

See Table 6.13 for more properties.

TABLE 6.13   Properties of Hypergeometric Distribution

Property	Expression	comments
Range of X	$x = \max(0, n - N + k), .., \min(k, n)$	Discrete, finite
Mean	$\mu = nk/N$	Real
Variance	$\sigma^2 = (nk/N)(1 - k/N)(N - n)/(N - 1)$	$= \mu(1 - k/N)(N - n)/$ $(N - 1), \sigma^2 < \mu$
Mode	$\lfloor(k + 1)(n + 1)/(N + 2)\rfloor$	
Skewness	$\gamma_1 = \dfrac{(N - 2k)(N - 2n)(N - 1)^{1/2}}{[nk(N - k)(N - n)]^{1/2}(N - 2)}$	
CV	$\{(N - k)(N - n)/[nk(N - 1)]\}^{1/2}$	
$\mathrm{Cov}(x_i, x_j)$	$np_i p_j \dfrac{N - n}{N - 1}$	
MD	$2 \sum\limits_{x=0}^{\mu}(nk - Nx)\dbinom{k}{x}\dbinom{N - k}{n - x} / \left[N\dbinom{N}{n}\right]$	$2\mu_2 * f_m[1+1/N]$
Factorial moments	$\mu_{(r)} = E[x_{(r)}] = (k)_r\dbinom{N - r}{n - r} / \dbinom{N}{n},$ $r = 1, 2, \ldots, \min(k, n)$	$(k)_r(n)_r/(N)_r$
MGF	$\dbinom{N - k}{n} / \dbinom{N}{n} {}_2F_1(-n, -k; N - k - n + 1; e^t)$	
PGF	$\dbinom{N - k}{n} / \dbinom{N}{n} {}_2F_1(-n, -k; N - k - n + 1; t)$	
Recurrence	$f(x)/f(x - 1) =$ $(n - x + 1)(k - x + 1)/[x(N - n - k + x)]$	

Symmetric when $N/2 = k$ or n. Write $(N - n)/(N - 1)$ as $1 - (n - 1)/(N - 1)$ to get another expression for variance.

Theorem 6.10 If X and Y are independent BINO(m, p) and BINO(n, p) random variables, then the distribution of $X|X + Y = n$ is hypergeometric and is independent of p.

Proof: Consider the random variable $Z = X + Y$. As the probability p is the same, this is distributed as BINO$(n + m, p)$. The conditional distribution of X given $Z = k$ is $P[X = x|Z = k] =$

$$P[X = x \cap Z = k]/P[Z = k] = P[X = x] * P[Y = k - x]/P[X + Y = k]. \quad (6.145)$$

As X and Y are independent, $X + Y \sim$ BINO$(m + n, p)$. Hence, we get

$$\binom{m}{x} p^x q^{m-x} \binom{n}{k - x} p^{k-x} q^{n-k+x} / \binom{m + n}{k} p^k q^{m+n-k}.$$

This reduces to $\dbinom{m}{x}\dbinom{n}{k-x} / \dbinom{m+n}{k}$. This obviously is independent of p. ∎

6.10.3 Approximations for Hypergeometric Distribution

Hypergeometric probabilities can be approximated by the binomial distribution. When N and k are large, $p = k/N$ is not near 0 or 1, and n is small with respect to both k and $N - k$, the HGD is approximately a BINO$(n, k/N)$. If k/N is small and n is large, the probability can be approximated using a Poisson distribution. Closed-form expressions for tail probabilities do not exist, except for particular values of the parameters. However in general

$$F_x(x; m, n, N) = F_x(y; N - m, N - n, N), \text{ where } y = N - m - n + x. \quad (6.146)$$

6.11 NEGATIVE HYPERGEOMETRIC DISTRIBUTION

This distribution is also called Markov–Polya distribution. The PDF is given by

$$f(x; a, b, n) = \binom{-a}{x}\binom{-b}{n-x} / \binom{-(a+b)}{n} \quad (6.147)$$

for $x = 0, 1, 2, \ldots, n$ where a, b, n are integers. The mean and variance are given by $\mu = E(X) = an/(a + b)$, variance $= \sigma^2 = abn(a + b + n)/[(a + b)^2(a + b + 1)]$. Replace $an/(a + b)$ on the RHS by μ, we get $\sigma^2 = \mu * b(a + b + n)/[(a + b)(a + b + 1)]$. Write the RHS as $\mu *[b/(a+b)*(a+b+n)/(a+b+1)]$. Beta binomial distribution discussed in the following is a special case when k and $(n - k)$ are integers.

6.12 BETA BINOMIAL DISTRIBUTION

This distribution can be obtained as the conditional distribution of binomial distribution in which the probability of success is distributed according to the beta law. Consider the binomial distributed random variable with PDF $b_x(n, p) = \binom{n}{x} p^x(1 - p)^{(n-x)}$, where p is distributed as $g_p(a, b) = (1/B(a, b))p^{a-1}(1 - p)^{b-1}$ for $0 \le p \le 1$. As p is a continuous random variable in the range $(0,1)$, we obtain the unconditional distribution of X by integrating out p from the joint probability distribution and using the expansion $B(a, b) = \frac{\Gamma(a)\Gamma(b)}{\Gamma(a+b)}$ as

$$f_x(n, a, b) = \int_0^1 f(x|p)g(p)dp = \binom{n}{x} / B(a, b) \int_{p=0}^1 p^{x+a-1}(1 - p)^{b+n-x-1}dp$$

$$= \binom{n}{x} \frac{\Gamma(a + b)}{\Gamma(a)\Gamma(b)} \frac{\Gamma(a + x)\Gamma(b + n - x)}{\Gamma(a + b + n)}. \quad (6.148)$$

By writing $\binom{n}{x} = n!/[x!(n - x)!] = \Gamma(n + 1)/[\Gamma(x + 1)\Gamma(n - x + 1)]$, this can also be written as $f_x(n, a, b) =$

$$\Gamma(n + 1)\Gamma(a+b)/[\Gamma(x+1)\Gamma(n - x+1)\Gamma(a)\Gamma(b)][\Gamma(a + x)\Gamma(b + n - x)/\Gamma(a + b + n)]. \quad (6.149)$$

This form is widely used in Bayesian analysis. The mean μ is most easily obtained from the conditional expectation as $E(X) = E[E(X|p)] = nE(p) = na/(a + b) = nP$, where $P = \frac{a}{(a+b)}$. The second raw moment $\mu'_2 = \mu * [n(1 + a) + b]/(a + b + 1)$, from which the variance follows as $\sigma^2 = nPQ + \frac{n(n-1)PQ}{a+b+1}$ where $Q = 1 - P$. This can also be written as $nPQ(a + b + n)/(a + b + 1)$. In Bayesian analysis, this is written as $n\pi(1 - \pi)[1 + (n - 1)\rho]$, where $\pi = P$ and $\rho = 1/(a + b + 1)$ is the pairwise correlation between the trials called overdispersion parameter. This form is obtained from the previous one by writing $a + b + n$ as $a + b + 1 + (n - 1)$ and dividing by the denominator $a + b + 1$ (see Table 6.14).

See Reference 153 for properties and generalizations.

6.13 LOGARITHMIC SERIES DISTRIBUTION

This is a special case of the left-truncated negative binomial distribution where the zero class has been omitted and the parameter k tends to one (Table 6.14).

Although log() is a continuous function, this is a discrete distribution with infinite support. It has PDF

$$f(x, p) = \begin{cases} q^x/[-x \log p] & \text{for } 0 < p < 1, x = 1, 2, \ldots \\ 0 & \text{elsewhere.} \end{cases}$$

TABLE 6.14 Properties of Beta Binomial Distribution

Property	Expression	Comments
Range of X	$x = 0, 1, \ldots, n$	Discrete, finite
Mean	$\mu = na/(a + b) = n * P$	Need not be integer
Variance(σ^2)	$nPQ + \dfrac{n(n - 1)PQ}{a + b + 1} =$ $nPQ(a + b + n)/(a + b + 1)$	$= \mu < \sigma^2$
Skewness γ_1	$(C + n)(b - a)/(D + 1)\sqrt{D/[nabC]}$	$C = a + b + n,$ $D = a + b + 1$
CV	$\sqrt{b(a + b + n)/[na(a + b + 1)]}$	$\sqrt{bC/[naD]}$
MD	$2\mu_2 * f_m[1 + 1/(n + 1)]$	
$E[X(X - 1) \cdots$ $(X - k + 1)]$	$k!/(1 - q)^{k+1} = k!/p^{k+1}$	
CDF	$[1 - B(n - k + b - 1, k + a + 1){}_3F_2(a, b, k)]/K$	$K = [B(a, b)B(n - $ $k, k + 2)(n + 1)]$
PGF	$\dbinom{n}{k} B(k + a, n - k + b)/B(a, b)$	
FMGF	${}_2F_1[a, -n, a + b, -t]$	

Variance is less than the mean for $(1/p - \mu) < 1$ or equivalently $p > 1/(1 + \mu)$.

An alternate parametrization is as

$$f(x, \theta) = \begin{cases} \alpha \ \theta^x / x & \text{for } \alpha = -[\log(1 - \theta)]^{-1}, 0 < \theta < 1, x = 1, 2, \dots \\ 0 & \text{elsewhere.} \end{cases}$$

6.13.1 Properties of Logarithmic Distribution

The logarithmic distribution has a single parameter p. The mean is $\mu = q/[-p\log(p)]$. As $0 < p < 1, \log(p)$ is negative, thereby canceling out negative sign. Variance is $\sigma^2 = -q(q + \log(p))/[(p \log(p))^2]$. In terms of μ, this is $\sigma^2 = \mu(1/p - \mu)$ or equivalently $\sigma^2 + \mu^2 = \mu/p$. Cross-multiply to get $p = \mu/(\sigma^2 + \mu^2)$. This shows that the variance is less than the mean for $(1/p - \mu) < 1$ or equivalently $p > 1/(1 + \mu)$. For the alternate representation (6.13), the mean is $\mu = a\theta/(1 - \theta)$ and variance is $\mu(1 - a\theta)/(1 - \theta)$. To fit the model, compute $\bar{x}$ and s^2 and find $\hat{p} = \bar{x}/(\bar{x}^2 + s^2)$. As the variance is +ve, $q < -\log(p)$. The factorial moments are easier to find than central moments. The kth factorial moment is given by

$$\mu_{(k)} = E[x(x - 1)..(x - k + 1)] = -\frac{(k - 1) \, !}{\log(p)}(q/p)^k. \tag{6.150}$$

The ChF is given by $\phi(t) = \ln (1 - qe^{it})/\ln (1 - q)$. As the values assumed by X are integers, it is used in those modeling situations involving counts. For instance, the number of items of a product purchased by a customer in a given period of time can be modeled by this distribution. See Table 6.15 for more properties.

6.14 MULTINOMIAL DISTRIBUTION

This distribution can be considered as a generalization of the binomial distribution with $n(>2)$ categories. The corresponding probabilities are denoted as p_i for the ith class such that $\sum_{i=1}^{n} p_i = 1$. We denote it by $MN(n, p_1, p_2, \dots, p_n)$. The PDF of a general multinomial distribution with k classes is

$$f(x; n, p_1, p_2, \dots, p_k) = \begin{cases} \frac{n \, !}{x_1 \, ! x_2 \, ! \cdots x_k \, !} p_1^{x_1} p_2^{x_2} \cdots p_k^{x_k} & \text{if } x_i = 0, 1, \dots, n \\ 0 & \text{elsewhere,} \end{cases}$$

where $x_1 + x_2 + \dots + x_k = n$ and $p_1 + p_2 + \dots + p_k = 1$. Using the product notation introduced in Chapter 1, this can be written as $(n!/ \prod_{i=1}^{k} x_i!) * \prod_{i=1}^{k} p_i^{x_i}$. This can also be written as $\begin{pmatrix} n \\ x_1, x_2, \dots, x_k \end{pmatrix}$, which is called the multinomial coefficient. As the p_i's are constrained as $\sum_{i=1}^{k} p_i = 1$, there are k parameters. For $k = 2$, this reduces to $BINO(n, p)$.

As in the case of binomial distribution, we could show that this distribution tends to the multivariate Poisson distribution:

TABLE 6.15   Properties of Logarithmic Distribution

Property	Expression	Comments
Range of X	$x = 0,1,...,\infty$	Discrete, infinite
Mean	$\mu = q/[-p\log(p)]$	$\alpha\theta/(1-\theta)$
Variance σ^2	$-q(q+\log(p))/(p\log(p))^2 = \mu(1/p - \mu)$	$= \mu((1-\theta)^{-1} - \mu)$, $\Rightarrow \mu < \sigma^2$
Mode	$1/\log(\theta)$	
Skewness	$\gamma_1 = [(1+\theta) - 3D + 2D^2]/[\sqrt{D}(1-D)^{3/2}$	$D = \alpha\theta$
Kurtosis	$(1 + 4\theta + \theta^2 - 4D(1+\theta) + 6D^2 - 3D^3/R$	$R = D(1-D)^2$
CV	$\sqrt{(1-\alpha\theta)}/\sqrt{\alpha\theta(1-\theta)^2}$	$(\alpha^{-1}\theta^{-1})^{1/2}$
Mean deviation	$2\alpha \sum_{k=1}^{\lfloor\mu\rfloor}(\mu-k)\theta^k/k$	
Moment recurrence	$\mu'_{r+1} = \theta[\partial/\partial\theta + \alpha/(1-\theta)]\mu'_r$	
Moment recurrence	$\mu_{r+1} = \theta\partial\mu_r/\partial\theta + r\mu_2\mu_{r-1}$	
Factorial moment	$\alpha\theta^r(r-1)!/(1-\theta)^r$	$(r-1)!(q/p)^r/-\log(p)$
PGF	$\log(1-\theta t)/\log(1-\theta)$	
MGF	$\log(1-\theta\exp(t))/\log(1-\theta)$	
Recurrence	$f(x+1;p)/f(x;p) = qx/(x+1)$	$(1-p)(1 - \dfrac{1}{x+1})$

Variance is less than the mean for $(1/p - \mu) < 1$ or equivalently $p > 1/(1+\mu)$.

Theorem 6.11 If n is large and p_i is small such that $np_i = \lambda_i$ remains a constant, the multinomial distribution approaches $e^{-(\lambda_1 + \lambda_2 + \cdots + \lambda_k)}\lambda_1^{n_1}\lambda_2^{n_2}\cdots\lambda_k^{n_k}/[n_1!n_2!\cdots n_k!]$.

Proof: The easiest way to prove this result is using pgf. As in the case of binomial distribution, it is easy to derive the pgf of multinomial as $(p_1t_1 + p_2t_2 + \cdots + p_kt_k)^n$. Now proceed as done in Section 6.5.8.   ∎

6.14.1   Properties of Multinomial Distribution

For each class, the means can be obtained using binomial distribution as $E(X_i) = np_i$, $\text{Var}(X_i) = np_iq_i$, and $\text{Cov}(X_i, X_j) = -np_ip_j$. As the covariance is negative, so is the correlation. This is because when one of them increases, the other must decrease due to the sum constraint on the X_i's. The ChF is given by $\phi(t) = [1 + \sum_{j=1}^m p_j(e^{it_j} - 1)]^n$. See Table 6.16 for more properties.

6.14.1.1   *Marginal and Conditional Distributions*   The marginal distributions are binomial that follows easily from the observation that the probabilities are obtained as terms in the expansion of $(p_1 + p_2 + \cdots + p_n)^N$. If marginal distribution of x_j is needed, put $p_j = p$ and the rest of the sum as $1-p$ (as $\sum_{i=1}^k p_i = 1, 1 - p = \sum_{i\neq j=1}^k p_i$). This results in the PGF of a binomial distribution.

TABLE 6.16   Properties of Multinomial Distribution

Property	Expression	Comments
Range of X	$x_i = 0, 1, \ldots, n$	Discrete, $\sum_{i=1}^{k} x_i = N$
Mean	$\mu = Np_i$	Need not be integer
Variance	$\sigma^2 = Np_i q_i$	$\mu > \sigma^2$
Covariance	$-np_i p_j$	$i \neq j$
Mode	$(x-1), x$ if $(n+1)*p_i$ is not integer	$x = (n+1)*p_i$ else
Skewness	$\gamma_1 = (1 - 2p_i)/\sqrt{np_i q_i}$	$= (q-p)/\sqrt{npq}$
Kurtosis	$\beta_2 = 3 + (1 - 6pq)/npq$	
$E[X(X-1)\cdots (X-k+1)]$	$n^{(r)} p^r$	
PGF	$\left[\sum_{j=1}^{m} p_j t_j\right]^n$	$= p^n(1+t)^n$ if each $p_i = p$
MGF	$\left[\sum_{j=1}^{m} p_j e^{t_j}\right]^n$	
ChF	$\phi(t) = \left[\sum_{j=1}^{m} p_j e^{it_j}\right]^n$	

Never symmetric, always leptokurtic. Satisfies the additivity property $\sum_{i=1}^{m}$ $MN(n_i, p_1, p_2, \ldots, p_m) = MN(\sum_{i=1}^{m} n_i, p_1, p_2, \ldots, p_m)$ if they are independent.

Conditional distributions of multinomials are more important as these are used in the expectation–maximization algorithms (EMAs) [22]. Let X_n be a multinomial distribution with k classes defined earlier. Suppose we have missing data in an experiment. For convenience, we assume that the first j components are observed, and $j + 1$ through k classes have missing data (unobserved). To derive the EMA for this type of problems, one needs to find the conditional distribution of X|observed variates. The conditional distribution of X_i given $X_j = n_j$ is binomial with parameters $n - n_j$ and probability $p_i/(1 - p_j)$.

As $X_{j+1}, X_{j+2}, \ldots, X_k$ are unobserved with respective probabilities $p_{j+1}, p_{j+2}, \ldots, p_k$, we write it using $P(A|B) = P(A \cap B)/P(B)$ as

$$P[X_{j+1} = m_{j+1}, \ldots, X_k = m_k | X_1 = m_1, \ldots, X_j = m_j]$$
$$= \frac{P[X_1 = m_1, \ldots, X_k = m_k]}{P[X_1 = m_1, \ldots, X_j = m_j]}. \tag{6.151}$$

Owing to the independence of the trials, this becomes

$$\frac{n!}{x_1! x_2! \cdots x_k!} \prod_{i=1}^{k} p_i^{x_i} \Big/ \frac{n!}{x_1! x_2! \cdots x_j!} \prod_{i=1}^{j} p_i^{x_i}. \tag{6.152}$$

Canceling out common terms, this can be simplified to a multinomial distribution. See Reference 154 for the mode of multinomial distribution,

◨ EXAMPLE 6.42   Human blood groups

Consider the human blood groups example in Chapter 5. Suppose that the percentage of people with the blood groups $\{A, B, O, \text{and } AB\}$ are 40, 12, 5, and 43, respectively. Find the probability that (i) in a group of 60 students, 30 or more are of blood group "A" and (ii) at least 4 persons have blood group O.

Solution 6.42 Using the frequency approach, we expect the probability of any person with blood group "A" as 0.40. Denote this as $p_1 = 0.40$. Similarly $p_2 = 0.12$, $p_3 = 0.05$, and $p_4 = 0.43$. This gives the PDF as

$$f(x) = 60!/[x_1!x_2!x_3!x_4!](0.40)^{x_1}(0.12)^{x_2}(0.05)^{x_3}(0.43)^{x_4} \qquad (6.153)$$

such that $x_1 + x_2 + x_3 + x_4 = 60$. Thus, the answer to (i) is $\sum_{x=30}^{60} \text{BINO}(60, 0.40)$, as the marginal distribution is binomial. (ii) Probability of "O" blood group is $5/100 = 1/20$. Thus, the answer is $1 - \sum_{i=0}^{3} \text{BINO}(60, 1/20) = 1 - \sum_{i=0}^{3} \binom{60}{i}(1/20)^i(19/20)^{60-i}$. ∎

6.15   SUMMARY

The collected data are either count or continuous number type. Several important discrete distributions encountered in probabilistic modeling are discussed and summarized in this chapter. Some of these are used in subsequent chapters. Sometimes, there are competing models (such as Poisson with small λ, geometric or logarithmic distributions that have striking similarities for some parameter values).

A manufactured item might meet engineering specifications (to be a quality item) or is a defective item otherwise. In a quality control inspection of sample items, the outcomes with respect to a specific item follow a Bernoulli distribution. The number of quality items in a random sample of n inspected items follows a binomial distribution. If the inspection of items is done until a defective item is encountered, then the number of items inspected until the termination of the inspection follows a geometric probability distribution. If a modification in the inspection process is made such that the inspection of items is continued until an accumulation of a specified number r of defective items, then the number of inspected items follows a negative binomial distribution.

Students and professionals are often interested in the tail probabilities of these distributions and approximations of it for power calculations. These tail probabilities can be obtained in closed form for some of the distributions. This in turn provides an alternative method to compute the mean deviation using the Power method introduced in Section 6.3 (p. 6–6). Several researchers have extended Abraham De Moivre's 1730 [130] result on the MD of a binomial distribution. The notable ones being by Bertrand

[131], Frisch [155], Kamat [156], Winkler [157], Diaconis and Zabell [132], Jogesh Babu and Rao [158], Pham-Gia et al. [159], Pham-Gia and Hung [160] (who also derives the distribution of sample mean absolute deviation), Egorychev et al. [161], and so on. In Section 6.3 (p. 6–6), we have provided a greatly simplified expression involving either the CDF (left-tail probabilities) or the SF (right-tail probabilities) when the mean μ is an integer or half integer.

EXERCISES

6.1 Mark as True or False

    a) For the binomial distribution mean is > variance

    b) Variance of a Bernoulli distribution lies between 0 and 1/4

    c) Poisson distribution is bimodal if λ is noninteger

    d) Poisson distribution satisfies the memory-less property

    e) Geometric distribution has infinite range

    f) Mean of a negative-binomial distribution is always greater than the variance

    g) The variance of BINO(n, p) and BINO(n, q) are the same

    h) Truncated discrete uniform distributions are of the same type.

6.2 Which of the following distributions have infinite range?
    (a) binomial (b) negative binomial (c) discrete uniform (d) hypergeometric

6.3 If the mean and variance of a binomial distribution are equal, what is the value of p?

6.4 If the mean and variance of a negative binomial distribution are equal, what is the value of p?

6.5 If $X \sim$ GEO(0.5), find $P[X \geq 3]$, and $P[X = 5]$.

6.6 For which distribution is $f(x)/f(x+k) = 1$ for all k in the range?.

6.7 Prove that the binomial distribution attains its maximum at $k = \lfloor (n+1)p \rfloor$. If k is an integer, then there are two maxima at $k = (n+1)p$ and $k = (n+1)p - 1 = np + p - 1 = np - q$.

6.8 If X and Y are independently and identically distributed geometric random variables, find the probability of each of the following:
(a) $P[X = Y]$ (b) $P[X \geq 2Y]$ (c) $P[X > Y]$ (d) $P[X|X + Y = (n + 1)] = 1/n$.

6.9 Using the binomial expansion, derive the following i) $\sum_{k=0}^{n} \binom{n}{k} = 2^n$ ii) $\sum_{k=0}^{n} (-1)^k \binom{n}{k} = 0$

6.10 If X is BINO(n, p) prove that $E(X/n) = p$, and Var$(X/n) = pq/n$.

6.11 Find the mean and variance of the distribution discussed in section 6.5.4, p. 204.

6.12 Prove that the mode of a negative binomial distribution NB(k, p) is $\lfloor (q/p)(k - 1) \rfloor$.

6.13 If $X_1, X_2, \ldots, X_n$ are independent BER(p) random variables, find the distribution of $U = \prod_{i=1}^{n}(1 + kX_i)$, where k is a constant.

6.14 Prove that variance of DUNI(N) is $\sigma^2 = \mu(\mu - 1)/3$ where $\mu = (N + 1)/2$. Hence or otherwise show that the variance is greater than the mean for $N > 7$.

6.15 Show that the coefficient of variation (CV) of a binomial distribution is CV $= \sqrt{q/(np)}$.

6.16 Show that the third moment μ_3 of a binomial distribution is npq $(q - p)$.

6.17 Prove that the variance of a binomial distribution is always less than the mean.

6.18 Find the covariance Cov(X_i, X_j) for multinomial distribution. Why is it negative?.

6.19 Can the Poisson approximation to BINO(n, p) be used when n is small? (say $n < 15$)?. If so, under what conditions?.

6.20 Prove that the NB(k, p) with PDF $\binom{x+k-1}{x} p^k q^x$ can be obtained using the expansion $(q - p)^{-k}$, where $q = 1 - p$.

6.21 The mean and variance can never be equal for which of the following distributions? (A) Binomial (B) Poisson (C) geometric (D) negative binomial.

6.22 What is the probability that a fair coin need to be flipped ten times to get the 5th head on 10th flip? 9th head on 10th flip?.

6.23 Find tail probabilities of binomial distributions using incomplete beta function. (i) Pr[$B(10, 0.4) \leq 6$], (ii) Pr[$B(22, 0.7) \geq 18$], (iii) Pr[$B(40, 0.2) \geq 35$].

6.24 Prove that a change of origin transformation simply displaces the discrete uniform distribution to the left or right with PDF $f(Y = a + x) = 1/N$, $y = a + 1, a + 2, \ldots, a + N$.

6.25 Which of the following discrete distributions is always leptokurtic?
(A) Binomial (B) Poisson (C) Geometric (D) Discrete uniform.

6.26 For which discrete distribution is the variance always greater than the mean? (a) Binomial (b) Poisson (c) Geometric (d) Discrete uniform.

6.27 Truncation never changes the skewness of which distribution?
(A) Binomial (B) Poisson (C) Geometric (D) Discrete uniform.

6.28 Which of the following discrete distributions is never symmetric?
(a) Binomial (b) Poisson (c) Geometric (d) Hypergeometric.

6.29 $Z = \min(X_1, X_2, \ldots, X_n)$ is identically distributed for which of the following discrete distributions?
(a) Binomial (b) Poisson (c) Geometric (d) Uniform.

6.30 For which distribution does the individual probabilities (divided by the first probability p) form a geometric progression?.

6.31 If X $\sim$ POIS(λ), prove that log(X) is approximately normal with mean $\log(\lambda)$ as $\lambda \to \infty$. What is the variance?.

6.32 Prove that the probability generating function (PGF) of DUNI(N) is $P_x(t) = (1 - t^n)/[n(1 - t)]$.

6.33 Prove that BINO$(n, \lambda/n) \to$ POIS(λ) as $n \to \infty$. What about limiting behavior of BINO$(n, 1 - \lambda/n)$?

6.34 Describe how you can approximate binomial probabilities using a Poisson distribution when p is not so small, but is near 1 (say 0.98).

6.35 Find the mode of negative binomial NBINO(k, p). Show that $\sigma^2 = \mu(1 + \mu/k)$.

6.36 If X is a discrete distribution such that $P(X = a) = p$ and $P(X = b) = 1 - p$, find distribution of $(X - a)/(b - a)$.

6.37 If $X \sim$ Geometric(p) find the distribution of $Y = \exp(x)$.

6.38 If X is discrete with support 1,2,... prove $E(X) = \sum_{k=1}^{\infty} \Pr[X \geq k]$.

6.39 If $X \sim$ Poisson(λ), find the probability that (i) X takes even values, (ii) X takes odd values.

6.40 If $X \sim$ Geometric(p) find the probability that (i) X takes even values, (ii) X takes odd values.

6.41 Let $X \sim$ Binomial(n, p) and $Y_k \sim$ Negative Binomial(k, p). Prove that $\Pr[X \geq k] = \Pr[Y_k \leq n]$.

6.42 Find x such that the binomial left-tail probabilities (say α) are (i) $B(20, 0.8)$ with $\alpha = 0.41145$ (ii) $B(10, 0.7)$ with $\alpha = 0.38278$?.

6.43 If $X \sim$ Binomial(n, p), find the covariance of $(X/n, (n - X)/n)$.

6.44 Obtain the mean and variance of zero-truncated geometric distribution.

6.45 What is the skewness of a discrete uniform distribution? Show that it is always platykurtic.

6.46 If $X \sim$ Geometric(p) with PDF $f(x; \lambda) = \frac{1}{1+\lambda}\left(\frac{\lambda}{1+\lambda}\right)^x$ for $x = 0, 1, 2, \ldots$ find the mean and variance.

6.47 Prove that truncating a DUNI(N) distribution results in another distribution of the same type as follows. If the truncation is at a single point on either extremes, the new distribution is DUNI($N - 1$). If the truncation is at both tails (one point each), the resulting distribution is DUNI($N - 2$). If k points are truncated at both tails, the new distribution is DUNI($N - 2k$).

6.48 If $X \sim$ BINO(n, p) where $p \to 1$ from below, find the limiting distribution of X when $n \to \infty$ and nq remains a constant.

6.49 Obtain the skewness coefficient of geometric distribution and argue that it is never symmetric.

6.50 If $X \sim$ GEO(p) where $p = e^{-\lambda}$ with PDF $f(x; \lambda) = e^{-\lambda}(1 - e^{-\lambda})^x$

for $x = 0, 1, 2, \ldots$ find the mean and variance.

6.51 Find the mean and variance of truncated NBINO(k, p) distribution $f(x; k, p) = \frac{1}{1-q^{-k}}\binom{k+x-1}{k-1}\frac{1}{(1+p)^k}\left(\frac{p}{q}\right)^x$, $x = 1, 2, 3 \ldots$.

6.52 Show that the truncated negative binomial distribution satisfies the recurrence $(1 + p)(k + 1)f_{k+1}(x) = p(n + k)f_k(x), k = 1, 2, \ldots$

6.53 Prove that the hypergeometric distribution $\binom{k}{x}\binom{N-k}{n-x} / \binom{N}{n}$ tends to the binomial distribution BINO(n, p) where $p = k/N$ as $N \to \infty$.

6.54 Let X denote the number of successes (or Heads) and Y denote the number of failures (or Tails) in n independent Bernoulli trials with the same probability of success p. Find the distribution of $U = (|X - Y| + 1)/2$ for n odd.

6.55 An urn contains 10 red and 6 blue balls. A sample of five balls is selected at random. Let Y denote the number of red balls in the sample. Find the density function of Y if sampling is with replacement and without replacement.

6.56 A company does tele marketing to sell its products. Three tele-operators X, Y, Z contact customers over the telephone and explains the company's products to them to get possible orders. The average success rate in selling at least one item out of 100 tele-contacts is 12 for X, 7 for Y, and 3 for Z with respective standard deviations 5, 2, and 1. Find the probability for each of the following: (a) X is able to get 20 or more sales orders out of 300 customers and (b) Y and Z together gets 30–60 orders.

6.57 Describe how to generate random numbers from negative binomial distribution using a random number generator for geometric distribution.

6.58 If X_i for $i = 1, 2, \ldots, n$ are IID geometric random variables, find the distribution of $Z_i = \min(X_1, X_2, \ldots, X_n)$.

6.59 Which of the discrete distributions satisfy: $P(X \geq s + t) | P(X \geq s) = P(X \geq t)$.

6.60 For which discrete distributions is the variance always less than the mean?

6.61 If $X_1, X_2, \ldots, X_m$ are independent BER(p_i), where each p_i is either equal to p or equal to q, find the distribution of $\sum_{i=1}^{m} X_i$.

6.62 If X and Y are independent BINO(n, p) find the distribution of $X | X + Y = n$. Find its mean and variance.

6.63 Prove that the variance of discrete uniform distribution is greater than the mean for $N > 7$.

6.64 Find mean and variance of

$$f(x; n, \mu) = \binom{n}{x} (\mu/n)^x ((n - \mu)/n)^{(n-x)}.$$

6.65 Find the PDF and CDF of the discrete uniform distribution when x takes the values $0, 1, \ldots, N$. Obtain the PGF, and the variance.

6.66 Find the mean deviation from the mean of a geometric distribution with PDF $f(x; p) = q^{x-1}p$, where $x = 1, \ldots$.

6.67 If X is distributed as $BINO(n,p)$ find $E(|X/n - p|)$ using power method.

6.68 What is the variance of (i) standard uniform distribution? (ii) truncated uniform distribution?

6.69 Prove that the maximum variance of a Bernoulli random variable is $1/4$ and that of a binomial distribution is $n/4$.

6.70 For which of the following values of p does the geometric distribution $q^x p$ tails off slowly? (a) $p = 0.1$ (b) $p = 0.5$ (c) $p = 0.8$ (d) $p = 0.9$

6.71 Obtain an expression for the mean deviation of a negative binomial distribution using incomplete beta function.

6.72 Prove that the mean deviation from the mean of the Poisson distribution is $2 * \exp(-\lambda)\lambda^{\lfloor\lambda\rfloor+1}/\lfloor\lambda\rfloor!$.

6.73 What is the probability distribution of $n - x$ failures in a $BINO(n,p)$? What is its variance?

6.74 If $X_i, i = 1, 2, \ldots, r$ are IID $GEO(p)$ random variables, prove that $\sum_{i=1}^{r} X_i \sim NBIN(r,p)$.

6.75 If X and Y are IID $GEO(p)$ find the distribution of (i) $X|(X + Y = 2n)$, (ii) $X|(X - Y = n)$.

6.76 What is the variance of limiting binomial variate with fixed n when $p \to 0$ from above or $q \to 1$ from below?

6.77 If X and Y are IID $GEO(p)$, find the distribution of $U = |X - Y|$, and its mean.

6.78 Prove that in a multinomial distribution $Cov(x_i, x_j) = -np_i p_j$, and
$$Corr(x_i, x_j) = -\left(\frac{p_i}{1-p_i}\frac{p_j}{1-p_j}\right)^{1/2}.$$

6.79 Describe independent Bernoulli trials. If $U = |X - Y|/2$ where X and Y are the number of success and failures, what is distribution of U if $p = 1/2$?

6.80 If a negative binomial distribution is defined as the number of trials needed to produce k successes in n trials, obtain a moment recurrence and find the mean.

6.81 If $X \sim geometric(p)$ find the probability that $X > \lceil |\log(p)| \rceil$

6.82 If X and Y are IID $GEO(p)$, show $P[X - Y = 0] = p^2/(1 - q^2) = p/(1 + q)$.

6.83 If $X \sim BINO(n,p)$ find the distribution of $Y = (X - np)^2$. Find its mean and mode.

6.84 If $X \sim NBINO(k,p)$ find a recurrence relation for moments using $f(x + 1, k, p) = q * f(x, k, p) * (x + k)/(x + 1)$.

6.85 Probability that high-rise structures in a city center will damage x other adjacent buildings after an earthquake is $POIS(0.008)$. Find the probability that it will damage three or more buildings in the vicinity.

6.86 If the mean and variance of a logarithmic distribution are $\mu = q/[-p\log(p)]$ and $-q(q + \log(p))/[(p\log(p))^2]$, prove that $p = \mu/(\sigma^2 + \mu^2)$. How can this be used to fit the distribution to data?

6.87 Wavelength W (in nanometer) and stopping potential V of a photo-electric surface are given in Table 6.17. Fit a logarithmic distribution and obtain the mean.

TABLE 6.17   Properties of Photo-electric Surface

W	360	400	440	480	540	580	600
V	1.45	1.12	0.9	0.6	0.35	0.24	0.18

6.88 It is given that the mean of a Binomial distribution is 6 and variance is 4.8. What is the probability of success p? What is n?

6.89 For which distribution does the change of origin transformation $Y = X + 1$ and left truncation at $X = 0$ result in the same law?

6.90 Prove that $\text{BINO}(n, p) \to \text{POIS}(\lambda)$ when $n \to \infty$ and $p^2 \to 0$ and np remains constant. Explain how to use this approximation when $p \to 1$.

6.91 Check if $f(x) = K * 2n/(n^2 - x^2)$ for $x = 1, 2, \ldots$ is a PDF for $K = 1/\pi \, \cot(\pi n)$

6.92 Is $f(x) = K/(n + x)^2$ for $x = $ integer $\in (-\infty, \infty)$, a PDF where $K = \pi^2/\sin^2(\pi n)$?

6.93 Prove that the geometric distribution $f(x; p) = q^{x-1} p$ can be obtained from $f(x; p) = q^x p$ by truncation at $x = 0$. How are the means and variances related?

6.94 The probability of success p of a $\text{BINO}(n, p)$ is given by a root of the quadratic equation $x^2 - x + \frac{cd}{2(c^2+d^2)} = 0$. Find μ. What are the conditions on c and d?

6.95 Check if $f(x) = K * 2^x \binom{n-x}{x}$ is a PDF for $x = 0, 1, \ldots \lfloor n/2 \rfloor$.

6.96 Check if $f(x) = K * (-1)^x \binom{n-x}{x} 2 \cos(c)^{n-2x}$ is a PDF for $x = 0, 1, \ldots \lfloor n/2 \rfloor$.

6.97 Use the Power method introduced in Section 6.3, page 6–6 to find the mean deviation of truncated binomial distribution (truncated at $x = 0$).

6.98 Use the Power method introduced in Section 6.3, page 6–6 to find the mean deviation of truncated Poisson distribution (truncated at $x = 0$).

6.99 The plumbing system of a high-rise building uses pipes from two sources. The probability of becoming defective in 2 years after installation is Poisson distributed with $\lambda = 0.003$ for both. If 120 pipes from source-one and 185 pipes from source-two are used, what is the probability that at least one pipe will leak in 2 years? What is the probability that at most 2 will leak?

6.100 A dyeing plant uses six steam boilers. The probability of any boiler exploding in 6 years time is Poisson distributed with $\lambda = 1/10,000$. What is the probability that none of the boilers will explode in 6 years?. If two new boilers with probability of exploding $\lambda = 1/12,000$ are added, find the probability that at most two boilers explode during 6 years from the time of installation.

6.101 Assume that a dam is built to hold 50,000 cubic feet of water. The exceedence in any year has a POIS(0.018). If the peak flow is independent yearly, find the probability of (i) at least three exceedences in a year and (ii) exactly two exceedences in a year.

6.102 A newly manufactured micro-chip is known to have 0.0001 probability of failing on any whole day (24 hours). Find the probability that (i) it will last at least 60 days, (ii) it will last between 50 and 100 days, and (iii) it will last at most 90 days (hint: use GEO(0.0001)).

6.103 A highway-patrolman is looking for speeding vehicles. It is known from past data that the total number of vehicles Y going beyond the set speed limit on a stretch of a highway during busy hours is $N * f(x; p)$, where $N =$ total number of vehicles passed during the time period and X has a logarithmic distribution with $f(x; p) = q^x / [-x \log p]$ with $p = 0.5$ and $x = 1, 2, \ldots$ represents the difference in speed in miles per hour beyond the permitted limit ($x =$ (actual speed-speed-limit) in miles per hour to nearest int). If 200 vehicles go above the speed limit in 1 hour, approximately how many vehicles are in the grace bracket $x \leq 5$? Approximately how many vehicles are ≥ 10 mph above the speed limit?. If an over-speed vehicle is stopped at random, what is the probability that it exceeds 10% of the speed limit?

6.104 A civil engineer wishes to test if adding a heated tri-chloride of aluminum to cement mixtures can improve the strength of high-rise structures to withstand powerful earthquakes and aftershocks. From laboratory tests, it is found that the probability of a crack developing in this type of cement is Poisson distributed with mean one in 6 thousand. If a building portfolio comprising 10 buildings in a city neighborhood is built using aluminum hardened cement, what is the probability that at least one building will develop cracks after an earthquake? What is the probability that at most two buildings develop cracks?

6.105 A variety of seed is experimented in a laboratory and is found to have a germination rate of 95% (95 out of 100 seeds will germinate). If 10,000 seeds are sawed in identical conditions at a field, what is the probability that (i) at least 99% will germinate and (ii) between 90% and 98% will germinate.

7

CONTINUOUS DISTRIBUTIONS

After finishing the chapter, students will be able to

- Understand various continuous distributions
- Describe basic properties of common continuous distributions
- Explain memory-less property of exponential distribution
- Utilize the limiting behavior of some distributions
- Comprehend the incomplete beta and gamma functions
- Apply continuous distributions in practical problems
- Use the Power method to find the MD of continuous distributions
- Explore the Power method in other applications

7.1 INTRODUCTION

Continuous distributions are encountered in many industrial experiments and research studies. For example, measurement of quantities (such as height, weight, length, temperature, conductivity, and resistance) on the ratio scale is continuous or quantitative data.

Definition 7.1 The variable that underlies quantitative data is called a continuous random variable, as they can take a continuum of possible values in a finite or infinite interval.

Statistics for Scientists and Engineers, First Edition. Ramalingam Shanmugam and Rajan Chattamvelli.
© 2015 John Wiley & Sons, Inc. Published 2015 by John Wiley & Sons, Inc.

This can be thought of as the limiting form of a point probability function, as the possible values of the underlying discrete random variable become more and more of fine granularity. Thus, the mark in an exam (say between 0 and 100) is assumed to be a continuous random variable, even if fractional marks are not permitted. In other words, even though marks are not measured at the finest possible granularity level of fractions, it can be modeled by a continuous law. If all students scored between say 50 and 100 in an exam, the observed range for that exam is of course $50 \leq x \leq 100$. This range may vary from exam to exam, so that the lower limit could differ from 50, and the upper limit of 100 is never realized (nobody got a perfect 100). As shown below, this range is in fact immaterial in several statistical procedures.

All continuous variables need not follow a statistical law. However, there are many phenomena that can be approximated by one of the statistical distributions such as the normal law, if not exact. For instance, errors in various measurements are assumed to be normally distributed with zero mean. Similarly, measurement variations in physical properties such as diameter, size of manufactured products, and exceedences of dams and reservoirs are assumed to follow a continuous statistical law. This is because they can vary in both directions from an ideal measurement or value called its central value. This chapter introduces the most common continuous univariate distributions. An extensive treatment requires entire volumes by itself. Our aim is to summarize the basic properties that are needed in subsequent chapters.

Before we proceed to discuss the popular distributions, we first derive a general method to find the mean deviation (MD) of continuous distributions. This result will be used extensively throughout the chapter to derive the MD of various distributions.

7.2   MEAN DEVIATION OF CONTINUOUS DISTRIBUTIONS

Finding the MD of continuous distributions is a laborious task, as it requires a lot of meticulous arithmetic work. It is also called the mean absolute deviation or L_1-*norm* from the mean. The MD is closely related to the Lorenz curve used in econometrics, Gini index and Pietra ratio used in economics and finance, and in reliability engineering. It is also used as an optimization model for hedging portfolio selection problems [162, 163], fuzzy multisensor object recognition [164], and minimizing job completion times on computer systems [165]. See also Jogesh Babu and Rao [158] for expansions involving the MD and Pham-Gia and Hung [160] for the sampling distribution of MD.

Johnson [125] surmised that the MD of some continuous distributions can be put in the form $2\mu_2 f_m$ where $\mu_2 = \sigma^2$ and f_m is the probability density expression evaluated at the integer part of the mean $m = \lfloor \mu \rfloor$. This holds good for exponential, normal, and χ^2 distributions. Kamat [156] generalized Johnson's result to several continuous distributions, (see Table 7.1). The multiplier is distribution specific (see the following discussion). The following theorem greatly simplifies the work and is very helpful to find the MD of a variety of univariate continuous distributions. It can easily be extended to the multivariate case and for other types of MDs such as mean deviation from the median and medoid.

TABLE 7.1   Summary Table of Expressions for MD

Name	Expression	Johnson's Conjecture
Bernoulli	$2pq$	$2\mu_2$
Binomial	$2npq\begin{pmatrix} n-1 \\ \mu-1 \end{pmatrix} p^{\mu-1}q^{n-\mu}$	$2\mu_2 f_m$
Negative binomial	$2c\begin{pmatrix} k+c-1 \\ c \end{pmatrix} q^c p^{k-1}$	$2\mu_2 f_m$
Poisson	$2*\exp(-\lambda)\lambda^{\lfloor\lambda\rfloor+1}/\lfloor\lambda\rfloor!$	$2\mu_2 f_m$
Geometric	$(2/q)\lfloor 1/p\rfloor(q^{\lfloor 1/p\rfloor})$	$2\mu_2 f_m$
Hypergeometric	$2\sum_{x=0}^{\mu}(nk-nx)\begin{pmatrix} k \\ x \end{pmatrix}\begin{pmatrix} N-k \\ n-x \end{pmatrix} / \left[N\begin{pmatrix} N \\ n \end{pmatrix}\right]$	$2\mu_2 * f_m[1+1/N]$
Beta-binomial	$2nPQ(a+b+n)/(a+b+1)$	$2\mu_2 * f_m[1+1/(n+1)]$
Discrete uniform	$(N^2-1)/(4N)$	$3\mu_2 f_m$
Logarithmic	$2\alpha\sum_{k=1}^{\lfloor\mu\rfloor}(\mu-k)\dfrac{\theta^k}{k}$	$2\mu_2 * f_m[1+1/(n+1)]$
Continuous uniform	$(b-a)/4$	$3\mu_2 f_m$
Exponential	$2/(e\lambda)$	$2\mu_2 * f_m$
Central chi-square	$e^{-n/2}n^{n/2+1}/[2^{n/2-1}\Gamma(n/2+1)]$	$2\mu_2 f_m$
Normal	$2\sigma*1/\sqrt{2\pi}=\sigma\sqrt{2/\pi}$	$2\mu_2 f_m$
Inverse Gaussian	$4\exp(2\lambda/\mu)\Phi(-2\sqrt{\lambda/\mu})$	
Double-exponential	b	$\mu_2 f_m$

See respective sections for other distributions.

Theorem 7.1 Power method to find the Mean Deviation

The MD of any continuous distribution that tails off to the left can be expressed in terms of the CDF as

$$MD = 2\int_{ll}^{\mu} F(x)dx, \tag{7.1}$$

where ll is the lower limit of the distribution, μ the arithmetic mean, and $F(x)$ the CDF.

Proof: By definition

$$E|X-\mu| = \int_{ll}^{ul} |x-\mu|f(x)dx, \tag{7.2}$$

where ll is the lower, and ul is the upper limit of the distribution. Split the range of integration from ll to μ, and μ to ul, and note that $|X-\mu| = \mu - X$ for $x < \mu$. This gives

$$E|X-\mu| = \int_{x=ll}^{\mu} (\mu-x)f(x)dx + \int_{x=\mu}^{ul} (x-\mu)f(x)dx. \tag{7.3}$$

∎

As $E(X) = \mu$, we can write $E(X - \mu) = 0$, where $E()$ is the expectation operator. Expand $E(X - \mu)$ as

$$E(X - \mu) = \int_{ll}^{ul} (x - \mu)f(x)dx = 0. \tag{7.4}$$

As done earlier, split the range of integration from ll to μ and μ to ul to get

$$E(X - \mu) = \int_{x=ll}^{\mu} (x - \mu)f(x)dx + \int_{x=\mu}^{ul} (x - \mu)f(x)dx = 0. \tag{7.5}$$

Substitute $\int_{x=\mu}^{ul}(x - \mu)f(x)dx = -\int_{x=ll}^{\mu}(x - \mu)f(x)dx$ in equation (7.3) to get

$$E|X - \mu| = \int_{x=ll}^{\mu} (\mu - x)f(x)dx - \int_{x=ll}^{\mu} (x - \mu)f(x)dx = 2 \int_{x=ll}^{\mu} (\mu - x)f(x)dx.$$

Split this into two integrals and integrate each of them to get

$$E|X - \mu| = 2 \left[\mu * F(\mu) - \int_{x=ll}^{\mu} xf(x)dx \right], \quad \text{if } F(ll) = 0. \tag{7.6}$$

Use integration-by-parts to evaluate the second expression. $\left\{ xF(x) \big|_{ll}^{\mu} - \int_{ll}^{\mu} F(x)dx \right\} = \mu * F(\mu) - ll * F(ll) - \int_{ll}^{\mu} F(x)dx$. The $\mu * F(\mu)$ terms cancel out leaving behind

$$E|X - \mu| = 2 \left[x F(x)\big|_{ll} + \int_{ll}^{\mu} F(x)dx \right]. \tag{7.7}$$

Here $x F(x)\big|_{ll} = ll * F(ll)$ means that we are to evaluate the limiting value of $x * F(x)$ at the lower limit of the distributions. For those distributions that extend to $-\infty$, this limit is obviously zero. If the lower limit of the distribution is either zero or it tails off to the limit, the first term in equation (7.7) is zero. If $F(x)$ contains expressions of the form $(x-ll)$, then also this term is zero. Similarly for distributions with range $x \geq 0$ for which the mode is not ll, $F(0) \to 0$ as $x \to 0$. If the distribution is symmetric, equation (7.7) becomes

$$E|X - \mu| = \left[\int_{ll}^{\mu} F(x)dx + \int_{\mu}^{ul} S(x)dx \right], \tag{7.8}$$

where $S(x) = 1 - F(x)$ is the survival function (SF). If the distribution tails off to the right, we evaluate this as

$$MD = 2 \int_{\mu}^{ul} S(x)dx. \tag{7.9}$$

Otherwise, we need to evaluate both the terms in equation (7.7). These situations are illustrated in the numerous MD examples throughout the chapter. Thus, we could write the aforementioned for distributions that tails off to one of the extremes as

$$MD = 2 \int_{ll}^{\mu} F(x)dx = 2 \int_{\mu}^{ul} S(x)dx. \tag{7.10}$$

This representation of MD in terms of CDF is extremely helpful when one needs to evaluate the MD using the CDF or SF. See the example on the MD of beta-I distribution, which is represented in terms of incomplete beta function in page 271.

In addition to finding the MD of continuous distributions, this formulation has other important applications in proving convergence of distributions and central limit theorems. Replace X on the LHS by $S = X_1 + X_2 + \cdots + X_n$. If X_i's are identically distributed continuous random variables, this has mean $n\mu$ so that the relationship becomes

$$E|X_1 + X_2 + \cdots + X_n - n\mu| = \left[\int_{ll}^{\mu} F(x)dx + \int_{\mu}^{ul} S(x)dx \right], \qquad (7.11)$$

where $F(x)$ and $S(x)$ are CDF and SF of S. Dividing both sides by n, we see that the LHS is the arithmetic mean $(X_1 + X_2 + \cdots + X_n)/n$ and RHS has an "n" in the denominator and $F(x)$, $S(x)$ are the CDF and SF of the mean rather than x. Taking the limit as $n \to \infty$, we see that the RHS tends to zero (because both the integrals are bounded for finite mean (see equation 7.13 given below) and the LHS converges to μ. This provides a simple and elegant proof for the asymptotic convergence of independent random variables, which can be extended to other cases. For example, if $g(X_i)$ is a continuous function of X_i with finite mean $v = g(\mu)$, replacing X_i by $g(X_i)$ in equation (7.11) provides a simple proof on the convergence of $g(X_i)$ to its mean asymptotically. Similarly, MD of functions of random variables can be easily obtained from equation (7.1) by replacing X with $Y = g(X)$, μ with $v = g(\mu)$, and $F(x)$ with the CDF of Y.

There are two other novel methods to find the MD. The first one uses generating functions (Chapter 9, Section 9.4, p. 381) to fetch a single coefficient of $t^{\mu-1}$ in the power series expansion of $(1 - t)^{-2} P_x(t)$, where $P_x(t)$ is the probability generating function. This works best for discrete distributions. The second method is using the inverse of distribution functions (Chapter 10, Section 10.4, p. 402). This works best for continuous distributions.

Similar expressions are available for the MD around the median as

$$E|X - \text{Median}| = \int_0^{\frac{1}{2}} (F^{-1}(1 - x) - F^{-1}(x))dx = \int_0^{\frac{1}{2}} (S^{-1}(x) - S^{-1}(1 - x))dx. \qquad (7.12)$$

Theorem 7.2 Variance of continuous distributions as tail areas
    Prove that the variance of a continuous distribution can be expressed in terms of tail areas.

Proof: We found earlier that MD $= 2 \int_{x=ll}^{\mu} F(x)dx = 2 \int_{x=\mu}^{ul} S(x)dx$, where $F(x)$ is the CDF and $S(x)$ is the SF. Equating Johnson's result that MD $= c \, \mu_2 f_m$ where $\mu_2 = \sigma^2$ and f_m is the probability mass evaluated at the integer part of the mean $m = \lfloor \mu \rfloor$ we get $\mu_2 * cf_m = 2 \int_{x=ll}^{\mu} F(x)dx$. Divide both sides by cf_m to get

$$\mu_2 = \sigma^2 = (2/(c * f_m)) \int_{x=ll}^{\mu} F(x)dx = (2/(c * f_m)) \int_{x=\mu}^{ul} S(x)dx. \qquad (7.13)$$

∎

An alternate expression given by Jones & Balakrishnan is

$$\sigma^2 = 2 \int_y \int_{-ll<x<y}^{ul} F(x)[1 - F(y)]dxdy, \qquad (7.14)$$

where ll and ul are the lower and upper limits [166]. See also Chapter 8 (p. 363). The constant multiplier ($c = 2$) proposed by Johnson may be different for some continuous distributions (e.g., for Laplace distribution $c = 1$) (see Table 7.1). Even in those situations, the above-mentioned result holds in general because the RHS of equation (7.13) simply get scaled by the constant (c).

7.2.1   Notion of Infinity

Another important point to remember in the study of continuous distributions is the notion of infinity (∞). Chapter 6 introduced several discrete distributions that extend to infinity. Examples are the Poisson, geometric, negative binomial, and logarithmic laws. All of them extend to $+\infty$. Here, ∞ is assumed to be a large *integer* (because discrete distributions take integer values; usually nonnegative). In this chapter, ∞ is assumed to be a large *real number* (because continuous distributions take real values). This difference is subtle but important because the majority of continuous distributions extend to infinity either at the positive end or both ends. In the discrete case, we write $x = 0, 1, \ldots, \infty$ (x actually assumes the value ∞), whereas in the continuous case we write it as $x < \infty$ (we seldom write $x \leq \infty$). For example, the range for standard normal distribution is written as $-\infty < z < \infty$. In the case of mixture distributions that have a discrete part and a continuous part like the noncentral χ^2 distribution, which is a Poisson-weighted sum of central χ^2 distributions, the continuous part takes precedence. This means that for such distributions the rule reverts to the continuous component, so that the discrete part (i.e., Poisson probabilities) assumes all values $< \infty$. If both components of a mixture distribution are discrete (e.g., noncentral negative binomial distribution), the rule reverts to the discrete case.

   This notion pertains only to the population variate values. If the parameter(s) of a discrete distribution takes any value on the real line, we write it as in the continuous case. For example, the parameter λ of Poisson distribution has range $-\infty < \lambda < \infty$. Sampling from such populations always gives us data values in a finite range. As shown below, this range can be fixed in terms of the mean and variance of the distributions. This is especially suitable for bell-shaped distributions.

7.3   CONTINUOUS UNIFORM DISTRIBUTION

As the name implies, this distribution assigns a constant probability to each point in a continuous interval. Thus, the range is always finite (and quite often small in practical applications). It is also called continuous rectangular or simply rectangular distribution. The PDF of continuous uniform distribution (CUNI(a, b)) is given by

$$f(u; a, b) = \begin{cases} 1/(b-a) & \text{for } a \leq u \leq b; \\ 0 & \text{otherwise.} \end{cases}$$

The CDF is obtained by integration from "a" to x as $F_X(x; a, b) = \int_a^x [1/(b - a)]dx = [1/(b - a)]x|_a^x = (x - a)/(b - a)$. Thus

$$F(x; a, b) = \begin{cases} 0 & \text{for } x < a; \\ (x - a)/(b - a) & \text{for } a \leq x \leq b; \\ 1 & \text{for } x > b. \end{cases}$$

Considered as an algebraic equation, $y = (x - a)/(b - a)$ represents a straight line with slope $1/(b - a)$ and intercept $a/(a - b)$. This line is defined only within the interval (a, b) (theoretically, a straight line extends to infinity in both directions). The slope is small when the range $(b - a)$ is large. The slope is large (line is steep) in the limiting case $b \to a$. Only the extremes of a sample $x_{(1)}$ and $x_{(n)}$ are sufficient to fit this distribution (Figures 7.1).

7.3.1 Properties of Continuous Uniform Distribution

This distribution has a special type of symmetry called *flat-symmetry*. Hence, all odd central moments except the first one are zeros. The median always coincides with the mean, and the mode can be any value within the range. As the probability is constant throughout the interval, the range is always finite (and quite often small). From equation (7.3), we see that a change of origin and scale transformation $y = (x - a)/(b - a)$ results in the standard uniform distribution. A uniform distribution defined in an interval $(c, c + \theta)$ has PDF $f(x; \theta) = 1/\theta$ for $c \leq x \leq c + \theta$. Take $c = 0$ to get the standard form $f(x; \theta) = 1/\theta, 0 < x < \theta$. This is the analog of the DUNI(N) with probability function $f(x; N) = 1/N, x = 0, 1, 2, \cdots, N - 1$ discussed in page 6-41 of Chapter 6.

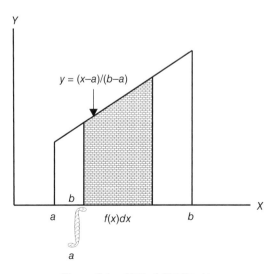

Figure 7.1 CDF of CUNI(a,b).

7.3.1.1 Moments and Generating Functions The moments are easy to find using the MGF. The mean is directly obtained as $\mu = [1/(b-a)] \int_a^b x\,dx = [1/(b-a)]\frac{x^2}{2}\,|_a^b = (b^2-a^2)/[2(b-a)] = (a+b)/2$. We will find higher order moments using the MGF. By definition

$$M_x(t) = E(e^{tx}) = \int_{x=a}^{b} [1/(b-a)]e^{tx}dx = [1/(b-a)]e^{tx}/t\,|_a^b = (e^{bt}-e^{at})/[(b-a)t].$$

The PGF is obtained from this as $P_x(t) = (t^b - t^a)/[(b-a)t]$.

To find the moments, we proceed as follows. Consider $e^{bt}/t = 1/t + b + b^2t/2! + \cdots + b^k t^{k-1}/k! + ..$ As $(1/t)$ is common in both e^{bt}/t and e^{at}/t, it cancels out. The second term is $(b-a)/(b-a) = 1$. Thus

$$(e^{bt}-e^{at})/[(b-a)t] = 1 + \frac{1}{b-a}\left(\sum_{k=2}^{\infty}[(b^k-a^k)/(b-a)]t^{k-1}/k!\right). \qquad (7.15)$$

If we differentiate equation (7.15) $(k-1)$ times with respect to t, all terms below the $(k-1)$th term will vanish (as they are derivatives of constants independent of t's) and all terms beyond the kth term will contain powers of t. Only the $(k-1)$th term is a constant with a $(k-1)!$ in the numerator, which cancels out with the $k!$ giving a k in the denominator. By taking the limit as $t \to 0$, we get

$$\mu'_{k-1} = (\partial^{k-1}/\partial t^{k-1})M_x(t)|_{t=0} = (b^k-a^k)/[(b-a)k]. \qquad (7.16)$$

Putting $k = 2$ gives $\mu_1 = (b+a)/2$. The second moment is obtained by putting $k = 3$ as $\mu'_2 = (b^3-a^3)/[3(b-a)] = (b^2-ab+a^2)/3$. From this, we get the second central moment as $\mu_2 = (b^2-ab+a^2)/3 - (a+b)^2/4$. Taking 12 as the LCM of 3 and 4, this simplifies to $\mu_2 = \sigma^2 = (b-a)^2/12$. See Table 7.2 for further properties.

7.3.1.2 Alternative Parametrization Write $\mu = (a+b)/2$ and $\sigma = (b-a)/(2\sqrt{3})$. Cross multiplying gives $(a+b) = 2\mu$ and $(b-a) = (2\sqrt{3})\sigma$. Add these two equations to get $b = \mu + \sqrt{3}\,\sigma$. Subtracting gives $a = \mu - \sqrt{3}\,\sigma$, from which $(b-a) = (2\sqrt{3})\sigma$. This allows us to write the PDF in the alternative form as

$$f(x, \mu, \sigma) = 1/(2\sqrt{3}\sigma), \mu - \sqrt{3}\sigma \le x \le \mu + \sqrt{3}\sigma. \qquad (7.17)$$

■ **EXAMPLE 7.1 Even moments of rectangular distribution**

Prove that the kth central moment is zero for k odd and is given by $\mu_k = (b-a)^k/[2^k(k+1)]$ for k even.

Solution 7.1 By definition $\mu_k = \frac{1}{b-a}\int_a^b (x - \frac{a+b}{2})^k dx$. Make the change of variable $y = x - (a+b)/2$. For $x = a$, we get $y = a - (a+b)/2 = (a-b)/2 = -(b-a)/2$. Similarly for $x = b$, we get $y = b - (a+b)/2 = (b-a)/2$. As the

TABLE 7.2 Properties of Continuous Uniform Distribution $U(a,b)$

Property	Expression	Comments
Range of X	$a \leq x \leq b$	Continuous; finite
Mean	$\mu = (a+b)/2$	Median $= (a+b)/2$
Variance	$\sigma^2 = (b-a)^2/12$	$\sigma^2 = (\mu^2 - ab)/3$
Skewness	$\gamma_1 = 0$	Special symmetry
Kurtosis	$\beta_2 = 9/5$	
Mean deviation	$E|X - \mu| = (b-a)/4$	$(\sqrt{3}/2)\sigma = 0.866\sigma$
CV	$(b-a)/[\sqrt{3}(a+b)]$	0.57735 if $a = 0, b = 1$
CDF	$(x-a)/(b-a)$	Line sloping up
SF	$(b-x)/(b-a)$	
Moments	$\mu'_r = (b^{r+1} - a^{r+1})/[(b-a)(r+1)]$	
Moments	$\mu_r = [(b-a)/2]^r/(r+1)$	r even
MGF	$(e^{bt} - e^{at})/[(b-a)t]$	
ChF	$(e^{ibt} - e^{iat})/[(b-a)it]$	
PGF	$(t^b - t^a)/[(b-a)\log(t)]$	

Standard uniform distribution results when $a = 0, b = 1$.

Jacobian is $\partial y/\partial x = 1$, the integral becomes $\mu_k = \frac{1}{b-a}\int_{-(b-a)/2}^{(b-a)/2} y^k dy$. When k is odd, this is an integral of an odd function in symmetric range, which is identically zero. For k even, we have $\mu_k = \frac{2}{b-a}\int_0^{(b-a)/2} y^k dy = \frac{2}{b-a}[y^{k+1}/(k+1)]|_0^{(b-a)/2} = (b-a)^k/[2^k(k+1)]$, as the constant $2/(b-a)$ cancels out. ∎

◨ **EXAMPLE 7.2 Mean deviation of rectangular distribution**

Find the MD of rectangular distribution.

Solution 7.2 By definition $E|X - \mu| = \int_a^b |x - \mu|/(b-a)dx$. Split the range of integration from "a" to μ and μ to "b" and note that $|X - \mu| = \mu - X$ for $x < \mu$. This gives

$$E|X - \mu| = \int_{x=a}^{\mu} (\mu - x)/(b-a)dx + \int_{x=\mu}^{b} (x - \mu)/(b-a)dx. \qquad (7.18)$$

Consider $\int_a^\mu \mu dx - \int_\mu^b \mu dx$. As $\mu = (a+b)/2$, this integral vanishes. What remains is

$$\frac{1}{b-a}\left(\int_\mu^b xdx - \int_a^\mu xdx\right) = \frac{1}{2(b-a)}(b^2 - \mu^2 - (\mu^2 - a^2))$$

$$= \frac{(a^2 + b^2 - 2\mu^2)}{2(b-a)}. \qquad (7.19)$$

Substitute the value $\mu = (a+b)/2$ and take 2 as a common denominator to get $\frac{1}{2(b-a)}(b-a)^2/2 = (b-a)/4$. Thus, the MD $E|X - \mu| = (b-a)/4$. ∎

Next we apply Theorem 7.1 (p. 257) to verify our result. As the rectangular distribution does not tail off to zero at the extremes, equation (7.1) seems to be not applicable. However, we know the CDF is $(x - a)/(b - a)$. If we substitute the lower limit is "a," in $(x - a)/(b - a)$, we get zero. Hence, Theorem 7.1 is applicable. This gives

$$\text{MD} = 2 \int_{\mu}^{\mu} F(x)dx = 2/(b - a) \int_{a}^{c} (x - a)dx \text{ where } c = \mu = (a + b)/2. \quad (7.20)$$

The integral $\int_{a}^{c}(x - a)dx$ is $(x - a)^2/2|_{a}^{c}$. The integral evaluated at the lower limit is obviously zero. As $c = (a + b)/2$, the upper limit evaluates to $(b - a)^2/8$. Substitute in equation (7.20). One $(b - a)$ cancels out and we get the MD as $(b - a)/4$. This tallies with the above-mentioned result.

7.3.2   Relationships with Other Distributions

Owing to its relationship with many other distributions, it is extensively used in computer generation of random variables. As mentioned earlier, a simple change of variable transformation $Y = (X - a)/(b - a)$ results in the standard uniform distribution $U(0, 1)$, usually denoted as $U(0, 1)$. If X is any continuous random variable with CDF $F(x)$, then $U = F(x) \sim U[0, 1]$. This property is utilized to generate random numbers from a distribution if the expression for its CDF involves simple or invertible arithmetic or transcendental functions. For example, the CDF of an exponential distribution (given below) is $F(x) = 1 - e^{-\lambda x}$. Equating to a random number u in the range [0,1] and solving for x, we get $1 - e^{-\lambda x} = u$ or $x = -\log(1 - u)/\lambda$. $U(0, 1)$ is a special case of BETA-I(a, b) when $a = b = 1$.

7.3.3   Applications

This distribution finds applications in many fields. It is used in nonparametric tests like Kolmogorov–Smirnov test. The rounding errors resulting from grouping data into classes uses a $U(0, 1)$ to obtain a correction factor known as Sheppard's correction. Quantization errors in audio coding use this distribution. It is also used in stratified sampling, nonrandom clustering, and so on. Random numbers for other distributions are easy to generate using $U[0, 1]$. Suppose we have a uniform random number generator between 0 and 1. The transformation $y = a + (b - a)x$ gives a random number in the interval $[a, b]$, where x is in [0,1] (if the random number generated is in [0,32767), we could use the mapping $a + (b - a)x/32767$ to get a random number in $[a, b]$).

▥ EXAMPLE 7.3   Estimating proportions

A jar contains a mixture of two liquids L_1 and L_2 that mixes well in each other (as water and wine or acid and water). All that is known is that "there is at most three times as much of one as the other." Find the probability that (i) $L_1/L_2 \leq 2$ and (ii) $L_1/L_2 \geq 1$.

Solution 7.3 The given condition is $\frac{1}{3} \leq L_1/L_2 \leq 3$. Let $U = L_1/L_2$. Assume that U is uniformly distributed in $[1/3, 3]$. As $3 - 1/3 = 8/3$, we take the density function as $f(x) = 3/8, \frac{1}{3} \leq x \leq 3$. The required answer for (i) is $P[U \leq 2] = \int_{1/3}^{2} f(x) \, dx = (3/8)*x|_{1/3}^{2} = (3/8) * (2 - 1/3) = 5/8$. (ii) $L_1/L_2 \geq 1 = \int_{1}^{3} f(x)dx = (3/8) * x|_{1}^{3} = 6/8 = 0.75$. ∎

7.4 EXPONENTIAL DISTRIBUTION

Exponential distribution ($EXP(\lambda)$) can be regarded as the continuous analog of geometric distribution. The PDF is given by

$$f(x; \lambda) = \begin{cases} \lambda e^{-\lambda x} & \text{for } x \geq 0, \lambda > 0; \\ 0 & \text{otherwise.} \end{cases}$$

When $\lambda = 1$, we get the standard exponential distribution $f(x) = e^{-x}$. Setting $\lambda = 1/\theta$ gives an alternative representation as $f(x, \theta) = \frac{1}{\theta}e^{-x/\theta}$. The CDF is given by

$$F(x) = \begin{cases} 1 - e^{-\lambda x}, & x \geq 0 \\ 0 & \text{otherwise.} \end{cases} \tag{7.21}$$

7.4.1 Properties of Exponential Distribution

This distribution has a single parameter, which is positive (Figure 7.2). Variance of this distribution is the square of the mean, as shown in the following. This means that

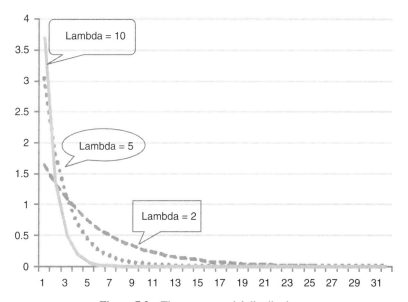

Figure 7.2   Three exponential distributions.

TABLE 7.3 Properties of Exponential Distribution ($\lambda e^{-\lambda x}$)

Property	Expression	Comments
Range of X	$x \geq 0$	Continuous
Mean	$\mu = 1/\lambda$	
Median	$\log(2)/\lambda$	
Variance	$\sigma^2 = 1/\lambda^2$	$\sigma^2 = \mu^2$
Skewness	$\gamma_1 = 2$	Never symmetric
Kurtosis	$\beta_2 = 9$	Always leptokurtic
Mean deviation	$E\|X - \mu\| = 2/(e\lambda)$	$2\mu_2 * f_m$
CV	1	
CDF	$1 - e^{-\lambda x}$	SF $= 1 - $ CDF $= e^{-\lambda x}$
Moments	$\mu'_r = 1/\lambda^r$	
MGF	$\lambda/(\lambda - t)$	
ChF	$\lambda/(\lambda - it)$	

Replace λ by $1/\lambda$ for the alternate parametrization.

when $\lambda \to 0$, the variance and kurtosis increases without limit. The SF is $1 - $ CDF $= e^{-\lambda x}$ (see Table 7.3).

7.4.2 Additivity Property

Several statistical distributions obey the additivity property. This information is useful while modeling data from two or more identical populations. The sum of k independent exponentially distributed random variables EXP(λ) has a gamma distribution with parameters k and λ. Symbolically, if X_i are EXP(λ), then $\sum_{i=1}^{k} X_i \sim$ GAMMA(k, λ). This is most easily proved using the MGF (see Table 7.4).

7.4.2.1 Moments and Generating Functions
The characteristic function is readily obtained by integration as $\phi_x(t; \lambda) =$

$$\int_0^\infty e^{itx} \lambda e^{-\lambda x} dx = \lambda \int_0^\infty e^{-(\lambda - it)x} dx = \frac{\lambda}{(\lambda - it)} = \frac{1}{1 - \frac{it}{\lambda}} = (1 - it/\lambda)^{-1}.$$

Expand as an infinite series using $(1 - x)^{-1} = 1 + x + x^2 + x^3 + \cdots$ to get

$$(1 - it/\lambda)^{-1} = 1 + it/\lambda + (it/\lambda)^2 + (it/\lambda)^3 + \cdots \tag{7.22}$$

From this, the mean and variance follows as $\mu = 1/\lambda$ and $\sigma^2 = 1/\lambda^2$. Alternately, the mean is given by $\mu = \lambda \int_0^\infty x e^{-\lambda x} dx$. Write the integral as $\int_0^\infty x^{2-1} e^{-\lambda x} dx$. Using gamma integral, this becomes $\mu = \lambda \Gamma(2)/\lambda^2$. One λ cancels out and we get $\mu = 1/\lambda$ as $\Gamma(2) = 1$. For the alternate parametrization $f(x, \theta) = \frac{1}{\theta} e^{-x/\theta}$, the mean $\mu = \theta$ and variance $\sigma^2 = \theta^2$.

TABLE 7.4   Summary Table of Additivity Property

Distribution	Parms-1	Parms-2	Combined	Conditions
Bernoulli	p	p	BINO$(2, p)$	Same p, independent
Binomial	n_1, p	n_2, p	$n_1 + n_2, p$	Same p, independent
Poisson	λ_1	λ_2	$\lambda_1 + \lambda_2$	Independent
Negative Bino.	r_1, p	r_2, p	$r_1 + r_2, p$	Same p, independent
Exponential	λ_1	λ_2	$\lambda_1 + \lambda_2$	Independent
Exponential	λ	λ	$\Gamma(2, \lambda)$	Same λ, independent
Gamma	$\Gamma(n_1, \lambda)$	$\Gamma(n_2, \lambda)$	$\Gamma(n_1 + n_2, \lambda)$	Same λ, independent
Normal	μ_1, σ_1^2	μ_2, σ_2^2	$\mu_1 + \mu_2, \sigma_1^2 + \sigma_2^2$	Independent
*Log-Normal	μ_1, σ_1^2	μ_2, σ_2^2	$\mu_1 + \mu_2, \sigma_1^2 + \sigma_2^2$	Product ($\prod$)
*BETA-I	a, b	$a + b, c$	$a, b + c$	Product ($\prod$)
Cauchy	θ	θ	2θ	Same θ, independent
Chi-square	$\chi_{n_1}^2$	$\chi_{n_2}^2$	$\chi_{n_1 + n_2}^2$	Independent
Unscaled F	(p, n)	(q, n)	$(p + q, n)$	Distribution of χ_p^2 / χ_n^2
Noncentral χ^2	$\chi_{n_1}^2(\lambda_1)$	$\chi_{n_2}^2(\lambda_2)$	$\chi_{n_1 + n_2}^2(\lambda_1 + \lambda_2)$	Independent

Note that for log-normal distributions (LNDs), it is the product and quotient that are identically distributed; and μ_i, σ_i^2 are the means of underlying normal variate. Multinomial satisfies MN$(n_1, p_1, p_2, \dots, p_m)$+ MN$(n_2, p_1, p_2, \dots, p_m) \sim$ MN$(n_1 + n_2, p_1, p_2, \dots, p_m)$ if they are independent.

The coefficients of skewness and kurtosis are 2 and 9, respectively. Hence, the distribution is always asymmetric and leptokurtic. Putting $Y = 1/X$ results in the inverse exponential distribution with PDF $f(y) = (\lambda/y^2)e^{-\lambda/y}$. See Table 7.3 for further properties.

EXAMPLE 7.4   Median of exponential distribution

Find the median of exponential distribution with PDF $f(x, \lambda) = \lambda e^{-\lambda x}$.

Solution 7.4 Let M be the median. Then $\int_M^\infty \lambda e^{-\lambda x} dx = 0.5$. This gives $-e^{-\lambda x}|_M^\infty = 1/2$, or equivalently $e^{-\lambda M} = 1/2$. Take log of both sides to get $-\lambda M = -\log(2)$ or $M = \log(2)/\lambda$ where the log is to the base e. ∎

EXAMPLE 7.5 $\Pr(X > \lambda/2)$, $\Pr(X > 1/\lambda)$ for EXP(λ) distribution

Show that $\Pr[X > \lambda/2]$ of the exponential distribution is $e^{-\lambda^2/2}$. What is the $\Pr[X > 1/\lambda]$?

Solution 7.5 As the SF is $e^{-\lambda x}$, $\Pr(X > \lambda/2)$ is easily seen to be the survival function evaluated for $x = \lambda/2$. This upon substitution becomes $e^{-\lambda^2/2}$. Putting $x = 1/\lambda$ in the SF, we get $e^{-1} = 1/e$. Thus, the mean $1/\lambda$ of an exponential distribution divides the total frequency in $(1 - \frac{1}{e}): \frac{1}{e}$ ratio. This is a characteristic property of exponential distribution. ∎

▌ **EXAMPLE 7.6   Lifetime of components**

The lifetime of a component is known to be exponentially distributed with mean $\lambda = 320$ hours. Find the probability that the component has failed in 340 hours, if it is known that it was in good working condition when time of operation was 325 hours.

Solution 7.6 Let X denote the lifetime. Then $X \sim$ EXP(1/320). Symbolically, this problem can be stated as $P[X < 340|X > 325]$. Using conditional probability, this is equivalent to $P[325 < X < 340]/P[X > 325]$. In terms of the PDF, this becomes $\int_{325}^{340} f(x)dx / \int_{325}^{\infty} f(x)dx$. Write the numerator as $\int_{325}^{\infty} f(x)dx - \int_{340}^{\infty} f(x)dx$, this becomes $1 - e^{-340/320} / e^{-325/320} = 1 - e^{-15/320} = 0.04579$.   ■

7.4.2.2   Relationship with Other Distributions It is a special case of gamma distribution with $m = 1$ (p. 283). If $X \sim$ EXP(λ) and b is a constant, then $Y = X^{1/b} \sim$ WEIB(λ, b) (p. 320). The difference of two IID exponential variates is Laplace distributed. It is also related to the $U(0, 1)$ distribution [167, 168] and power-law distribution, which is a discrete analog of this distribution [169].

▌ **EXAMPLE 7.7   Memory-less property**

Prove that the exponential distribution has memory-less property $P(X \geq s + t)|P(X \geq s) = P(X \geq t)$ for $s, t \geq 0$.

Solution 7.7 Consider P(X≥s+t)∩P(X≥s)/P(X≥s). The numerator simplifies to $P(X \geq s + t) = \lambda \int_{x=(s+t)}^{\infty} e^{-\lambda x} dx = e^{-\lambda(s+t)}$ using $e^{-\infty} = 0$. The denominator is $\lambda \int_{x=s}^{\infty} e^{-\lambda x} dx$. This simplifies to $e^{-\lambda s}$. Taking the ratio of these gives $e^{-\lambda(s+t)} / e^{-\lambda s} = e^{-\lambda t}$, which is the RHS.   ■

7.4.2.3   Applications This distribution is used to model random proportions and life-time of devices and structures. It has applications in reliability theory and waiting times in queuing theory. For example, the expected life length of a new light bulb can be assumed to follow an exponential distribution with parameter $\lambda = 1/500$ hours so that the life time is given by $f(x) = (1/500)(e^{-x/500})$.

Other examples include modeling: (i) Lifetime of destructive devices that are (more or less) continuously or regularly in use, such as light bulbs and tubes, electronic chips. (ii) Lifetime of nondestructive or reusable devices until next repair work, electronic devices such as computer monitors and LCD screens, microwaves, electrical appliances such as refrigerators, and lifetime of automobile tires. Time until the arrival of the next event (such as next telephone call and emergency call) or time until next customer to an office or business.

▌ **EXAMPLE 7.8   Mean deviation of exponential distribution**

Find the mean deviation of the exponential distribution $f(x, \lambda) = \lambda e^{-\lambda x}$.

Solution 7.8 We apply Theorem 7.1 (page 267) to find the MD. As the exponential distribution does not tail off to zero at the lower limit (i.e., at 0), equation (7.1) seems like not applicable. We know that the CDF is $1 - e^{-\lambda x}$. If we apply L'Hospital's rule once on $x * F(x)$ we get $x \exp(-\lambda x) + (1 - \exp(-\lambda x))$. As both terms $\to 0$ as $x \to 0$, the $\lim_{x \to 0} x * F(x) = 0$, and the Theorem 7.1 becomes applicable. This gives

$$MD = 2 \int_0^{1/\lambda} (1 - e^{-\lambda x}) dx. \tag{7.23}$$

Split this into two integrals and evaluate each to get

$$MD = 2[1/\lambda + (1/\lambda)e^{-1} - (1/\lambda)] = 2/(e\lambda) = 2\mu_2 * f_m, \tag{7.24}$$

where $f_m = \lambda e^{-1} = \lambda/e$. Alternatively, use the SF() version as the exponential distribution tails off to the upper limit (Table 7.3). ∎

7.4.2.4 *General form*   The general form of the exponential distribution is given by

$$f(x) = \lambda e^{-\lambda(x-\delta)}, \qquad x \geq 0, \qquad \lambda > 0, \qquad x \geq \delta. \tag{7.25}$$

The corresponding characteristic function is

$$\phi(t) = \frac{e^{i\delta t}}{1 - it/\lambda}. \tag{7.26}$$

7.5   BETA DISTRIBUTION

The beta distribution is widely used in statistics owing to its close relationship with other continuous distributions. It is also used in Bayesian models with unknown probabilities, in order statistics and reliability analysis. It is used to model the proportion of fat (by weight) in processed or canned food and percentage of impurities in some manufactured products such as food items, cosmetics, and laboratory chemicals. In Bayesian analysis, the prior distribution is assumed to be the beta for binomial proportions. Important distributions belonging to the beta family are discussed in the following. These include type I and type II beta distributions. We will use the respective notations Beta-I(*a*,*b*) and Beta-II(*a*,*b*). Beta distributions with three or more parameters are also briefly mentioned.

7.5.1   Type-I Beta Distribution

This is also called the standard beta distribution. The PDF of Beta-I(a, b) is given by

$$f_x(a, b) = x^{a-1}(1 - x)^{b-1}/B(a, b), \tag{7.27}$$

where $0 < x < 1$, and $B(a, b)$ is the complete beta function (CBF). Particular values for a and b results in a variety of distributional shapes.

7.5.2 Properties of Type-I Beta Distribution

This distribution has two parameters, both of which are positive real numbers. The range of x is between 0 and 1. The variance is always bounded, irrespective of the parameter values. Put $y = 1 - x$ in the above to get the well-known symmetry relationship $f_x(a, b) = f_y(b, a)$ or in terms of tail areas $I_x(a, b) = 1 - I_{1-x}(b, a)$, where $I_x(a, b)$ is described below (p. 277). If $a = b$, the distribution is symmetric about $X = 1/2$. If $a = b = 1$, it reduces to uniform (rectangular) distribution. When $a = b = 1/2$, this distribution reduces to the arc-sine distribution of first kind (Section 7.8, p. 279). If $b = 1$ and $a \neq 1$, it reduces to power-series distribution $f(x; a) = ax^{a-1}$ using the result $\Gamma(a + 1) = a * \Gamma(a)$. Put $a = \alpha + 1, b = \beta + 1$ to get an alternate form

$$f_x(\alpha, \beta) = x^\alpha (1 - x)^\beta / B(\alpha + 1, \beta + 1). \tag{7.28}$$

7.5.2.1 Moments and Generating Functions The moments are easy to find using beta integral. The kth moment can be obtained as

$$\mu'_k = \frac{1}{B(a, b)} \int_0^1 x^{a+k-1}(1 - x)^{b-1} dx = B(a + k, b)/B(a, b) = \frac{\Gamma(a + b)\Gamma(a + k)}{\Gamma(a + b + k)\Gamma(a)}.$$

In terms of rising factorials, this becomes $\mu'_k = a^{[k]}/(a + b)^{[k]}$. The mean is obtained by putting $k = 1$ as $\mu = a/(a + b) = 1 - b/(a + b)$. This has the interpretation that increasing the parameter "a" by keeping "b" fixed moves the mean to the right (toward 1). Put $k = 2$ to get the second moment as $a(a + 1)/[(a + b)(a + b + 1)]$. The variance is $\sigma^2 = ab/[(a + b)^2(a + b + 1)]$. This is symmetric in the parameters and increasing both "a" and "b" together decreases the variance. If $a > 1$ and $b > 1$, there exist a single mode at $(a - 1)/(a + b - 2)$. The characteristic function is

$$\phi(t) = \frac{1}{B(a, b)} \int_0^1 e^{itx} x^{a-1}(1 - x)^{b-1} dx = {}_1F_1(a, a + b; it), \tag{7.29}$$

where ${}_1F_1(a, a + b; it)$ is the confluent hypergeometric function. The kth central moment can be obtained as follows:

$$\mu_k = \frac{1}{B(a, b)} \int_0^1 (x - a/(a + b))^k x^{a-1}(1 - x)^{b-1} dx$$

$$= (-a/(a + b))^k {}_2F_1(-k, a, a + b, (a + b)/a), \tag{7.30}$$

where ${}_2F_1(a, b, c; x)$ is the hypergeometric function. The coefficient of skewness is $\gamma_1 = 2(b - a)\sqrt{a + b + 1}/[\sqrt{ab}(a + b + 2)]$. Mean deviation about the mean is given by

$$E|X - \mu| = 2a^a b^b / [B(a, b)(a + b)^{a+b+1}]. \tag{7.31}$$

See Table 7.5 for further properties (Figures 7.3 and 7.4).

TABLE 7.5 Properties of Beta-I Distribution

Property	Expression	Comments
Range	$0 \leq x \leq 1$	Continuous
Mean	$\mu = a/(a+b)$	$= 1 - b/(a+b)$
Variance	$ab/[(a+b)^2(a+b+1)] = \mu(1-\mu)/$ $(a+b+1)$	$\Rightarrow \mu > \sigma^2$
Mode	$(a-1)/(a+b-2)$	$a > 1, b > 1$
CV	$(b/[a(c+1)])^{1/2}$	$c = a + b$
Skewness	$\gamma_1 = 2(b-a)\sqrt{a+b+1}/[\sqrt{ab}$ $(a+b+2)]$	
Kurtosis	$\beta_2 = 3c(c+1)(a+1)(2b-a)/$ $[ab(c+2)(c+3)]$	$c = a + b$
Mean deviation	$E\|X - \mu\| = 2a^a b^b/[B(a,b)(a+b)^{a+b+1}]$	$2c[I_c(a,b) - I_c(a+1,b)],$ $c = a/(a+b)$
Moments	$\mu'_r = \prod_{i=0}^{r-1}(a+i)/(a+b+i)$	$a^{(r)}/(a+b)^{(r)}$
Moments	$\mu_r = (-c)^r {}_2F_1(a, -r, a+b, 1/c)$	$c = a/(a+b)$
ChF	$\dfrac{\Gamma(a+b)}{\Gamma(a)} \sum_{j=0}^{\infty} \dfrac{\Gamma(a+j)(it)^j}{\Gamma(a+b+j)\Gamma(1+j)}$	${}_1F_1(a, a+b; it)$
Additivity	$\sum_{i=1}^m \text{BETA}(a_i, b) = B\left(\sum_{i=1}^m a_i, b\right)$	Independent
Recurrence	$f(x; a+1, b)/f(x; a, b) = (1 + b/a)x$	
	$f(x; a, b+1)/f(x; a, b) = (1 + a/b)(1 - x)$	
U-shaped	$a < 1$ and $b < 1$	
J-shaped	$(a-1) * (b-1) < 0$	
Tail area	$I_x(a,b) = \dfrac{1}{B(a,b)} \displaystyle\int_0^x t^{a-1}(1-t)^{b-1}dt$	I = Incomplete beta

Symmetric when $a = b$. ${}_2F_1()$ is hypergeometric function that is related to the incomplete beta function as $I_x(a,b) = [f(x; a+1, b)/(a+b)]*_2F_1(1-b, 1; a+1; -x/(1-x))$, where $f(x; a+1, b)$ is the density of BETA-I. It can also be represented using the Euler identity ${}_2F_1(a, -r, a+b, 1/c) = (1 - 1/c)^{b+r} {}_2F_1(b, a+b+r, a+b; 1/c)$ [170].

◼ EXAMPLE 7.9   Mean deviation of beta distribution

Find the mean deviation of the beta distribution using Theorem 7.1.

Solution 7.9 As the beta distribution does not tail off to the lower or upper limits for some parameter values (e.g., $a = b = 0.25$), equation (7.1) seems like not applicable. We know that the CDF is $I_x(a,b)$. As done in the case of exponential distribution, using L'Hospital's rule, it is easy to show that $x * F(x) \to 0$, so that the Theorem 7.1 is applicable. This gives MD $= 2\int_0^c I_x(a,b)dx$, where

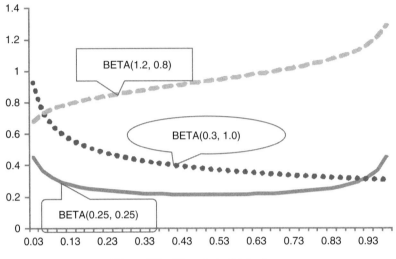

Figure 7.3   Three beta distributions.

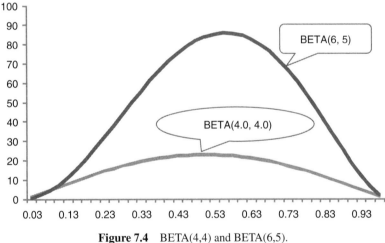

Figure 7.4   BETA(4,4) and BETA(6,5).

$c = a/(a + b)$ is the mean. Taking $u = I_x(a, b)$ and $dv = dx$, this becomes

$$\text{MD} = 2 \left[x I_x(a, b) dx |_0^c - \int_0^c x g_x(a, b) dx \right] = 2 \left[c * I_c(a, b) - \int_0^c x g_x(a, b) dx \right].$$

Write $x * g_x(a, b)$ as $x^a(1 - x)^{b-1}/B(a, b)$. Multiply numerator and denominator by $B(a + 1, b)$ and write $B(a + 1, b)/B(a, b)$ as $a/(a + b)$ to get

TABLE 7.6   Mean Deviation of Beta-I using Equations (7.31) and (7.32)

a	b	$a/(a+b)$	Equation (7.32)	Equation (7.31)	a	b	$a/(a+b)$	Equation (7.32)	Equation (7.31)
1	1	0.50000	0.25000	0.25000	5	5	0.50000	0.12305	0.12305
1	2	0.33333	0.19753	0.19753	10	2	0.83333	0.08225	0.08225
1	5	0.16667	0.11163	0.11163	10	5	0.66667	0.09525	0.09525
2	2	0.50000	0.18750	0.18750	8	8	0.50000	0.09819	0.09819
2	5	0.28571	0.13010	0.13010	10	20	0.33333	0.06801	0.06801
2	10	0.16667	0.08225	0.08225	30	30	0.50000	0.05129	0.05129
5	2	0.71429	0.13010	0.13010	30	10	0.75000	0.05414	0.05414

First and second columns are the shape parameters and third column is $c = a/(a+b)$. Fourth column finds the MD using equation (7.32) and fifth column using equation (7.31).

$2 \int_0^c x * g_x(a,b) dx = 2a/(a+b) I_c(a+1,b)$. This gives

$$MD = 2c[I_c(a,b) - I_c(a+1,b)], \quad \text{where } c = a/(a+b). \tag{7.32}$$

∎

This can be simplified using a result in Reference 4 as

$$I_c(a,b) - I_c(a+1,b) = (1-c)^2/(b-1) \, g_c(a+1,b-1), \tag{7.33}$$

where $g(\ )$ is the PDF of BETA-I. This gives $MD = 2c(1-c)^2/(b-1) \, g_c(a+1,b-1)$. Substitute for c and simplify to get the above-mentioned form (7.31). Alternately, use

$$I_c(a,b) - I_c(a+1,b) = c^a(1-c)^b/[a \, B(a,b)], \tag{7.34}$$

to get $MD = 2bc/[(a+b)(a+b+1)]g_c(x;a,b)$, where $c = a/(a+b)$. Results are compared in Table 7.6.

⬛ EXAMPLE 7.10   Mean versus variance of BETA-I

Prove that the variance can never equal the mean of a beta-I distribution.

Solution 7.10 We know that the variance of BETA-I can be represented in terms of the mean as $\mu(1-\mu)/(a+b+1)$. Assume the contrary that the variance can be equal to the mean. Put $\mu = x$ in the above to get $x(1-x)/(a+b+1) = x$. This simplifies to $-x^2 = (a+b)x$. As the mean cannot be zero (as "a" cannot be zero), there is no solution possible. Hence, the variance of BETA-I is always less than the mean. Alternatively, divide the variance by the mean and argue that as $\mu \in (0,1)$, $1-\mu$ is always less than 1, showing that the ratio is <1, which implies that $\sigma^2 < \mu$. ∎

7.5.3 Type-II Beta Distribution

Beta distribution of the second kind (also called Type-II beta distribution or inverted beta distribution (IBD)) is obtained from the above by the transformation $Y = X/(1 - X)$ or equivalently $X = Y/(1 + Y)$. When $x \to 0, y \to 0$, and when $x \to 1, y \to \infty$. Hence, the range of Y is from 0 to ∞. The PDF is given by

$$f_y(a, b) = y^{a-1}/[B(a, b)(1 + y)^{a+b}], \quad y > 0, \quad a, b > 0. \tag{7.35}$$

The BETA-I distribution is used to model random experiments or occurrences that vary between two finite limits, which are mapped to the (0,1) range, whereas BETA-II is used when upper limit is infinite.

7.5.4 Properties of Type-II Beta Distribution

This is a special case of the unscaled F distribution (distribution of χ_m^2/χ_n^2) or an F with the same degrees of freedom. In other words, put $Y = (m/n) * X$ in F distribution to get BETA-II distribution. If Y is BETA-II(a,b) then $1/Y$ is BETA-II(b, a). This means that $1/[X/(1 - X)] = (1 - X)/X$ is also beta distributed (see the following discussion).

7.5.4.1 Moments and Generating Functions
The mean and variance are $\mu = a/(b - 1)$ and $\sigma^2 = a(a + b - 1)/[(b - 1)^2(b - 2)]$ for $b > 2$. Consider $E(Y^k)$

$$\int_0^\infty y^k f_y(a, b) dy = \int_0^\infty y^{a+k-1}/[B(a, b)(1 + y)^{a+b}] dy. \tag{7.36}$$

Put $x = y/(1 + y)$ so that $y = x/(1 - x), (1 + y) = 1/(1 - x)$, and $dy/dx = [(1 - x) - x(-1)]/(1 - x)^2$. This simplifies to $1/(1 - x)^2$. The range of X is [0,1]. Hence, equation (7.36) becomes

$$(1/B(a, b)) \int_0^\infty y^{a+k-1}/(1 + y)^{a+b} dy = (1/B(a, b)) \int_0^1 x^{a+k-1}(1 - x)^{b-k-1} dx. \tag{7.37}$$

This is $B(a + k, b - k)/B(a, b)$. Put $k = 1$ to get the mean as $\Gamma(a + 1)\Gamma(b - 1)$ $\Gamma(a + b)/[\Gamma(a)\Gamma(b)\Gamma(a + b)]$. Write $\Gamma(a + 1) = a\Gamma(a)$ in the numerator and $\Gamma(b) = (b - 1)\Gamma(b - 1)$ in the denominator and cancel out common factors to get $\mu = a/(b - 1)$. Put $k = 2$ to get the second moment as $B(a + 2, b - 2)/B(a, b) =$ $\Gamma(a + 2)\Gamma(b - 2)\Gamma(a + b)/[\Gamma(a)\Gamma(b)\Gamma(a + b)] = a(a + 1)/[(b - 1)(b - 2)]$. From this, the variance is obtained as $a(a + 1)/[(b - 1)(b - 2)] - a^2/(b - 1)^2$. Take $\mu = a/(b - 1)$ as a common factor. This can now be written as $\mu(\frac{a+1}{b-2} - \mu)$. Substitute for μ inside the bracket and take $(b - 1)(b - 2)$ as common denominator. The numerator simplifies to $b - a + 2a - 1 = (a + b - 1)$. Hence, the variance becomes $\sigma^2 = a(a + b - 1)/[(b - 1)^2(b - 2)]$. As $(a + 1)/(b - 2) - \mu = (a + b)/[(b - 1)(b - 2)]$, this expression is valid for $b > 2$. Unlike the BETA-I distribution whose variance is always bounded, the variance of BETA-II can be

TABLE 7.7   Properties of Beta-II Distribution

Property	Expression	Comments
Range of X	$0 \le x < \infty$	Continuous
Mean	$\mu = a/(b-1)$	
Variance	$\sigma^2 = a(a+b-1)/[(b-1)^2(b-2)] = \mu(\dfrac{a+1}{b-2} - \mu)$	
Mode	$(a-1)/(b+1)$	$a > 1$
$E[X/(1-X)]^k$	$\dfrac{\Gamma(a+k)\Gamma(b-k)}{\Gamma(a)\Gamma(b)}$	
Skewness	$\gamma_1 = 2(b-a)\sqrt{a+b+1}/[\sqrt{ab}(a+b+2)]$	
Kurtosis	$\beta_2 = 3c(c+1)(a+1)(2b-a)/[ab(c+2)(c+3)]$	$c = a+b$
Mean deviation	$E|X - \mu| = 2\int_0^{a/(b-1)} I_{y/(1+y)}(a,b)dy$	
Moments	$\mu'_r = B(a+r, b-r)/B(a,b)$	$a^{(r)}/b_{(r)}$
Moments	$\mu_r = (-c)^r {}_2F_1(a, -r, a+b, 1/c)$	$c = a/(a+b)$
ChF	${}_1F_1(a, a+b; it) = \dfrac{\Gamma(a+b)}{\Gamma(a)}\sum_{j=0}^{\infty}\dfrac{\Gamma(a+j)(it)^j}{\Gamma(a+b+j)\Gamma(1+j)}$	
Additivity	$\sum_{i=1}^{m} BETA(a_i, b) = BETA\left(\sum_{i=1}^{m}(a_i, b)\right)$	Independent
Recurrence	$f(x; a+1, b)/f(x; a, b) = (1 + b/a)(x/(1+x))$	
	$f(x; a, b+1)/f(x; a, b) = (1 + a/b) * 1/(1+x)$	
U-shaped	$a < 1$ and $b < 1$	
J-shaped	$(a-1) * (b-1) < 0$	
Tail area	$I_x(a,b) = \dfrac{1}{B(a,b)}\int_0^x t^{a-1}(1-t)^{b-1}dt$	$x = y/(1+y)$

Beta distribution of the first kind (also called Type-I beta distribution) is obtained by the transformation $X = Y/(1+Y)$.

increased arbitrarily by keeping b constant (say near 2^+) and letting $a \to \infty$. It can also be decreased arbitrarily when $(a+1)/(b-2)$ tends to $\mu = a/(b-1)$. The expectation of $[X/(1-X)]^k$ is easy to compute in terms of complete gamma function as $E[X/(1-X)]^k = \frac{\Gamma(a+k)\Gamma(b-k)}{\Gamma(a)\Gamma(b)}$. See Table 7.7 for further properties.

⬛ EXAMPLE 7.11   The mode of BETA-II distribution

Prove that the mode of BETA-II distribution is $(a-1)/(b+1)$.

Solution 7.11 Differentiate equation (7.35) (without constant multiplier) with respect to y to get

$$f'(y) = [(1+y)^{a+b}(a-1)y^{a-2} - y^{a-1}(a+b)(1+y)^{a+b-1}]/(1+y)^{2(a+b)}. \quad (7.38)$$

Equate the numerator to zero and solve for y to get $y[a + b - a + 1] = (a - 1)$ or $y = (a - 1)/(b + 1)$. It is left as an exercise to verify that the second derivative is $-$ve for this value of y. ∎

7.5.5 Relationship with Other Distributions

Put $a = b = 1$ to get Beta(1,1), which is identical to $U(0, 1)$. If X is beta1(a, b) then $(1 - X)/X$ is beta2(b, a), and $X/(1 - X)$ is beta2(a, b). If X and Y are independent gamma random variables $\Gamma(a, \lambda)$ and $\Gamma(b, \lambda)$, then $X/(X + Y)$ is BETA(a, b) (see Exercise 7.26). As gamma and χ^2 are related, this result can also be stated in terms of normal variates as follows. If X and Y are independent normal variates, then $Z = X^2/(X^2 + Y^2)$ is beta distributed. In addition, if $X_1, X_2, \cdots, X_k$ are IID $N(0, 1)$ and $Z_1 = X_1^2/(X_1^2 + X_2^2), Z_2 = (X_1^2 + X_2^2)/(X_1^2 + X_2^2 + X_3^2)$, and so on, $Z_j = \sum_{i=1}^{j} X_j^2 / \sum_{i=1}^{j+1} X_j^2$, then each of them are BETA-I distributed, as also the product of any consecutive set of Z_j's are beta distributed [167, 171]. The logistic distribution and type II beta distribution are related as $Y = -\ln(X)$. If X is BETA-I(a, b) then $Y = \ln(X/(1 - X))$ has a generalized logistic distribution [172, 173]. Dirichlet distribution is a generalization of beta distribution. Order statistic from uniform distribution is beta distributed. In general, jth highest order statistic from a uniform distribution is BETA-I$(j, n - j + 1)$. See Reference 174 for the beta-generalized exponential distribution, Reference 167 for relationships among various statistical distributions, Reference 133 for normal approximations, and Reference 175 for new properties of this distribution.

As the random variable takes values in [0,1], any CDF can be substituted for x to get a variety of new distributions [22]. For instance, put $x = \Phi(x)$, the CDF of a normal variate to get the beta-normal distribution with PDF

$$f(x; a, b) = (1/B[a, b]) \ \phi(x)[\Phi(x)]^{a-1}[1 - \Phi(x)]^{b-1}, \tag{7.39}$$

where $B(a, b)$ is the CBF, $\phi(x)$ is the PDF and $\Phi(x)$ is the CDF of normal distribution, so that the range is now extended to $-\infty < x < \infty$.

7.6 THE INCOMPLETE BETA FUNCTION

The incomplete beta function (IBF) denoted by $I_x(a, b)$ or $I(x; a, b)$ has several applications in statistics and engineering. It is used in wind velocity modeling [176], flood water modeling, and soil erosion modeling. It is used to compute Bartlett's statistic for testing homogeneity of variances when unequal samples are drawn from normal populations [177] and in several tests involving likelihood ratio criterion [178]. It is also used in computing the power function of nested tests in linear models [179], approximating the distribution of largest roots in multivariate inference, and detecting two outliers in the same direction in a linear model [180]. Its applications to traffic accident proneness are discussed by Haight [181].

7.6.1 Tail Areas Using IBF

Tail areas of several statistical distributions are related to the beta CDF as discussed below. The survival function of a binomial distribution BINO(n, p) is related to the left-tail areas of BETA-I distribution as

$$\sum_{x=a}^{n} \binom{n}{x} p^x q^{n-x} = I_p(a, n - a + 1). \tag{7.40}$$

Using the relationship (7.47), the CDF becomes

$$\sum_{x=0}^{a-1} \binom{n}{x} p^x q^{n-x} = I_q(n - a + 1, a). \tag{7.41}$$

When both a and b are integers, this has a compact representation as

$$I_x(a, b) = 1 - \sum_{k=0}^{a-1} \binom{a+b-1}{k} x^k (1 - x)^{a+b-1-k}. \tag{7.42}$$

The survival function of negative binomial distribution is related as follows

$$\sum_{x=a}^{n} \binom{n+x-1}{x} p^n q^x = I_q(a, n) = 1 - I_p(n, a). \tag{7.43}$$

The relationship between the CDF of central F distribution and the IBF is

$$F_{m,n}(x) = I_y(m/2, n/2), \tag{7.44}$$

where (m, n) are the numerator and denominator degrees of freedom (DoF) and $y = mx/(n + mx)$. Similarly, Student's T CDF is evaluated as

$$T_n(t) = \frac{1}{2}\left[1 + \text{sign(t)} I_x\left(\frac{1}{2}, \frac{n}{2}\right)\right] = \frac{1}{2}\left\{1 + \text{sign}(t) \left[1 - I_y\left(\frac{n}{2}, \frac{1}{2}\right)\right]\right\} \tag{7.45}$$

where $x = t^2/(n + t^2)$ and $y = n/(n + t^2)$, and sign(t) $= +1$ if $t > 0$, -1 if $t < 0$ and is $= 0$ for $t = 0$.

The IBF is related to the tail areas of binomial, negative binomial, Student's T, and central F distributions [182]. It is also related to the confluent hypergeometric function, generalized logistic distribution, the distribution of order statistics from uniform populations, and the Hotelling's T^2 statistic. The hypergeometric function can be approximated using the IBF also [183]. The Dirichlet (and its inverse) distribution can be expressed in terms of the IBF [184]. It is related to the cumulative distribution function (CDF) of noncentral distributions [7, 185–192] and the sample multiple correlation coefficient [193, 194]. For instance, the CDF of singly noncentral beta [179, 195], singly type-II noncentral beta, doubly noncentral beta [4], noncentral T [188, 196], noncentral F [5, 197, 198], and the sample multiple correlation coefficient [199, 200] could all be evaluated as infinite mixtures of the IBF.

Definition 7.2 The IBF is the left-tail area of the beta distribution

$$I_x(a,b) = (1/B(a,b)) \int_0^x t^{a-1}(1-t)^{b-1} dt, \quad (a,b > 0) \text{ and } 0 \le x \le 1, \quad (7.46)$$

where $B(a,b)$ is the CBF. Obviously, $I_0(a,b) = 0$ and $I_1(a,b) = 1$. Replace x by $(1 - x)$ and swap a and b to get a symmetric relationship.

$$I_x(a,b) = 1 - I_{1-x}(b,a). \quad (7.47)$$

This symmetry among the tail areas was extended by Chattamvelli [196] to noncentral beta, noncentral Fisher's Z, and doubly noncentral distributions.

If the CDF (left-tail area) of a type-II noncentral beta distribution is denoted as $J_x(a,b,\lambda)$, then the tail areas are related as $J_x(a,b,\lambda) = 1 - I_{1-x}(b,a,\lambda)$, where $I_y(b,a,\lambda)$ is the CDF of a type-I noncentral beta distribution. We write $I_x(b)$ or $I(x;b)$ for the symmetric IBF $I_x(b,b)$ [201, 202]. The parameters of an IBF can be any positive real number. Simplified expressions exist when either of the parameters is an integer or a half-integer. These representations have a broad range of applications to evaluating or approximating other related distributions and test statistics mentioned earlier.

The IBF has representations in terms of other special functions and orthogonal polynomials [183, 202–205]. For example, it could be expressed in terms of hypergeometric series in the following form:

$$I_x(a,b) = \frac{x^a(1-x)^{b-1}}{aB(a,b)} \, {}_2F_1(1-b,1;a+1;-x/(1-x)), \quad (7.48)$$

where ${}_2F_1$ denotes the hypergeometric series.

7.6.2 Tables

Many tables for the IBF are available. See, for example, Soper [206], Pearson [207], Aroian [203], Majumder and Bhattacharjee [208], and Boston and Battiste [209].

Random variate generation from beta distribution is accomplished using the relationship between the beta and gamma distributions. Hence, if two random numbers are generated from $\Gamma(1,a)$ and $\Gamma(1,b)$, where $a < b$, then the beta variate is given by $B(a,b) = \Gamma(1,a)/[\Gamma(1,a) + \Gamma(1,b)]$ [22].

7.7 GENERAL BETA DISTRIBUTION

General three parameter beta distribution is given by

$$f_x(a,b,c) = (x/c)^{a-1}(1-x/c)^{b-1}/c\,B(a,b). \quad (7.49)$$

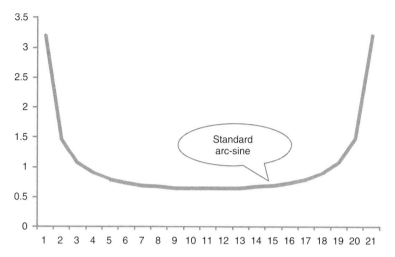

Figure 7.5   Standard arc-sine distribution.

The four-parameter beta distribution is obtained from the above-mentioned representation by the transformation $y = (x - a)/(b - a)$ to get the PDF

$$f_x(a, b, c, d) = \frac{\Gamma(c + d)}{\Gamma(c)\Gamma(d)(b - a)^{c+d-1}}(x - a)^{c-1}(b - x)^{d-1}. \tag{7.50}$$

This has mean $(ad + bc)/(c + d)$ and variance $\sigma^2 = cd(b - a)^2/[(c + d + 1)(c + d)^2]$. The location parameters are "a,""b" and scale parameters are c and d. Coefficient of skewness is $2cd(d - c)/[(c + d)^2(c + d)^{(3)}[cd/((c + d)(c + d)^{(2)})]]$, where $(c + d)^{(k)}$ is raising Pochhammer notation with $(c + d)^{(3)} = (c + d)(c + d + 1)(c + d + 2)$. The mode is $\frac{a(d-1)+b(c-1)}{(c+d-2)}$ for c not 1 and d not 1. See References 210 and 134.

7.8   ARC-SINE DISTRIBUTION

This is a special case of the beta distribution when $a = 1/2, b = 1/2$. The *standard* arc-sine distribution (SASD) of first kind has support $0 < x < 1$, is U-shaped and symmetric around $x = 1/2$ (see Figure 7.5). Its PDF is given by

$$f(x) = \begin{cases} \dfrac{1}{\pi\sqrt{x(1 - x)}} & \text{for } \ 0 < x < 1 \\[2mm] 0 & \text{elsewhere.} \end{cases}$$

To prove that this is indeed a PDF, put $x = \sin^2(\theta)$ so that $dx = 2\sin(\theta)\cos(\theta)d\theta$ and $1 - x = 1 - \sin^2(\theta) = \cos^2(\theta)$. The denominator $\sqrt{x(1 - x)}$ becomes

$\sqrt{\sin^2(\theta)\cos^2(\theta)} = \sin(\theta)\cos(\theta)$. When $x = 0, \theta = 0$ and when $x = 1, \theta = \pi/2$.

$$\int_0^{+1} \frac{1}{\pi\sqrt{x(1-x)}} dx = \int_0^{\pi/2} (2/\pi)\sin(\theta)\cos(\theta)d\theta/[\sin(\theta)\cos(\theta)] \qquad (7.51)$$

$$= (2/\pi)\int_0^{\pi/2} d\theta = (2/\pi)\theta|_0^{\pi/2} = 1.$$

This shows that the above is indeed a PDF. Another form of the distribution called arc-sine distribution of second kind has support $-1 < x < 1$ with PDF given by (see Exercise 7.74, p. 331)

$$f(x) = \begin{cases} \dfrac{1}{\pi\sqrt{(1-x^2)}} & \text{for } -1 < x < 1 \\ 0 & \text{elsewhere.} \end{cases}$$

To prove that this is a PDF, integrate over the range to get $\int_{-1}^{+1} 1/\sqrt{(1-x^2)}dx = \sin^{-1}x|_{-1}^{+1} = (3\pi/2) - (\pi/2) = \pi$. The π cancels out, showing that this is indeed a PDF.

▉ EXAMPLE 7.12   Moments of arc-sine distribution

Find the kth moment of arc-sine distribution of second kind

Solution 7.12 The kth moment is $(\frac{1}{\pi})\int_{-1}^{+1} x^k/\sqrt{(1-x^2)}dx$. Put $x = \sin(\theta)$ so that $dx = \cos(\theta)d\theta$ and $\sqrt{(1-x^2)} = \sqrt{1-\sin^2(\theta)} = \cos(\theta)$. Thus $\mu_k = \frac{1}{\pi}\int_{-\pi/2}^{\pi/2} \sin^k\theta d\theta$. As $\sin(\theta)$ is an odd function, the integral vanishes when k is odd. When k is even, the integral becomes $(\frac{1}{\pi})\int_{-\pi/2}^{\pi/2} \sin^k\theta d\theta = (\frac{2}{\pi})\int_0^{\pi/2} \sin^p\theta d\theta$ where $p = 2k$. Using integration by parts, this reduces to $(\frac{2}{\pi})(\sqrt{\pi}/2)\Gamma((p+1)/2)/\Gamma(p/2+1) = (1/\sqrt{\pi})\Gamma((2k+1)/2)/\Gamma(k+1)$. For $k = 1$, this becomes $(1/\sqrt{\pi})\Gamma(3/2) = 1/2$ using $\Gamma(1/2) = \sqrt{\pi}$.   ▉

The two-parameter arc-sine distribution (ASD) is obtained by putting $y = (x-a)/b$ in the above as

$$f(x; a, b) = \begin{cases} \dfrac{b}{\pi\sqrt{(x-a)(a+b-x)}} & \text{for } a < x < b \\ 0 & \text{elsewhere.} \end{cases}$$

It has mean = median = $(a+b)/2$, variance $(b-a)^2/8$, skewness = 0, and excess kurtosis $-3/2$.

TABLE 7.8   Properties of Arc-Sine Distribution

Property	Expression	Comments
Range of X	$0 \leq x \leq 1$	SASD-I; Continuous
Mean	$\mu = 1/2 = 0.50$	
Median	0.50	Mode $\in \{0, 1\}$
Variance	$\sigma^2 = 1/8$	0.125
Skewness	$\gamma_1 = 0$	Symmetric
Kurtosis	$\beta_2 = 3/2$	Always platykurtic
Mean deviation	$E\lvert X - \mu\rvert = 1/8$	
CV	$1/\sqrt{2}$	
CDF	$\frac{2}{\pi}\sin^{-1}(\sqrt{x})$	$\sin^{-1}(x) + \pi/2$
Moments	$(1/\pi)B(k + 0.5, 0.5)$	$\mu_{2k} = \binom{2k}{k}(1/2)^{2k}$
MGF	$e^{t/2}I_0(t/2)$	Modified Bessel function
ChF	$e^{-t/2}I_0(it/2)$	$_1F_1(1/2, 1; it)$

Third column of CDF and moments line are for the other parametrization.

7.8.1   Properties of Arc-Sine Distribution

The SASD-I is a special case of beta type-I distribution. Put $Y = X - \frac{1}{2}$ to get

$$f(y) = \frac{1}{\pi\sqrt{(y + 1/2)(1/2 - y)}}, \quad -1/2 \leq y \leq 1/2. \tag{7.52}$$

As $(1/2 + y)(1/2 - y) = (1/4 - y^2)$, the PDF becomes $f(y) = (2/\pi)1/\sqrt{(1 - 4y^2)}$, for $-1/2 \leq y \leq 1/2$. The CDF of SASD-I is

$$F(x) = \frac{2}{\pi}\sin^{-1}(\sqrt{x}), \quad 0 \leq x \leq 1. \tag{7.53}$$

The mean is 0.5 and variance is 0.125 for the SASD. As the distribution is symmetric, coefficient of skewness is zero. The kurtosis coefficient is $\beta_2 = 3/2$. Hence, it is always platykurtic. Note that the density is maximum when x is near 0 or 1 with the center as a cusp (U-shaped). Hence, there are two modes (bimodal) that are symmetrically placed in the tails. This is the reason why it is platykurtic. The central moments of arc-sine distribution of second kind is $\mu_{2k} = \binom{2k}{k}(1/2)^{2k}$. The MGF is $M_x(t) = e^{t/2}I_0(t/2)$, where $I_0(x)$ is the modified Bessel function of first kind. The two parameter ASD satisfies an interesting property:–If $X \sim \text{ASD}(a, b)$ then $cX + d \sim \text{ASD}(c * a + d, c * b + d)$. See References 121 and 134 for other relationships and applications. See Table 7.8 for further properties.

■ EXAMPLE 7.13   Mean of arc-sine distributions

Find the mean of arc-sine distribution of first kind.

Solution 7.13 $E(X) = (1/\pi) \int_0^1 x/\sqrt{x(1-x)}dx$. Put $x = \sin^2(\theta)$ as before so that $E(X) = (2/\pi) \int_0^{\pi/2} \sin^2(\theta)d\theta$. Put $\sin^2(\theta) = (1 - \cos(2\theta))/2$ and integrate to get $2/\pi[(1/2)\theta|_0^{\pi/2} - (1/4)\sin(2\theta)|_0^{\pi/2}] = 2/\pi[\pi/4 - 0] = 1/2$.   ∎

This distribution is related to the beta distribution when $a = 1/2, b = 1/2$.

■ EXAMPLE 7.14   Mean deviation of arc-sine distribution

Find the mean deviation of the arc-sine distribution using Theorem 7.1.

Solution 7.14 We have seen earlier that the CDF is $\frac{2}{\pi}\sin^{-1}(\sqrt{x})$. As the mean is 0.5, we get the MD using Theorem 7.1 as

$$MD = 2 \int_{ll}^{\mu} F(x)dx = \frac{4}{\pi} \int_0^{.5} \sin^{-1}(\sqrt{x})dx. \qquad (7.54)$$

Put $\sqrt{x} = t$ so that $dx = 2tdt$. Adjust the upper limit of integration as $c = \sqrt{0.5}$. This gives $MD = \frac{8}{\pi} \int_0^c t\sin^{-1}(t)dt$. Now use $\int t \sin^{-1}(t)dt = (2t^2 - 1)/4\sin^{-1}(t) + t\sqrt{1 - t^2}/4$. The first term evaluates to zero, and we get the MD as 1/8.   ∎

7.9   GAMMA DISTRIBUTION

The two parameter gamma distribution can be considered as a generalization of the exponential distribution. Its PDF is given by

$$f_x(\lambda, m) = \lambda^m x^{m-1} e^{-\lambda x}/\Gamma(m), \qquad x \geq 0, m > 0, \ \lambda > 0. \qquad (7.55)$$

When $m = 1$, this reduces to the exponential distribution. Hence, it is considered a generalization of the exponential distribution. For $m = 1/2$, we get $f_x(\lambda, m) = \sqrt{\lambda/\pi x} \ e^{-\lambda x}$. The parameter λ is called scale parameter and m is the shape parameter (see figure 7.8). A reparametrization as

$$f(x; \lambda, m) = e^{-x/\lambda}x^{m-1}/[\lambda^m \Gamma(m)] \qquad (7.56)$$

also exist. A change of scale transformation $Y = \lambda X$ (so that $dy = \lambda dx$) in equation (7.55) gives $f_y(m) = y^{m-1}e^{-y}/\Gamma(m)$ given below. As this form is easier to work with, it is extensively tabulated.

7.9.1 Properties of Gamma Distribution

There are two parameters, both of which are real numbers. For $\lambda = 1$, we get the one-parameter gamma distribution with PDF

$$f(x; m) = x^{m-1}e^{-x}/\Gamma(m) \text{ for } x > 0. \tag{7.57}$$

For m an integer, this is called Erlang distribution. The coefficient of skewness and kurtosis are $1/\sqrt{m}$ and $3(1 + 2/m)$, which are both independent of λ. This distribution is always leptokurtic.

7.9.1.1 Additivity Property This distribution can be obtained as the sum of m independent exponential variates with parameter λ, resulting in GAMMA(m, λ). If X and Y are two independent gamma random variables with the same scale parameter λ and shape parameters m_1 and m_2, respectively, their sum $X + Y$ is distributed as gamma with the same scale parameter and $m_1 + m_2$ as shape parameter. This result can be generalized to any number of independent gamma variates as "the sum of m independent gamma variates with shape parameters m_i and the same scale parameter λ is distributed as Gamma($\sum_i m_i, \lambda$), see Table 7.4 in page 269."

7.9.1.2 Moments and Generating Functions The raw moments are easy to find using gamma integral. Consider

$$E(X^k) = \int_0^\infty \lambda^m x^k x^{m-1}e^{-\lambda x}./\Gamma(m) \tag{7.58}$$

Using gamma integral, this becomes $\Gamma(k + m)/\lambda^{k+m}$. From this, we get the mean as $\mu = m/\lambda$ and variance $\sigma^2 = m/\lambda^2 = \mu/\lambda$. This shows that the variance is more than the mean for $\lambda < 1$ and vice versa. The characteristic function is

$$\phi(t) = (\lambda^m/\Gamma(m)) \int_0^\infty x^{m-1}e^{-x(\lambda-it)}dx. \tag{7.59}$$

Put $Y = (\lambda - it)X$, so that $dy = (\lambda - it)dx$. The range of integration remains the same and we get

$$\phi(t) = (\lambda^m/[\Gamma(m)(\lambda - it)^m]) \int_0^\infty y^{m-1}e^{-y}dy = (\lambda/(\lambda - it))^m = (1 - it/\lambda)^{-m} \tag{7.60}$$

for $t < \lambda$. By expanding this as an infinite series (see Chapter 6), we get $(1 - it/\lambda)^{-m} =$

$$\sum_{k=0}^\infty \binom{m+k-1}{k}(-it/\lambda)^k = 1 + (m/\lambda)it + m(m+1)/[1*2](it/\lambda)^2 + \cdots. \tag{7.61}$$

7.9.2   Relationships with Other Distributions

The χ^2 distribution is a special case of gamma distribution as $\chi_n^2 = \text{GAMMA}(n/2, 1/2)$. Symbolically, if $X_1, X_2, \cdots, X_n$ are independent standard normal random variables, $Y = X_1^2 + \cdots + X_n^2 \sim \text{GAMMA}(n/2, 1/2)$. If $X_1 \sim \text{Gamma}(a, b)$ and $X_2 \sim \text{Gamma}(c, d)$, are independent, then $X_1/(X_1 + X_2)$ is distributed as BETA-I. Inverse gamma distribution is obtained by a simple change of variable $Y = 1/X$ as

$$f_y(\lambda, m) = \lambda^m y^{-(m+1)} e^{-\lambda/y}/\Gamma(m), \quad y \geq 0, m > 0, \ \lambda > 0. \tag{7.62}$$

Log-gamma distribution is the analog of LND in the Gamma case. The PDF is given by

$$f(x) = a^b/\Gamma(b)(\ln x)^{b-1} x^{-a-1}. \quad a > 1, b > 0. \tag{7.63}$$

Boltzmann distribution in engineering is related to the gamma law. If the quantized energies of a molecule in an ensemble are $E_1, E_2, \cdots, E_n$, the probability that a molecule has energy E_i is given by $C \exp(-E_i/(KT))$, where K is the Boltzmann constant (gas constant divided by Avogadro number) and T is the absolute temperature. The sum of the energies is gamma distributed when $E_i's$ are independent. See Table 7.9 for further properties. Estimating the median of gamma distributions is discussed in Reference 211.

TABLE 7.9   Properties of Gamma Distribution $(\lambda^m x^{m-1} e^{-\lambda x}/\Gamma(m))$

Property	Expression	Comments
Range of X	$x \geq 0$	Continuous
Mean	$\mu = m/\lambda$	
Median	$\log(2)/\lambda$	
Mode	$(m-1)/\lambda$	
Variance	$\sigma^2 = m/\lambda^2$	$\sigma^2 = \mu^2/m = \mu/\lambda$
Skewness	$\gamma_1 = 2/\sqrt{m}$	
Kurtosis	$\beta_2 = 3(1 + 2/m)$	Always leptokurtic
Mean deviation	$E\lvert X - \mu\rvert = 2m^m e^{-m} \lambda/\Gamma(m)$	$2\lambda^m/\Gamma(m) \int_0^{m/\lambda} \Gamma(x; \lambda, m) dx$
CV	$1/\sqrt{m}$	
CDF	$\Gamma(x; \lambda, m)$	$\frac{\lambda^m}{\Gamma(m)} \int_0^x e^{-\lambda y} y^{m-1} dy$
Moments	$\mu'_r = \Gamma(k+m)/\lambda^{k+m}$	
MGF	$[\lambda/(\lambda - t)]^m$	
ChF	$[\lambda/(\lambda - it)]^m$	

Replace λ by $1/\lambda$ for the alternate parametrization. $\sigma^2 > \mu^2$ when $m < 1$. $\frac{\lambda^m}{\Gamma(m)} \int_0^x e^{-\lambda y} y^{m-1} dy$ is called the incomplete gamma function. See References 211 and 212 for median estimates.

7.9.3   Incomplete Gamma Function (IGF)

Definition 7.3 The left-tail area of gamma distribution is called the incomplete gamma function. It is given by

$$P(x; \lambda, m) = \frac{\lambda^m}{\Gamma(m)} \int_0^x e^{-\lambda y} y^{m-1} dy. \tag{7.64}$$

As there are two parameters, a simple change of scale transformation mentioned earlier can be made. This gives one-parameter gamma distribution. The function $\Gamma(m) = \int_0^\infty x^{m-1} e^{-x} dx$ is called the complete gamma function. When m is an integer, the above-mentioned integral becomes $\Gamma(m) = (m-1)!$ When m is not an integer, we get the recurrence $\Gamma(m) = (m-1) * \Gamma(m-1)$ using integration by parts. The integral with and without the normalizing constant is denoted as $\gamma(x; m) =$

$$\int_0^x y^{m-1} e^{-y} dy = \Gamma(m) - \Gamma(x; m) \text{ and } P(x, m) = \frac{1}{\Gamma(m)} \int_0^x y^{m-1} e^{-y} dy. \tag{7.65}$$

These satisfy the recurrence

$$\gamma(x; m+1) = m * \gamma(x; m) - x^m e^{-x} \text{ and } P(x, m+1) = P(x; m) - \frac{x^m e^{-x}}{\Gamma(m+1)}.$$

Put $y = x^2$ and $m = 1$ to get $\gamma(1, x^2) = \int_0^{\sqrt{y}} e^{-y^2} dy$. An approximate relation with erfc() is available as $\gamma(x, m)/\Gamma(m) \sim 0.5 * \text{erfc}(-K\sqrt{m/2})$, where $K^2 = 2(x/m - 1 - \ln(x/m))$.

The CDF of gamma is a sum of Poisson probabilities when the shape parameter is an integer: $-F_x(m, p) = 1 - \sum_{k=0}^{p-1} e^{-\lambda x} (\lambda x)^k / k!$. The corresponding right-tail area is denoted as $P(x; m, p) = \frac{m^p}{\Gamma(p)} \int_x^\infty e^{-mt} t^{p-1} dt$. Both of these are extensively tabulated. For example, Pearson [207] tabulated the function $I(x, p) = \frac{1}{\Gamma(p+1)} \int_0^{x\sqrt{p+1}} e^{-t} t^p dt$. It has a representation in terms of confluent hypergeometric functions as $\gamma(x; 1, p) = (x^p/p) e^{-x} {}_1F_1(1, p+1; x) = (x^p/p) {}_1F_1(p, p+1; -x)$ and error function as $\gamma(x^2; 1, 1/2) = \text{erf}(x)$. See References 121, 213–216 for other properties and relationships.

7.10   COSINE DISTRIBUTION

The PDF is given by

$$f(x; a, b) = \frac{1}{2b} \cos((x - a)/2b), \quad a - \pi b/2 \leq x \leq a + \pi b/2, \ b > 0 \tag{7.66}$$

$$F(x) = \frac{1}{2} \left[1 + \sin((x - a)/(2b)) \right]. \tag{7.67}$$

TABLE 7.10   Properties of Cosine Distribution ($\frac{1}{4b}\cos((x-a)/2b)$)

Property	Expression	Comments
Range of X	$a - \pi b \leq x \leq a + \pi b$	Continuous; finite
Mean	$\mu = a$	
Median	a	Mode $= a$
Variance	$\sigma^2 = b^2(\pi^2 - 8)$	$cv = \dfrac{b\sqrt{\pi^2 - 8}}{a}$
Skewness	$\gamma_1 = 0$	Symmetric
Kurtosis	$\beta_2 = 9$	Always leptokurtic
MD	$b(\pi - 2)$	
CDF	$\dfrac{1}{2}\left[1 + \sin\dfrac{x-a}{2b}\right]$	

As this distribution is symmetric, the mean, median, and mode coincide at $x = a$, which is the location parameter; and skewness is zero (Table 7.10). The variance depends only on b and is given by $\sigma^2 = b^2[\pi^2 - 8]$. Random sample generation: Generate $U(-1, 1)$, then transform $x = a + 2b \sin^{-1}(2U - 1)$ if $u = [0, 1]$.

7.11   THE NORMAL DISTRIBUTION

The normal distribution is perhaps the most widely studied distribution in statistics. It is known by the name Gaussian distribution in Engineering in honor of the German mathematician Carl Friedrich Gauss (1777–1855). It has two parameters, which are by convention denoted as μ and σ to indicate that they capture the location (mean) and scale information. The PDF is

$$\phi(x; \mu, \sigma) = \frac{1}{\sigma\sqrt{2\pi}} e^{-\frac{1}{2}\left(\frac{x-\mu}{\sigma}\right)^2}, \quad -\infty < x < \infty, \ -\infty < \mu < \infty, \ \sigma > 0. \quad (7.68)$$

It is denoted by $N(\mu, \sigma^2)$, where the first parameter is always the population mean and second parameter is the population variance. Some authors use the notation $N(\mu, \sigma)$, where the second parameter is the population standard deviation and $Z(0, 1)$ for a standard normal distribution. Even if the mean is zero, the first parameter should be specified. Thus, $N(0, \sigma^2)$ denotes a normal distribution with zero mean.

Any normal distribution (with arbitrary μ and σ) can be converted into the standard normal form $N(0, 1)$ using the transformation $Z = (X - \mu)/\sigma$. This is called "standard normalization". It is applicable for approximation due to central limit theorem, even in nonnormal situations when the sample size is quite large. The reverse transformation is $X = Z\sigma + \mu$. This shows that from the table of standard normal distribution, we could obtain the tail areas of any other normal distribution.

⬛ **EXAMPLE 7.15   Probability of normal deviates**

The radius of a batch of pipes is known to be normally distributed with mean 0.5 inch and variance 0.009. What proportions of a batch of 132 pipes have radius more than 2 standard deviations in the higher side?

Solution 7.15 As radius $\sim N(0.5, 0.009)$, standard deviation is 0.0948683. Standard normalize it to get $Z = (X - 0.5)/0.0948683$. Area above 2 standard deviations for $N(0, 1)$ is $1 - 0.9772 = 0.0228$. Thus, in a batch of 132 pipes, we expect $132*0.0228 = \lfloor 3.0096 \rfloor = 3$ pipes to have radius more than two standard deviations.   ■

7.11.1   Properties of Normal Distribution

The general normal distribution has two parameters, the second of which is positive (the first parameter by convention is the mean μ). The distribution is symmetric about the mean with relatively shorter tails than Cauchy and T distributions (see Figures 7.6 and 7.7). When the mean $\mu = 0$ and the variance $\sigma^2 = 1$, it is called the standard normal distribution, which is denoted by $Z(0, 1)$ or simply by Z. The corresponding PDF and CDF are denoted by $\phi(x)$ and $\Phi(x)$. Owing to symmetry $\Phi(-c) = 1 - \Phi(c)$ for $c > 0$ and $\Phi(0) = 1/2$, so that median = mean = mode with modal value $1/[\sigma\sqrt{2\pi}]$. If $c < d$, the area from c to d can be expressed as $\Phi(d) - \Phi(c)$.

7.11.1.1 Moments and Generating Functions As the distribution is symmetric, all odd central moments are zeros. The even moments are given by

$$\mu_{2k} = (2k!)(\sigma^2)^k/[2^k k!].   \tag{7.69}$$

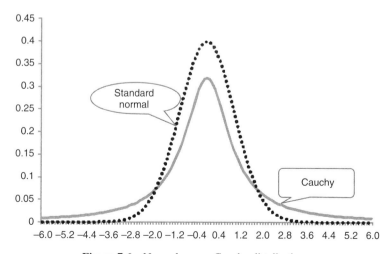

Figure 7.6 Normal versus Cauchy distributions.

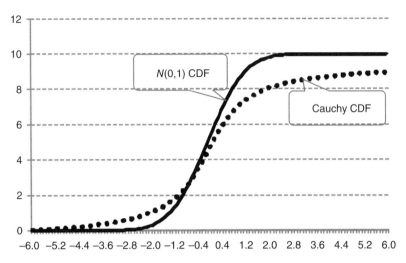

Figure 7.7 $N(0, 1)$ versus Cauchy CDF.

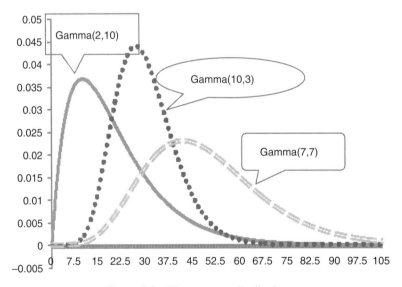

Figure 7.8   Three gamma distributions.

This can easily be proved using gamma integrals (see Exercise 7.34 in p. 285). The mean deviation is $\sigma\sqrt{2/\pi}$. The MGF is easily obtained as

$$M_x(t) = E(e^{tx}) = \int_{-\infty}^{\infty} \frac{1}{\sigma\sqrt{2\pi}} e^{tx} e^{-\frac{1}{2}((x-\mu)/\sigma)^2} dx. \tag{7.70}$$

Put $z = (x - \mu)/\sigma$ in the above so that $dz = dx/\sigma$ and $x = \mu + \sigma z$, to get

$$M_x(t) = (1/\sqrt{2\pi}) \int e^{t(\mu+\sigma z)} e^{-z^2/2} dz = e^{t\mu}/\sqrt{2\pi} \int e^{t\sigma z - \frac{1}{2}z^2} dz. \tag{7.71}$$

TABLE 7.11 **Properties of Normal Distribution**$(\frac{1}{\sigma\sqrt{2\pi}}e^{-\frac{1}{2}(\frac{x-\mu}{\sigma})^2})$

Property	Expression	Comments
Range of X	$-\infty < x < \infty$	Continuous; infinite
Mean	μ	
Median	μ	Mode $= \mu$
Variance	σ^2	
Skewness	$\gamma_1 = 0$	Symmetric
Kurtosis	$\beta_2 = 3$	For N(0,1)
Mean deviation	$E\lvert X - \mu \rvert = \sigma\sqrt{2/\pi}$	$2\int_{-\infty}^{\mu}\Phi(x)dx$
CDF	$\Phi((x-\mu)/\sigma)$	
Absolute Moments	$E(\lvert X \rvert^r) = (r-1)!!\sigma^r$ for $r = 2k$	$\sqrt{2/\pi}\,2^k k!\sigma^r$ for $r = 2k+1$
Moments	$\mu_{2r} = \sigma^{2r}(2r)!/[2^r\, r!]$	Even
MGF	$\exp(t\mu + \frac{1}{2}t^2\sigma^2)$	
ChF	$\exp(it\mu - \frac{1}{2}t^2\sigma^2)$	
Additivity IID	$X \sim N(\mu_1, \sigma_1^2)$ & $Y \sim N(\mu_2, \sigma_2^2)$	$X \pm Y \sim N(\mu_1 \pm \mu_2, \sigma_1^2 + \sigma_2^2)$

$\mathrm{erf}(z) = \dfrac{1}{\sqrt{\pi}}\int_0^z e^{-x^2/2}dx = \dfrac{2}{\sqrt{\pi}}\int_0^z e^{-t^2}dt.$

Write the exponent as $-\frac{1}{2}(z-t\sigma)^2 + \frac{1}{2}t^2\sigma^2$. As $e^{\frac{1}{2}t^2\sigma^2}$ is constant, take it outside the integral to get

$$e^{t\mu+\frac{1}{2}t^2\sigma^2}(1/\sqrt{2\pi})\int_{-\infty}^{\infty}e^{-(z-t\sigma)^2/2}dz. \tag{7.72}$$

As the integral evaluates to one, we get the desired result $M_x(t) = e^{t\mu+\frac{1}{2}t^2\sigma^2}$. See Table 7.11 for further properties.

The normal distribution is the basis of many procedures in statistical inference. These include confidence intervals for unknown parameters, prediction intervals for future observations, tests of various hypotheses, and estimation of parameters. Given below is a theorem about the mean of a random sample from a normal population, which is used in CI construction and tests about the mean.

Theorem 7.3 The sample mean $\bar{x}_n$ of any sample of size $n \geq 2$ from a normal population $N(\mu, \sigma^2)$ is itself normally distributed as $N(\mu, \sigma^2/n)$.

Proof: The easiest way to prove this result is using the MGF (or ChF) as follows. Let $M_x(t)$ be the MGF of a normal distribution (Chapter 8). Then $M_{\bar{x}}(t) = [M_x(t/n)]^n = [e^{\mu t/n + \frac{1}{2}(t/n)^2\sigma^2}]^n = e^{n\mu t/n + \frac{1}{2}n\sigma^2(t/n)^2} = e^{\mu t + \frac{1}{2}t^2(\sigma^2/n)}$, which is the MGF of a normal distribution with mean μ and variance σ^2/n. This proves the result. ∎

▉ **EXAMPLE 7.16   Mean deviation of normal distribution**

Find the mean deviation of the normal distribution using Theorem 7.1.

Solution 7.16 Let $X \sim N(\mu, \sigma^2)$. As the normal distribution tails off to zero at the lower and upper limits, equation (7.1) is applicable. This gives

$$\text{MD} = 2 \int_{\mu}^{\mu} F(x)dx = 2 \int_{-\infty}^{\mu} \Phi(x)dx. \tag{7.73}$$

Put $z = (x - \mu)/\sigma$, so that $dx = \sigma dz$. The lower limit remains the same, but the upper limit in equation (7.73) becomes 0. Thus, we get

$$\text{MD} = 2\sigma \int_{-\infty}^{0} \Phi(z)dz. \tag{7.74}$$

This integral can readily be evaluated using integration by parts. Put $u = \Phi(z)$ and $dv = dz$, so that $du = \phi(z)$. This gives

$$\text{MD} = 2\sigma \left[z\Phi(z) \mid_{-\infty}^{0} - \int_{-\infty}^{0} z\phi(z)dz \right] = 2\sigma \int_{0}^{\infty} z\phi(z)dz. \tag{7.75}$$

Substitute for $\phi(z) = (1/\sqrt{2\pi})\exp(-z^2/2)$ and use $\int_{0}^{\infty} z^{2n+1} \exp(-z^2/2)dz = n!2^n$. Then the integral becomes $1/\sqrt{2\pi}$. Apply the constant 2σ to get the

$$\text{MD} = 2\sigma * 1/\sqrt{2\pi} = \sigma\sqrt{2/\pi}. \tag{7.76}$$

■

7.11.2   Transformations to Normality

Normality of parent population is a fundamental assumption in many statistical procedures. For example, error terms in logistic and multiple regression are assumed to be normally distributed with zero mean. Normality tests of the sample can reveal whether the data came from a normal distribution or not. When the data are not normally distributed, a simple transformation may sometimes transform it to nearly normal form. For instance, count data are usually transformed using the square root transformation, and proportions are transformed using logit transformation as $y = \frac{1}{2} \log(p/q)$ where $q = 1 - p$.

7.11.3   Functions of Normal Variates

Any linear combinations of normal variates are normally distributed. Symbolically, if $X_1, X_2, \cdots, X_k$ are IID $N(\mu_i, \sigma_i^2)$, then $Y = \mp c_1 X_1 \mp c_2 X_2 \mp \cdots \mp c_k X_k = \sum_{i=1}^{k} \mp c_i X_i$ is normally distributed with mean $\sum_{i=1}^{k} \mp c_i \mu_i$ and variance $\sum_{i=1}^{k} c_i^2 \sigma_i^2$.

If X and Y are IID normal variates with zero means, then $U = XY/\sqrt{X^2 + Y^2}$ is normally distributed [217, 218]. In addition, if $\sigma_x^2 = \sigma_y^2$, then $(X^2 - Y^2)/(X^2 + Y^2)$ is

TABLE 7.12   Area Under Normal Variates

Sigma Level	Area As %	Outside Range (ppm)	Sigma Level	Area As %	Outside Range (ppm)
$\mp 1\sigma$	68.26	317,400	$\mp 1.5\sigma$	86.64	133,600
$\mp 2\sigma$	95.44	45,600	$\mp 2.5\sigma$	98.76	12,400
$\mp 3\sigma$	99.74	2600	$\mp 3.5\sigma$	99.96	400
$\mp 4\sigma$	99.99366	63.40	$\mp 5\sigma$	99.9999426	0.574
$\mp 6\sigma$	99.9999998	0.002	$>6\sigma$	99.99999999	Negligible

also normally distributed. Product of independent normal variates has a Bessel-type III distribution. The square of a normal variate is gamma distributed (of which χ^2 is a special case). In general, if $X_1, X_2, \cdots, X_k$ are IID $N(0, \sigma^2)$, then $\sum_{i=1}^{k} X_i^2/\sigma^2$ is χ_k^2 distributed. As shown in page 276, if $Z_1 = X_1^2/(X_1^2 + X_2^2), Z_2 = (X_1^2 + X_2^2)/(X_1^2 + X_2^2 + X_3^2)$, and so on, $Z_j = \sum_{j=1}^{j} X_i^2 / \sum_{i=1}^{j+1} X_i^2$, then each of them are BETA-I distributed, as also the product of any consecutive set of Z_j's are beta distributed [171]. Normal distribution is also related to Student's T, Snedecor's F, and Fisher's Z distributions [219].

■ **EXAMPLE 7.17   Probability $P(X \leq Y)$ for two populations $N(\mu, \sigma_i^2)$**

If X and Y are independently distributed as $N(\mu, \sigma_i^2)$, for $i = 1, 2$ find the probability $P[X \leq Y]$.

Solution 7.17 As X and Y are IID, the distribution of $X - Y$ is $N(0, \sigma_1^2 + \sigma_2^2)$. Hence $P[X \leq Y] = P[X - Y \leq 0] = \Phi(0) = 0.5$, irrespective of the variances. ■

7.11.4   Relation to Other Distributions

If X and Y are independent normal random variables, then X/Y is Cauchy distributed. This is proved in Chapter 10. If X is chi distributed (i.e., $\sqrt{\chi_m^2}$) and Y is independently distributed as $BETA((b-1)/2, (b-1)/2)$, then the product $(2 * Y - 1) * X$ is distributed as $N(0, 1)$ [220]. There are many other distributions that tend to the normal distribution under appropriate limits. For example, the Binomial distribution tends to the normal curve when the sample size n becomes large. The convergence is more rapid when $p \to 1/2$ and n is large. The lognormal and normal distributions (Section 7.14, p. 297) are related as follows: If $Y = \log(X)$ is normally distributed, then X has LND. The LND can be obtained from a normal law using the transformation $X = e^Y$ (p. 298).

7.11.4.1   Tail Areas   The CDF of standard normal distribution is given by

$$F_z(z_0) = \Phi(z_0) = \frac{1}{\sqrt{2\pi}} \int_{-\infty}^{z_0} e^{-z^2/2} dz. \qquad (7.77)$$

This can be approximated as $\Phi(z) = 0.5 + (1/\sqrt{2\pi})(z - z^3/6 + z^5/40 - \dots = 0.5 + (1/\sqrt{2\pi}) \sum_{k=0}^{\infty} (-1)^k z^{2k+1}/[(2k+1)k!2^k]$. Because of the symmetry of the normal curve, the CDF is usually tabulated from 0 to some specified value x. The area of the normal curve from 0 to z is called the error function

$$\text{erf}(z) = \frac{1}{\sqrt{\pi}} \int_0^z e^{-x^2/2} dx = \frac{2}{\sqrt{\pi}} \int_0^z e^{-t^2} dt. \tag{7.78}$$

Using a simple change of variable, this can be expressed in terms of the incomplete gamma integral as $\sqrt{\pi}\gamma(1/2, z^2)$. The complement of the above-mentioned integral is denoted by $\text{erfc}(z)$. The $\text{erf}(z)$ can be expressed in terms of confluent hypergeometric functions as

$$\text{erf}(z) = \frac{2z}{\sqrt{\pi}} {}_1F_1(1/2, 3/2, -z^2) = \frac{2z}{\sqrt{\pi}} e^{-z^2} {}_1F_1(1, 3/2, z^2). \tag{7.79}$$

When the ordinate is in the extreme tails, another approximation as $\Phi(-x) =$

$$\frac{\phi(x)}{x}(1 - 1/x^2 + 3/x^4 - \dots + (-1)^n 3 * 5 * \dots (2n-1)/x^{2n}) \tag{7.80}$$

can be used. Replace $-x$ by $+x$ to get an analogous expression for right-tail areas. Scaled functions of the form $C * e^{-dx^2}$ are quite accurate to approximate the tail probabilities. The error function $\text{erf}(z)$ has an infinite series expansion as

$$\text{erf}(z) = \frac{2}{\sqrt{\pi}} \int_0^z e^{-t^2} dt = \frac{2}{\sqrt{\pi}} e^{-z^2} \sum_{k=0}^{\infty} 2^k z^{2k+1}/[1.3. \cdots (2k+1)]. \tag{7.81}$$

This series is rapidly convergent for small z values. See References 134 and 221.

7.11.4.2 Additivity Property

As mentioned in Section 7.11.3, linear combinations of IID normal variates are normally distributed. If $X \sim N(\mu_1, \sigma_1^2)$ and $Y \sim N(\mu_2, \sigma_2^2)$ are independent, then $X \pm Y \sim N(\mu_1 \pm \mu_2, \sigma_1^2 + \sigma_2^2)$. This is an important point in practical experiments, where results from two or more processes that are independent and normally distributed need to be combined.

■ **EXAMPLE 7.18   Linear combination of IID normal variates**

$X \sim N(10, 3)$, $Y \sim N(15, 6)$, and $Z \sim N(9, 2.5)$, find the mean and variance of the following functions: (i) $U = X - 2Y + 3Z$ and (ii) $V = 2X - 1.2Y - Z$.

Solution 7.18 Use the linear combination property to get $E(U) = 10 - 2 * 15 + 3 * 9 = 37 - 30 = 7$, $\text{Var}(U) = 3 + 6 + 2.5 = 11.5$ so that $U \sim N(7, 49.5)$ In the second case, $E(V) = 2 * 10 - 1.2 * 15 - 9 = -7$ and $\text{Var}(V) = 4 * 3 + 4 * 6 + 9 * 3.5 = 23.14$. ■

7.11.5   Algorithms

Random variate generation: If x is uniform in the interval [0,1] then $z = \sqrt{-2\,\log x}\,\cos(2\pi x)$ is distributed as standard normal. This is known as Box–Muller method [22, 222].

◼ EXAMPLE 7.19   Truncated normal distribution PDF

Prove that the PDF of an asymmetrically truncated normal distribution with truncation points a and b is

$$f(x; \mu, \sigma, a, b) = \frac{1}{\sigma}\phi((x - \mu)/\sigma)/[\Phi((b - \mu)/\sigma) - \Phi((a - \mu)/\sigma)] \text{ for } a < x < b.$$

Solution 7.19 As the truncation point is asymmetric, the area enclosed is $\int_a^b \phi((x - \mu)/\sigma)dx$. Put $z = (x - \mu)/\sigma$, so that $dz = dx/\sigma$. The limits are changed as $(a - \mu)/\sigma$ and $(b - \mu)/\sigma$. Thus

$$\int_a^b \phi((x - \mu)/\sigma)dx = \sigma \int_{(a-\mu)/\sigma}^{(b-\mu)/\sigma} \phi(z)dz = \sigma[\Phi((b - \mu)/\sigma) - \Phi((a - \mu)/\sigma)].$$

$$(7.82)$$

Dividing by this quantity gives the PDF as desired.   ◼

7.12   CAUCHY DISTRIBUTION

The Cauchy distribution is named after the French mathematician A. L. Cauchy(1789–1857), although it was known to Fermat and Newton much earlier. It is symmetric, unimodal and has the general PDF

$$f(x; a, b) = 1/[b\pi[1 + (x - a)^2/b^2]] \qquad a, b > 0, \quad -\infty < x < \infty. \tag{7.83}$$

The location parameter is "a" and scale parameter is "b." The standard Cauchy distribution (SCD) is obtained from the above by putting $a = 0$ and $b = 1$:–

$$f(x) = \frac{1}{\pi}\frac{1}{1 + x^2}, \quad -\infty < x < \infty. \tag{7.84}$$

The CDF of SCD is

$$F(x; a, b) = 1/2 + (1/\pi)\tan^{-1}x \tag{7.85}$$

and that of general Cauchy distribution is given by $F_x(a, b) =$

$$\frac{1}{\pi}\int_{-\infty}^x \frac{dx}{[b\pi[1 + (x - a)^2/b^2]]} = \frac{1}{2} + \text{sign}(x - a)\,(1/\pi)\tan^{-1}\left(\frac{x - a}{b}\right). \tag{7.86}$$

From this, the inverse CDF follows as $a + b[\tan(\pi(u - 0.5))]$.

TABLE 7.13   Properties of Cauchy Distribution $(1/[b\pi[1 + (x - a)^2/b^2]])$

Property	Expression	Comments
Range of X	$-\infty < x < \infty$	Continuous; infinite
Mean	$\mu =$ does not exist	
Median	a	Mode $= a$
Variance	$\sigma^2 =$ does not exist	
Skewness	$\gamma_1 = 0$	Symmetric
Kurtosis	$\beta_2 = 9$	Always leptokurtic
Mean deviation	Does not exist	
CDF	$\dfrac{1}{2} + \dfrac{1}{\pi} \tan^{-1}((x - a)/b)$	
Moments	Does not exist	
$Q_1 = a - b$	$Q_3 = a + b$	
ChF	$\exp(ita - |t|b)$	

7.12.1   Properties of Cauchy Distribution

As the integral $\int_{-\infty}^{\infty} x/(1 + x^2)dx$ does not exist, the mean is undefined. The limiting value $\frac{1}{\pi} \underset{R \to \infty}{Lt} \int_{-R}^{R} \frac{x}{1+x^2} dx$ is zero. The characteristic function is

$$\phi(t) = \frac{1}{\pi} \int_{-\infty}^{\infty} \frac{e^{itx}}{1 + x^2} dx = \frac{2}{\pi} \int_{0}^{\infty} \frac{\cos(tx)}{1 + x^2} dx = e^{-|t|}. \tag{7.87}$$

Median and mode of the general Cauchy distribution coincide at $x = a$, with modal value $1/(b\pi)$. If the distribution is truncated in both tails at $x = c$, the resulting PDF is $\frac{1}{2 \tan^{-1}(c)} \frac{1}{1+x^2}$ for $-c \leq x \leq +c$. This has mean 0 and variance $c/\tan^{-1}(c) - 1$. See Table 7.13 for further properties.

7.12.2   Functions of Cauchy Variate

If $X_1, X_2, \cdots X_n$ are independent Cauchy distributed random variables, then $\overline{X} = (X_1 + X_2 + \cdots + X_n)/n$ is also distributed as Cauchy. An implication of this result is that the Cauchy mean does not obey the central limit theorem or the law of large numbers (the law of large numbers states that S_n/n for SCD converges to μ in probability, and the CLT states that the distribution of S_n/n tends to $N(0, 1)$ as $n \to \infty$).

Cauchy distribution is related to the uniform distribution $U(0, 1)$ through the tangent function $\tan(\pi\theta)$. More precisely, if $U \sim U(0, 1)$, then $\tan(\pi U)$ has a Cauchy distribution. This allows us to find tangent functions of Cauchy distributed variates like $2X/(1 - X^2), (3X - X^3)/(1 - 3X^2), (4X - X^4)/(1 - 6X^2 + X^4)$, and so on, which are, respectively, $\tan(2X), \tan(3X), \tan(4X)$, and so on. These are all Cauchy distributed [223–225]. If X and Y are IID Cauchy variates, $(X - Y)/(1 - XY)$ and $(X - 1)/(X + 1)$ are identically distributed. It satisfies an additivity property: If $X_1 \sim \text{Cauchy}(0, b_1)$ and $X_2 \sim \text{Cauchy}(0, b_2)$ are independent, then $X_1 + X_2 \sim \text{Cauchy}(0, b_1 + b_2)$.

7.12.3   Relation to Other Distributions

The Student's T distribution with 1 DoF is identically Cauchy distribution. If Z is Cauchy distributed, then $(a + bZ)$ is Cauchy distributed.

Many other similar relationships can be found in Arnold [224].

7.13   INVERSE GAUSSIAN DISTRIBUTION

This is also called Wald's distribution (see Figure 7.9). The PDF takes many forms

$$f(x, \mu, \lambda) = \sqrt{\lambda/2\pi x^3} \; \exp\{ - \frac{\lambda}{2\mu^2 x}(x - \mu)^2\}, \; \text{ for } x > 0, \mu, \lambda > 0. \qquad (7.88)$$

Expand $(x - \mu)^2$ as a quadratic and divide each term by μx to get

$$f(x, \mu, \lambda) = \sqrt{\lambda/2\pi x^3} \; \exp\{ - \frac{\lambda}{2\mu}(x/\mu - 2 + \mu/x)\}, \; \text{ for } x > 0, \mu, \lambda > 0. \quad (7.89)$$

Take the constant in the exponent as a separate multiplier and put $\lambda/\mu = \delta$ to get another form

$$f(x, \mu, \delta\mu) = \sqrt{\delta\mu/2\pi x^3} \; \exp(\delta) \; \exp\{ - \frac{\delta}{2}(x/\mu + \mu/x)\}, \; \text{ for } x > 0, \mu, \delta > 0. \qquad (7.90)$$

The CDF is expressible in terms of the standard normal CDF as

$$F(x, \mu, \lambda) = \Phi\left(\sqrt{\frac{\lambda}{x}} \left[\frac{x}{\mu} - 1 \right] \right) + e^{2\lambda/\mu} \Phi\left(-\sqrt{\frac{\lambda}{x}} \left[\frac{x}{\mu} + 1 \right] \right). \qquad (7.91)$$

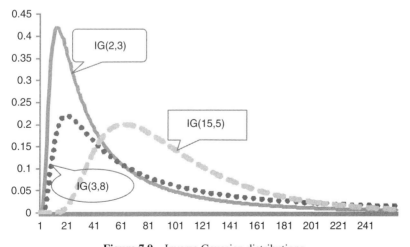

Figure 7.9   Inverse Gaussian distributions.

7.13.0.1   Properties of IGD   Multiply equation (7.90) by e^{itx} and integrate over the range to get

$$\phi(t) = \exp\left(\frac{\lambda}{\mu}\{1 - (1 - (2i\mu^2 t/\lambda)^{1/2}\}\right) = \exp(\delta\{1 - (1 - (2i\mu t/\delta)^{1/2}\}). \quad (7.92)$$

The mean is μ, but the mode depends on both μ, λ as $\mu((1 + 9/(4\delta^2))^{1/2} - 3/(2\delta))$ where $\delta = \lambda/\mu$. The variance also depends on both μ, λ as $\sigma^2 = \mu^3/\lambda = \mu^2/\delta$. The coefficient of skewness is $3\sqrt{\mu/\lambda}$ so that a practitioner can choose between a variety of distributional shapes. The mean deviation is given by

$$\text{MD} = 2\int_0^\mu \{\Phi(\sqrt{\lambda/x}[x/\mu - 1]) + e^{2\lambda/\mu}\Phi(-\sqrt{\lambda/x}[x/\mu + 1])\}dx. \quad (7.93)$$

This can be simplified as given in table 7.14. Linear combinations of IGD are IGD distributed (see summary table 7.4 (p. 297)). In particular, if X_i's are IGD(μ_i, λ_i) then $\sum_{i=1}^n \lambda_i/\mu_i^2 X_i$ is IGD$(\sum_{i=1}^n \lambda_i/\mu_i, (\sum_{i=1}^n \lambda_i/\mu_i)^2)$. The kurtosis is $3 + 15(\mu/\lambda)$ showing that it is always leptokurtic. Writing the exponent as $(\frac{x-\mu}{\mu})^2 = (\frac{x}{\mu} - 1)^2$ and letting $\mu \to \infty$ this becomes $(-1)^2 = 1$. The resulting distribution is called one-parameter IGD:–

$$f(x, \lambda) = \sqrt{\lambda/2\pi x^3} \exp\{-\lambda/(2x)\}. \quad (7.94)$$

7.13.1   Relation to Other Distributions

If $X \sim$ IGD(μ, λ) then $Y = \lambda(X - \mu)^2/(\mu^2 X)$ has chi-square distribution. If λ is held constant and $\mu \to \infty$, IGD(μ, λ) approaches a gamma distribution GAMMA$(\lambda/2, 1/2)$. When $\mu = 1$, the CDF can be expressed in terms of standard normal CDF as

$$F_x(\delta) = \Phi((x - 1)\sqrt{\delta/x}) + e^{2\delta} \Phi(-(x + 1)\sqrt{\delta/x}). \quad (7.95)$$

See Reference 226 for other approximations.

The moments and inverse moments are related as $E(X/\mu)^{-r} = E(X/\mu)^{r+1}$, where negative index denotes inverse moments. See Table 7.14 for further properties.

7.14   LOGNORMAL DISTRIBUTION

LND arises in a variety of applications. For example, rare-earth elements and radioactivity, micro-organisms in closed boundary regions, solute mobility in plant cuticles, pesticide distribution in farm lands, time between infection and appearance of symptoms in certain diseases, file sizes on hard disks, and so on follow approximately the LND. It also has applications in insurance and economics. It is the widely used parametric model in mining engineering for low-concentration mineral deposits.

It is obtained from the normal distribution $\frac{1}{\sqrt{2\pi}}e^{-x^2/2}$ using the transformation $y = e^x$ or equivalently $x = \log(y)$. This means that the transformed variate is lognormally

TABLE 7.14   **Properties of IGD ($\sqrt{\delta\mu/2\pi x^3}\ \exp(\delta)\ \exp\{-\frac{\delta}{2}(x/\mu + \mu/x)\}$)**

Property	Expression	Comments
Range of X	$x \geq 0$ continuous;	Infinite
Mean	μ	
Median	$\log(2)/\lambda$	
Variance	$\sigma^2 = \mu^3/\lambda = \mu^2/\delta$	$\delta = \lambda/\mu$
Mode	$\mu((1 + 9/(4\delta^2))^{1/2} - 3/(2\delta))$	symmetric tail areas are used in 6-sigma rule (table 7.12)
Skewness	$\gamma_1 = 3/\sqrt{\delta}$	
Kurtosis	$\beta_2 = 3 + 15/\delta$	Always leptokurtic
Mean deviation	$E\|X - \mu\| = 4\exp(2\lambda/\mu)\Phi(-2\sqrt{\lambda/\mu})$	
CV	$1/\sqrt{\delta}$	$\delta = \lambda/\mu$
CDF	$\Phi\left(\sqrt{\dfrac{\lambda}{x}}\left[\dfrac{x}{\mu} - 1\right]\right) + e^{2\lambda/\mu}\Phi\left(-\sqrt{\dfrac{\lambda}{x}}\left[\dfrac{x}{\mu} + 1\right]\right)$	
Cumulants	$\kappa_r = 1.3.5\cdots(2r-3)\mu^{2r-1}/\lambda^{r-1}$	$r \geq 2$
MGF	$\exp(\delta(1 - (1 - 2\mu^2 t/\lambda)^{1/2}))$	
CGF	$\delta(1 - [1 + 2\mu^2 it/\lambda]^{1/2})$	
ChF	$\exp(\delta(1 - (1 - 2\mu^2 it/\lambda)^{1/2}))$	
Additivity	$X_i \sim \mathrm{IG}(\mu, \lambda) \Rightarrow \sum_i X_i \sim \mathrm{IG}(n\mu, n^2\lambda)$	
	$X_i \sim \mathrm{IG}(\mu_i, \mu_i^2) \Rightarrow \sum_i X_i \sim \mathrm{IG}\left(\sum_i \mu, 2\left(\sum_i \mu^2\right)\right)$	
	$X_i \sim \mathrm{IG}(\mu, \lambda) \Rightarrow \bar{x} \sim \mathrm{IG}(\mu, n\lambda)$	

Approaches normality as $\lambda \to \infty$, otherwise it is skewed. The cumulant generating function of $\mathrm{IG}(\mu, \lambda)$ is the inverse of the CGF of normal distribution.

distributed. It is important to remember that if X is normally distributed, $\log(X)$ is not lognormal (a normal variate extends from $-\infty$ to ∞, but logarithm is undefined for negative argument). This gives $\partial x/\partial y = 1/y$, so that the PDF of standard lognormal distribution becomes

$$f_y(0, 1) = \frac{1}{\sqrt{2\pi}\ y} e^{-(\ln y)^2/2}, \ 0 \leq y < \infty. \tag{7.96}$$

The general form of the lognormal distribution is easily obtained as

$$f_y(\mu, \sigma^2) = \frac{1}{\sqrt{2\pi}\ \sigma y} e^{-(\ln y - \mu)^2/(2\sigma^2)}. \tag{7.97}$$

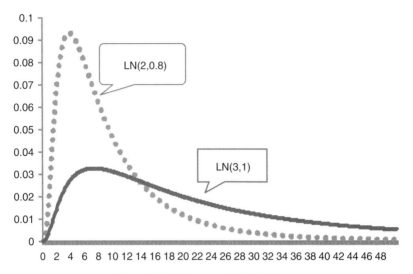

Figure 7.10   Lognormal distributions.

Here, μ and σ^2 are not the mean and variance of lognormal distribution but that of the underlying normal law (from which LND is obtained by the transformation $y = e^x$). Tail probabilities can be easily evaluated using the CDF of a normal distribution. For instance, if $Y \sim \text{lognormal}(0,1)$ then $P[Y > y_0] = \Pr[Z > \ln(y_0)] = 1 - \Phi(\ln(y_0))$.

7.14.1   Properties of Lognormal Distribution

This distribution and IG distribution are somewhat similar shaped for small parameter values (Figure 7.10). The CDF can be expressed in terms of erf() function as

$$F(x) = \frac{1}{2}\text{erfc}((\mu - \ln (x))/\sigma\sqrt{2}) = \Phi((\ln (x) - \mu)/\sigma). \qquad (7.98)$$

From this, it is easy to show that the area from the mode to the mean of an LND is $\Phi(\sigma) - \Phi(-\sigma)$, where $\Phi()$ denotes the CDF of standard normal. This result can be used to characterize lognormal distributions.

The quantiles of standard normal and lognormal are related as $Q_p(x) = \exp(\mu + \sigma Z_p(z))$, where Z_p denotes the corresponding quantile of standard normal variate. Replace p by $p + 1$ and divide by the above-mentioned expression to get

$$Q_{p+1}(x) = Q_p(x)\exp(\sigma(Z_{p+1}(z) - Z_p(z))). \qquad (7.99)$$

The sum of several independent LNDs can be *approximated* by a scaled LND. A first-order approximation can be obtained by equating the moments of linear combination with target lognormal distribution as done by Patnaik [198] for noncentral χ^2 distribution. As the cumulants of LND are more tractable, we could equate the cumulants and obtain a reasonable approximation. There are many other approaches for this purpose. See, for example, References 227–229, and so on.

📖 EXAMPLE 7.20   Mode of lognormal distribution

Prove that LND is unimodal with the mode at $\exp(\mu - \sigma^2)$. What is the modal value?

Solution 7.20 Consider the PDF (7.97). To find the maximum, we take log first, as the maximum of $f(x)$ and $\log(f(x))$ are the same. This gives $\log(f_y(\mu, \sigma^2)) = K - \log(y) - (\log y - \mu)^2/(2\sigma^2)$, where K is a constant. Differentiate with respect to y and equate to zero to get $-1/y - (\log y - \mu)/(y\sigma^2) = 0$. Cross-multiply and solve for y to get $(\log y - \mu) = -\sigma^2$ or equivalently $\log(y) = \mu - \sigma^2$. Exponentiate both sides to get the result $y = \exp(\mu - \sigma^2)$. Put the value in equation (7.97) to get the modal value $\dfrac{1}{\sqrt{2\pi}\sigma \exp(\mu-\sigma^2)} e^{-(\mu-\sigma^2-\mu)^2/(2\sigma^2)}$. This simplifies to $\dfrac{1}{\sqrt{2\pi}\sigma \exp(\mu-\sigma^2)} e^{-\sigma^2/2}$.   ∎

7.14.2   Moments

The mean is $\mu = e^{\mu + \frac{1}{2}\sigma^2}$ and variance $\sigma^2 = e^{2\mu+\sigma^2}(e^{\sigma^2} - 1) = e^{2\mu}\omega(\omega - 1)$ where $\omega = e^{\sigma^2}$. The ratio μ/σ^2 simplifies to $(e^{\sigma^2} - 1)$. While the variance of the general normal distribution is given by a single-scale parameter σ^2, the variance of lognormal distribution depends on both the location and scale parameters μ and σ^2. As this distribution in the "logarithmic scale" reduces to the normal law, many of the additive properties of the normal distribution have multiplicative analogs for the LND. For example, the additive form of the central limit theorem that asserts that the mean of a random sample tends to normality for increasing values of n can be stated for LN() as follows: If $X_1, X_2, \cdots, X_n$ are independent lognormal random variables with finite $E(\log(X_i))$, then $Z = (\log(S_n) - n * E[\log(X_i)])/(n * \text{Var}(\log(X_i)))^{1/2}$ asymptotically approaches normality, where S_n is the product of the $X_i's$. See Table 7.15 for further properties.

📖 EXAMPLE 7.21   Mean deviation of lognormal distribution

Find the mean deviation of LND using Theorem 7.1.

Solution 7.21 Let X~LN(μ, σ^2). As the lognormal distribution tails off to zero at the lower and upper limits, equation (7.1) is applicable. This gives

$$\text{MD} = 2\int_{ll}^{\mu'} F(x)dx = 2\int_0^c \Phi((\ln(x) - \mu)/\sigma)dx, \quad \text{where } c = e^{\mu+\frac{1}{2}\sigma^2}. \quad (7.100)$$

Put $z = ((\ln(x) - \mu - \sigma^2/2)$, so that $dx = e^{z+\mu+\sigma^2/2}dz$. The lower limit in equation (7.100) becomes $-\infty$, and the upper limit is 0 because $\ln(c) = \mu + \frac{1}{2}\sigma^2$. Thus, we get MD =

$$2\int_{-\infty}^{0} \Phi(z/\sigma + \sigma/2)e^{z+\mu+\sigma^2/2}dz = 2e^{\mu+\sigma^2/2}\int_{-\infty}^{0} e^z\Phi(z/\sigma + \sigma/2)dz. \quad (7.101)$$

TABLE 7.15   Properties of Lognormal Distribution

Property	Expression	Comments
Range of X	$0 \leq x < \infty$	Continuous
Mean	$\mu = e^{\mu + \frac{1}{2}\sigma^2}$	Median $= e^\mu$
Mode	$\exp(\mu - \sigma^2)$	Mode<median<mean
Variance	$e^{2\mu + \sigma^2}(e^{\sigma^2} - 1)$	$e^{2\mu}\omega(\omega - 1)$ where $\omega = e^{\sigma^2}$
Skewness	$\gamma_1 = \sqrt{\omega - 1}(\omega + 2)$	Approximately symmetry as $\sigma \to 0$
Kurtosis (β_2)	$(\omega - 1)[\omega^2(\omega + 3) + 6(\omega + 1)]$	
Mean deviation	$2e^{\mu + \sigma^2/2}[2\Phi(\sigma/2) - 1]$	$2e^{\mu + \sigma^2/2}\mathrm{erf}(\sigma/(2\sqrt{2}))$
CV	$\sqrt{(\omega - 1)}$	
CDF	$\int_0^{\log(x)} \frac{1}{\sigma\sqrt{2\pi}} e^{-\frac{1}{2}(\frac{x-\mu}{\sigma})^2} dx$	
Moments	$\mu_k' = \exp(k\mu + k^2\sigma^2/2)$	
*Log-normal ($\prod$)	$(\mu_1, \sigma_1^2), (\mu_2, \sigma_2^2)$	$\mu_1 + \mu_2, \sigma_1^2 + \sigma_2^2$ product XY
*Log-normal (ratio)	$(\mu_1, \sigma_1^2), (\mu_2, \sigma_2^2)$	$\mu_1 - \mu_2, \sigma_1^2 + \sigma_2^2$ ratio X/Y

The product and ratio of two independent log normal variates are log normal with the parameters as shown. Similarly, the geometric mean of n IID lognormal variates is lognormally distributed. The mean-median-mode inequality is mode<median<mean.

Take $u = \Phi(z/\sigma + \sigma/2)$ and $dv = e^z dz$ so that $v = e^z$, and $du = (1/\sigma) \phi(z/\sigma + \sigma/2)$. Apply integration by parts to equation (7.101) to get MD =

$$2e^{\mu + \sigma^2/2} \left[\Phi(z/\sigma + \sigma/2)e^z \mid_{-\infty}^0 - \int_{-\infty}^0 e^z(1/\sigma) \phi(z/\sigma + \sigma/2)dz \right]. \quad (7.102)$$

The first expression in equation (7.102) is $\Phi(\sigma/2)$ as $\Phi(-\infty) = e^{-\infty} = 0$. To evaluate the second expression, we use $-\int_b^a f()dx = \int_a^b f()dx$, expand $\phi()$ and write it as

$$\int_0^\infty e^z(1/\sigma) \phi(z/\sigma + \sigma/2)dz = (1/\sigma\sqrt{2\pi}) \int_0^\infty \exp(z - \frac{1}{2}(z/\sigma + \sigma/2)^2)dz.$$

Expand the quadratic and combine the exponent as $z - \frac{1}{2}(z/\sigma + \sigma/2)^2 = -\frac{1}{2}(z^2/\sigma^2 - z + \sigma^2/4) = -\frac{1}{2}(z/\sigma - \sigma/2)^2$. This gives the above integral as

$$(1/\sigma\sqrt{2\pi}) \int_0^\infty \exp(-\frac{1}{2}(z/\sigma - \sigma/2)^2)dz. \quad (7.103)$$

Now put $(z/\sigma - \sigma/2) = v$ so that $dz = \sigma dv$. The upper limit remains the same, but the lower limit becomes $-\sigma/2$. Upon substitution, the σ cancels out from the numerator and denominator and equation (7.103) becomes

$$(1/\sqrt{2\pi}) \int_{-\sigma/2}^{\infty} \exp(-v^2/2)dv = 1 - \Phi(-\sigma/2). \qquad (7.104)$$

Substitute in equation (7.102) to get the MD as

$$\text{MD} = 2\,e^{\mu+\sigma^2/2}[\Phi(\sigma/2) + 1 - \Phi(-\sigma/2)]. \qquad (7.105)$$

Divide the area under the normal curve from $-\infty$ to $-\sigma/2, -\sigma/2$ to $+\sigma/2$, and from $+\sigma/2$ to $+\infty$. We notice that as the total area is unity, the expression (7.105) is simply the middle area from $-\sigma/2$ to $+\sigma/2$. Hence, it becomes $2* \Phi(\sigma/2) - 1$. Substitute for the bracketed expression to get the MD as

$$\text{MD} = 2\,e^{\mu+\sigma^2/2}[2 * \Phi(\sigma/2) - 1]. \qquad (7.106)$$

∎

▣ EXAMPLE 7.22   Geometric mean of IID lognormal variates

If $X_1, X_2, \cdots, X_n$ are independent lognormal random variables $\text{LN}(\mu, \sigma^2)$, find the distribution of the GM $= (X_1 * \cdots * X_n)^{1/n}$.

Solution 7.22 As X_i is LND, $\log(X_i)$ are normally distributed. Taking log gives $Y = \log(\text{GM}) = (\log(X_1) + \cdots + \log(X_n))/n$. Each component in this expression is normal $N(\mu, \sigma^2)$, so that Y is $N(\mu, \sigma^2/n)$. Taking the inverse transformation $X = e^y$ shows that GM is lognormal $\text{LN}(\mu, \sigma^2/n)$. ∎

7.14.2.1 Partial Expectation of Lognormal Distribution The partial expectation of LND has applications in economics, finance, and insurance. It is defined as

$$g(k) = \int_k^{\infty} xf(x; \mu, \sigma^2)dx = e^{\mu+\sigma^2/2}[\Phi([\mu + \sigma^2 - \ln(k)]/\sigma)]. \qquad (7.107)$$

Consider the survival function form (7.9) of MD as

$$E|x - \mu| = 2\int_{\mu}^{ul} S(x)dx \qquad (7.108)$$

Take $u = S(x)$, and $dv = dx$ so that $du = -f(x)$, and we get

$$E|x - \mu| = 2\left(\left[xS(x) \,|_{\mu}^{\infty}\right] + \int_{\mu}^{\infty} xf(x)dx \right). \qquad (7.109)$$

Using L'Hospital's rule, the first expression inside the bracket reduces to $-\mu S(\mu)$. Divide throughout by 2 and rearrange equation (7.109) to get

$$\int_{\mu}^{\infty} xf(x)dx = E|x - \mu|/2 + \mu S(\mu). \tag{7.110}$$

Depending on whether $k < \mu$ or $k > \mu$, the integral between them can be expressed in terms of $\Phi()$. This shows that the partial expectation of LND is related to the MD through the SF value at μ.

7.14.3   Fitting Lognormal Distribution

We have seen earlier that the mean $E(X) = e^{\mu + \frac{1}{2}\sigma^2}$ and Variance $V(X) = e^{2\mu + \sigma^2}(e^{\sigma^2} - 1)$. Take log and solve for μ and σ^2 to get $\mu = \ln(E(X)) - 0.5*\ln(1 + (\text{Var}(X)/E(X)^2))$, and $\sigma^2 = \ln(1 + (\text{Var}(X)/E(X)^2))$. If the sample size is sufficiently large, we could replace $E(X)$ by the sample mean $\bar{x}_n$, and Var(X) by s_n^2, and obtain estimates of the unknown parameters.

7.15   PARETO DISTRIBUTION

This distribution is named after the Italian economist Vilfredo Pareto (1848–1923), who studied the income distribution of populace during his lifetime. The PDF is $ck^c x^{-(c+1)}$ where $x \geq k, c > 0$ are constants [230]. For income and wealth distributions, the constant c is greater than 1 (and near 2.0 in developed countries).

7.15.1   Properties of Pareto Distribution

The survival function of this distribution takes the simple form $S(x) = (k/x)^c$. The median is given by $k2^{1/c}$ and coefficient of variation is $1/\sqrt{c(c-2)}$, which is independent of k.

7.15.1.1   Moments and Generating Functions   The ordinary and inverse moments are easy to find. Moments higher than c do not exist.

■ **EXAMPLE 7.23   Moments of Pareto distribution**

Prove that the rth moment of Pareto distribution is $\mu_r = c * k^r/(c-r)$ for $r < c$.

Solution 7.23   As the range of x is from k to ∞, we have

$$\mu_r' = ck^c \int_k^{\infty} x^r/x^{c+1} dx = ck^c \int_k^{\infty} x^{r-c-1} dx = ck^c/(r-c)x^{r-c}||_k^{\infty}. \tag{7.111}$$

As the integrand is a power of x, it converges for $r < c$ to get $ck^c/(r - c)(0 - k^{r-c})$. The k^c cancels out giving $\mu_r = c * k^r/(c - r)$. Take c as a common factor from denominator to get $\mu_r = k^r(1 - r/c)^{-1}$. This gives the recurrence relation $\mu_{r+1} = \mu_r * k * (c - r)/(c - r - 1)$. From this, the mean and variance are easily obtained as $\mu = kc/(c - 1)$, $\sigma^2 = k^2c/[(c - 1)^2(c - 2)]$. The generalized Pareto distributions are obtained by change of origin and scale transformation. This has CDF $1 - (k/(x + b))^c$ and mode k. Median is $k * 2^{1/c}$.   ∎

◧ EXAMPLE 7.24   Mean deviation of Pareto distribution

Find the mean deviation of the Pareto distribution $f(x, k, c) = ck^c x^{-(c+1)}$ where $x \geq k$.

Solution 7.24 We apply Theorem 7.1 (p. 256) to find the MD. Note that the Pareto distribution is defined for $x \geq k$. At $x = k$, the functional value is $1/k$. As the PDF does not tail off to zero at the lower limit (i.e., at k), equation (7.1) seems like inapplicable. We know that the CDF is $1 - (k/x)^c$. If we apply L'Hospital's rule once on $x * F(x)$, we get $1 [1 - (k/x)^c] + x[ck^c x^{-c-1}]$. The first term $\to 0$ as $x \to k$, whereas the second term tends to c. We need to use the equation (7.7). However, the term $x F(x)|_{ll=k} \to 0$, so that it reduces to

$$\text{MD} = 2 \int_{ll}^{\mu} F(x)dx = 2 \int_{k}^{d} [1 - (k/x)^c]dx, \quad \text{where } d = kc/(c - 1). \quad (7.112)$$

Separate into two terms and integrate each term to get $\text{MD} = 2\{[kc/(c - 1) - k] - k^c \int_k^d x^{-c}dx\}$. The expression inside the square bracket simplifies to $k/(c - 1)$ and the integral simplifies to $(k/(1 - c))(c - 1)^{c-1}/c^{c-1} + k/(1 - c)$. The term $k/(1 - c)$ cancels with first term $k/(c - 1)$. Take c outside from the second expression to get $\text{MD} = 2k(1 - (1/c)^{c-1})/(c-1)$. See table 7.16 for further properties.   ∎

7.15.2   Relation to Other Distributions

This distribution is related to exponential distribution as follows: If Y is exponentially distributed, then $X = k * \exp(Y/c)$ has Pareto distribution. The zipf distribution is the discrete analog of Pareto distribution. As $c \to \infty$, the PDF approaches Dirac's δ function. Left truncation results in Pareto distributions. The sum of the logarithm of several independent scaled Pareto distributions has a gamma distribution. See References 122, 230–233 for further properties.

7.15.3   Algorithms

As the CDF is $1 - (k/x)^c$, it is easy to generate random numbers using the inverse-CDF method. Let u be a uniform random number. Equate $u = 1 - (k/x)^c$ and solve for x to get $(1 - u)^{1/c} = k/x$ or $x = k/(1 - u)^{1/c}$.

TABLE 7.16   Properties of Pareto Distribution ($f(x;k,c) = ck^c x^{-(c+1)}$)

Property	Expression	Comments
Range of X	$k < x < \infty$	Continuous
Mean	$\mu = kc/(c-1)$	$k[1 + 1/(c-1)]$
Median	$k2^{1/c}$	Mode $= k$
Variance	$\sigma^2 = k^2 c/[(c-1)^2(c-2)]$	$= \mu^2/[c(c-2)]$
Skewness	$\gamma_1 = 2[(c-2)/c]^{1/2}(1+c)/(c-3)$	Valid for $c > 3$
Kurtosis	$\beta_2 = 6(c^3 + c^2 - 6c - 2)/[c(c-3)(c-4)]$	Valid for $c > 4$
Mean deviation	$E\lvert X - \mu \rvert = 2k(1 - c^{-1})^{c-1}/(c-1)$	
CV	$1/\sqrt{c(c-2)}$	$c > 2$
SF	$(k/x)^c$	
CDF	$1 - (k/x)^c$	
Moments	$\mu_r = c * k^r/(c-r)$	
MGF	$c(-kt)^c \Gamma(-c, -kt)$	For $t < 0$
ChF	$c(-ikt)^c \Gamma(-c, -ikt)$	

Put $x = y - b$ to get a three-parameter version. Note that the expression for skewness is valid for $c > 3$, and it is never symmetric. A symmetric Pareto distribution can be defined by replacing x by $\lvert x \rvert$, changing the range as $\lvert x \rvert > k$ and adjusting the normalizing constant. β_2 given here is the excess kurtosis, $\Gamma()$ is incomplete gamma function.

7.16   DOUBLE EXPONENTIAL DISTRIBUTION

This distribution, invented by Pierre Laplace in 1774, has many applications in quality control, error modeling (called Laplacian noise), and inventory control (especially of slow moving items). It is also called Laplace distribution.

$$f_x(a, b) = (1/2b) \exp(-\lvert x - a \rvert/b), \quad -\infty < x < \infty, \quad -\infty < a < \infty, b > 0, \quad (7.113)$$

where "a" is the location parameter and b is the scale parameter. The standard form is obtained by putting $a = 0, b = 1$ as $f(z) = e^{-\lvert z \rvert}$.

7.16.0.1   Properties of Double Exponential Distribution   This distribution is symmetric around $x = a$ (see Figure 7.11). Hence β_1 and all odd moments are zeros. The mean and variance are $\mu = $ a, and $\sigma^2 = 2b^2$. When $a = 0$, the resulting distribution and that of two IID exponential(b) distributions are the same.

■   EXAMPLE 7.25   Even moments of double exponential

Prove that the even moments of double exponential are given by $\mu_{2k} = (2k)! b^{2k}$.

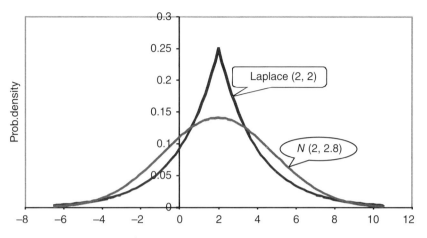

Figure 7.11 Laplace distributions.

Solution 7.25 As the distribution is symmetric, the odd moments are all zeros. The even moment is given by $\mu_{2k} =$

$$\int_{-\infty}^{\infty} (x-\mu)^{2k} \frac{1}{2b} \exp(-|x-a|/b)dx = \frac{1}{2b}[\int_{-\infty}^{\mu} (x-\mu)^{2k} \exp(-|x-a|/b)dx$$

$$+ \int_{\mu}^{\infty} (x-\mu)^{2k} \exp(-|x-a|/b)dx].$$

As $\mu = a$, put $y = (x-a)$ and change the limits accordingly.

$$\mu_{2k} = \frac{1}{2b}[\int_{-\infty}^{0} y^{2k} \exp(y/b)dx + \int_{0}^{\infty} y^{2k} \exp(-y/b)dy]. \qquad (7.114)$$

Put $-y = t$ in the first integral. Then, it becomes the second integral. Hence $\mu_{2k} = 2\int_{0}^{\infty} y^{2k} \exp(-y/b)dy$. Write y^{2k} as $y^{(2k+1)-1}$. Using gamma integral, this becomes $\Gamma(2k+1)b^{2k}$. As $\Gamma(2k+1) = (2k)!$, the result follows. ∎

By integrating with respect to x, we get the CDF as

$$F_x(a,b) = (1/2)[1 + \text{sign}(x-a)(1 - \exp(-|x-a|/b)], \qquad -\infty < x < \infty. \quad (7.115)$$

The standard form of the Laplace distribution is obtained by putting $a = 0$ and $b = 1$ in the above.

$$(1/2)\exp(-|x|), \qquad -\infty < x < \infty. \qquad (7.116)$$

The characteristic function is obtained easily as follows:

$$\phi(t) = \frac{1}{2b}\int_{-\infty}^{\infty} e^{itx}e^{(-|x-a|/b)}dx. \qquad (7.117)$$

By splitting the range of integration from $-\infty$ to a and from a to ∞ and changing the variable as $y = (x - a)/b$, we get

$$\phi(t) = \frac{e^{iat}}{1 + b^2 t^2} \tag{7.118}$$

from which the mean and variance can be obtained easily as $\mu = a$ and $\sigma^2 = 2b^2$. For $a = 0, b = \pm 1$, this becomes $\phi(t) = 1/(1 + t^2)$, which shows that Laplace and Cauchy distributions are related through characteristic functions. See Table 7.17 for further properties.

■ **EXAMPLE 7.26   Mean deviation of Laplace distribution**

Find the mean deviation of the Laplace distribution using Theorem 7.1.

Solution 7.26 Let X~Laplace(a, b). As the Laplace distribution tails off to zero at the lower and upper limits, equation (7.1) is applicable. This gives

$$MD = 2 \int_{ll}^{a} F(x)dx = 2 \int_{-\infty}^{a} \frac{1}{2} \exp(-(a - x)/b)dx, \tag{7.119}$$

(see Table 7.17, the CDF line). Put $z = (a - x)/b$, so that $dx = -bdz$. When $x = a, z = 0$; but when $x = -\infty$, z becomes $+\infty$. Cancel out the 2 to get

$$MD = -b \int_{\infty}^{0} e^{-z}dz = b \int_{0}^{\infty} e^{-z}dz = -be^{-z} \mid_{0}^{\infty} = -b[0 - 1] = b. \tag{7.120}$$

This shows that the mean deviation is b.   ■

TABLE 7.17   Properties of Double Exponential Distribution

Property	Expression	Comments
Range of X	$-\infty < x < \infty$	Continuous
Mean	$\mu = a$	
Median	a	Mode $= a$
Variance	$\sigma^2 = 2b^2$	
Skewness	$\gamma_1 = 0$	Always symmetric
Kurtosis	$\beta_2 = 6$	
Mean deviation	$E\lvert X - \mu \rvert = b$	
CV	$\sqrt{2}b/a$	
CDF	$\frac{1}{2}\exp(-(a - x)/b)$	$x < a$
	$1 - \frac{1}{2}\exp(-(x - a)/b)$	$x > a$
Moments	$\mu_r = \Gamma(r + 1)b^r$	r even
MGF	$\exp(at)/(1 - b^2 t^2)$	
ChF	$\exp(iat)/(1 + b^2 t^2)$	

7.16.1 Relation to Other Distributions

The standard Laplace distribution is the distribution of the difference of two independent exponential variates. In general, if X_i are two independent $EXP(\lambda)$ variates, then $Y = X_1 - X_2$ has a Laplace distribution. The ratio of two IID Laplace variates $L(0, b_1)$ and $L(0, b_2)$ is a central $F(2, 2)$ variate.

7.17 CENTRAL $\chi 2$ DISTRIBUTION

This distribution has a long history dating back to 1838 when Bienayme obtained it as the limiting form of multinomial distribution (see Figure 7.12). It was used by Karl Pearson for contingency table analysis during 1900. It is also used in testing goodness of fit between observed data and predicted model and in constructing confidence intervals for sample variance.

If $X_1, X_2, \ldots, X_n$ are independent standard normal random variables, the distribution of $Y = X_1^2 + \cdots + X_n^2$ is called the central χ^2 distribution. It has only one parameter called DoF (n) with PDF

$$f_x(n) = x^{n/2-1}e^{-x/2}/[2^{n/2}\Gamma(n/2)]. \tag{7.121}$$

It is a special case of the Gamma distribution GAMMA($n/2, 1/2$).

The distribution of $\sqrt{\chi^2}$ is known as chi distribution and has PDF

$$f_x(n) = x^{n-1}e^{-x^2/2}/[2^{n/2}\Gamma(n/2)]. \tag{7.122}$$

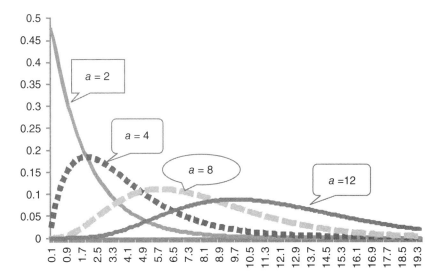

Figure 7.12 Chi-square distribution.

7.17.1   Properties of Central χ^2 Distribution

Integrating equation (7.121) with respect to x, we get the CDF as

$$F_n(x) = (1/[2^{n/2}\Gamma(n/2)]) \int_0^x y^{n/2-1} e^{-y/2} dy. \tag{7.123}$$

Make the change of variable $u = y/2$ to get the above in terms of incomplete gamma function as $F_n(x) = \gamma(n/2, x/2)/\Gamma(n/2) = P(n/2, x/2)$. It satisfies the recurrence $F_n(x) - F_{n-2}(x) = -2 * f_n(x)$.

7.17.1.1   *Moments and Generating Functions*   The MGF is easily obtained as

$$M_x(t) = E(e^{tx}) = \int_0^\infty e^{tx} x^{n/2-1} e^{-x/2}/[2^{n/2}\Gamma(n/2)] dx \tag{7.124}$$

$$= 1/[2^{n/2}\Gamma(n/2)] \int_0^\infty e^{-x[1-2t]/2} x^{n/2-1} dx = (1-2t)^{-n/2}. \tag{7.125}$$

From this, the kth moment is easily obtained as $\mu'_k = 2^k\Gamma(n/2 + k)/\Gamma(n/2)$. The mean is n and variance is $2n$. See Table 7.18 for further properties.

TABLE 7.18   Properties of χ^2 Distribution $x^{n/2-1} e^{-x/2}/[2^{n/2}\Gamma(n/2)]$

Property	Expression	Comments
Range of X	$0 \le x < \infty$	Continuous
Mean	$\mu = n$	
Variance	$\sigma^2 = 2n = 2\mu$	$\sigma^2 > \mu$
Mode	$n - 2 = \mu - 2$	$n > 2$
Median	$n - (2/3) = \mu - 2/3$	Approximately for large n
CV	$(2/n)^{1/2}$	
Skewness	$\gamma_1 = 2\sqrt{2/n}$	$= 2^{3/2} n^{-1/2}$
Kurtosis	$\beta_2 = 3 + 12/n$	Always leptokurtic
Mean deviation	$e^{-\frac{n}{2}} n^{\frac{n}{2}+1} / \left[2^{\frac{n}{2}-1}\Gamma\left(\frac{n}{2}+1\right) \right]$	$= 2\int_0^n P(n/2, x/2) dx$
Moments	$\mu'_r = 2^r\Gamma(r + n/2)/\Gamma(n/2)$	
MGF	$(1 - 2t)^{-n/2}$	$t < 1/2$
ChF	$(1 - 2it)^{-n/2}$	
CGF	$-(n/2)\log(1-2it)$	
Additivity	$\chi_m^2 + \chi_n^2 = \chi_{m+n}^2$	Independent
Recurrence	$f_{n+2}(x)/f_n(x) = x/n$	$F_n(x) - F_{n-2}(x) = -2f_n(x)$
Approximation	$(\chi_n^2/n)^{1/3}$	$N((1 - 2/(9n)), 2/(9n))$
Tail probability	$1 - P(n/2, x/2)$	P = Regularized gamma fn

The mean-median-mode inequality is mode<median<mean.

■ **EXAMPLE 7.27   Mean deviation of central χ^2 distribution**

Find the mean deviation of the central χ^2 distribution with PDF $f_x(n) = x^{n/2-1}e^{-x/2}/[2^{n/2}\Gamma(n/2)]$.

Solution 7.27 We apply Theorem 7.1 (p. 257) to find the MD. As the χ^2 distribution does not tail off to zero at the lower limit (i.e., at 0) for $n < 3$, equation (7.1) seems like not applicable. We know that the CDF is $P(n/2, x/2)$. If we apply L'Hospital's rule once on $x * F(x)$, we find that it $\to 0$ as $x \to 0$. As the $\lim_{x \to 0} x * F(x) = 0$, and the Theorem 7.1 becomes applicable. This gives $MD = 2\int_0^n P(n/2, x/2)dx$. Use integration by parts by taking $u = P(n/2, x/2), dv = dx$ to get

$$MD = 2nP(n/2, n/2) - 2/\Gamma(n/2)\int_0^n x^{n/2}e^{-x/2}dx. \qquad (7.126)$$

Put $x/2 = u$ so that $dx = 2du$ to get

$$MD = 2nP(n/2, n/2) - 4/\Gamma(n/2)\int_0^{n/2} u^n e^{-u}du. \qquad (7.127)$$

Multiply the numerator and denominator of the integral by $(n/2)$ and write $(n/2) * \Gamma(n/2) = \Gamma(n/2 + 1)$. This gives

$$MD = 2n[P(n/2, n/2) - P(n/2 + 1, n/2)]. \qquad (7.128)$$

Now use $P(a, x) - P(a + 1, x) = e^{-x}x^a/\Gamma(a + 1)$ (Abramowitz & Stegun eq. 6.5.21) with $a = x = n/2$ to get

$$MD = 2ne^{-n/2}(n/2)^{n/2}/\Gamma(n/2 + 1). \qquad (7.129)$$

This simplifies to

$$MD = e^{-n/2}n^{n/2+1}/[2^{n/2-1}\Gamma(n/2 + 1)]. \qquad (7.130)$$

■

See References 234 and 235 for properties including the median of χ^2 and Reference 236 for critical values.

7.17.1.2   Additivity Property This distribution in particular and its noncentral version in general satisfies a reproductive property given below. If $X \sim \chi_m^2$ and $Y \sim \chi_n^2$ are independent, then $X + Y \sim \chi_{m+n}^2$. This result was proved by Helmert [238]. See Table 7.4.

7.17.1.3 *Approximations* Wilson and Hilferty proved that $(\chi_n^2/n)^{1/3}$ is approximately normal with mean $(1 - 2/(9n))$ and variance $2/(9n)$. This allows the CDF to be expressed in terms of standard normal CDF as

$$F_n(x) = \Phi(((x/n)^{1/3} - 1 + 2/(9n))/\sqrt{2/(9n)}). \tag{7.131}$$

This has been extended by many researchers [239, 240].

7.17.2   Relationships with Other Distributions

If $X \sim \chi_n^2$, then $c * X$ for $c > 0$ is GAMMA$(n/2, 2c)$. It is related to the $U = U(0, 1)$ as $-2 \log(U) \sim \chi_2^2$. If $X_i \sim$ Laplace(a, b), $\sum_{i=1}^{n} 2|X_i - a|/b$ is distributed as χ_{2n}^2. If $X \sim$ Rayleigh(1), then $X^2 \sim \chi_2^2$. The SF of a chi-square distribution is related to the Poisson CDF (Section 6.9 in p. 229). It is also related to the normal, beta, T, and F distributions (see References 239, 241–245).

7.18   STUDENT'S *T* DISTRIBUTION

This distribution was obtained by William Gosset [246] as the distribution of the ratio $Z/\sqrt{\chi^2(n)/n}$ where Z is a standard normal variate and Z and $\chi^2(n)$ are independent. A derivation is given in Chapter 11. It is frequently encountered in small sample statistical inference when population variance is unknown. It is used in tests for the means, in testing the significance of correlation coefficients, in constructing confidence intervals, and so on.

It has a single parameter n called the DoF (df), which was described in Chapter 3. Theoretically, n need not be an integer. As n represents the sample size adjusted for "loss of information," it is always an integer in statistical inference. The PDF is given by

$$f(t) = K(1 + t^2/n)^{-(n+1)/2}, \tag{7.132}$$

where $K = \Gamma((n + 1)/2)/[\sqrt{n\pi}\ \Gamma(n/2)]$. As it is an even function of t, it is always symmetric around $t = 0$. The more general form is obtained by a change of origin and scale transformation $t = (y - \mu)/c$.

7.18.1   Properties of Student's *T* Distribution

It is always symmetric, unimodal with mode $t = 0$. The modal value is $(\Gamma(n + 1)/2)/[\sqrt{n\pi}\Gamma(n/2)]$. The mean $\mu = 0$ if $n > 1$ and does not exist otherwise. It has a single parameter n, which controls both the spread and peakedness. For higher values of n, the flatness in the tails decreases, and the peakedness increases. Eventually, it coincides with the standard normal distribution for large n. The variance is $n/(n - 2)$ if $n > 2$.

The distribution is concave upward for $|t| < -\sqrt{(n/(n + 2))}$ and concave downward otherwise. For $n > 4$, the kurtosis coefficient is given by $\beta_2 = 3 + \frac{6}{n-4}$, showing that it is always leptokurtic. Write the PDF in equation (7.132) as

$$f(t) = K(1 + t^2/n)^{1/2} * (1 + t^2/n)^{-n/2}. \text{ Let } n \to \infty \text{ and use } \underset{x \to \infty}{Lt}(1 + a/x)^{-x} = e^{-a}$$

to get $f(t) = Ke^{-t^2/2}$ as $(1 + t^2/n)^{1/2}$ will tend to one. It can be shown that $K = \Gamma((n+1)/2)/[\sqrt{n\pi}\ \Gamma(n/2)] \to 1/\sqrt{2\pi}$ as $n \to \infty$. This shows that the limiting distribution is a standard normal. By factoring the PDF into two asymmetric products, Jones & Faddy obtained the skew-t distributions

$$f(t; a, b) = K\left\{1 + t/\sqrt{C + t^2}\right\}^{a+\frac{1}{2}}\left\{1 - t/\sqrt{C + t^2}\right\}^{b+\frac{1}{2}}, \tag{7.133}$$

where K is given by $1/K = C^{1/2}2^{C-1}B(a, b)$ and $C = (a + b)$ [247]. See References 146, 248–251 for tail areas, Reference 252 for the distribution of the difference of two t-variables, and Reference 253 for applications to statistical tests.

▮ EXAMPLE 7.28 Mean deviation of Student's T distribution

Find the mean deviation of Student's T distribution using Theorem 7.1.

Solution 7.28 Let $K = \Gamma((n + 1)/2)/[\sqrt{n\pi}\ \Gamma(n/2)]$. As the Student's T distribution is symmetric around zero, the MD is given by

$$MD = K \int_{-\infty}^{\infty} |t|(1 + t^2/n)^{-(n+1)/2}dt. \tag{7.134}$$

Split the integral from $-\infty$ to 0; and 0 to ∞. As $|x| = -x$ when $x < 0$, the first integral becomes $- \int_{-\infty}^{0} tf(t)dt = \int_{0}^{\infty} tf(t)dt$. Hence

$$MD = 2K \int_{0}^{\infty} t(1 + t^2/n)^{-(n+1)/2}dt. \tag{7.135}$$

Put $t^2 = n\tan^2(\theta)$ so that $t = \sqrt{n}\ \tan(\theta)$, and $dt = \sqrt{n}\ \sec^2(\theta)d\theta$. The limits of integration become 0 to $\pi/2$ and we get

$$MD = 2Kn \int_{0}^{\pi/2} \tan(\theta)\sec^{-(n+1)}(\theta)\ \sec^2(\theta)d\theta. \tag{7.136}$$

Using $\sec(\theta) = 1/\cos(\theta)$, this becomes

$$MD = 2Kn \int_{0}^{\pi/2} \sin(\theta)\cos^{(n-2)}(\theta)d\theta. \tag{7.137}$$

Put $\cos(\theta) = t$ so that $\sin(\theta)d\theta = -dt$, and the limits are changed as 1 to 0; and we get

$$MD = -2Kn \int_{1}^{0} t^{n-2}dt = 2Kn \int_{0}^{1} t^{n-2}dt. \tag{7.138}$$

TABLE 7.19 Properties of T Distribution $(1/[\sqrt{n}\,B(\frac{1}{2},\frac{n}{2})](1+\frac{t^2}{n})^{-(n+1)/2})$

Property	Expression	Comments
Range of T	$-\infty < t < \infty$	Infinite
Mean	$\mu = 0$	
Median	0	Mode $= 0$
Variance	$\sigma^2 = n/(n-2) = 1 + 2/(n-2)$	$n > 2$
Skewness	$\gamma_1 = 0$	Symmetric
Kurtosis	$\beta_2 = 3(n-2)/(n-4) = 3 + 6/(n-4)$ for $n > 4$	Always leptokurtic
Mean deviation	$\sqrt{n/\pi}\,\Gamma((n-1)/2)/\Gamma(n/2)$	$\displaystyle\int_0^\infty I_{n/(n+t^2)}\left(\frac{n}{2},\frac{1}{2}\right)dt$
CDF	$F_n(t_0) = 1 - \dfrac{1}{2}I_{n/(n+t^2)}(n/2,1/2)$	I = Incomplete beta
Moments	$\mu_k = n^{k/2}\dfrac{\Gamma\left(\dfrac{k+1}{2}\right)\Gamma\left(\dfrac{n-k}{2}\right)}{\sqrt{\pi}\,\Gamma(n/2)}$	
ChF	$\exp(-|t\sqrt{n}|)S_n(|t\sqrt{n}|)$	

See equation (7.155), page 7-91.

As the integral evaluates to $1/(n-1)$, MD $= 2\sqrt{n}/[(n-1)B(1/2,n/2)]$. Expand the complete beta function $B(1/2,n/2) = \Gamma(1/2)\Gamma(n/2)/\Gamma((n+1)/2)$ and write $\Gamma((n+1)/2) = ((n-1)/2)\Gamma((n-1)/2)$. One $(n-1)$ cancels out from numerator and denominator giving the alternative expression $\sqrt{n/\pi}\,\Gamma((n-1)/2)/\Gamma(n/2)$.

Next we apply the Theorem 7.1 to find the MD. As the Student's T distribution tails off to zero at the lower and upper limits, equation (7.1) is applicable. This gives

$$\text{MD} = 2\int_{\text{ll}}^{\mu} F_n(t)dt = 2\int_{\mu}^{\text{ul}} S_n(t)dt = 2\int_{t=0}^{\infty} S_n(t)dt. \tag{7.139}$$

As the SF can be expressed in terms of the IBF as

$$1 - S_n(t) = F_n(t) = \frac{1}{2}[1 + \text{sign}(t)\,I_y(1/2,n/2)], \tag{7.140}$$

where $y = t^2/(n+t^2)$, and $\text{sign}(t) = -1$ for $t < 0$ (see Table 7.19, the CDF line), equation (7.139) becomes

$$\text{MD} = 2 * \int_0^\infty \frac{1}{2}[1 - I_y(1/2,n/2)]dt. \tag{7.141}$$

Using $1 - I_y(1/2,n/2)dt = I_{1-y}(n/2,1/2)$ where $1 - y = n/(n+t^2)$, this becomes

$$\text{MD} = \int_0^\infty I_{1-y}(n/2,1/2)dt. \tag{7.142}$$

To evaluate this integral, take $u = I_{1-y}(n/2, 1/2)$ and $dv = dt$ so that $v = t$. Use the chain rule of differentiation to get

$$du = (\partial/\partial t)I_{1-y}(n/2, 1/2) = (\partial/\partial y)I_{1-y}(n/2, 1/2) * (\partial y/\partial t). \tag{7.143}$$

Differentiate $1 - y = n/(n + t^2)$ to get $-\partial y/\partial t = -2nt/(n + t^2)^2$. In addition, $\frac{\partial}{\partial y}I_{1-y}(n/2, 1/2) = g_{1-y}(n/2, 1/2)$ where $g()$ is the PDF of BETA-I. Integrate equation (7.143) by parts to get

$$t * [I_{1-y}(n/2, 1/2)] \mid_0^\infty + \int_0^\infty t * g_{1-y}(n/2, 1/2)(2nt/(n + t^2)^2)dt. \tag{7.144}$$

The first term is zero using L'Hospital's rule. Take $2n$ outside the integral to get

$$\text{MD} = 2n \int_0^\infty [t^2/(n + t^2)^2] * g_{1-y}(n/2, 1/2)dt. \tag{7.145}$$

Put $v = n/(n + t^2)$, and $1 - v = t^2/(n + t^2)$. This gives $t = \sqrt{n}((1 - v)/v)^{1/2}$, and $dv = -2nt/(n + t^2)^2 dt$. Write $[t^2/(n + t^2)^2] = [t^2/(n + t^2)] * 1/(n + t^2) = v(1 - v)/n$, and $dv = (-2/\sqrt{n})v^{1/2}(1 - v)^{3/2}$. Expand $g_{1-y}(n/2, 1/2)$. The n cancels out from the numerator and denominator, and equation (7.145) becomes

$$(\sqrt{n}/B(n/2, 1/2)) \int_0^1 v^{(n-3)/2}(1 - v)^0 dv. \tag{7.146}$$

This simplifies to $(\sqrt{n}/B(n/2, 1/2))/[(n-1)/2] = 2\sqrt{n}/[(n - 1) * B(n/2, 1/2)]$, which is the same expression obtained earlier.   ∎

7.18.2 Relation to Other Distributions

For $n = 1$, it reduces to the Cauchy distribution. If X and Y are IID χ^2-distributed random variables with the same DoF, then $(\sqrt{n}/2)(X - Y)/\sqrt{XY}$ is Student's T distributed [254]. If X is an F variates with $n\,df$, then $T = \frac{\sqrt{n}}{2}(\sqrt{X} - 1/\sqrt{X})$ is Student's $T(n)$. The analog of log normal to normal distribution is the log-Student's T distribution as $f(y, n) =$

$$\Gamma\left(\frac{n+1}{2}\right) / \left\{\sqrt{\pi n}\Gamma(n/2) \, y\left(1 + \frac{1}{n}(\log y)^2\right)^{(n+1)/2}\right\}, \tag{7.147}$$

where $y = \log(T)$ has a Student's T distribution (p. 311).

7.18.2.1 Tail Areas

The CDF of a Student's T random variable is encountered frequently in small sample statistical inference. For example, it is used in tests for the means, testing the significance of correlation coefficients, and constructing confidence intervals for means. The area under Student's distribution from $-t$ to $+t$ is of special interest in finding two-sided confidence intervals and tests of hypothesis.

We denote it as $T(-t : t|n)$ or $T_n(-t : t)$.

$$T_n(-t : t) = \frac{1}{\sqrt{n}B(\frac{1}{2},\frac{n}{2})} \int_{-t}^{t} (1 + x^2/n)^{-(n+1)/2} dx. \qquad (7.148)$$

This integral can be converted into an IBF by the transformation $y = n/(n + x^2)$ giving

$$T_n(-t : t) = 1 - I_{n/(n+t^2)}(n/2, 1/2). \qquad (7.149)$$

Owing to the symmetry, the area from $\pm t$ to the mode $(x = 0)$ is

$$T_n(-t : 0) = T_n(0 : t) = \frac{1}{2} - \frac{1}{2}I_{n/(n+t^2)}(n/2, 1/2). \qquad (7.150)$$

The CDF (area from $-\infty$ to t) is given by

$$F_n(t_0) = \int_{-\infty}^{t_0} f(t)dt = \int_{-\infty}^{0} f(t)dt + \int_{0}^{t_0} f(t)dt. \qquad (7.151)$$

Owing to symmetry, the first integral evaluates to 1/2. Represent the second integral using equation (7.150) to get

$$F_n(t_0) = 1 - \frac{1}{2}I_y(n/2, 1/2), \qquad (7.152)$$

where $y = n/(n + t_0^2)$, and $I(x; a, b)$ is the IBF.

For even degrees of freedom, the CDF of Student's T distribution can be obtained as $F_n(t) =:$

$$\frac{1}{2}(1 + \sqrt{(x/\pi)}) \sum_{i=0}^{\frac{n}{2}-1} (1 - x)^i \Gamma(i + 1/2)/\Gamma(i + 1)), \text{ where } x = t^2/(n + t^2).$$

The special cases $n = 2$ and $n = 4$ are $F_2(t) = \frac{1}{2}(1 + t/\sqrt{(2 + t^2)})$ and $F_4(t) = \frac{1}{2}[1 + (1 + 2/(4 + t^2))t/\sqrt{(4 + t^2)}]$. As mentioned earlier, this reduces to Cauchy CDF for $n = 1$ as $\frac{1}{2} + \frac{1}{\pi}\text{sign}(t) \tan^{-1}(t)$. For $n = 3, 5$, similar expressions exist (see Reference 121).

7.18.2.2 Moments and Generating Functions
As this distribution is symmetric, all odd moments vanish. The even moments are given by

$$\mu_k = n^{k/2} \frac{\Gamma\left(\frac{k+1}{2}\right)\Gamma\left(\frac{n-k}{2}\right)}{\sqrt{\pi}\Gamma(n/2)}. \qquad (7.153)$$

This satisfies the first-order recurrence

$$(n - k)\mu_k = n(k - 1)\mu_{k-2}, \qquad (7.154)$$

repeated application of which gives a closed-form expression (see Table 7.19). The characteristic function is given by

$$\phi(t) = K * \int_{-\infty}^{\infty} e^{itx}/(1 + x^2/n)^{(n+1)/2} dx, \tag{7.155}$$

where we have used the variable x instead of t owing to the dummy variable in the ChF [255]. Upon putting $x^2/n = y^2$, this becomes

$$K\sqrt{n} * \int_{-\infty}^{\infty} \cos(ty\sqrt{n})/(1 + y^2)^{(n+1)/2} dy. \tag{7.156}$$

If n is odd ($=2m + 1$), this reduces to $\exp(-|t\sqrt{n}|)S_n(|t\sqrt{n}|)$ where S is a polynomial of degree $n - 1$ that satisfies the recurrence $S_{m+3}(t) = S_{m+1}(t) + t^2/(m^2 - 1)S_{m-1}(t)$. See Table 7.19 for further properties.

7.19   SNEDECOR'S *F* DISTRIBUTION

This distribution, named after G.W.Snedecor [256], is used extensively in ANOVA and related procedures. This is due to the normality assumption of the population from which sample came, so that the null distribution of the test statistic has an F distribution. It is also used in computing the power of various statistical tests that employ the sample variance.

This is the distribution of the ratio of two independent scaled χ^2 variates $F = (\chi^2(m)/m)/(\chi^2(n)/n) = \frac{n}{m}(\chi_m^2/\chi_n^2)$ with PDF

$$f(x; m, n) = \frac{\Gamma((m + n)/2)m^{m/2}n^{n/2}}{\Gamma(m/2)\Gamma(n/2)} \frac{x^{m/2-1}}{(n + mx)^{(m+n)/2}}, \quad 0 < x < \infty. \tag{7.157}$$

A derivation is given in Chapter 11. The unscaled F distribution is the distribution of the ratio $\chi^2(m)/\chi^2(n)$, which is BETA-II $(m/2, n/2)$.

7.19.1   Properties of *F* Distribution

As both the numerator and denominator variates in the definition are χ^2, this distribution is defined for $x > 0$. Owing to symmetry, $1/F$ has exactly identical distribution with the DoF reversed. The parameters m and n are integers in practical applications. Theoretically, the distribution is defined for noninteger DoF values as well. The distribution of $Z = (1/2) \log(F)$ is more tractable, as it converges to normality faster than F itself. As the χ^2 distribution is a special case of gamma distribution, the ratio of two properly scaled independent gamma variates has an F distribution [257]. The F distribution has a long right tail and is skewed to the right for small parameter values. Several recurrences satisfied by the density, distribution functions, and moments can be found in Reference 129.

7.19.1.1 Moments and Generating Functions

The mean is undefined when $n < 2$, but it is $n/(n-2) = 1 + 2/(n-2)$ for $n > 2$. This does not depend on the numerator DoF parameter m. Although the distribution has infinite range, the mean (center of mass) is bounded by 3, and rapidly approaches 1 as n becomes large. The variance is $\sigma^2 = \frac{2n^2(m+n-2)}{m(n-2)^2(n-4)}$, which is defined for $n > 4$. This in terms of the mean is $2\mu^2/(n-4) * ((m+n-2)/m)$. The mode is $[(m-2)/m] * [n/(n+2)]$. As $n/(n-2)$ is >1 and $n/(n+2)$ is <1, the mode is less than the mean. As n becomes large, the mean tends to 1 but the mode tends to $(m-2)/m$. Similarly, the skewness coefficient is undefined for $n \leq 6$ (all of these conditions are on n and not on m). For $n > 6$, the skewness coefficient is $\beta_1 = \frac{2(2m+n-2)\sqrt{2(n-4)}}{\sqrt{m(n-6)}\sqrt{m+n-2}}$. The characteristic function of F variate is (see Reference 258)

$$\phi(t) = \Gamma((m+n)/2)/\Gamma(n/2)\psi(m/2, 1 - n/2, -itn/m), \qquad (7.158)$$

where $\psi(m/2, 1 - n/2, -itn/m)$ is the confluent hypergeometric function of type-2. A double infinite sum for it is as follows (see References 259 and 260)

$$\phi(m, n; t) = \frac{1}{B(m/2, n/2)} \sum_{i=0}^{\infty} \sum_{j=0}^{\infty} \frac{(it)^i}{[i!(i+j+m/2)]} \binom{i+j-n/2}{j}, \qquad (7.159)$$

which is valid for n even. See Table 7.20 for further properties.

7.19.2 Relation to Other Distributions

As mentioned earlier, if $X \sim F(m, n)$, then $Y = 1/X$ is $F(n, m)$. As the T distribution is the ratio of a standard normal to the square root of an independent scaled χ_n^2 random variate, the square of T is F distributed with 1 and $n\, df$. If X and Y are independent F variates with the same df, then $T = \frac{\sqrt{n}}{2}(\sqrt{X} - \sqrt{Y})$ is Student's $T(n)$ [254]. Tail area of binomial distribution is related to the F distribution as

$$\sum_{x=0}^{k} \binom{n}{x} p^x q^{n-x} = 1 - F_y(2(k+1), 2(n-k)), \quad \text{where } y = p(n-k)/(q(1+k)). \qquad (7.160)$$

As the denominator DoF $n \to \infty$, the variate $m * X$ approaches a χ_m^2 distribution.

7.19.2.1 Tail Areas

Integrating from 0 to $+x$ gives the CDF of a Snedecor's F distribution with (m, n) DoF using equation (7.47) as

$$F_{(X)}(x; m, n) = I_{n/(n+mx)}(n/2, m/2) = 1 - I_{mx/(n+mx)}(m/2, n/2). \qquad (7.161)$$

The tail areas are related as $F(x; m, n) = 1/F(1 - x; n, m)$. The special cases are $F(x; 1, 1) = (\frac{2}{\pi})\tan^{-1}(\sqrt{mx/n})$, $F(x; 1, 2) = \sqrt{mx/(n+mx)}$, $F(x; 2, 1) = 1 - \frac{\sqrt{n}}{\sqrt{(n+mx)}}$, and $F(x; 2, 2) = mx/(n+mx)$. See References 261–264 for further properties and References 236 and 265 for percentile points.

TABLE 7.20   Properties of F Distribution $\left(\frac{\Gamma((m+n)/2)m^{m/2}n^{n/2}}{\Gamma(m/2)\Gamma(n/2)} \frac{x^{m/2-1}}{(n+mx)^{(m+n)/2}} \right)$

Property	Expression	Comments
Range	$0 \le x \le \infty$	Infinite
Mean	$\mu = n/(n-2)$	$= 1 + 2/(n-2)$
Variance	$\sigma^2 = 2\mu^2(m+n-2)/[m(n-4)]$	$= \mu > \sigma^2$
Mode	$n(m-2)/[m(n+2)]$	$m > 2$
CV	$(2(m+n-2)/[m(n-4)])^{1/2}$	
Skewness	$(2m+n-2)[8(n-4)]^{1/2}/[\sqrt{m}(n-6)(n+m-2)^{1/2}]$	γ_1
Kurtosis	$\beta_2 > 3$	
MD	$E\|X - \mu\| = 2 \int_0^{n/(n-2)} I_{mx/(n+mx)}(m/2, n/2)dx$	
Moments	$\mu'_r = (n/m)^r \Gamma(m/2+r)\Gamma(n/2-r)/[\Gamma(m/2)\Gamma(n/2)]$	
ChF	$[\Gamma((m+n)/2)/\Gamma(n/2)]\,\Psi(m/2, 1-n/2; -nit/m)$	
Additivity	$\sum_{i=1}^m F(m_i, n) = F\left(\sum_{i=1}^m m_i, n\right)$	Unscaled IID F
Tail area	$I_{n/(n+mx)}(n/2, m/2) = 1 - I_{mx/(n+mx)}(m/2, n/2)$	$\Pr(F \le x)$

I is the incomplete beta function. Additivity is for unscaled F with the same denominator DoF.

7.20   FISHER'S Z DISTRIBUTION

This distribution is obtained as a transformation from the F distribution as $Z = \frac{1}{2}\log(F)$. It is also called logarithmic F distribution. A derivation is given in Chapter 11. As the range of F is from 0 to ∞, the range of Z is $-\infty$ to ∞ and the PDF is obtained directly from the previous PDF as

$$f_z(m, n) = \frac{2m^{m/2}n^{n/2}}{B(m/2, n/2)} \frac{e^{mz}}{(n + me^{2z})^{(m+n)/2}}, \qquad (7.162)$$

where $B(m/2, n/2)$ is the CBF. The unnormalized Z distribution results when F is replaced by the unnormalized F (which is BETA-II). If the F-distribution is noncentral, the corresponding Z is singly noncentral. If both chi squares in the F-distribution are noncentral, the corresponding Z is called doubly noncentral [4, 266].

7.20.1   Properties of Fisher's Z Distribution

Left tail areas can be expressed in terms of IBF as follows

$$Z_{m,n}(x) = P[\frac{1}{2}\log(F_{m,n}(x)) = P[F_{m,n}(x)] \le e^{2x} = I_c(m/2, n/2), \qquad (7.163)$$

where $c = e^{2x}/(n + me^{2x})$. The CDF satisfies the symmetry relationship $Z_c(m, n) = 1 - Z_{-c}(n, m)$. See Table 7.21 for further properties (Figures 7.13 and 7.14).

TABLE 7.21 Properties of Fisher's Z $\left(\frac{2m^{m/2}n^{n/2}}{B(m/2,n/2)}\frac{e^{mz}}{(n+me^{2z})^{(m+n)/2}}\right)$

Property	Expression	Comments
Range of Z	$-\infty \le z \le \infty$	Infinite
Mean	$\mu \simeq (m-n)/(2mn)$	mode $z = 0$
Variance	$\sigma^2 \simeq (m+n)/2mn$	$\Rightarrow \mu < \sigma^2$
ChF	$(n/m)^{it/2}\Gamma((n-it)/2)\Gamma((m+it)/2)/$ $[\Gamma(m/2)\Gamma(n/2)]$	
Tail area	$I_c(m/2, n/2),\ c = m\ e^{2x}/(n+me^{2x})$	
Symmetry relation	$Z(x; m, n) = 1 - Z(-x; n, m)$	

$I_x(m, n)$ is the incomplete beta function.

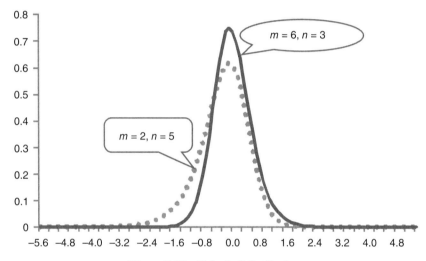

Figure 7.13 Fisher's Z distribution.

7.20.1.1 *Moments*

The characteristic function is $(n/m)^{it/2}\Gamma((n-it)/2)\Gamma((m+it)/2)/[\Gamma(m/2)\Gamma(n/2)]$. Using the derivatives of gamma function, the first two moments are $\mu = (m-n)/[2mn] = (1/n - 1/m)/2$ and $\mu_2 = (m+n)/[2mn] = (1/n + 1/m)/2$ approximately. The cumulants are easier to find in terms of digamma function [60, 267, 268].

7.20.1.2 *Relationship with Other Distributions*

When both the parameters $\to \infty, Z \to N(\frac{1}{2}\frac{m-n}{mn}, \frac{1}{2}\frac{m+n}{mn})$. Convergence of Z to normality is faster than the convergence of F distribution. If $X \sim Z(m, n)$ then $\exp(2Z) \sim F(m, n)$. The transformation $V = (N/(N+1))^{1/2}(Z/b)$ is approximately distributed as T_N, where $N = m + n - 1, b^2 = \frac{1}{2}(1/m + 1/n)$ and T_N is Student's T distribution.

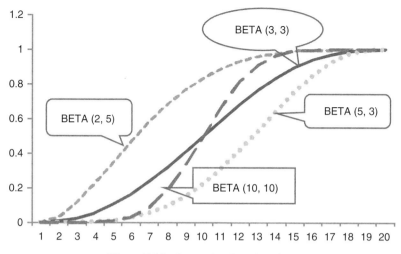

Figure 7.14 Incomplete beta function.

7.21 WEIBULL DISTRIBUTION

This distribution is named after the Swedish physicist W. Weibull (1887–1979), who invented it in 1939 in connection with strength of materials, although it was known to Rosen and Rammler [270]. It finds applications in reliability theory, quality control, strength of materials, and so on [271, 272]. It is used in the design of wind turbines, model wind speed distributions, fading channels in wireless communications, describe the size of particles in motion (such as raindrops), or those being grinded, milled, crushed, or subjected to external pressure (for which another choice is the lognormal law). The one-parameter Weibull distribution has PDF

$$f(x, a) = ax^{a-1}e^{-x^a}, \quad x > 0. \tag{7.164}$$

The two-parameter Weibull distribution is obtained from the above by a simple transformation $x = y/b$ as $f(y, a, b) = \frac{a}{b}(y/b)^{a-1}e^{-(y/b)^a}$, where b is the scale and a is the shape parameter. Mode is $b(\frac{a-1}{a})^{1/a}$, median is $b(\log 2)^{1/a}$. This reduces to the above form when $b = 1$. We denote it as WEIB(a, b). It is easy to see that if $[(X - a)/b]^c$ has an exponential distribution, then X has a general Weibull distribution. The corresponding CDF is easily found as $F(x) = 1 - e^{-x^a}$, from which the quantile function can be obtained as $Q(u) = [-\log(1 - u)]^{1/a}$.

7.21.1 Properties of Weibull Distribution

The mode of the distribution is 0 for $a < 1$. When $a = 1$, we get the exponential distribution. It is also related to the Rayleigh distribution. If $X \sim$ WEIB(a, b) then

$\log(X)$ has extreme value distribution. For $a > 1$, the mode is at $(1 - \frac{1}{a})^{1/a}$. This tends to the limit 1 as $a \to \infty$. The three-parameter Weibull model has CDF

$$F_X(a, b, c) = 1 - \exp\left(-\left(\frac{x-c}{a}\right)^b\right) \quad \text{for } x \geq c. \tag{7.165}$$

Put $c = 0$ to get two-parameter version earlier.

7.21.1.1 Moments
The kth moment is given by $E(x^k) = \int_0^\infty x^k f(x)dx = b^k \Gamma(k/a + 1)$. From this, we get the mean as $\mu = b\Gamma(\frac{1}{a} + 1)$. The second moment is $b^2\Gamma(\frac{2}{a} + 1)$, from this we get the variance as $\sigma^2 = b^2[\Gamma(2/a + 1) - \Gamma(1/a + 1)^2]$. Put $b = 1$ to get the variance of standard form (7.164). See Table 7.22 for further properties.

■ **EXAMPLE 7.29   Mean deviation of Weibull distribution**

Find the mean deviation of the Weibull distribution using Theorem 7.1.

TABLE 7.22   Properties of Weibull Distribution

Property	Expression	Comments
Range of X	$0 \leq x < \infty$	Continuous
Mean	$\mu = b\Gamma(1 + \frac{1}{a})$	
Variance	$\sigma^2 = b^2\left[\Gamma(1 + \frac{2}{a}) - \Gamma(1 + \frac{1}{a})^2\right]$	$= b^2\Gamma(1 + \frac{2}{a}) - \mu^2$
Mode	$b(1 - \frac{1}{a})^{1/a}, a > 1$	$\to b$ as $a \to \infty$
Median	$b(\log 2)^{1/a}$	
CV	$\left(\Gamma(1 + \frac{2}{a})/\Gamma(1 + \frac{1}{a})^2 - 1\right)^{1/2}$	$\to 0$ as $a \to$ large
Skewness	$\gamma_1 = (\Gamma(1 + \frac{3}{a})b^3 - 3\mu\sigma^2 - \mu^3)/\sigma^3$	$=$
Kurtosis β_2	$3c(c + 1)(a + 1)(2b - a)/[ab(c + 2)(c + 3)]$	$c = a + b$
Moments	$\mu'_r = b^r\Gamma(1 + r/a)$	
MGF	$\sum_{k=0}^\infty (bt)^k \Gamma(1 + k/a)/k!$	
ChF	$\sum_{k=0}^\infty (bit)^k \Gamma(1 + k/a)/k!$	
Additivity	$\sum_{i=1}^m Y(a, b_i) = Y(a, \prod_{i=1}^m b_i)$	IID $Y = \log(\text{WEIB})$
Recurrence	$(1 + \frac{1}{a})(x/b)\exp(-(x/b)^a[1 - x/b])$	$f(x; a + 1, b)/f(x; a, b)$
Tail area	$\exp(-(x/b)^a)$	

The mode of WEIB$(a, b) \to b$ as a becomes large ($0.9974386\,b$ for $a = 20$, $0.999596\,b$ for $a = 50$), but median $\to b$ much slower ($0.981841\,b$ for $a = 20$, $0.992697\,b$ for $a = 50$, $0.9995964\,b$ for $a = 908$). The MFG and ChF of logarithm of the Weibull variate are more tractable, $E[\exp(\ln(X)t)] = b^t\Gamma(1 + t/a)$. A Weibull plot can reveal if the data came from this distribution. For this, plot $\ln(x)$ along the X-axis and $\ln(-\ln(1-F(x)))$ along the Y-axis, where F is the empirical CDF obtained from random sample. A straight line indicates Weibull parent population.

Solution 7.29 We apply Theorem 7.1 (page 7-5) to find the MD. As the Weibull distribution does not tail off to the lower limit for some parameter values (e.g., $b < 1$), equation (7.1) seems like not applicable. We know that the CDF is $1 - \exp(-(\frac{x-c}{a})^b)$. As done in the case of exponential distribution, using L'Hospital's rule, it is easy to show that $x * F(x) \to 0$, so that the Theorem 7.1 is applicable. This gives

$$\text{MD} = 2 \int_0^m \left[1 - \exp\left(-\left(\frac{x-c}{a}\right)^b \right) \right] dx, \tag{7.166}$$

where $m = b\Gamma(1 + \frac{1}{a})$ is the mean. Split this into two integrals and integrate the first term to get $2b\Gamma(1 + \frac{1}{a})$. The second integral is $-2\int_0^m \exp(-(\frac{x-c}{a})^b)dx$. Expand $\exp(-(\frac{x-c}{a})^b)$ as an infinite series and integrate term by term to get $-2\sum_{k=0}^{\infty} (-1)^k/k! \int_0^m (\frac{x-c}{a})^{bk}dx$. This simplifies to $2\sum_{k=0}^{\infty} (-1)^{k+1}/[k!(bk+1)a^{bk}][(m-c)^{bk+1} - (-c)^{bk+1}]$. Now combine with the first term to get the MD.

∎

7.21.1.2 Relationship with Other Distributions If $[(X-a)/b]^c$ has an exponential distribution, then X has a general Weibull distribution. It is also related to the uniform distribution $U(0,1)$ as follows: –if $X \sim U(0,1)$ then $Y = (-\ln(X)/a)^{1/b} \sim$ WEIB(a,b). See Reference 273 discrete Weibull distribution.

7.21.2 Random Numbers

As $F(x) = 1 - e^{-(x/b)^a}$, we could generate random numbers using uniform pseudo-random numbers in [0,1] as $u = 1 - e^{-(x/b)^a}$, which on rearrangement becomes $x = b(-\log(u))^{1/a}$. Notice that the log function takes negative values for the argument in [0, 1]. Hence, the $-\log(u)$ maps it into the positive interval.

7.22 RAYLEIGH DISTRIBUTION

This distribution is named after the British physicist Rayleigh (1842–1919). It finds applications in reliability theory and communication systems. See Reference 274 for an application of Rayleigh distribution to wind turbine modeling. Stability of the Rayleigh distribution is discussed in Reference 275. See References 276 and 277 for the distribution of the product of two IID Rayleigh random variables. This distribution can be considered as the distribution of the radial distance of a point on the bivariate normal surface (with zero means) from the origin. In other words, it is the distribution of $\sqrt{X^2 + Y^2}$, where (X, Y) have a joint bivariate normal distribution. It is a special case of χ-distribution. The PDF is

$$f_x(x; a) = x/a^2 \, e^{-x^2/(2a^2)}, \quad x \geq 0. \tag{7.167}$$

The corresponding CDF can be expressed in terms of scaled normal PDF as given below. An alternate parametrization can be obtained by the linear transformation $y = (\sqrt{b}/[a\sqrt{2}])x$, so that $dx = a\sqrt{2}/\sqrt{b}dy$ and we get the PDF as

$$f_y(y; b) = 2(y/b)e^{-y^2/b}, \quad y \geq 0. \tag{7.168}$$

In general, if $X_1, X_2, \cdots, X_n$ are independent normal random variables $N(0, \sigma^2)$, the distribution of $X = (X_1^2 + \cdots + X_n^2)^{1/2}$ is given by

$$f(x; n, \sigma) = \frac{2}{(2\sigma^2)^{n/2}\Gamma(n/2)}x^{n-1}e^{-x^2/2\sigma^2} \quad \text{for } x > 0. \tag{7.169}$$

See Table 7.23 for further properties. The CDF is given by

$$F_x(x, a) = 1 - \exp(-x^2/(2a^2)), \quad x \geq 0. \tag{7.170}$$

■ EXAMPLE 7.30   Mean deviation of Rayleigh distribution

Find the mean deviation of the Rayleigh distribution with PDF $f_x(x; a) = (x/a^2) \, e^{-x^2/(2a^2)}$.

TABLE 7.23   Properties of Rayleigh Distribution $(x/a^2) \, e^{-x^2/(2a^2)}$

Property	Expression	Comments
Range of X	$0 \leq x < \infty$	Continuous
Mean	$\mu = a\sqrt{\pi/2}$	
Variance	$\sigma^2 = (2 - \pi/2)a^2 = 2a^2 - \mu^2 = \mu^2(4/\pi - 1)$	$= 0.27324\mu^2$
Mode	a	
Median	$a\sqrt{\ln(4)}$	$1.17741\,x$
Skewness	$\gamma_1 = 2(\pi - 3)\sqrt{\pi}/(4 - \pi)^{3/2}$	$\simeq 0.63111$
Kurtosis	$\beta_2 = (32 - 3\pi^2)/(4 - \pi)^2$	
Mean deviation	$2[a\sqrt{\pi/2} - P(\pi/4, 1/2)]$	$P = $ incomplete Γ
Quartiles	$Q_1 = 0.75853\,a$	$Q_3 = 1.66551\,a$
CV	$\sqrt{4/\pi - 1}$	0.522723
Moments	$\mu'_r = 2^{r/2}a^r\Gamma(r/2 + 1)$	
MGF	$1 + bt\exp(b^2t^2/2)\sqrt{\pi/2}[\mathrm{erfc}(bt/\sqrt{2}) + 1]$	
ChF	$1 - bit\exp(-b^2t^2/2)\sqrt{\pi/2}[\mathrm{erfc}(bt/\sqrt{2}) - i]$	
Tail area	$\Pr[X > x] = \exp(-x^2/2a^2)$	

The ratio $\sigma/\mu = 0.5227232$ shows that $\mu > \sigma$. The mean-median-mode inequality is mode <median<mean.

Solution 7.30 We apply Theorem 7.1 (p. 256) to find the MD. As the Rayleigh distribution does not tail off to zero at the lower limit (i.e., at 0), equation (7.1) seems like not applicable. We know that the CDF is $1 - \exp(-x^2/2a^2)$. If we apply L'Hospital's rule once on $x * F(x)$, we find that it $\to 0$ as $x \to 0$. As the $\lim_{x \to 0} x * F(x) = 0$, Theorem 7.1 becomes applicable. This gives

$$MD = 2 \int_0^m [1 - \exp(-x^2/2a^2)]dx \text{ where } m = a\sqrt{\pi/2}. \tag{7.171}$$

Split the integral into two parts. The first one integrates to $2m$. The second one is $-2\int_0^m \exp(-x^2/2a^2)dx$. Put $y = x^2/(2a^2)$ so that $dy = x/a^2dx$. The upper limit of integration becomes $m^2/(2a^2) = \pi/4$. We get

$$MD = 2m - 2 \int_0^{\pi/4} y^{\frac{1}{2}-1}e^{-y}dy = 2[m - P(\pi/4, 1/2)], \tag{7.172}$$

where $P()$ is the incomplete gamma integral. Put the value of m to get the MD. ∎

7.22.1   Properties of Rayleigh Distribution

The standard Rayleigh distribution is obtained by putting $a = 1$. As the skewness is 0.63111, it is always positively skewed. Variance $\sigma^2 = 0.27324\mu^2$ shows that $\sigma^2 < \mu^2$ or equivalently $\sigma/\mu = 0.5227232$. See Reference 278 for an application to the distance between pairs of points in wireless networks.

7.22.1.1   Moments and Generating Functions   Ordinary moments can be obtained in terms of gamma function as $E[X^k] =$

$$\frac{2}{(2\sigma^2)^{n/2}\Gamma(n/2)} \int_0^\infty x^{n+k-1}e^{-x^2/(2\sigma^2)}dx = \frac{2^{k/2}\sigma^k\Gamma((n+k)/2)}{\Gamma(n/2)}, \tag{7.173}$$

from which we get the mean as $\mu = \sigma\sqrt{2}\Gamma((n+1)/2)/\Gamma(n/2)$ and $\sigma^2 = (4-\pi)a^2/2$.

7.22.1.2   Relationship with Other Distributions   If X is Rayleigh(1), then X^2 is χ_2^2. It is related to $U(0, 1)$ as $X = a(-2\ln(1-U))^{1/2}$. If X_i is Rayleigh(b), then $\sum_i X_i^2$ is gamma distributed.

7.23   CHI-DISTRIBUTION

If $X_1, X_2, \cdots, X_n$ are independent standard normal random variables, the distribution of $Y = (X_1^2 + X_2^2 + \cdots + X_n^2)^{1/2}$, that is, $\sqrt{\chi_n^2}$ is called a chi-distribution with $n df$. The PDF is easily obtained as

$$f(x; n) = x^{n-1}e^{-x^2/2}/[2^{n/2-1}\Gamma(n/2)]. \tag{7.174}$$

7.23.1   Properties of Chi-Distribution

This distribution reduces to the half-normal or folded normal distribution for $n = 1$, the Rayleigh distribution for $n = 2$, and the Maxwell distribution for $n = 3$. The rth moment is $\mu_r = 2^{r/2}\Gamma((n+r)/2)/\Gamma(n/2)$. The mean is $\sqrt{2}\Gamma((n+1)/2)/\Gamma(n/2)$.

EXAMPLE 7.31   Mode of chi-distribution

Find the mode of chi-distribution with $n > 1 df$.

Solution 7.31 Take log of equation (7.174) to get $\log(f(x)) = k + (n - 1)\log(x) - x^2/2$. Differentiate with respect to x to obtain the RHS as $(n-1)/x - x$. Equating to zero results in $(n-1) = x^2$ so that $x = \sqrt{n-1}$ is the solution. As the second derivative is $-(n-1)/x^2 - 1$, which is negative for $n > 1$, this is indeed the mode.   ∎

7.24   MAXWELL DISTRIBUTION

This distribution is frequently encountered in engineering. It is named after the Scottish physicist James C. Maxwell (1831–1879). The PDF is given by

$$f_x(x; a) = \sqrt{2/\pi} x^2 e^{-x^2/(2a^2)}/a^3, \quad x > 0. \tag{7.175}$$

Here, "a" is a scale parameter. Put $y = x/a$ to get the standard Maxwell distribution.

7.24.1   Properties of Maxwell Distribution

The Maxwell and Rayleigh distributions are surprisingly similar shaped for small parameter values (Figures 7.15 and 7.16). It has an alternate parametrization known as Maxwell's velocity distribution that represents the velocity of a gas molecule as

$$f(x; a, k, T) = 4\pi x^2 (a/2\pi kT)^{3/2} \exp(-ax^2/(2kT)), \tag{7.176}$$

where a = molecular weight, T = absolute temperature, and k is the Boltzmann constant. Mean $\mu = \sqrt{8/(\pi a)}$ and variance $\sigma^2 = (3 - 8/\pi)/a$. The mean velocity of a gaseous molecule at room temperature can then be estimated as $\bar{x} = [8kT/(a\pi)]^{1/2}$. Integration of equation (7.175) allows us to write the CDF in either of the following formats:–

$$2\gamma(3/2, x^2/(2a^2))/\sqrt{\pi} = \text{erf}(x/(a\sqrt{2})) - (x/a)\sqrt{2/\pi} \exp(-x^2/(2a^2)). \tag{7.177}$$

The MD is easily obtained using the power method as

$$MD = (4/\sqrt{\pi}) \int_0^{2a/\sqrt{\pi/2}} \gamma(3/2, x^2/(2a^2)) dx. \tag{7.178}$$

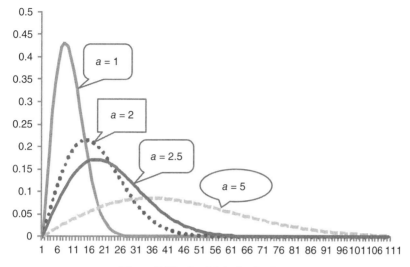

Figure 7.15   Rayleigh distributions.

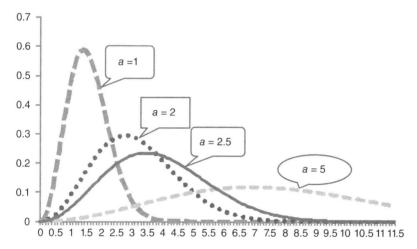

Figure 7.16   Maxwell distributions.

This distribution is a special case of the chi-distribution that has PDF

$$f(x) = x^{n-1}e^{-x^2/(2\sigma^2)}/[2^{n/2-1}\sigma^n\Gamma(n/2)]. \tag{7.179}$$

Put $n = 3$ in equation (7.179) to get equation (7.175). See Table 7.24 for further properties.

TABLE 7.24 **Properties of Maxwell Distribution** $\sqrt{2/\pi}x^2 \exp(-x^2/(2a^2))/a^3$

Property	Expression	Comments
Range of X	$0 \le x < \infty$	Continuous
Mean	$\mu = 2a\sqrt{2/\pi}$	
$E(X^2)$	$3a^2$	
Variance	$\sigma^2 = (3 - 8/\pi)\,a^2$	
Mode	$a\sqrt{2}$	
Skewness	$2a\sqrt{2}(16 - 5\pi)/(3\pi - 8)^{3/2}$	a* to 0.4857 approximately
Mean deviation	$4/\sqrt{\pi}\displaystyle\int_0^{\mu}\gamma(3/2, x^2/2a^2)dx$	$4/\sqrt{\pi}\displaystyle\int_0^{2a\sqrt{2/\pi}}\gamma(3/2, x^2/2a^2)dx$
Moments	$\mu_r' = 2^{r/2+1}a^r\Gamma((r+3)/2)/\sqrt{\pi}$	
CDF	$2\,\gamma(3/2, x^2/(2a^2))/\sqrt{\pi}$	
MGF	$a\,t\,\sqrt{2/\pi}+2a\,(1+t^2)e^{t^2/2}\,\Phi(t)$	

Note that there are two parametrizations for Maxwell distribution.

7.25 SUMMARY

Statistical distributions play an important role in data modeling in various fields, including psychology, education, various branches of engineering, medical sciences, management, and the worldwide web. A single distribution suffices for most modeling situations. A linear combination of homogeneous models (such as normals with different means) is sometimes used. A researcher has to choose the most appropriate model depending on the data to be modeled at hand. A simple data plot can quite often reveal the most appropriate distribution that fits it well.

Statistical properties such as the mean, variance, cumulative probability function, median, and mean deviation are obtained in summary format for some commonly employed continuous distribution such as uniform, normal (Gaussian), exponential, gamma, Weibull, and lognormal among others. The exponential distribution has no memory, whereas other continuous distributions in the above-mentioned list do retain a level of memory.

Ever since the landmark paper of Abraham De Moivre in 1730 [130], numerous research work had gone into finding the mean deviation of common distributions. See, for example, References 134 and 221, and so on. In Reference 166, the authors gave several expressions involving the integral of distribution function that pertains to higher order moments and moments of spacings. In Section 7.2, we gave an easy method to find the MD of continuous distributions and demonstrated its use throughout this chapter.

Extensive bibliographies exist for each distribution. See Balakrishnan and Nevzorov [121], Evans et al. [122], Johnson, Kotz and Balakrishnan [60], and Hazewinkel [279] for theoretical discussions and properties of these distributions. This chapter gave a bird's eye view of the main results. Separate volumes are available exclusively for some of these distributions. For example, see Reference 230 for Pareto distributions, Reference 272 for Weibull distributions, References 221 and 280 for normal and related distributions, and Reference 281 for LND. Numerous application papers have also appeared recently.

EXERCISES

7.1 Mark as True or False

a) $F[1 - \alpha; m, n]$ is always $>1/F[\alpha; n, M]$, where F is CDF of Snedecor's F

b) BETA-I and BETA-II are related as $Y = \ln(X)$

c) The MD of a distribution in the range $[a, b]$ is always within the range

d) The central limit theorem is applicable to the Cauchy mean

e) Geometric mean of lognormal distributions is lognormal distributed

f) Maxwell distribution is a special case of chi-distribution

g) Truncated $U(a, b)$ distribution has the same skewness

h) Variance of exponential distribution is square of the mean.

7.2 Find the unknown K and verify if each of the following is a PDF:
(a) $f(x) = Kx^2(1 - x)$, for $0 < x < 1$, (b) $f(x) = K/x^a$, for $x > 1, a > 0$.

7.3 Prove that $B(a + 1, b) = [a/(a + b)] B(a, b)$, where $B(a, b)$ denotes the complete beta function (CBF). What is the value of $B(0.5, 0.5)$?

7.4 If $X \sim EXP(\lambda)$ and $Y \sim EXP(\mu)$ prove that $P(X < Y) = \lambda/(\lambda + \mu)$.

7.5 If $X \sim EXP(\lambda)$ find the probability that $\Pr(X - 1/\lambda) < 1$.

7.6 Find C if $f(x, y, z) = Cz(x + y)$ for $0 < x < 1, 0 < y < 1, 0 < z < 1$ is a PDF.

7.7 If $X \sim EXP(\lambda)$, find the PDF of floor(X) and ceil(X).

7.8 Prove that mode <median<mean for lognormal distribution.

7.9 Prove that for the lognormal distribution mean/median $= \exp(\sigma^2/2)$.

7.10 What is the range of lognormal distribution? If X is normal, is $\log(X)$ lognormal distributed?.

7.11 For what values of the parameters p and q is the beta distribution BETA-I(p, q) U-shaped?

7.12 Prove that the normal distribution $N(\mu, \sigma^2)$ has points of inflection at $x = \mu \pm \sigma$. How does it change by the change of scale transformation $y = x/\sigma$

7.13 For which of the following distributions is the variance the square of the mean?. (a) exponential (b) beta (c) normal (d) Student's T

7.14 Prove that area from $1/\lambda$ to λ of an exponential distribution $f(x; \lambda) = \lambda e^{-\lambda x}$ is $\frac{1}{e} - e^{-\lambda^2}$ if $\lambda > 1$ and $e^{-\lambda^2} - \frac{1}{e}$ otherwise.

7.15 Find the area from 0 to λ and from λ to ∞ of the exponential distribution $\lambda e^{-\lambda x}$. Hence or otherwise obtain the area from $1/\lambda$ to λ.

7.16 Express the areas of standard normal in terms of error function $\mathrm{erf}(z)$ (i) from $-c$ to $+c$ and (ii) $-c$ to ∞.

7.17 If $X \sim$ Beta-I(a, b), find the distribution of $Y = (1 - X)/X$ and obtain its mean and variance. Find the ordinary moments.

7.18 Prove that the mode of an F distribution is $((m - 2)/m)(n/(n + 2))$ for $m > 2$.

7.19 If X_i, $i = 1, 2,...,n$ are IID $U(0, 1)$, find the distribution of $S = \sum_{i=1}^{n} X_i$.

7.20 Prove that the mean of an exponential distribution divides the area in $(1 - \frac{1}{e}):\frac{1}{e}$ ratio.

7.21 Prove that the characteristic function of general Cauchy distribution is $\exp(itb-|t|a)$.

7.22 What are the central moments of a rectangular distribution CUNI(a, b)?. Obtain μ_2.

7.23 Prove that $a = k - c$ and $b = k + c$ in CUNI(a, b) results in $f(x; c) = 1/(2c)$ for $a-c \le x \le a+c$. Find its mean.

7.24 Which of the following distributions is always asymmetric and leptokurtic. (a) exponential (b) beta (c) normal (d) Student's T.

7.25 If $X \sim U(0, 1)$, find the distribution of $Y = -\log(1 - X)$. Hence prove that $P[Y|X > x_0] = 1 - \log(1 - x_0)$ characterizes the $U(0, 1)$ distribution.

7.26 If X and Y are independent gamma random variables $\Gamma(a, \lambda)$ and $\Gamma(b, \lambda)$, then prove that $X/(X + Y)$ is BETA(a, b).

7.27 What transformation to use to obtain standard arc-sine distribution from $b/[\pi \sqrt{(x - a)(a + b - x)}]$ for $a < x < b$?.

7.28 Prove that $\mu = a/(b - 1)$ and $\sigma^2 = a(a + b - 1)/[(b - 1)^2(b - 2)]$ $(b > 2)$ for the BETA-II distribution with PDF $f_y(a, b) = y^{a-1}/[B(a, b)(1 + y)^{a+b}]$.

7.29 Consider a distribution defined as $f(x; p) = pq^{(x-1)/2}$, where $x > 0$, $p > 0$, $q > 0$. Find the normalizing constant, the mean and variance.

7.30 Find the kth moment of arc-sine distribution of first kind given in equation (7.8), page 7-40, and obtain the μ and σ^2.

7.31 Find the kth moment of the two parameter arc-sine distribution $b/\pi \sqrt{(x - a)(a + b - x)}$.

7.32 Find first two moments of the distribution $f_x(\lambda, m) = \sqrt{\lambda/\pi x}\; e^{-\lambda x}$ with range $x \ge 0$.

7.33 Find the constant C of the distribution $f(y) = C/\sqrt{(1 - 4y^2)}$, for $-1/2 \le y \le 1/2$. Find μ and σ^2.

7.34 Prove that the rth moment, for r even, of $N(0, 1)$ is $\mu_r = 2^{r/2}\Gamma((r + 1)/2)/\sqrt{\pi}$.

7.35 Prove that the distribution of the sum of n independent Cauchy variates is $f(x) = \frac{1}{\pi} \frac{n}{n^2+x^2}$.

7.36 Show that the PDF of square root of a half-standard normal variate (for $z > 0$) is $f(x) = \sqrt{2/\pi}e^{-x^4/2}$, $x > 0$.

7.37 Prove that if $X \sim \text{WEIB}(p, q)$, then $Y = (X/q)^p$ has an exponential distribution.

7.38 If X, Y are IID $\sim U[0, 1]$, find the distribution of $-2\log(XY)$.

7.39 If $X \sim \text{Cauchy}$ find the distribution of $Y = (X - (1/X))/2$.

7.40 A left-truncated exponential distribution with truncation point c has PDF $f(x; \lambda) = \lambda e^{-\lambda x}/[1 - e^{-c}]$ for $x > c$. Obtain the mean and variance.

7.41 Obtain the PDF for a symmetrically both-side truncated Cauchy distribution with the truncation point θ. Does the variance exist?

7.42 Prove that the harmonic mean of n IID Cauchy variates is Cauchy distributed.

7.43 If $X \sim \text{Beta} - I(a, b)$, find the distribution of $1/X$, $(1 - X)/X$, $X/(1 - X)$.

7.44 If X is distributed as $N(0, 1)$, find the distribution of $Y = \exp(X)$.

7.45 Prove that the median and mode of the log-normal distributions are e^μ and $e^{\mu - \sigma^2}$.

7.46 For $N(\mu, \sigma^2)$ distribution, 68.26 of the area lies in the interval $(\mu - \sigma, \mu + \sigma)$, and so on. What are the corresponding intervals for log-normal distribution?

7.47 Verify whether $f(x; c, d) = (1 + x)^{c-1}(1 - x)^{d-1}/[2^{c+d-1}B(c, d)]$ is a PDF for $-1 < x < 1$, where $B(c, d)$ is the complete beta function.

7.48 Consider the Lindley distribution with PDF $f(x) = \theta^2 \frac{1+x}{1+\theta}e^{-\theta x}m^x$, for $x > 0, \theta > 0$. Prove that the mgf is $\frac{\theta^2}{\theta+1}\frac{\theta+t+1}{(\theta+t)^2}$. Find the mean and variance.

7.49 If $X \sim \text{Cauchy}(\mu, \sigma)$ with PDF $f(x; \mu, \sigma) = 1/[\sigma\pi\left(1 + (\frac{x-\mu}{\sigma})^2\right)]$ prove that (i) $2X/(1 - X^2)$ is identically distributed (ii) $1/X \sim \text{Cauchy}(\mu/(\mu^2 + \sigma^2), \sigma/(\mu^2 + \sigma^2))$.

7.50 Prove that the mean and median of Pareto law $f(x; c) = c/x^{c+1}$ are $c/(c - 1)$ and $2^{1/c}$. Find the corresponding mean of power-law using $Y = 1/X$.

7.51 If $X \sim \text{Cauchy}(\mu, \sigma)$ with PDF $f(x; \mu, \sigma) = 1/[\sigma\pi\left(1 + (\frac{x-\mu}{\sigma})^2\right)]$ find the PDF of each of (i) $1/(1 + X^2)$ and (ii) $X^2/(1 + X^2)$.

7.52 If X has a lognormal distribution, find the distribution of $Y = X^n$ for $n \geq 2$.

7.53 Prove that $P[X > nc] = P[X > c]^n$ for an $\text{EXP}(\lambda)$. Evaluate $P[X > c]/\frac{\partial}{\partial c}P[X > c]$.

7.54 Show that for a lognormal distribution, $\mu'_{k+1}/\mu'_k = \exp(\mu + \frac{\sigma^2}{2}(2k + 1))$.

7.55 Verify whether the log-normal distribution satisfies $\text{GM}^2 = \text{AM*HM}$, where GM, AM and HM are the geometric, arithmetic and harmonic means, respectively.

7.56 Prove that the third moment of beta distribution is $\mu_3 = 2(b - a)\mu_2/[(a + b)(a + b + 2)]$.

7.57 Prove that the harmonic mean of n IID Cauchy random variables has PDF $f(x) = n/[\pi(n^2 + x^2)]$.

7.58 Prove that the lognormal distribution is unimodal with mode $\exp(\mu - \sigma^2)$. What is the modal value?

7.59 Prove that the variance of Student's T with n DoF is $1 + (1/(n/2 - 1))$ if $n > 2$, which $\to 1$ as $n \to \infty$.

7.60 If X is CUNI(a, b) find the distribution of $Y = (2X - (a + b))/(b - a)$

7.61 Prove that BETA-I(a, b) can be approximated by $N(0, 1)$ when $a, b \to \infty$ and a/b is constant.

7.62 If $X \sim$ EXP(λ), prove that $c * e^x$ has Pareto distribution.

7.63 Prove that difference of two IID EXP(λ) variates is Laplace$(0, \frac{1}{\lambda})$.

7.64 Show that the constant C of the generalized Cauchy distribution with PDF $f(x; m) = C/(1 + x^2)^m$ is $C = \Gamma(m)/[\sqrt{\pi}\Gamma(m - 1/2)]$.

7.65 Prove that $f(x)/[1 - F(x)]$ is a constant for the exponential distribution, irrespective of its parametric forms.

7.66 Prove that the mean tends to the mode from above for a BETA-II distribution by taking the ratio of mode to the mean as $(1 - \frac{1}{a})(1 - \frac{2}{b+1})$. Show that they coincide as both a, b becomes large.

7.67 If X_i for $i = 1, 2, \cdots, n$ are independent EXP(λ_i) variates where $\lambda_i \neq \lambda_j$ for $i \neq j$, the sum $S = \sum_{i=1}^{n} X_i$ is distributed as

$\sum_{i=1}^{n} C(i, j, n)\lambda_i \exp(-\lambda_i x)$, where $C(i, j, n) = \prod_{j \neq i=1}^{n} \lambda_j/(\lambda_j - \lambda_i)$.

7.68 Prove that the mean deviation of CUNI(a, b) is $\sqrt{3}\sigma/2$, where σ is the standard deviation.

7.69 If X and Y are IID random variables with $f(x) = 1/[\pi\sqrt{(2/b^2) - x^2}]$, where $|x| < \sqrt{2}/|b|$, prove that $(X + Y)/b$ has the same distribution as XY.

7.70 The weight of baggage compartment of a small aircraft is normally distributed with mean 1800~kg and variance 20,64~kg. Find the probability that the baggage compartment weighs (i) $\geq$ 1942~kg, (ii) <1516~kg, and (iii) between 1374 and 2226 kg.

7.71 If X and Y are independent F variates with the same numerator and denominator degrees of freedom n, then prove that $T = \frac{\sqrt{n}}{2}(\sqrt{X} - \sqrt{Y})$ is Student's T distributed with n degrees of freedom.

7.72 Prove that the ratio of variances of Student's T with $(n + 1)$ and n DoF is $(1 + 1/n) * (1 - 1/(n - 1))$ if $n > 2$.

7.73 Prove that the ratio of modal values of Student's T with n and $(n + 2)$ DoF is $(1 - 1/(n + 1)) * \sqrt{(1 + 2/n)}$ if $n > 2$.

7.74 Find first two moments of the arc-sine distribution of second kind with PDF $f(y) = \frac{1}{\pi\sqrt{(1-y^2)}}$ for $-1 < y < 1$.

7.75 What does $B(m/2, n/2) - I_y(m/2, n/2)$ where $y = m/(m +$

nx), $B()$ is the complete beta function and $I_y(a, b)$ is the unskilled IBF represent.

7.76 Show that as $n \to \infty K = \Gamma((n + 1)/2)/[\sqrt{n\pi}\ \Gamma(n/2)] \to 1/\sqrt{2\pi}$.

7.77 Prove that the modal value of lognormal distribution is
$$\frac{1}{\sqrt{2\pi}\sigma \exp(\mu - \sigma^2)} e^{-\sigma^2/2}.$$

7.78 Let X be distributed as lognormal $LN(\mu, \sigma)$. Prove that the mean is $\psi = \exp(\mu + \frac{1}{2}\sigma^2)$. If a change of scale transformation $Y = X/\psi$ is applied, prove that the $(r+1)$th mean of Y is $\exp(\frac{1}{2}\sigma^2 r(r+1)) = E(Y^{-r})$, the rth inverse moment.

7.79 The fraction (by weight) of impurities in a kitchen cleaning liquid has a Beta-I distribution with known parameter $a = 2$. If the average fraction of impurities is 0.18, find the parameter b. What is the variance for the impurities?.

7.80 If the shape parameter of a GAMMA(m, n) is an integer, prove that the survival function can be expressed as a Poisson sum.

7.81 Show that the ordinary moments of lognormal can be found using $\mu'_{r+1} = \mu'_r * \exp(\mu + \sigma^2(r + \frac{1}{2}))$.

7.82 Prove that the quartiles Q_1 and Q_3 of generalized Cauchy distribution are $a - b$ and $a + b$, respectively.

7.83 Prove that the quartiles Q_1 and Q_3 of $U[a, b]$ are given by $x_p = a + p(b - a)$ where $p = 0.25$ for Q_1 and 0.75 for Q_3.

7.84 Prove that area from the median to the mode of lognormal distribution is $\Phi(0) - \Phi(-\sigma)$ where $\Phi()$ is the normal CDF.

7.85 Prove that the area up to the mode of lognormal distribution is $\Phi(-\sigma)$ where $\Phi()$ is the CDF of underlying normal.

7.86 Prove that the mean-median-mode inequality of lognormal distribution is mode<median<mean.

7.87 Prove that the mean–median-mode inequality of central χ^2 distribution is mode<median<mean.

7.88 Find the mean and variance of Laplacian distribution with PDF $f(x; \sigma) = (1/\sigma\sqrt{2}) \exp (-\sqrt{2}|x|/\sigma)$, $-\infty < x < \infty$.

7.89 Prove that the MD of lognormal distribution is $2e^{\mu + \sigma^2/2}[2\Phi (\sigma/2) - 1]$, where $\Phi(x)$ is the CDF of standard normal.

7.90 For which distribution is the mean asymmetric and the variance symmetric in the parameters? (a) negative binomial (b) Beta-I(a, b) (c) binomial (d) hypergeometric

7.91 Find the mean deviation of gamma distribution using the incomplete gamma function (IGF) and Theorem 7.1 as MD $= 2 \int_0^{m/\lambda} F(x)dx$.

7.92 If X is BETA-I(a, b), find the distribution of $X/(1 - X)$ and $(1 - X)/X$.

7.93 If X is BETA-II(a, b), find the distribution of $X/(1 + X)$ and $1//(1 + X)$.

7.94 If $Y \sim T_n$, find the distribution of $(1/2) + (1/2)*y/\sqrt{n + y^2}$.

7.95 If $X \sim N(0, 1)$ prove that the distribution of X^2 is $\phi(\sqrt{x})/\sqrt{x}$.

7.96 The distribution of age (x) of patients to a clinic is given by the frequency function $f(x) = 12x(100 - x)^2/100^4$ for $0 \le x \le 100$. Prove that the modal age is 33.33, and the mean age is 40. Show that the standard deviation is 20 approximately. Find the approximate number of patients between two standard deviations from the mean age. Is the age distribution positively or negatively skewed?

7.97 Prove that the quantiles of exponential distribution is given by $x_p = -\ln(1 - p)/\lambda$. Find an upper bound to $P|X - \mu| \ge k\sigma$ using Chebychev's inequality. Find the hazard rate function $\rho(t) = f(t)/SF(t)$, where $SF()$ is the survival function.

7.98 If $F(x)$ is the CDF of a continuous random variable defined on $[0,1]$, prove the following: –(i) $F(x)$ is $U(0, 1)$ distributed, (ii) $-\ln(F(x))$ has standard exponential distribution, and (iii) $-2\ln(F(x))$ has a chi-square distribution with two DoF.

7.99 Chloride content in kilogram per cubic meter at a distance x (in inches) from the surface of thick concrete floors is distributed as $f(x) = K[1 - erf(x/(2\sqrt{tD}))]$, where t is the time in years and D is the diffusion coefficient (cm^2/sec). Determine K if time is taken as 20 years. Find total chloride up to 3 inches from the surface.

7.100 The annual maintenance cost in 1000 of a high-rise office complex is BETA-I$(1.2, 0.8)$ distributed where $x = (1 - 1/t)$, t being the age (in years) of the building. Find the (i) maintenance cost (in000) for the fifth year, (ii) total maintenance costs for 10 years, and (iii) maintenance cost for first 3 years.

8

MATHEMATICAL EXPECTATION

After finishing the chapter, students will be able to

- Understand the meaning of mathematical expectation
- Find the expectation of sums and functions of random variables
- Derive moments (ordinary, central, factorial) as expected values
- Interpret variance and covariance as expected values
- Explain conditional expectation and independence
- Apply the concepts learned to practical problems

8.1 MEANING OF EXPECTATION

Many location measures were introduced in Chapter 2. These measures concisely summarize the information in a sample as a single number (for univariate data). Analogous measures are needed to succinctly summarize the characteristics of statistical populations. The *population* and *sample space* were defined in Chapter 5. In most of the discussions later, the functional form of the population is known precisely. However, theoretically the concept is valid even in those situations where the exact form is either unknown or is partially known. The concept of the mathematical expectation (or simply called expectation) relies on a *random variable* defined below.

8.2 RANDOM VARIABLE

The concept of random variable is of prime importance in mathematical expectation. Discrete random variables was defined in Section 6.1 of Chapter 6

Definition 8.1 A random variable is a function defined on the sample space of a random experiment that maps each possible outcome of the sample space to real numbers such that the associated probabilities sum to one.

This concept is easy to understand for discrete random variables as the number of points in the sample space are countably finite. Any number of random variables can be defined on a given sample space. These may be related or independent (see Figure 8.3). In every random experiment, there are some numerical values of interest. For example, consider the toss of a die. The possible outcomes are the faces numbered $\{1,2,3,4,5,6\}$, each with probability $1/6$. If X denotes the face that turns up, we express it mathematically as $f(x) = 1/6$ for $x = 1, 2, \ldots 6$. What we have done is to simply assign a mathematical function to each outcome of a random experiment. This is the most common way to define a discrete random variable. For example, $f(x) = q^x p$ is a mathematically defined random variable. There is one more way to define discrete random variables. It is called complete enumeration. Consider the random variable $p(1) = 0.2, p(2) = 0.6, p(3) = 0.2$. Here, x takes three values $\{1,2,3\}$. As the probabilities add up to one, it is a well-defined random variable. This can also be written as $p(x = 1) = 0.2, p(x = 2) = 0.6, p(x = 3) = 0.2$ for a univariate random variable. This notation can be extended to bivariate and higher-dimensional random variables. It is better suited when the sample space is of small size. The individual probabilities can also be defined using a recurrence relation.

Definition 8.2 A probability mass function (PMF) defined on the discrete sample space of a random experiment is a mapping that can be represented as an ordered pair $\{x, f(x)\}$ if for each possible outcome x of the sample space, the following three conditions are satisfied: (i) $f(x) \geq 0 \forall x$ values, (ii) $\sum_x f(x) = 1$, and (iii) $P(X = x) = f(x)$ unambiguously.

Note that in the case of continuous random variables, it is the area in an infinitesimal interval $\int_{x-\epsilon}^{x+\epsilon} f(x) dx$, where $\epsilon = dx/2$ that represents the value of the variable at x (Figure 8.4). This means that $\Pr(x = c) = 0$ for a fixed c. Thus, we have the following definition.

Definition 8.3 A probability density function (PDF) defined on the continuous sample space of a random experiment is a mapping that can be represented as $\{x, \int_{-dx/2}^{dx/2} f(x) dx\}$ satisfying the following conditions: (i) $f(x) \geq 0 \forall x$ values, (ii) $\int_{x=-\infty}^{\infty} f(x) dx = 1$, and (iii) $P(a < X < b) = \int_a^b f(x) dx$, see Figure 8.1.

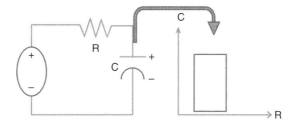

Figure 8.1   Continuous density function takes any value in a range.

⬛ **EXAMPLE 8.1   Check if $f(x)$ is a PDF**

Verify whether a function defined as $f(x; c) = 2cx \exp(-cx^2)$ over $[0, \infty)$ is a PDF.

Solution 8.1 As $\exp(-cx^2)$ takes nonnegative values, the above is positive for $c > 0$, proving (i). Integrate over the range of x to get $2c \int_0^\infty x \exp(-cx^2)dx$. Put $x^2 = t$. The range of integration is the same and we get $c \int_0^\infty \exp(-ct)dt = c[\frac{\exp(-ct)}{-c}|_0^\infty] = 1$, proving condition (ii). Integrate from 0 to y to get the CDF (cumulative distribution function) as $c[\frac{\exp(-ct)}{-c}|_0^{y^2}] = 1 - \exp(-cy^2)$. Now consider $P(a < X < b) = \int_a^b 2cx \exp(-cx^2)dx = 2c \int_a^b x \exp(-cx^2)dx = c \int_{a^2}^{b^2} \exp(-ct)dt$ using the transformation $x^2 = t$. This integrates to $c[\frac{\exp(-ct)}{-c}|_{a^2}^{b^2}] = \exp(-ca^2) - \exp(-cb^2)$. Putting $x = b^2$ and $x = a^2$ in the CDF, we get $(1 - \exp(-cb^2)) - (1 - \exp(-ca^2)) = \exp(-ca^2) - \exp(-cb^2)$, which is the same as above. This proves condition (iii). Hence it is a PDF for $c > 0$.   ⬛

8.2.1   Cumulative Distribution Function

The cumulative probabilities are computed by summing individual probabilities from the lowest possible x value to a higher number (see figure 8.2). The implicit assumption here is that the random variable is arranged in increasing order of possible values of the outcomes. Symbolically $F_x(x) = \Pr[X \leq x]$. Using the summation notation introduced in Chapter 1, this can be written as $F_x(x) = \sum_{k \leq x} p(k)$. For $x = 1, F(x) = p(x) = p(1)$. For $x = 2, F(x) = p(1) + p(2)$, and so on. The CDF is a jump (or step) function if X is discrete (Figure 8.2). In this case, we call it a cumulative probability function. Analogously the right tail probabilities is called the survival function $S(x)$. They are related as $F(x) = 1 - S(x)$. If X is continuous, we use integration instead of summation. Thus, irrespective of the nature of the random variable, we can define the cumulative distribution function as the probability that the random variable X takes values less than or equal to x, where x is a specified number within the range. Obviously the CDF is an increasing function of x. This means that $F(x) - F(x - 1) = p(x)$ for discrete random variables. In the case of continuous random variables, we have (i) $F(x)$ is a nondecreasing function of x, (ii) $\lim_{x \to ll} F(x) = 0$ (i.e., $F(ll) = 0$), (iii) $\lim_{x \to ul} F(x) = 1$ (i.e., $F(ul) = 1$), (iv) $\frac{\partial}{\partial x} F(x) = f(x)$, where ll and

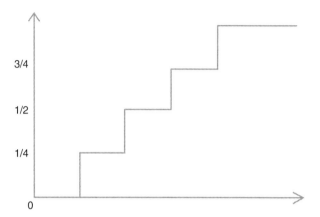

Figure 8.2 Discrete distribution function is a step function.

ul are the lower and upper limits, and (v) $F(x)$ is continuous function of x on the right, with countable number of discontinuities, if any. Property (i) follows trivially due to our implicit assumption that the outcomes of the random experiment are arranged in ascending order of their values. This means that if x_1 is strictly less than x_2, then all sample points that are part of the left tail up to x_1 are automatically part of the left tail up to x_2. Thus, the sum of probabilities up to x_1 is strictly less than that up to x_2. Other properties follow easily due to the increasing nature of $F(x)$, which must eventually equal 1.

As the CDF accumulates probabilities from the left tail, it easily follows that for $b > a$,

$$F(b) = \begin{cases} F(a) + \displaystyle\sum_{k=a+1}^{b} p(k) & \text{if } X \text{ is discrete;} \\[2ex] F(a) + \displaystyle\int_{a}^{b} f(x)dx & \text{if } X \text{ is continuous.} \end{cases}$$

This gives $\Pr[a < X \le b] = F(b) - F(a)$ (Table 8.1). The PDF of a continuous distribution can be obtained from the CDF by differentiation with respect to the random variable, and that of a discrete distribution can be obtained from the CDF by differencing. Symbolically,

$$f(x) = \lim_{\Delta x \to 0} [F(x + \Delta x) - F(x)]/\Delta x = \partial F(x)/\partial x, \tag{8.1}$$

provided the limit exists.

■ **EXAMPLE 8.2 Check if $F(x)$ is a CDF**

Verify whether a function defined as $F(x; c, d) = (1/(d - c))[dx^c - cx^d]$ over $[0, 1]$ is a CDF.

TABLE 8.1   Comparison of Discrete and Continuous Random Variables

Property	Discrete	Continuous	Comment
PMF $f(x)$	Probability mass function	Probability density function	
CDF $F_X(x)$	$\sum_{k=lo}^{x} f(k)$	$\int_{u=lo}^{x} f(u)du$	lo is the lower limit
Sum	$\sum_{x=lo}^{hi} f(x) = 1$	$\int_{x=lo}^{hi} f(x)dx = 1$	lo = lower, ul = upper limit
Partial sum	$\sum_{k=a}^{b} f(x) = F(b) - F(a)$	$\int_{a}^{b} f(x)dx = F(b) - F(a)$	
$f(x)$ from $F(x)$	$f(x) = F(x) - F(x-1)$	$f(x) = \partial F(x)/\partial x$	$f(lo) = F(lo)$ discrete
$x_1 < x_2$	$F(x_1) \le F(x_2)$	$F(x_1) < F(x_2)$	

For the continuous case, $f(x = c) = 0$ for fixed c. It is the area in an infinitesimal interval $\int_{x-\varepsilon}^{x+\varepsilon} f(x)dx$, where $\varepsilon = dx/2$ that represents the value of the variable at x.

Solution 8.2 Differentiate with respect to x to get $f(x; c, d) = (1/(d - c))[cdx^{c-1} - cdx^{d-1}]$. Take cd as a common factor to get $f(x; c, d) = (cd/(d - c))[x^{c-1} - x^{d-1}]$. As $c < d$ and $0 \le x \le 1$, the expression inside the square bracket is always positive. This shows that $F(x)$ is a nondecreasing function of x. As $ll = 0, F(ll) = F(0) = (1/(d - c))[0 - 0] = 0$, showing that $\lim_{x \to ll} F(x) = 0$. As $ul = 1, F(ul) = F(1) = (1/(d - c))[d - c] = 1$, showing that $\lim_{x \to ul} F(x) = 1$. Obviously, $\int_0^1 f(x; c, d)dx = (cd/(d - c))[x^c/c - cx^d/d]|_0^1 = (cd/(d - c))[1/c - 1/d] = 1$. Here, $F(x)$ does not have discontinuities in the interval $[0, 1]$. Hence all the conditions mentioned above are satisfied, proving that $F(x)$ is indeed a CDF. ∎

8.2.2   Expected Value

Measures of location introduced in Chapter 2 refer to a sample. The expected value can be thought of as the arithmetic mean (weighted average) of random variables or populations, where the weights are the probabilities. The mean has a much more broader meaning, as it could be associated with any numeric quantity which may or may not be random. The concept of "expected value" appeared for the first time in the works of Christian Huygens around 1657.

Definition 8.4 Mathematical expectation is used to concisely quantify the mean value of an event, an experiment, a random variable, or a function of it in a population.

It is also called the expected or average value of the argument, or the center of mass in physical sciences. It can be any real number within the range for real-valued random variables. The expected value of integer-valued random variables need not be an integer. It is so called because (i) population may be unknown, (ii) population could be un-enumerable, (iii) it could be a function of the unknown parameters of the population, and (iv) it is just the *expected value* (and not the exact value) of the population under study. It is defined for discrete as well as continuous random variables, and any well-defined functions of them. It is a scalar for univariate populations, and a vector for multivariate random variables. It is denoted as $E(X)$, $E[X]$, or $E\,X$, where E is the *expectation operator*, followed by the argument. The argument (also called the operand, which is usually an expression or a function in capital letters) of "E" can be any well-defined function of X, including integer and fractional powers of X, or conditional distributions. It can also be another expectation expression. Thus, the "E" operator can be recursively nested (Section 8.4, p. 358). In the rest of the chapter, we will simply call it the expected value. In the particular case when the argument is X itself, it is called the population mean.

Definition 8.5 The expected value of a univariate random variable is defined as:

$$
E(x) = \begin{cases} \displaystyle\sum_{k=-\infty}^{\infty} x_k p_k & \text{if } X \text{ is discrete;} \\[2ex] \displaystyle\int_{-\infty}^{\infty} x f(x)\,dx = \int_{-\infty}^{\infty} x\,dF(x) & \text{if } X \text{ is continuous.} \end{cases}
$$

Whenever this sum is *absolutely convergent*, we call it the expected value of X. This means that $\int |x|dF(x) < \infty$ in the continuous case, and $\sum_x |x| f(x) < \infty$ in the discrete case. Although we have indicated the range as $-\infty$ to ∞, the actual range depends on the random variable. For example, as the binomial distribution takes values between 0 and a positive integer n, all expected values of binomial random variables use the summation range from 0 to n. Similarly, as the Poisson and geometric distributions take values between 0 and ∞, the summation is carried out in this range only. Hence the very first step in finding the expectation is to figure out the exact range (Figures 8.3, 8.4).

As p'_ks are probabilities that sum to 1 ($\sum_k p_k = 1$, $\int f(x)dx = 1$), the above sum will almost always converge when $\sum_{k=-\infty}^{\infty} |x_k| p_k < \infty$ or $\int_{-\infty}^{\infty} |x| f(x)dx < \infty$. On occasion this sum may diverge, in which case we say that the expected value does not exist. This seldom happens for random variables defined over a finite range.

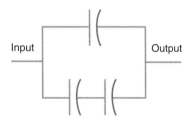

Figure 8.3   Parallel circuit.

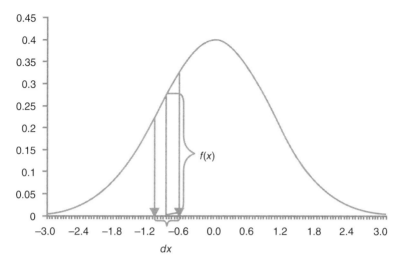

Figure 8.4   Continuous density function.

As mentioned earlier, $E(X)$ is a weighted average of all possible values that the variable can take with corresponding probabilities as the weights. In other words, the notation $E(X)$ can be considered as an *expectation operator*, operating upon the entire range of its argument (which is not shown, but is understood from context). The argument of this expectation operator is quite often a random variable, with the implied meaning that it is the centroid of the argument. As discussed below, the argument could also be any well-defined function of the random variable (e.g., $E(X^2), E(X^{-k})$). If all weights (p_k or $f(x_k)$) are equal (and the range of X is finite), the expected value reduces to the ordinary average (arithmetic mean) of all possible values. For instance, let X take values in the range 1–n and let $p_k = 1/n$ (so that $\sum_k p_k = 1$) then $E(X) = \sum_k p_k x_k = \sum_k k/n = \frac{1}{n}\sum_k k = n(n+1)/[2n] = (n+1)/2$.

⬛ **EXAMPLE 8.3   Expected value of $c/x^2, x \geq 1$**

Find the expected value of the random variable X with PDF $f(x) = c/x^2, x \geq 1$, where c is the normalizing constant.

Solution 8.3 As the total probability must be one, $c\int_1^\infty 1/x^2 dx = 1 \Rightarrow -c[1/x|_1^\infty] = 1$ giving $c = 1$. The expected value is $E(X) = \int_1^\infty (1/x)dx = [\log (x)|_1^\infty] = \infty$. Hence $E(X)$ does not exist. ∎

In many of the examples given below, we are interested in the expected value of random experiments. In such situations, $E(X)$ can be considered as the average value of the experiment, if it is repeated under identical conditions a large number of times.

EXAMPLE 8.4   Promotional coupons in cereals pack

A cereals manufacturer offers a promotional coupon with a new brand of cereals pack. Two types of coupons (that carry either 1 point or 2 points) are printed, and exactly one of them is put in each pack. Probability that a customer will find a 1-point coupon is p, and a 2-points coupon is $q = 1 - p$. If a customer purchases n packs of the cereal, what is the expected number of points earned?

Solution 8.4 As each pack contains a coupon, the minimum score is n and the maximum score is $2n$. These two scores can happen in one way each with respective probabilities p^n and q^n. The customer can score $(n + 1)$ points in n ways (exactly one cereal pack contains a 2-points coupon, and the rest $(n - 1)$ packs contain 1-point coupons) with probability $p^{n-1}q$, so that the probability of score $n + 1$ is $\binom{n}{1}p^{n-1}q$, and so on. Hence the expected score is found by summing the points earned multiplied by the corresponding probabilities as $E(X) = (n * p^n + (n + 1)\binom{n}{1}p^{n-1}q + \ldots + 2n * q^n)$. Separate the first term from each, and use binomial expansion for $(p + q)^n = 1$ to simplify the above as $n + \binom{n}{1}p^{n-1}q + 2\binom{n}{2}p^{n-2}q^2 + \cdots + nq^n$. The "$n+$" term here is a guarantee that the minimum score is n. Using the identity 2 (p. 188) as $i * \binom{n}{i} = n * \binom{n-1}{i-1}$ this can be further simplified as $n + nq(p + q)^{n-1} = n + nq = n(1 + q)$. The maximum score of $2n$ occurs when $q \to 1$. If an equal number of coupons are printed, $q = 1/2$, so that the expected score is $3n/2$. ∎

EXAMPLE 8.5   Coin tossing game

Consider a simple game in which a fair coin is tossed. You win \$100 if the Head turns up. If it is a Tail that turns up, you lose \$90. What is your expected loss or gain in one toss? What is the expected value in n (>2) tosses? Does the expected value converge when a sufficiently large number of trials are conducted?

Solution 8.5 As the coin is fair, $P(\text{Head}) = P(\text{Tail}) = 1/2$. Thus, the expected value in one toss $= (1/2) * 100 - (1/2) * 90 = 50 - 45 = 5$, which is a gain. If this game is repeated n times, our expected gain is $5 * n$. This expected value is divergent as $n \to \infty$.

The expected value can also be a negative number. In the above example, if the winning amount is 90 and losing amount is 100, the expected value is negative. Another example is given below: ∎

EXAMPLE 8.6   Roll of a die

Consider the experiment of rolling a fair die once. If the number that comes on top is an even integer, you win c units of money. If the number on top is odd, you lose $2*c$ units of money. What is your expected gain or loss?

TABLE 8.2   Number of Chicken Hatched in 10 Days

Eggs	42	50	49	50	45	47	48	49	50	48
Chicken	37	43	47	44	40	45	44	42	48	45

Solution 8.6 If the number on top is 2, 4, or 6, the winning amount is c. If it is 1, 3, or 5, the losing amount is $2c$. As the die is fair, each of them have equal probability 1/6. Let X denote the loss or gain. Then $E(X) = -2c * P(X = 1) + c * P(X = 2) - 2c * P(X = 3) + c * P(X = 4) - 2c * P(X = 5) + c * P(X = 6) = (1/6)[-6c + 3c] = -(3/6)c = -c/2$, which is a negative number for $c > 0$.   ∎

▣ EXAMPLE 8.7   Egg hatching

A farmer hatches between 40 and 50 eggs every week. Total number of chickens hatched are given in Table 8.2. Find the expected number of chickens that will be hatched next week if n eggs are kept for hatching.

Solution 8.7 We assume that the eggs are uniformly sampled from a hypothetical population with a constant probability p of hatching. From Table 8.2, we get the probabilities of hatching as $p1 = 37/42 = 0.881, p2 = 43/50 = 0.86, p3 = 0.9592, p4 = 0.88, p5 = 0.8889, p6 = 0.9575, p7 = 0.9167, p8 = 0.8571, p9 = 48/50 = 0.96$, and $p10 = 45/48 = 0.9375$. The expected probability of hatching is found as $E(p) = [0.881 + 0.86 + 0.9592 + .. + 0.9375]/10 = 9.09778/10 = 0.909778$. If n eggs are kept for hatching, the expected number of chickens hatched is $0.909778*n$. The expected values for $n = 40$–50 are [36.39,37.30,38.21,39.12,40.03,40.94,41.85, 42.76, 43.67,44.58,45.49]. As the answer must be a whole integer, we could round-off the above values to the nearest integer.   ∎

8.2.3   Range for Summation or Integration

We have taken the general range for X anywhere on the real line in the above definition. Range of X depends upon the random variable. For example, Poisson and Geometric distributions assume integer values in 0 to ∞, whereas binomial distribution has range 0–n. Thus, the range for summation (or integration) collapses to the range of the random variable or event involved (see Figure 8.4).

▣ EXAMPLE 8.8   Expected value of marks

A multiple choice exam comprises of 50 questions, each with 5 answers. All correctly marked answers score 1 mark, and all wrong answers get a penalty of 1/4 mark. What is the expected number of marks obtained by a student who guesses n questions?

Solution 8.8 Assume that the student has actually obtained x correct and $(n - x)$ wrong answers. Then the marks obtained is $x * 1 - (n - x) * (1/4) = (5/4) * x - n/4 = (5x - n)/4 = Y$ (say). As X_i can take the possible values $0, 1, \ldots, n$ we get the expected value as $E(X) = 1 * 1/5 + 0 * 4/5 = 1/5$. Now $E(Y) = (5E(X) - n)/4 = (5x(n/5) - n)/4 = 0$. Hence the answer is zero. As it is an expected value, the actual score earned when all n questions are attempted could fluctuate around the expected value. ∎

8.2.4 Expectation Using Distribution Functions

Some of the statistical distributions have simple expressions for the CDF. Examples are the exponential, logistic, and extreme value (Gumbel) distributions. The expected value could be found in terms of the distribution functions as follows:

Theorem 8.1 If X is discrete, then $E(X) = \sum_k P(X \geq k)$.

Proof: Without loss of generality, assume that X takes the values 1,2,.. Then

$$E(X) = \sum_k kP(X = k) = 1 * P(X = 1) + 2 * P(X = 2) + 3 * P(X = 3) + \cdots .$$

$$(8.2)$$

Split $2 * P(X = 2)$ as $P(X = 2) + P(X = 2)$, and so on and sum alike terms to get

$$E(X) = [P(X = 1) + P(X = 2) + \cdots] + [P(X = 2) + P(X = 3) + \cdots] + \cdots$$

$$= P(X \geq 1) + P(X \geq 2) + \cdots = \sum_k P(X \geq k). \qquad (8.3)$$

As $E(X \pm c) = E(X) \pm c$ (Section 8.3 in page 348), the result follows for arbitrary range of X. ∎

🔖 EXAMPLE 8.9   Closed form summation of Incomplete Beta Function

Prove that

$$\sum_{k=0}^{n} I_p(k, n - k + 1) = np, \quad \text{where } I_x(a, b) = \frac{1}{B(a, b)} \int_0^x t^{a-1}(1 - t)^{b-1} dt. \qquad (8.4)$$

is the Incomplete Beta Function (IBF) defined in Chapter 7 (p. 277).

Solution 8.9 We have seen in Chapter 6 (p. 208) that if $X \sim \text{BINO}(n, p)$,

$$\sum_{x=k}^{n} \binom{n}{x} p^x q^{n-x} = I_p(k, n - k + 1). \qquad (8.5)$$

Using Theorem 8.1, we have

$$E(X) = \sum_k P(X \geq k) = \sum_{k=0}^{n} I_p(k, n - k + 1) = np. \qquad (8.6)$$

because from Chapter 6, we know that the mean of $BINO(n, p)$ is np. Note that the IBF is a continuous function of p in equation (8.6), which is being summed. The LHS gives exact result only for small n values, as error could propagate for large n values. For example, if $n = 8$ and $p = 0.9$ (equation 8.6) gives 7.2, which is exact. For $n = 12$ and $p = 0.5$, equation (8.6) gives 5.999756, whereas $np = 6$. Use symmetry relation to get a similar expression

$$\sum_{k=0}^{n} I_{1-p}(n - k + 1, k) = n - np = nq, \tag{8.7}$$

which is better suited for $p > 0.5$. Hence the closed form expression (on the RHS) in equations (8.6) and (8.7) is extremely useful to evaluate sums on the LHS, especially when n is large. ∎

EXAMPLE 8.10   Closed form for infinite summation of IBF

Prove that

$$\sum_{j=0}^{\infty} I_q(j + 1, k) = kq/p, \quad \text{where } q = 1 - p. \tag{8.8}$$

Solution 8.10 We have seen in Chapter 6 (p. 228) that if $X \sim NBINO(k, p)$, the lower tail probabilities are found as

$$\sum_{x=0}^{c} \binom{x + k - 1}{x} p^k q^x = I_p(k, c + 1). \tag{8.9}$$

Upper tail probability (SF) is obtained by subtraction as $\sum_{x=c+1}^{\infty} \binom{x+k-1}{x} p^k q^x = 1 - I_p(k, c + 1)$. Using Theorem 8.1, we have

$$E(X) = \sum_{j} P(X \geq j) = \sum_{j=0}^{\infty} [1 - I_p(k, j + 1)] = \sum_{j=0}^{\infty} I_q(j + 1, k) = kq/p. \tag{8.10}$$

because the mean of $NBINO(k, p)$ is kq/p. Using symmetry relation, a similar expression follows easily as

$$\sum_{j=0}^{\infty} I_p(k, j + 1) = 1 - kq/p = 1 + k - k/p. \tag{8.11}$$

Although an infinite sum, rapid convergence of equation (8.10) occurs for p values in the right tail (> 0.5, especially for p above 0.80 or equivalently $q < 0.20$) so that it can be truncated at $j = 2k$, giving the LHS as $\sum_{j=0}^{2k} I_p(k, j + 1)$ for $p > 0.5$. However, the RHS provides a simple and elegant expression, which avoids the expensive evaluation of IBF. ∎

■ **EXAMPLE 8.11** $E(X)$ **in terms of** $F(x)$ **for continuous variates**

If X is a continuous random variable with CDF $F(x)$, then

$$E(X) = \int_0^\infty [1 - F(x)]dx - \int_{-\infty}^0 F(x)dx. \tag{8.12}$$

Solution 8.11 Consider $\int_0^\infty [1 - F(x)]dx$. As $[1 - F(x)]$ is the survival function (right tail area), it can be written as $[1 - F(x)] = \int_x^\infty f(t)dt$. Substitute in the RHS above to get $\int_0^\infty [1 - F(x)]dx = \int_{x=0}^\infty \int_{t=x}^\infty f(t)dtdx$. Change the order of integration to get $\int_{t=0}^\infty \int_{x=0}^t f(t)dxdt$. As $f(t)$ is a constant while integrating with respect to x, the inner integral simplifies to $t * f(t)$. Thus, the RHS becomes $\int_{t=0}^\infty t * f(t)dt$, which is the LHS. Similarly $\int_{-\infty}^0 F(x)dx = -\int_{-\infty}^0 t * f(t)dt$. Combine both results to get the desired answer. ■

The quantiles of a distribution was defined in Chapter 3. This includes the quartiles, deciles, and percentiles. We can express the quantiles using the CDF as follows: The kth quantile is that value of x for which $F(x) = k/100$. In the case of quartiles, the divisor is 4 so that we get 3 quartiles that divide the total area into 25, 50, and 75 points.

Lemma 1 If X is non-negative, prove that $E(X) = \int_0^\infty [1 - F(x)]dx$.
Proof: By definition,

$$E(X) = \int_0^\infty xf(x)dx = \int_0^\infty xdF(x) = \lim_{t\to\infty} \int_0^t xdF(x)$$

$$= \lim_{t\to\infty} \left[xF(x)|_0^t - \int_0^t F(x)dx \right] = \lim_{t\to\infty} \left[tF(t) - 0 - \int_0^t F(x)dx \right]. \tag{8.13}$$

Letting t tend to infinity, this becomes $F(\infty)\lim_{t\to\infty} \int_0^t dx - \int_0^t F(x)dx$. Put $F(\infty) = 1$ to get $\lim_{t\to\infty} \int_0^t [1 - F(x)]dx$. Now let t tend to infinity to get the final results as $\int_0^\infty [1 - F(x)]dx$.

■ **EXAMPLE 8.12** $E(X)$ **and** $E(X^2)$ **of a Poisson distribution**

Find $E(X)$ and $E(X^2)$ for a Poisson random variable.

Solution 8.12 Let $X \sim$ POIS(λ). Then $E(X) = \sum_{x=0}^\infty xe^{-\lambda}\lambda^x/x! = e^{-\lambda}\lambda \sum_{x=1}^\infty \lambda^{x-1}/(x-1)! = e^{-\lambda}\lambda e^\lambda = \lambda$. To find $E(X^2)$, write $X^2 = X * (X-1) + X$. Then $E(X^2) = \sum_{x=0}^\infty x(x-1)e^{-\lambda}\lambda^x/x! + \sum_{x=0}^\infty xe^{-\lambda}\lambda^x/x!$. The $x(x-1)$ in the first sum cancels out with $x!$ in the denominator giving $(x-2)!$. Thus, the first term becomes $\lambda^2 e^{-\lambda} \sum_{x=2}^\infty \lambda^{x-2}/(x-2)!$. Putting $x - 2 = y$, the summation reduces to e^λ giving $\lambda^2 e^{-\lambda}e^\lambda = \lambda^2$. Substitute for the second sum from above to get

$E(X^2) = \lambda^2 + \lambda$. From Chapter 6, page 234, we know that the tail probabilities of a Poisson distribution is related to the incomplete gamma function as follows:

$$F(r) = P[x > r] = \sum_{x=r+1}^{\infty} e^{-\lambda} \lambda^x / x! = \frac{1}{\Gamma(r+1)} \int_0^\lambda e^{-x} x^r dx = P_{r+1}(x). \quad (8.14)$$

∎

Substitute for $P[x > r] = P_{r+1}(x)$ using Section 8.1 to get $E(X) = \sum_k P(X \geq k) = \sum_r P_{r+1}(x)$.

Mathematical expectation can also be defined on events associated with a random variable. Consider the event Y that a Poisson random variable X takes even values. The possible values of Y are $0, 2, 4, \ldots, \infty$. Then $E(Y) = \sum_{y \text{ even}} y e^{-\lambda} \lambda^y / y!$. Put $u = y/2$, so that u takes all integer values starting with 0. As before $E(U) = \sum_{u=0}^{\infty} u e^{-\lambda} \lambda^u / u! = \lambda$, so that $E(Y) = 2\lambda$ (see Example 6.39 in p. 233).

▣ EXAMPLE 8.13

If X is a continuous random variable, and d is a constant, find the unknowns a,b,c such that $E[X - d|X > d] = \int_a^b c[1 - F(x)]dx$, where $b > a$.

Solution 8.13 Consider the probability $Y = P[X - d|X > d]$. Then $P[Y \leq y] = P[X - d|X > d \leq y]$. This probability is the same as $P[X - d \leq y|X > d]$. As X is continuous, this can be written as $P[d < X < y + d|X > d]$. Using the conditional probability formula $P[A|B] = P[A \cap B]/P[B]$ this becomes $P[d < X < y + d]/P[X > d]$. The numerator can be written in terms of unconditional CDF of X as $F_X(y + d) - F_X(d)$ if d is positive, and $F_X(d) - F_X(y - d)$ if d is negative. The denominator being the survival function can be written as $1 - F_X(d)$. Now apply equation 8.10 on the expectation of Y as $E[Y] = \int_0^\infty [1 - G_Y(y)]dy$. Substitute for $G_Y(y)$ from the above to get $E[Y] = \int_0^\infty [1 - [F_X(y + d) - F_X(d)]/[1 - F_X(d)]]dy$. This simplifies to $E[Y] = \int_0^\infty [1 - [F_X(y + d)]/[1 - F_X(d)]dy$. Apply a change of variable transformation $x = y + d$ to get $E[Y] = \int_d^\infty [1 - [F_X(y)]/[1 - F_X(d)]dx$. From this, we see that $a = d, b = \infty$, and $c = 1/[1 - F_X(d)]$. ∎

▣ EXAMPLE 8.14 Variance in terms of $F(X)$

If X is continuous random variable, the variance of X can be expressed as $\sigma_x^2 = \int_0^\infty 2x[1 - F_X(x) + F_X(-x)]dx - \mu_X^2$.

Solution 8.14 As $\text{Var}(X) = E[X^2] - (E[X])^2 = E[X^2] - \mu_x^2$. Substitute for $F(x)$ as in Example 8–11 and proceed as above to get the result. ∎

▣ EXAMPLE 8.15 Expectation of Integer part of Exponential

If $X \sim \text{EXP}(\lambda)$ find $E[\lfloor X \rfloor]$, where $\lfloor X \rfloor$ denotes the integer part of X.

Solution 8.15 We have seen in Chapter 6, page 218, that the integer part of X has a geometric distribution with $q = \exp(-\lambda)$. Using the above method, the problem reduces to finding the mean of GEO(p), where $p = 1 - q = [1 - \exp(-\lambda)]$. Hence $E(Y) = E(\lfloor X \rfloor)$ is $\exp(-\lambda)/[1 - \exp(-\lambda)]$.   ∎

8.3   EXPECTATION OF FUNCTIONS OF RANDOM VARIABLES

In many practical applications, we have to work with simple mathematical functions of random variables. A possible method is to first find the distribution of these functions and then find its expected value. However, the following theorem gives us a simple method to find expected values of functions of random variables without either deriving their distributions or knowing about the exact distributions.

There is another way to find $E[g(x)]$ if $g(x)$ has a well-defined distribution. Instead of using the original PDF inside the summation or integration, we could find the distribution of $Y = g(X)$, and find $E[Y]$ of this distribution. As an example, let $X \sim N(0, 1)$ and $Y = X^2$. We wish to find $\mu_2' = E[X^2] = \int_{-\infty}^{\infty} x^2 \phi(x) dx$, where $\phi(\ )$ is the standard normal PDF. We know that $Y = X^2$ has a central χ^2 distribution with 1 degree of freedom, whose expected value is 1. Hence $E[X^2]$ is also 1. This technique may not always work. In the above example, if we wanted $E[X^2 - 2X + 3]$, we need to resort to the first approach because $X^2 - 2X + 3$ does not have a simple distribution.

8.3.1   Properties of Expectations

Let X and Y be any two random variables, discrete or continuous, univariate or multivariate. In the following discussion, it is assumed that $E(X)$ and $E(Y)$ exist (they are finite).

Theorem 8.2 The expected value of a constant is constant.

Proof: The proof follows trivially because the constant can be taken outside the summation (discrete case) or integration (continuous case) and what remains is either the summation or integration of probabilities that evaluates to a 1.0. Symbolically, $E(c) = c$. Here, c is a scalar constant for univariate distributions, and a constant vector for multivariate distributions. Symbolically, $E(c) = \sum_k c p_{x=k} = c \sum_k p_{x=k} = c$, for the discrete case. If X is continuous, $E(c) = \int_x c p(x) dx = c \int_x p(x) dx = c$.   ∎

Theorem 8.3 The expected value of linear function $c * X$ is c times the expected value of X, where c is a nonzero constant and the expected value exists.

Proof: As above, the constant can be taken outside the summation (discrete case) or integration (continuous case) and what remains is either the summation or integration of X that evaluates to $E(X)$. Applying the multiplier c gives the result as $c * E(X)$.   ∎

Theorem 8.4 Prove that expected value of linear combination $E(a * X + b) = a * E(X) + b$ for any random variable X, and nonzero constant a.

Proof: Let X be discrete, and take values $x_1, x_2, \ldots, x_\infty$. From the definition of expected values, $E(aX + b) = \sum_k (ax_k + b)p_{x_k} = a\sum_k x_k p_{x_k} + b\sum_k p_{x_k} = aE(X) + b$ because $\sum_k p_{x_k} = 1$. If X is continuous, $E(aX + b) = \int (ax + b)p(x)dx = a \int xp(x)dx + b \int p(x)dx = aE(X) + b$. We have not made any assumption on the distribution of the random variable X in this theorem, but only the existence of the first moment. ∎

Corollary 1 $E(c - X) = c - E(X)$. This follows by writing $(c - X)$ as $(-1 * X + c)$ and applying the above theorem with $a = -1$, and $b = c$. Putting $c = 0$, we get $E(-X) = -E(X)$.

▣ EXAMPLE 8.16 $E(n - x)$ of a Binomial

If X has a binomial distribution with parameters n and p (BINO(n, p)), find $E(n - X)$, of a binomial distribution.

Solution 8.16 Write $n - X$ as $Y = (-1) * X + n$, and apply Theorem 8.4 to get $E(Y) = (-1) * E(X) + E(n)$. Substitute $E(X) = np$, and use $E(c) = c$ to get $E(Y) = -np + n = n(1 - p) = nq$. ∎

Theorem 8.5 If X and Y are two random variables, $E(X \pm Y) = E(X) \pm E(Y) = E(Y \pm X)$ and $E(aX \pm bY) = aE(X) \pm bE(Y)$.

Proof: The sum of two random variables X and Y makes sense only when they are compatible random variables. The first result follows trivially by distributing the summation or integration over the individual components (connected by + or −). The second result can be proved using the fact that $E(cX) = cE(X)$, twice. This is called the linearity property of expectation. ∎

8.3.1.1 *Expected Value of Independent Random Variables* We defined independent random variables in Chapter 5 as $P(XY) = P(X) * P(Y)$.

This result is defined in terms of probabilities. As expected values can be considered as functions of random variables with probabilities as weights, we could get analogous results in terms of expected values. Two outcomes are independent if knowing the outcome of one does not change the probabilities of the outcomes of the other. When two events are independent, we find the probability of both events happening by multiplying their individual probabilities.

Definition 8.6 If X and Y are two *independent* random variables, then $E(XY) = E(X) * E(Y)$.

Let X and Y be discrete. Then $E(XY) = \sum_k (x_k y_k)p_{x_k,y_k} = \sum_k x_k y_k p_{x_k} p_{y_k}$ using the above theorem on the independence of probabilities. Pairing x_k with p_{x_k}, this becomes $\sum_k (x_k p_{x_k})(y_k p_{y_k})$. Hence, if X and Y are independent, then $E[XY] = E[X]E[Y]$.

Theorem 8.6 If X and Y are two *independent* random variables, then $P(X \leq x, Y \leq y) = P(X \leq x) * \Pr(Y \leq y)$.

Proof: Let both be discrete. Write the LHS as

$$P[X \leq c, Y \leq d] = \sum_{x \leq c} \sum_{y \leq d} p(x, y) = \sum_{x \leq c} \sum_{y \leq d} p(x)p(y). \tag{8.15}$$

∎

Using the properties of summation mentioned in Chapter 1, split this into two products as

$$P[X \leq c, Y \leq d] = \sum_{x \leq c} p(x) \sum_{y \leq d} p(y) = P[X \leq c] * P[Y \leq d] \tag{8.16}$$

due to the independence of X and Y.

Theorem 8.7 Expected value of a linear combination of scaled functions is the linear combination of expected value of the functions with respective scaling factors. Symbolically, $E(c_1 g_1(x) + c_2 g_2(x) + \cdots +) = c_1 E(g_1(x)) + c_2 E(g_2(x)) + \cdots +$, where the constants c_i's are any real numbers.

Proof: This result follows trivially using theorem and corollaries above. Putting $c_1 = +1$ and $c_2 = -1$, we get $E([g_1(x) - g_2(x)]) = E(g_1(x)) - E(g_2(x))$. ∎

Definition 8.7 If X and Y are two *independent* random variables, and $g(x), h(y)$ are everywhere continuous functions of X and Y defined on the range of X and Y, then $E(g(X) * h(Y)) = E(g(X)) * E(h(Y))$ if the expectations on the RHS exist.

The proof follows exactly as done above using summation in the discrete case and integration in the continuous case.

■ **EXAMPLE 8.17** $E(X^2)$ using $E(X(X - 1))$

Prove that $E[x^2]$ can be found easily using $E[x(x - 1)]$ and $E[x]$ when the denominator of the random variable involves factorials. Find a similar method to find $E[x^3]$. Use this technique to find the expected values of a Poisson random variable.

Solution 8.17 Write $x^2 = x(x - 1) + x$. Use the above result to break the expected value of RHS into two terms to get $E[x^2] = E[x(x - 1)] + E[x]$. Write $x^3 = x(x - 1)(x - 2) + 3x(x - 1) + x$. Take expectation to get $E(x^3) = E[x(x - 1)(x - 2)] + 3E[x(x - 1)] + E[x]$. For a Poisson random variable, $E[x(x - 1)(x - 2)] = \sum_{x=0}^{\infty} x(x - 1)(x - 2) \exp(-\lambda)\lambda^x/x! = \sum_{x=3}^{\infty} \exp(-\lambda)\lambda^x/(x - 3)! = \lambda^3$. Similarly, $E[x(x - 1)] = \lambda^2$. Put these values in the above expression to get $E(x^3) = \lambda^3 + 3\lambda^2 + \lambda = \lambda(\lambda^2 + 3\lambda + 1) = \lambda[(\lambda + 1)^2 + \lambda]$. ∎

Theorem 8.8 If $X_1, X_2, \ldots, X_m$ are independent compatible random variables, then $E(c_1 X_1 + c_2 X_2 + \cdots + c_m X_m) = \sum_i c_i E(X_i)$.

Proof: This follows easily by repeated application of the above result. If $X_1, X_2, \ldots, X_m$ are m compatible random variables, then $E(X_1 + X_2 + \cdots + X_m) = E(X_1) + E(X_2) + \cdots + E(X_m)$, provided each of the expectations exist. ∎

■ EXAMPLE 8.18   Sum of Bernoulli random variables

If $X \sim \text{BER}(p_1)$ and $Y \sim \text{BER}(p_2)$, then $E(X + Y) = p_1 + p_2$.

Solution 8.18 As X and Y are compatible, we apply the above theorem to get the result. This theorem can be extended to any number of random variables.   ■

Theorem 8.9 If $X_1, X_2, \ldots, X_m$ are $\text{BER}(p_i)$, then $E(X_1 + X_2 + \cdots + X_m) = p_1 + p_2 + \cdots + p_m = \sum_{i=1}^{m} p_i$.

Proof: The proof follows easily by taking the expectation term by term. If the probability of success is equal (the same probability p for each of them), then $E(X_1 + X_2 + \cdots + X_m) = mp$.   ■

Theorem 8.10 If $X_1, X_2, \ldots, X_n$ are random variables, each with mean μ, the mean of $\overline{X}_n = (X_1 + X_2 + \cdots + X_n)/n$ is also μ.

Proof: Take $1/n$ as a constant on the RHS and apply the above theorem to get $E[\overline{X}_n] = (1/n) * (E[X_1] + E[X_2] + \cdots + E[X_n]) = (1/n) * n\mu = \mu$.   ■

Theorem 8.11 If $X \leq Y$, then $E[X] \leq E[Y]$.

Proof: For simplicity assume that X and Y have the same range. Consider $Z = Y - X$. As $X \leq Y$, Z is always positive. Hence $E[Z] \geq 0$. This means that $E(Y - X) \geq 0$, or equivalently $E[Y] \geq E[X]$.   ■

As noted above, the sum on the left makes sense only when the random variables are compatible. Any of the "+" can also be replaced by a "−" with the corresponding sign changed on the RHS accordingly.

■ EXAMPLE 8.19   Expected value of functions of Binomial

If X has a binomial distribution with parameters n and p ($\text{BINO}(n, p)$), then (i) $E(X/n) = p$ and (ii) $E(n - X)^2 = nq[n - p(n - 1)] = nq + n(n - 1)q^2$.

Solution 8.19 The first result follows trivially from the above by replacing c with $1/n$, and taking $1/n$ as a constant outside the expectation operator. For (ii) expand the quadratic as $E(n - X)^2 = E(n^2 - 2nX + X^2)$. Take term by term expectation to get $E(n^2) - 2nE(X) + E(X^2))$. Now apply above theorems to get $n^2 - 2n * np + (np + n(n - 1)p^2)$. This simplifies to $nq[n - p(n - 1)]$. Write p as $1 - q$ so that $[n - p(n - 1)] = n - (1 - q)(n - 1)$. The n cancels out giving $1 + q(n - 1)$. Substitute in the above to get $E(n - X)^2 = nq[1 + (n - 1)q] = nq + n(n - 1)q^2$. This result can also be obtained directly from the observation that $n - X$ has $\text{BINO}(n, q)$ (Example 6.11 1, p. 203) so that its second moment is $nq + n(n - 1)q^2$.

■

■ **EXAMPLE 8.20   Expected heads in flips of four coins**

What is the expected number of heads in four flips of a fair coin?

Solution 8.20 Let X be the number of heads. It can take values $0, 1, 2, 3, 4$ with probabilities $P(X = 0) = P(\text{"TTTT"}) = q^4, P(X = 1) = \binom{4}{1} pq^3 = 4pq^3, P(X = 2) = \binom{4}{2} p^2 q^2 = 6p^2 q^2, P(X = 3) = \binom{4}{3} p^3 q = 4p^3 q$, and $P(X = 4) = P (\text{"HH-HH"}) = p^4$. Hence $E(X) = 0 * q^4 + 1 * 4pq^3 + 2 * 6p^2 q^2 + 3 * 4p^3 q + 4 * p^4 = 4[pq^3 + 3p^2 q^2 + 3p^3 q + p^4]$. As the coin is fair, $p = 1/2$. Thus, the required expected number is $\frac{4}{2^4}[1 + 3 + 3 + 1] = 8/4 = 2 = np$.   ■

Theorem 8.12 Prove that $|E[XY]| \leq \sqrt{E[X^2]} \sqrt{E[Y^2]}$.

Proof: Consider the expression $(aX + Y/a)^2 = a^2 X^2 + Y^2/a^2 + 2 * a * (1/a) * XY = a^2 X^2 + Y^2/a^2 + 2 * XY$. As the LHS being a square is always nonnegative, this can be written as $\pm 2XY \leq a^2 X^2 + Y^2/a^2$. Take expectation of both sides to get $\pm 2E[XY] \leq a^2 E[X^2] + (1/a^2)E[Y^2]$. If $E[X^2] > 0$, take $a^2 = \sqrt{E[Y^2]}/\sqrt{E[X^2]}$ to get the RHS as $2 * \sqrt{E[X^2]} \sqrt{E[Y^2]}$. Cancel out 2 from both sides to get the result.   ■

Corollary 2 $|E[X]| \leq \sqrt{E[X^2]}$.

This follows easily by taking $Y = X$.

8.3.2   Expectation of Continuous Functions

Some applications involve functions of random variables. Examples are fractional powers of X, integer powers of X, exponential, logarithmic and trigonometric functions, and other transcendental functions.

■ **EXAMPLE 8.21   Expected value of exp(λX)**

If X is BINO(n, p) find expected value of $\exp(\lambda X)$, where λ is a nonzero constant.

Solution 8.21 By definition $E[\exp (\lambda X)] = \sum_{k=0}^{n} \exp(\lambda k) \binom{n}{k} p^k q^{n-k}$. Combine $\exp(\lambda k)$ with p^k and write this as $\sum_{k=0}^{n} \binom{n}{k} (p \exp(\lambda))^k q^{n-k}$. This simplifies to $(q + pe^\lambda)^n$.   ■

Corollary 3 If $g(x)$ is undefined for at least one value of X, then $E(g(x))$ does not exist. For instance, the first inverse moment $E(1/X)$ is undefined for all random variables that assume a nonzero value for $x = 0$ (i.e., $f(x = 0) \neq 0$).

■ **EXAMPLE 8.22 $E(1/X) \geq 1/E(X)$**

Prove that $E(1/X) \geq 1/E(X)$ for positively defined random variables X.

Solution 8.22 Let μ be the mean of X. Then $E[(X - \mu)(1/X) - 1/\mu] = E[(X - \mu)(\mu - X)/\mu X = -E[(X - \mu)^2/(\mu X) \le 0$. ∎

Definition 8.8 If the function $g(x)$ is everywhere continuous in the range of X, then

$$E(g(x)) = \begin{cases} \displaystyle\sum_{x=-\infty}^{\infty} g(x)f(x) & \text{if } X \text{ is discrete;} \\ \displaystyle\int_{-\infty}^{\infty} g(x)f(x)dx & \text{if } X \text{ is continuous.} \end{cases}$$

Theorem 8.13 $E(g(X, Y)) = \underset{\text{over } x}{E} \left[\underset{\text{over } y}{E} (g(X, Y)|X) \right]$

EXAMPLE 8.23   Expected value of a function

If X is EXP(λ), find $E[e^{-x/2}]$.

Solution 8.23 As X is EXP(λ), $f(x; \lambda) = (1/\lambda)\exp(-x/\lambda)$.
   Hence $E[e^{-x/2}] = \int_0^\infty e^{-x/2}(1/\lambda)\exp(-x/\lambda)dx$. Take the constant outside the integral and combine the exponents to get $E[e^{-x/2}] = (1/\lambda) \int_0^\infty \exp(-x(\frac{1}{2} + \frac{1}{\lambda}))dx$. This evaluates to $2/(\lambda + 2)$. ∎

EXAMPLE 8.24 $E((-1)^x)$ of a Poisson Distribution

If X is POIS(λ), find $E[(-1)^x]$.

Solution 8.24 This follows easily as $E[(-1)^x] = \sum_{x=0}^\infty (-1)^x \exp(-\lambda)\lambda^x/x!$. Using the above theorem, we take the constant outside the summation to get

$$E[(-1)^x] = \exp(-\lambda) \sum_{x=0}^{\infty} (-\lambda)^x/x! = \exp(-\lambda)\exp(-\lambda) = \exp(-2\lambda). \qquad (8.17)$$

∎

Corollary 4 Expected value of a scaled function is the scaling factor times the expected value of the function. Symbolically, $E(c * g(x)) = c * E(g(x))$.

EXAMPLE 8.25   Moments of geometric distribution $q^{x/2}p$

Find the mean and variance of a distribution defined as

$$f(x; p) = \begin{cases} q^{x/2}p & \text{if } x \text{ ranges from } 0, 2, 4, 6, \ldots, \infty \\ 0 & \text{elsewhere.} \end{cases}$$

Solution 8.25 Put $Y = X/2$ to get the standard form. Take expectation of both sides to get $E(Y) = E(X)/2 = q/p$, so that $E(X) = 2q/p$. Similarly, $V(X) = 4$ $V(Y) = 4q/p^2$. ∎

8.3.3 Variance as Expected Value

The variance is a measure of spread in the population. This is captured in a single parameter for normal, Laplace, Gumbel, and some other distributions, but is a function of two or more parameters for others.

Definition 8.9 Variance of a random variable is $\sigma_x^2 = E[(X - E(X))]^2 = E(X^2) - E(X)^2$.

Here, $E(X)$ is the population mean, which we denote by μ_x. If $\mu_x = 0$, the population variance takes the simple form $\sigma_x^2 = E(X^2)$. The above can be expressed as $E[(X - E(X))]^2 = \sum_k (x_k - \mu)^2 p_X(x_k)$ when X is discrete, and $\int_x (x - \mu)^2 f(x) dx = \int_x (x - \mu)^2 dF(x)$ when X is continuous.

8.3.3.1 Properties of Variance

1. $\mathrm{Var}(X) = E[X^2] - E[X]^2$.
   The proof follows trivially by expanding $(x - \mu)^2 = x^2 - 2\mu x + \mu^2$, then taking expectation term by term, and using $E[X] = \mu$ in the middle term (Table 8.4).

2. The variance of independent random variables are additive. Symbolically $V(X + Y) = V(X) + V(Y)$.
   This is known as the additivity property, which is valid for any number of independent random variables. We prove it for two variates X and Y. By definition, $\mathrm{Var}(X + Y) = E[(X + Y) - E[X + Y]]^2$. Use $E[X + Y] = E[X] + E[Y]$ in the inner expectation and combine with $(X + Y)$ to get RHS as $E[(X - E[X]) + (Y - E[Y])]^2$. Expand as a quadratic to get $E[(X - E[X])^2 + (Y - E[Y])^2 + 2(X - E[X])(Y - E[Y])]$. Now take term by term expectation and use $E(X - E[X])(Y - E[Y]) = 0$ (as X and Y are independent, $E(XY) = E(X) * E(Y)$ so that $E(X - E[X])(Y - E[Y]) = 0$) to get the result.

3. $\mathrm{Var}(c * X) = c^2 * \mathrm{Var}(X)$.
   By definition $\mathrm{Var}(c * X) = E[cX - E(cX)]^2$. As c is a constant, it can be taken outside the expectation to get $c^2 * E[X - E(X)]^2$.

4. $\mathrm{Var}(c * X \pm b) = c^2 * \mathrm{Var}(X)$.
   By definition $\mathrm{Var}(c * X + b) = E[cX + b - E(cX + b)]^2$. The $+b$ and $-b$ cancels out giving $E[cX - E(cX)]^2 = c^2 * \mathrm{Var}(X)$. Replace b by $-b$ to get a similar result. The above two results allow us to find the variance of any linear combination of random variables by finding the $\mathrm{Var}(X)$ just once, and doing simple arithmetic with the constants to get the desired result.

5. $\mathrm{Var}(X \pm b) = \mathrm{Var}(X)$.
   By definition $\mathrm{Var}(X + b) = E[X + b - E(X + b)]^2$. The $+b$ and $-b$ cancels out giving $E[X - E(X)]^2 = \mathrm{Var}(X)$. Replace b by $-b$ to get a similar result. This result shows that a change of origin transformation does not affect the variance.

6. $VAR(\sum_{j=1}^{m} X_j) = \sum_{j=1}^{m} VAR(X_j)$ if X_j's are independent.
   This can be proved by induction using the above result 1 (see Table 8.4).

Theorem 8.14 If $X_1, X_2, \ldots, X_n$ are random variables, each of which are pair-wise uncorrelated with the same mean μ and variance σ^2, then the variance of $\overline{X}_n = \sigma^2/n$.

Proof: Consider $Var(\overline{X}_n) = Var((X_1 + X_2 + \cdots + X_n)/n) = 1/n^2 * [Var(X_1 + X_2 + \cdots + X_n)] = 1/n^2 * n\sigma^2 = \sigma^2/n$. ∎

EXAMPLE 8.26   Variance of $Y = n - X$ of binomial distribution

If X has a binomial distribution with parameters n and p, derive the variance of $Y = n - X$.

Solution 8.26 This is already derived in Chapter 6 (p. 203). Here we use the above property to derive it. $Var(Y) = Var(n - X) = Var(n) + (-1)^2 Var(X)$. As the variance of a constant is zero, the RHS simplifies to $Var(X)$. Hence $V(X) = Var(Y) = npq$. This is obtainable directly because $n - X \sim BINO(n, q)$. ∎

EXAMPLE 8.27   Variance of points earned in cereals coupon

In the cereals coupon example 8.4 in page 342 find the variance on the number of points earned.

Solution 8.27 Let X_i denote the event associated with ith packet. Then X_i takes the value 1 with probability p and 2 with probability $1 - p$ so that the expected value is $1.p + 2.(1 - p) = 2 - p$. Write this as $1 + (1 - p) = 1 + q$. If $X \geq 1$ packets are bought, $E(X) = X_1 + X_2 + \cdots + X_n = n(1 + q)$. $E(X^2) = 1^2 * p + 2^2 * q = p + 4q = 1 + 3q$ using $p + q = 1$. From this $V(X_i) = E(X_i^2) - E(X_i)^2 = 1 + 3q - (1 + q)^2 = 3q - 2q - q^2 = q - q^2 = pq$. $V(X) = V(X_1 + X_2 + \cdots + X_n) = npq$. ∎

8.3.4   Covariance as Expected Value

Covariance is a nonstandardized measure of the dependency between the variables involved. We denote it by $Cov(X, Y)$. The order of the variables X and Y is unimportant, as it is symmetric in the variables involved.

Definition 8.10 Covariance of two random variables X and Y is $Cov(X, Y) = E(XY) - E(X)E(Y) = E(X - E[X])(Y - E[Y]) = E(Y - E[Y])(X - E[X])$.

8.3.4.1 Properties of Covariance Covariance satisfies several interesting properties listed below. It is assumed that both X and Y are quantitative.

1. The covariance of two independent random variables is zero
   This follows from the above definition because $E(XY) = E(X) * E(Y)$ when X and Y are independent.

2. The covariance of a random variable with an independent linear combination is additive.
   Symbolically, $\text{Cov}(X + Y, Z) = \text{Cov}(X, Z) + \text{Cov}(Y, Z)$. By definition

$$\text{LHS} = \text{Cov}(X + Y, Z) = E(X + Y)Z - E(X + Y)E(Z). \qquad (8.18)$$

As Z is independent of $X + Y$, $E(X + Y)Z = E(XZ) + E(YZ)$. Also, $E(X + Y) = E(X) + E(Y)$. Substitute in RHS to get $E(XZ) + E(YZ) - [E(X) + E(Y)]E(Z)$. Rewrite this as $[E(XZ) - E(X)E(Z)] + [E(YZ) - E(Y)E(Z)] = \text{Cov}(X, Z) + \text{Cov}(Y, Z)$. Similar results can be derived for $\text{Cov}(X - Y, Z) = \text{Cov}(X, Z) - \text{Cov}(Y, Z)$ (if Z is independent of $X - Y$), $\text{Cov}(X, Y + Z) = \text{Cov}(X, Y) + \text{Cov}(X, Z)$ (if X is independent of $Y + Z$), and $\text{Cov}(X, Y - Z) = \text{Cov}(X, Y) - \text{Cov}(X, Z)$ (if X is independent of $Y - Z$).

3. $\text{Cov}(X, Y) = E(XY)$ when either or both of $E(X)$ or $E(Y) = 0$.
   This follows easily from the definition $\text{Cov}(X, Y) = E(XY) - E(X) * E(Y)$.

4. $\text{Cov}(X, Y) = -\mu_x * \mu_y$ when X and Y are orthogonal.
   This follows from the fact that $E(XY) = 0$ under orthogonality.

5. $\text{Cov}(a * X, b * Y \pm c * Z) = ab * \text{Cov}(X, Y) \pm ac * \text{Cov}(X, Z)$.
   $\text{LHS} = E[a * X - aE(X)][(b * Y \pm c * Z) - (b * E(Y) \pm c * E(Z))]$. Split the second expression into two and combine with the first expression to get $E[a * X - aE(X)][b * Y - E(b * Y)] \pm E[a * X - aE(X)][c * Z - E(c * Z)]$. Take out a, b as constants from the first; and c,d as constants from the second expression to get the result. If $c = 0$, we get $\text{Cov}(a * X, b * Y) = ab * \text{Cov}(X, Y)$.

6. $\text{Cov}((X - a)/c, (Y - b)/d) = \text{Cov}(X, Y)/(cd)$.
   As the change of origin transformation does not affect the covariance, the LHS is equal to $\text{Cov}(X/c, Y/d)$. Now apply above result with $a = 1/c, b = 1/d$ to get the result.

7. If $U = a * X + b * Y$ and $V = c * X + d * Y$, where a, b, c, and d are nonzero constants, then $\text{Cov}(U, V) = a * c\sigma_x^2 + b * d\sigma_y^2 + (a * d + b * c)\text{Cov}(X, Y)$. This result allows us to find the covariance of two arbitrary linear combinations. The proof follows exactly in the same way as above.

8. If $U = \cos(\theta) * X - \sin(\theta) * Y$ and $V = \sin(\theta) * X + \cos(\theta) * Y$, then $\text{Cov}(U, V) = \cos(\theta) * \sin(\theta)[\sigma_x^2 - \sigma_y^2] + [\cos^2(\theta) - \sin^2(\theta)]\text{Cov}(X, Y)$. As θ is a constant, take $a = \cos(\theta), b = -\sin\theta$, and so on and apply the above theorem.

9. $\text{COV}(\sum_{j=1}^{m} X_j, \sum_{k=1}^{n} Y_k) = \sum_{j=1}^{m} \sum_{k=1}^{n} \text{COV}(X_j, Y_k)$ if the X's and Y's are independent. This allows us to find the covariance of sums of random variables.

10. $\text{VAR}(\sum_{j=1}^{m} X_j) = \text{COV}(\sum_{j=1}^{m} X_j, \sum_{j=1}^{m} X_j) = \sum_{j=1}^{m} \sum_{k=1}^{m} \text{COV}(X_j, X_k) = \sum_{j=1}^{m} \text{COV}(X_j, X_j) + \sum_{j=1}^{m} \sum_{k \neq j=1}^{m} \text{COV}(X_j, X_k) = \sum_{j=1}^{m} \text{VAR}(X_j) + 2 \sum_{j=1}^{m} \sum_{k<j} \text{COV}(X_j, X_k)$.

Theorem 8.15 If X_i and Y_i are pair-wise independent, and $U = c_1 * X_1 + c_2 * X_2 + \cdots + c_n * X_n$ and $V = d_1 * Y_1 + d_2 * Y_2 + \cdots + d_n * Y_n$, then $\text{COV}(U, V) = \sum_{k=1}^{n} c_k * d_k * \text{COV}(X_k, Y_k)$.

Proof: Consider $U * V = \sum_{i=1}^{n} \sum_{j=1}^{n} c_i d_j X_i Y_j$. Separate the indexvar into two groups as $i = j$ and $i \neq j$ and write this as $\sum_{k=1}^{n} c_k * d_k * X_k Y_k + \sum_{i=1}^{n} \sum_{j \neq i=1}^{n} c_i d_j X_i Y_j$. Take covariance of both sides to get $\text{Cov}(U, V) = \text{Cov}[\sum_{k=1}^{n} c_k * d_k * X_k Y_k] + \text{Cov}[\sum_{i=1}^{n} \sum_{j \neq i=1}^{n} c_i d_j X_i Y_j]$. As X_i and Y_i are pair-wise independent, the second sum is zero (by using the Theorem 8.8). Taking covariance inside the summation in the first term gives the result. ∎

8.3.5 Moments as Expected Values

The arithmetic mean of a random variable is the first *raw* or *uncentered* moment, which is denoted as $\mu = E(X)$. We call $E(X^k)$ as the kth raw moment and denote it as m_k; and $E((X - \mu)^k)$ as the kth central moment μ_k. Here, μ is called the pivot. Theoretically, the pivot can be any nonzero constant, so long as the expected value exists. By expanding $(X - \mu)^k$ using binomial theorem, it is possible to express the central moments in terms of raw moments as follows:

$$\mu_k = \sum_{j=0}^{k} \binom{k}{j} (-\mu)^{k-j} m_j. \tag{8.19}$$

When k is even, $(X - \mu)^k = (\mu - X)^k$ so that $\mu_k = \sum_{j=0}^{k} (-1)^{k-j} \binom{k}{j} \mu^j m_{k-j}$.

Lemma 2 The change of origin and scale transformation yields $\mu_r(c * X + b) = c^r * \mu_r(X)$.

By definition $\mu_r(c * X + b) = E[cX + b - E(cX + b)]^r$. The $+b$ and $-b$ cancels out giving $E[cX - E(cX)]^r = c^r * \mu_r(X)$. Replace b by $-b$ to get a similar result.

Lemma 3 Prove that $E[cX + b]^r = \mu_r'(c * X + b) = \sum_{k=0}^{r} \binom{r}{k} c^{r-k} b^k E[X^{r-k}]$, where r is a positive integer.

Here, μ_r' denotes the raw moments given by $E[cX + b]^r$. As r is a positive integer, we could expand this as a power series to get $E[\sum_{k=0}^{r} \binom{r}{k} b^k (cX)^{r-k}]$. Take c^{r-k} as a constant outside and operate the expectation on each term to get the result.

8.4 CONDITIONAL EXPECTATIONS

Conditional expectation is a useful concept that defines the expected value of a random variable or function thereof by conditioning one or more dependent variables. Conditional expectation can also be defined in terms of conditional density functions. The conditional expectation considers a non-null subset of random variables by fixing some other random variables as constant.

$$E[Y|X = x] = \begin{cases} \displaystyle\sum_{x=-\infty}^{\infty} yf_y(Y|X = x) & \text{if } Y \text{ is discrete;} \\ \displaystyle\int_{y=-\infty}^{\infty} yf_y(Y|X = x)dy & \text{if } Y \text{ is continuous.} \end{cases}$$

Thus, the conditional expected value of Y for a given value of $X = x$ is the mean of Y computed relative to the conditional distribution, which is a function of x.

Theorem 8.16 Show that $E(Y) = E[E(Y|X)]$ if X and Y are independent.

Proof: For simplicity assume that X and Y are continuous. Consider the RHS. $E[E(Y|X)] = \int_{x=-\infty}^{\infty} E[Y|X = x] f_X(x)dx$. Here we have expanded the outer expectation operator. Next, expand the inner expectation operator to get

$$E[E(Y|X)] = \int_{x=-\infty}^{\infty} \left(\int_{y=-\infty}^{\infty} y f_{X,Y}(x, y)/f_X(x)dy \right) f_X(x)dx.$$

As $f_X(x)$ inside the inner integral is a constant while integrating with respect to y, this cancels out from the numerator and denominator to get

$$\int_{x=-\infty}^{\infty} \int_{y=-\infty}^{\infty} y f_{X,Y}(x, y)dxdy.$$

As X and Y are independent, the density function $f_{X,Y}(x, y)$ factorizes into $f_X(x) * f_Y(y)$. Integrate out $f(x)$ over its entire range to unity, and the remaining expression becomes $\int_{y=-\infty}^{\infty} f_Y(y)dy = E[Y]$, which is the LHS. ∎

Theorem 8.17 Show that $E(XY) = E[E(XY|X)]$.

Proof: As before, assume that X and Y are continuous. Consider the RHS.

$$E[XY|X = x] = \int_{y=-\infty}^{\infty} xy f_{X,Y}(x, y)/f_X(x)dy \qquad (8.20)$$

As x is a constant inside the integral over y, this becomes

$$x \int_{y=-\infty}^{\infty} y f_{X,Y}(x, y)/f_X(x)dy.$$

This integral was shown above as $E[Y|X = x]$. Expand the outer expectation operator on the RHS as $E[E(XY|X)] = \int_{x=-\infty}^{\infty} E[XY|X = x]f_X(x)dx$. Next, expand the inner integral also to get

$$E[E(XY|X)] = \int_{x=-\infty}^{\infty} \left(\int_{y=-\infty}^{\infty} y \, xy \, f_{X,Y}(x, y)/f_X(x)dy \right) f_X(x)dx.$$

As $f_X(x)$ inside the inner integral is a constant while integrating with respect to y, this cancels out from the numerator and denominator to get

$$E[E(XY|X)] = \int_{x=-\infty}^{\infty} \int_{y=-\infty}^{\infty} xy \, f_{X,Y}(x, y) dx dy = E(XY).$$

∎

Lemma 4 Show that $E(Y + Z|X) = E(Y|X) + E(Z|X)$.

Proof: We will prove the result for the continuous case. By the above definition, $E(Y + Z|X) = \int_{y=-\infty}^{\infty} (y + z) f_y(Y|X = x) dy = \int_{y=-\infty}^{\infty} y f_y(Y|X = x) dy + \int_{y=-\infty}^{\infty} z f_y(Y|X = x) dy = E(Y|X) + E(Z|X)$.

∎

■ EXAMPLE 8.28   Expected number of devices in working condition

The lifetime of a device in years is distributed as $EXP(\lambda)$, where $\lambda = 1/8$. If n such devices are put together in a satellite, find the following: (i) probability that half or more of the devices are in good working condition after 5 years. (ii) Expected number of devices in working condition after 8 years.

Solution 8.28 Put $t = 5$ to get the probability that any device is working after 5 years as $\lambda \exp(-5 * \lambda) = 0.0669$. Probability that it is not working is $1 - \lambda \exp(-5\lambda) = 0.9331$. As there are n such devices, probability that at least half of the devices are working is $\sum_{k=n/2}^{n} \binom{n}{k} [0.0669.]^k [0.9331]^{n-k}$. Using the relationship between binomial SF and IBF, this can be written as $I_c(n/2, n/2 + 1)$, where $c = 0.0669$. For case (ii), we need to find the expected value after 8 years. The probability of good working condition is $\lambda \exp(-8\lambda)$. The number of devices in working condition is a binomial variate, so that the expected value is $np = n\lambda * \exp(-8\lambda) = 0.04598493 * n$.

∎

■ EXAMPLE 8.29   Mean of noncentral beta distribution

Find the mean of noncentral beta distribution using conditional expectation.

Solution 8.29 The noncentral beta distribution is an infinite sum of Poisson weighted central beta distributions [7]. Depending on whether the central beta distribution is of first or second kind, there exist noncentral beta distribution (NCB) of two kinds [4]. Symbolically, NCB of first kind has CDF $I_x(a, b; \lambda) \equiv I_x(a + N, b)$, where $N \sim P(\frac{\lambda}{2})$ has a Poisson distribution. Hence conditional on N, the random variable X has a central beta distribution of first kind. From this, an expression for the mean is easily obtained as follows:

$$E(X) = E[E(X|N)] = E[(a + N)/(a + b + N)], \tag{8.21}$$

where we have used the fact that the mean of a beta distribution of first kind is $a/(a + b)$. Write the numerator $(a + N)$ as $(a + b + N) - b$ and simplify to get the

RHS as $1 - b * E[1/(a + b + N)]$. As N is Poisson distributed, the expression in the bracket is the first inverse moment of displaced Poisson distribution, which is given as $E\left(\frac{1}{A+N}\right) = \frac{e^{-\lambda}}{A} \, _1F_1[A, A + 1; \lambda]$, where $A = a + b$, and $_1F_1[A, A + 1; \lambda]$ is the confluent hypergeometric function. ∎

As the numerator and the denominator are dependent, we use the following formula:

$$E[X/Y] \simeq \mu_x/\mu_y \quad [1 + \text{Var}(Y)/\mu_y^2 - \text{Cov}(X, Y)/(\mu_x \mu_y)],$$

where μ_x and μ_y denote the mean of X and Y, respectively. Here, $\mu_x = E[b] = b, \mu_y = E(a + b + N) = (a + b + \lambda/2), \text{Cov}(X, Y) = 0, \text{Var}(X) = 0, \text{Var}(Y) = \text{Var}(a + b + N) = \text{Var}(N)$ (using Section 5, p. 354) $= \lambda/2$. Hence equation (8.21) becomes

$$\mu = E(X) \simeq 1 - (b/C)[1 + \lambda/(2C^2)], \tag{8.22}$$

where $C = a + b + \frac{\lambda}{2}$. Some approximations are given in Table 8.3, where the actual mean is an infinite sum as

$$\mu = \sum_{k=0}^{\infty} e^{-\lambda/2}(\lambda/2)^k/k![(a + k)/(a + b + k)]. \tag{8.23}$$

The difference between actual and approximate values are given in the last column. The results are quite good for increasing noncentrality parameter (λ) values. The biggest advantage of equation (8.22) is that it takes only 10 arithmetic operations (including the computation of C once), whereas equation (8.23) takes a large number of operations when λ is large. See Reference 282 for integral representations of moments.

■ **EXAMPLE 8.30   Mean of noncentral χ^2 distribution**

Find the mean of noncentral chi-square distribution using conditional expectation.

TABLE 8.3   Mean of Noncentral Beta Using Equation (8.22)

a	b	λ	$C = a + b + \lambda/2$	$1 - \dfrac{b}{C} * (1 + \dfrac{\lambda}{2C^2})$	Actual Mean	Difference
1	2	0.5	3.25	0.37005007	0.37300047	0.00295
3	3	2	7	0.56268220	0.56340118	0.000719
10	2	0.5	12.25	0.836462698	0.836482166	0.00002
2	10	4	14	0.278425656	0.278725707	0.00030
10	10	6	23	0.562751705	0.562078005	−0.00067
5	5	20	20	0.743750000	0.743599032	−0.00015
12	10	30	37	0.726768405	0.726752438	−0.00002
20	10	40	50	0.798400000	0.798393998	−0.00001

Solution 8.30 Let Y be distributed as noncentral chi-square. As this is a Poisson weighted central chi-square distribution we write $Y \sim \chi^2_{n+2N}$, where N has Poisson distribution with parameter $\lambda/2$. From this we get $E(N) = \lambda/2$.

$$E(Y) = E[E(Y|N)] = E(n + 2N) = n + \lambda. \tag{8.24}$$

∎

■ **EXAMPLE 8.31   Mean of noncentral F distribution**

Find the mean of noncentral F distribution.

Solution 8.31 The noncentral F distribution is the distribution of the scaled ratio of a noncentral $\chi^2(\lambda)$ over an independent central χ^2 distribution. Symbolically $F(p, q, \lambda) = (q/p)\, \chi^2_{p+2N}/\chi^2_q$, where p and q are the DoF of numerator and denominator χ^2, and λ is the noncentrality parameter.   ∎

This may be written as $Z\, F(p, q, \lambda) \sim \dfrac{(p + 2N)\chi^2_{p + 2N}/(p+2N)}{p\chi^2_q/q}$. Conditional on N, the noncentral F distribution is a multiple of central F distribution [129, 283]. We write this as $F(p, q, \lambda) \sim \dfrac{p + 2N}{p} F_{p+2N,q}$. The moments follow by the same argument as

$$E(Z) = E[E(Z|N)] = E\left[\frac{p + 2N}{p}\ \frac{q}{q - 2}\right]$$

$$= \frac{q}{q - 2}\ \frac{p + \lambda}{p}, q > 2,$$

where we have used the fact that the mean of central $F(p, q)$ is $q/(q - 2)$.

8.4.1   Conditional Variances

The variance of a random variable, conditionally on another variable or on the count of IID (independent identically distributed) random variables occur in several applications. The conditional variance of Y for a given X is $\mathrm{var}(Y|X) = E\{[Y - E(Y|X)]^2 |X\}$. Expanding the quadratic and taking term by term expectation results in $\mathrm{Var}(Y|X) = E(Y^2|X) - [E(Y|X)]^2$ (see Table 8.4).

Theorem 8.18 Let N be an integer valued random variable that takes values ≥ 1. Let $X_1, X_2, \ldots X_N$ be N IID random variables. Define $S_N = X_1 + X_2 + \cdots + X_N$. Then (i) $E(S) = E(X)E(N)$, provided the expectations exist, (ii) $P_{S_N}(t) = P_N(P_X(t))$, and $M_{S_N}(t) = M_N(K_X(t)) = (K_X(t))^N$ due to independence.

Proof: Assume that N is fixed. Then $E(S) = E(X_1 + X_2 + \cdots + X_N) = E(X_1) + E(X_2) + \cdots + E(X_N)$. As each of the X_i's are IID, the above becomes $E(S) = NE(X)$. Now allow N to vary in its range and take the expectation of both sides to get $E(S) = E(N)E(X)$ (because $E(E(S)) = E(S)$ and $E(E(X)) = E(X)$).   ∎

To prove $P_{S_N}(t) = P_N(P_X(t))$, we proceed as above and assume that N is held constant. Then $P_{S_N}(t) = E(t^{S_N}) = E(t^{X_1+X_2+\cdots+X_N}) = E(t^{X_1})E(t^{X_2})\cdots E(t^{X_N}) = E(t^X)^N = [P_X(t)]^N$. Taking expectation of both sides, we get the desired result. Next consider $M_{S_N}(t) = E(e^{tS_N}) = E(e^{t[X_1+X_2+\cdots+X_N]}) = M_{X_1}(t) * M_{X_2}(t) * \cdots * M_{X_N}(t) = [M_X(t)]^N$. Taking log of both sides, we get $K_{S_N}(t) = \log(M_{S_N}(t)) = N * \log(M_X(t)) = NK_X(t)$. Now allow N to vary to get the result.

8.4.2 Law of Conditional Variances

Theorem 8.19 The unconditional variance can be expressed in terms of conditional variances as $V(X) = E[V(X|Y)] + V[E[X|Y]]$ where $V(X) = Variance(X)$, assuming that the variances exist.

Proof: Subtract and add $E[X|Y]$, and write $X - E[X] = (X - E[X|Y]) + (E[X|Y] - E[X])$. Square both sides and take expectation of each term to get

$$E[X - E[X]]^2 = V(X) = E(X - E[X \mid Y])^2 + E(E[X \mid Y] - E[X])^2$$
$$+ 2E(X - E[X \mid Y])(E[X \mid Y] - E[X]) = (1) + (2) + (3) \text{ (say)}.$$
$$(8.25)$$

As $E(E[X|Y]) = E(X)$, the last term (3) is zero. Substitute $E(X) = E(E[X|Y])$ in the second term $E(E[X \mid Y] - E[X])^2$ to get $(2) = E(E[X \mid Y] - E(E[X|Y]))^2$. As this is the expectation of the squared deviation of $E[X \mid Y]$ from its mean, it is $\text{Var}(E[X|Y])$. Symbolically, $(2) = \text{Var}(E[X|Y])$. ∎

Using the law of total expectation we have $V(X|Y) = E[X^2|Y] - E[X|Y]^2$. Take expectation of both sides to get $E[V(X|Y)] = E\{E[X^2|Y]\} - E\{E[X|Y]^2\}$. Write the first term $E(X - E[X \mid Y])^2$ in equation (8.25) as $E\{E(X - E[X \mid Y])^2|Y\}$. Expand the quadratic and take term by term expectation to get $E\{E[X^2|Y]\} - 2E\{E(X)E[X|Y]\} + E\{E[X|Y]^2\}$. Substitute $E(X) = E(E[X|Y])$ in the second term, and cancel out the third term. This reduces to $E\{E[X^2|Y]\} - E\{E[X|Y]^2\}$ showing that it is the expected value of $V[X|Y]$. Symbolically, $(1) = E[V(X|Y)]$. Substitute for (1) and (2) in equation (8.25) to get the result.

⬛ **EXAMPLE 8.32 Variance of noncentral chi-square**

Find the variance of noncentral chi-square distribution using conditional expectation.

Solution 8.32 We know that the noncentral chi-square distribution is a Poisson weighted linear combination of independent central chi-square distributions. This allows us to write it as $Y \sim X + 2N$, where conditional on N, X is a central chi-square distribution. For convenience let the DoF of central chi-square be denoted by p and N has POIS($\lambda/2$). Then

TABLE 8.4 **Summary Table of Expressions for Variance**

Using	Expression	Comment
Definition	$EX^2 - E(X)^2$	$E(X)^2, E(X)$ finite
$\text{Var}(Y\|X)$	$E(Y^2\|X) - [E(Y\|X)]^2$	Conditional on X
$V(X)$	$E[V(X\|Y)] + V[E[X\|Y]]$	Unconditional and conditional
CDF	$\int_0^\infty 2x[1 - F_X(x) + F_X(-x)]dx - \mu_X^2$	
$P_x(t)$	$P_x''(1) + P_x'(1) - [P_x'(1)]^2$	PGF
$M_x(t)$	$M_x''(0) - [M_x'(0)]^2$	MGF
$K_x(t)$	$K_x''(0)$	$K_x'(0) = \mu$
$\ln(F_X(x))$	$[\ln(F_X(x))]'\|_{x=1} + [\ln(F_X(x))]''\|_{x=1}$	F is CDF
$\phi_x(t)$	$\phi_x''(0) - [\phi_x'(0)]^2$	ChF
$\text{FM}_x(t)$	$\text{FM}_x''(0) + \text{FM}_x'(0) - [\text{FM}_x'(0)]^2$	FMGF

See Section 9.2.1 in page 9–9 and Wilf [284].

$$V(Y) = V[E(Y|N)] + E[V(Y|N)] = V(p + 2N) + E(2p + 4N)$$

$$= 4V(N) + 2p + 4E(N) = 2p + 4\lambda, \qquad (8.26)$$

where we have used the facts that $V(c + b * X) = b^2 V(X)$ and $E(X) = V(X) = \lambda$ for a Poisson distribution [7, 283] (Table 8.4). ∎

8.5 INVERSE MOMENTS

The definition of ordinary moments can be extended to the case where the order is a negative integer as follows:

$$E(1/x^k) = \begin{cases} \displaystyle\sum_{j=-\infty}^{\infty} x_j^{-k} p_j & \text{if } X \text{ is discrete}; \\ \displaystyle\int_{-\infty}^{\infty} x^{-k} f(x)dx & \text{if } X \text{ is continuous.} \end{cases}$$

A necessary condition for the existence of the first inverse moment is that $f(0) = 0$. For instance, the Poisson distribution has $p(x = 0) = e^{-\lambda}\lambda^0/0! = e^{-\lambda}$, which is nonzero $\forall\lambda$. Hence the first inverse moment does not exist. However, there are a large number of distributions that satisfy the necessary condition. Examples are chi-square (and gamma), Snedecor's F, beta, and Weibull distributions. The exponent k is an integer in most of the applications of inverse moments. However, inverse moments could also be defined for fractional k (called fractional inverse moments).

■ EXAMPLE 8.33 Inverse moment of central χ^2 distribution

Find first inverse moment of central χ^2 distribution.

Solution 8.33 By definition, $E(1/X) =$

$$K \int_0^\infty \left(\frac{1}{x}\right) x^{n/2-1} e^{-x/2} dx = K \int_0^\infty x^{n/2-2} e^{-x/2} dx = K\Gamma(n/2-1)2^{n/2-1}, \quad (8.27)$$

where $K = 1/(2^{n/2}\Gamma(n/2))$. This simplifies to $1/(n-2)$. ∎

8.6   INCOMPLETE MOMENTS

Ordinary and central moments discussed above are defined for the entire range of the random variable X. There are several applications when the summation or integration is carried out partially over the range of X. The omitted range can either be in the left tail or in the right tail. We define the first incomplete moment as $E_I(X) = \sum_{x=k}^\infty xf(x)$.

8.7   DISTANCES AS EXPECTED VALUES

Statistical distances can be expressed as expected values. Consider two real-valued random variables X and Y. The k-norm distance between them is $D_k(X, Y) = \|X - Y\|^k = [E(|Y - X|^k)]^{1/k}$. This is also called the k-metric. It satisfies the following properties:

(i) $D_k(X, Y) \geq 0$, (ii) $D_k(X, Y) = 0$ iff $X = Y$, (iii) $D_k(X, Y) + D_k(Y, Z) \geq D_k(X, Z)$ (triangle inequality).

Particular values of k give various distances such as Euclidean metric, Manhattan metric, and so on [22]. The sample analogs of these distances are used in cluster analysis as dissimilarity metrics. The above definition can be extended from scalar random variables to vectors and matrices. For instance, if X is an $m \times n$ matrix of real-valued random variables, where X_{ij} denotes the (i, j)th entry, we define $E(X)$ as that matrix whose (i, j) entry is $E[X_{ij}]$, provided the individual expectations exist. Using matrix commutativity, associativity, and so on with respect to addition, we could obtain the following results:

(i) $E(X + Y) = E(X) + E(Y)$ if X and Y are compatible matrices, (ii) $E(AX) = AE(X)$ if A is a scalar $m \times n$ matrix and X has as many rows as columns of A matrix (i.e., A is $n \times p$), and (iii) if X and Y are independent, then $E(XY) = E(X)E(Y)$ (Table 8.5).

8.7.1   Chebychev Inequality

This is a useful result connecting the expected value of a function of a random variable and the tail area of it. Let X be a random variable and $g(x)$ be a nonnegative function of it. Then the right tail area of $g(X)$ is related to its expected value as $P[g(X) \geq c] \leq E[g(X)]/c$.

TABLE 8.5   Summary of Mathematical Expectation

Function	Name	Conditions
$E[X^k]$	kth raw moment	k is real, finite
$E[X^{-k}]$	kth inverse moment	k real, $f(x = 0) = 0$
$E[(X - \mu)^k]$	kth central moment	k is real, finite
$E\lvert X - \mu \rvert$	Mean deviation (about mean)	μ finite
$E[X(X - 1) \cdots (X - k + 1)]$	kth falling factorial moment	$k \geq 1$ is real, finite
$E[X(X + 1) \cdots (X + k - 1)]$	kth raising factorial moment	$k \geq 1$ is real, finite
$EX^2 - E(X)^2$	Variance	$E(X)^2, E(X)$ finite
$E[(X - \mu_x)(Y - \mu_y)]$	Covariance	μ_x, μ_y finite
$E(c_1 g_1(x) + c_2 g_2(x) + \cdots)$	Linear combination	c_i's $\neq 0$
$E(e^{tx})$	Moment generating function	$-\varepsilon < t < \varepsilon, \varepsilon > 0$
$E(e^{itx})$	Characteristic function	$-\varepsilon < t < \varepsilon, \varepsilon > 0$
$E(t^x)$	Probability gen. function	$-\varepsilon < t < \varepsilon, \varepsilon > 0$
$E[(1 + t)^x]$	Falling factorial mgf	$-\varepsilon < t < \varepsilon, \varepsilon > 0$
$E[(1 - t)^{-x}]$	Raising factorial mgf	$-\varepsilon < t < \varepsilon, \varepsilon > 0$
$E(XY)^2 \leq E(X^2)E(Y^2)$	Cauchy–Schwartz inequality	
$E(g(x)) \geq g(E(x))$	Jensen's inequality	$g(x)$ is convex

See Section 9.2.1 in page 9–9.

8.8   SUMMARY

This chapter introduced the basic ideas and rules of both the mathematical expectation and conditional expectation, see Table 8.5. Mathematical expectation plays an important role in digital signal processing, actuarial sciences, astronomy, and many other fields. For example, the average energy $\omega(t)$ of a periodic or random signal in the time domain is represented for continuous signals as $\omega(t) = \int_{-\infty}^{\infty} f(t)dt$, from which the average power of the signal over a time period t_1 to t_2 is given by

$$E[P] = \frac{1}{t_2 - t_1} \int_{t_1}^{t_2} f_1(t)f_2(t)dt = \frac{1}{T} \int_0^T f^2(t)dt, \qquad (8.28)$$

if $f_1(t) = f_2(t)$, where $f(t)$ represents the signal value as a time-varying function. As the spectra of periodic signals are more revealing in the frequency domain, most DSP applications use one of the frequency transforms such as Fourier transform, cosine transform, wavelet transform, and so on under the assumption that $\int_{-\infty}^{\infty} \lvert x_T(t) \rvert dt < \infty$ where $\lvert x_T(t) \rvert$ emphasizes that it is a random variable in the time-domain. The average power in the frequency domain can then be represented by expected value as $E[\lvert X_T(f) \rvert] = \frac{1}{2T} \int_{-\infty}^{\infty} \lvert X_t(f) \rvert^2 df$. As $T \to \infty$, $E[\lvert X_T(f) \rvert]$ will stabilize for stationary processes and signals, resulting in power spectral density of the signal. See references 285, 286 for further examples.

EXERCISES

8.1 Mark as True or False

a) Expected value of a random variable always exist

b) Expected value is unchanged by a change of scale transformation

c) Chebychev inequality can provide an upper bound on expected values

d) Variance of a distribution defined in $[0, 1]$ can be > 1

e) $E(1/X) \geq 1/E(X)$ for all random variables X.

f) $|E[X]| \leq \sqrt{E[X^2]}$ for all random variables X

g) If $X \leq Y$, then $E[X] \leq E[Y]$

h) Expectation "E" is a linear and monotone operator.

8.2 A——assigns a value to each element of the sample space: (a) generating function (b) random variable (c) cumulant (d) expected value.

8.3 If $X_1, X_2, \ldots, X_n$ are IID, and $Y = \sum_i X_i$ prove that $M_Y(t) = \prod_i M_{X_i}(t)$.

8.4 If $p(x=0)=q^2, p(x=1)=2pq, p(x=2)=p^2$, find the CDF and PGF

8.5 Express $\Pr[c - \delta < x < c + \delta]$ in terms of CDF. For what value of c is this area maximum for bell-shaped curves?.

8.6 The current flow I through a resistor fluctuates according to arcsin law. Find the expected value of the Power $= R * I^2$, where R is the resistance (given).

8.7 Prove that $c = E(X)$ minimizes the expression $E(x - c)^2$

8.8 Prove that $c = \text{Median}(X)$ minimizes the expression $E|x - c|$.

8.9 Show that all cumulants except the first one vanish for a symmetric distribution.

8.10 If X is a negative random variable (values of x are always < 0), prove that $E(X) = \int_0^\infty -F(-x)dx$.

8.11 If X is uniformly distributed in $[a, b]$ find the expected value of $\exp(-\lambda X)$.

8.12 If X is a nonnegative discrete random variable, prove that $E(X^2) = \sum_{k=1}^\infty (2k + 1)P(X > k)$

8.13 Prove that $[E(XY)]^2 \leq E(X)^2 * E(Y)^2$ if X and Y are real-valued.

8.14 If $X > 0$, prove that $E[1/X] \geq 1/E[X]$, if each expectation exists.

8.15 If $f(x; a, b) = Cx^{a-1}e^{-(x/b)^a}$ prove that $C = a/b^a$. Find the CDF and MGF, and obtain the first two moments where $x > 0$.

8.16 Consider a discrete random variable $p(x = k) = 4/(\pi^2 k^2)$ for $k = 1, 2, \ldots$. Verify whether the expected value exists.

8.17 If $X \sim \text{BETA}(a, b)$, where $a < b$, find $E[[X(1 - X)]^k]$

8.18 Prove that $\text{COV}(X, Y - Z) = \text{COV}(X, Y) - \text{COV}(X, Z)$.

8.19 For the Geometric distribution with PDF $f(x) = q^{x-1}p$, find $E(X)$

and $E(X^2)$ and deduce Var(X). What is the MD?

8.20 Consider a game in which a fair die (marked 1–6) is thrown and the player losses k dollars if the top point k is odd, and gains 2k dollars if it is even. Find $E(X), V(X)$.

8.21 For any continuous distribution, prove that $E|X - c|$ is minimum when c is the median.

8.22 If $f(x, y) = x + y$ for $0 < x < 1, 0 < y < 1$, find $E(Y|X)$ and $E(X|Y)$.

8.23 Let X be a random variable that denotes the sum of the numbers that shows up when two dice are thrown. Define the PDF of X and find its expected value $E(X)$. Does $E(E(X))$ exist?

8.24 Find the normalizing constant K in $f(x) = K/(x + c)^{n+1}$, where n is an integer and c is a real constant. Prove that all ordinary moments of order up to $n - 1$ are non-existent for this distribution.

8.25 When is $Cov(X, Y) = -\mu_x \mu_y$?

8.26 If X,Y are IID $\sim EXP(\lambda)$, find $E(\frac{\sqrt{X}+\sqrt{Y}}{2})^2$

8.27 Prove that $M_{ax}(t) = M_x(at)$ and $M_{ax+b}(t) = e^{bt}M_x(at)$. Deduce that $M_{(x-\mu)/\sigma}(t) = e^{-\mu t/\sigma}M_x(t/\sigma)$.

8.28 Prove that the expected value of MD is $E(\text{MD}) = \delta\sqrt{1 - \frac{1}{n}}$, where n is the sample size and δ is the population MD.

8.29 If $f(x) = Kx\exp(-x)$ for $x \geq 0$, find $K, E(X)$, and the MGF.

8.30 If $f(x) = K\exp(-|x|)$ for $-\infty < x < \infty$, find $K, E(X)$, and the MGF.

8.31 Show that $E(cY|X) = c * E(Y|X)$, where c is a constant.

8.32 If $X \sim EXP(\lambda)$, find $E[\sqrt{X}]$

8.33 If $f(x) = \binom{x-1}{r-1}p^r q^{x-r}, x = r, r + 1, \ldots$, find the MGF and derive the first two moments. What is the survival function?

8.34 Prove that $\sigma^2 = E(X(X - 1)) + E(X) - [E(X)]^2$. Apply it to find the variance of Geometric and Poisson distributions.

8.35 For the Poisson distribution (POIS(λ)), prove that

$$\sum_{x=0}^{|\lambda|} e^{-\lambda}\lambda^x/x! \begin{cases} \leq \dfrac{1}{2} & \text{if } \lambda \text{ is large;} \\[2mm] \geq \dfrac{1}{e} & \text{otherwise.} \end{cases}$$

Use this result to derive the first incomplete moment $\sum_{x=0}^{|\lambda|} xe^{-\lambda}\lambda^x/x!$.

8.36 Prove that $E(X - k)^2 = \sigma^2 + [E(X) - k]^2$

8.37 Find $E(\frac{1\pm X}{1+Y})$ and $E(\frac{1\pm X^2}{1+Y^2})$ for the random variable in Q 8.22.

8.38 Find $E((1 + \frac{x}{n})^n)$ for the binomial distribution with n trials, where x is the number of successes. What is the limit of this expectation as $n \to \infty$?

8.39 Find the MGF of logarithmic series distribution $f(x) = c\theta^x/x, x = 1, 2, \ldots$ where $c = -1/\log(1-\theta)$, and $0 < \theta < 1$. Prove that rth factorial moment $\mu_{(r)} = c(r-1)![\theta(1-\theta)]^r$.

8.40 Find the expected value for a binomial distribution BINO(n, p).

8.41 If $X \sim \text{BETA}(a, b)$, prove that $E(X) = a/(a+b)$.

8.42 If $X \sim \text{BETA}(a, b-a)$, prove that $M_x(t) = {}_1F_1(a, b; x) = \frac{\Gamma(b)}{\Gamma(a)\Gamma(b-a)} \int_0^1 e^{xt}t^{a-1}(1-t)^{b-a-1}dt$ is the confluent hypergeometric function. Hence show that $E(X) = a/b$.

8.43 Find K to make the following functions a PDF. Then find $E(X)$ and $E(X^2)$ (a) $f(x) = Kx^2(1-x)$ for $0 < x < 1$, (b) $K/x^{a+1}, a > 1, x > 0$, (c) $f(x) = K(X^2+1)$ for $x \in \{-2, -1, 0, 1, 2\}$.

8.44 If $f(x) = 1/\pi$ for $0 < x < \pi$, show that $E[\sin(x)] = 2/\pi$.

8.45 If $X > 0$, find constants a, b, c such that $E[X - t | X > t] = c \int_a^b [1 - F(x)]dx$.

8.46 Prove that $E[X^2] = \sum_{k=0}^{\infty}(2k+1)P[X > k]$

8.47 What is the expected value of an indicator variable?

8.48 If $X \sim \text{BINO}(n, p)$ find $E(\frac{x-1}{x+1})$.

8.49 Find $\sum_{x=0}^{np} \binom{n}{x} p^x q^{n-x}$.

8.50 The PDF of a discrete random variable is given by $f(x) = K(|x| + 1)$ for $x = -3, -2, -1, 0, 1, 2, 3$. Find K and the CDF. Evaluate $F(2)$ and $P[X \geq 0]$.

8.51 The PDF of a discrete random variable is given by $f(x) = cx^2$ for $x = \{1, 2, 3\}$. Find the mean and variance.

8.52 Verify if $f(x) = \frac{1}{\sqrt{2}\sigma}e^{-|x|/(\sigma/\sqrt{2})}$ is a PDF. Find the MGF and $E(X)$.

8.53 If $f(x) = Kx(x+1)$ for $x = 1, 2, 3, 4$; find the $E[X]$ and $P[X \geq 2]$.

8.54 If $f(x) = K/2^{x-1}$ for $x = 1, 2, 3, 4$; find K, $E[X]$ and the probability that $X \geq 2$.

8.55 Find the mean and variance for the distribution $f(x) = (1 - \frac{\mu}{n})^n \binom{n}{x}(\frac{\mu}{n-\mu})^x, x = 0, 1, \ldots n$.

8.56 If $f(x, y) = C(x+y)$ for $0 < x < y < 1$, find C. What is the value of $E(X^2 + Y^2)$?

8.57 If X and Y are independent random variables with $E(X) = -3$ and $E(Y) = 5$, find $E(2X-3)(Y+5)$.

8.58 If $\phi(x)$ is a real-valued, monotonic function of a positive random variable X, prove that $E[\phi(x)] = \phi(0) + \int_0^\infty [1 - F(x)]\partial\phi(x)/\partial x$. Hence derive that $E[X^n] = n\int_0^\infty x^{n-1}[1 - F(x)]dx$.

8.59 What are the conditions for a function to be a moment generating function? Are the following functions true MGF? (a) $e^{a(t-1)+b(t-1)^2}$, (b) $e^{a(t-1)/(1-bt)}$, (c) $e^{a(t-1)+b(t^2-1)}$, (d) $e^{|(t-1)/(t^2-1)|}$.

8.60 Suppose an urn contains m red balls and n blue balls. If r balls are drawn with replacement, what is expected number of blue balls drawn?

8.61 Suppose an urn contains n coins numbered $1 - n$. If r coins are drawn without replacement, what is the expected sum of the numbers?

8.62 If X and Y are IID distributed as lognormal (μ, σ^2), (i) find $E[XY]$ and (ii) approximate value of $E[XY \log(XY)]$.

8.63 If X is a discrete symmetric random variable $(P[X = k] = P[X = -k])$, find expected value of $\sin(\pi X)$.

8.64 Prove that $\mathrm{COV}(\overline{X}, X_i - \overline{X}) = 0$ for any random sample.

8.65 Prove that $\mathrm{COV}(\overline{X}, X_i) = \sigma^2/n$ for any random sample.

8.66 Use $x^2 = x(x - 1) + x$ and $x^3 = x(x - 1)(x - 2) + 3x(x - 1) + x$ to

find the expected values of a Poisson random variable. What is the expression to find $E[x^4]$?

8.67 Find the mean and variance of the distributions: $(i) f(x, n) = (n/2) \sin (nx), 0 \le x \le \pi/n, n > 0$ is real; $(ii) f(x, n) = (n/2) \cos(nx), -\pi/2n \le x \le \pi/2n, n > 0$ is real.

8.68 Prove that $\sigma^2_{Y|X} = \sum_k y_k^2 f(y|x) - \mu^2_{Y|X}$.

8.69 If the second derivative of $h(x)$ is positive, prove $E[h(x)] \ge h(E[x])$.

8.70 If X is continuous, prove that $\mu'_r = \int_0^\infty rx^{r-1}[1 - F_X(x) + (-1)^r F_X(-x)]dx = E(X^r)$.

8.71 Find the expected values of x^2 and x^3 in a random experiment of tossing a fair die.

8.72 Prove that the factorial moments for the following distributions are as given:

BINO$(n, p) : E[X(X - 1)..(X - r + 1)] = n_{(r)}p^r$.

HYPG$(N, n, p) : n_{(r)}Np_{(r)}/N_{(r)}$ if $f(x) = \binom{Np}{k} \binom{Nq}{n-k} / \binom{N}{n}$

GEOM$(p) : r!q^{r-1}/p^r$

NBINO$(n, p) : r(r + 1)..(r + s)(q/p)^s$.

8.73 If $X_1, X_2, \ldots, Xn$ are independent identical random variables with the same Mean μ, and same variance σ^2, find the expected value and variance of the arithmetic mean of $x'_i s'$.

8.74 If X is a nonnegative continuous random variable, and $Y = \exp (X^2)$ find the expected value of Y. Find $E(y)$ if y is $U(0.1)$.

8.75 Find the MGF and first three moments of the geometric distribution $f(x) = (1 - e^{-\lambda})e^{-\lambda x}$.

8.76 If X is a real-valued continuous random variable, prove that $E[X^2] = 2 \int_0^\infty x[1 - F(x) + F(-x)]dx$

8.77 The number of MMS messages arriving in Emily's cell phone between 9 AM and 5 PM is Poisson distributed with $\lambda = 1$ for a 10-min interval. What is the expected number of MMS messages received in 1 h? What is the total expected number of messages she will receive between 1 PM and 5 PM?

8.78 The mean-excess function of a variate is defined as $\Delta[X] = (1/S(x)) \sum_u^\infty (1 - F(x))$ if X is discrete, and $\Delta[X] = (1/S(x)) \int_u^\infty (1 - F(x))dx$ if X is continuous, where $S(x) = 1 - F(x)$ is the survival function. Find $\Delta[X]$ of Poisson and exponential variates.

8.79 Suppose $X_1, X_2, \ldots, X_n$ are IID random variables, with $E(X_i) = \mu$ and $\text{VAR}(X_i) = \sigma^2$. Define $Y = X_1 + X_2 + \cdots + X_N$, where N is another random variable independent of X. Prove that $E(Y|N) = N\mu$, and $\text{Var}(Y|N) = N\sigma^2$. Use $E(Y) = E[E(Y|N)]$ to show that $E(Y) = \nu\mu$, where $\nu = E(N)$. Show that the unconditional variance of Y is $\nu\sigma^2 + \mu^2\delta^2$ where δ^2 is variance of N.

8.80 When a cell phone is powered on, it is registered with a base station. Each base station has a "cell" which is the coverage region (say a circular or square region) around it. When the caller moves from place to place, they may move out of one region and into an adjacent region. The phone company automatically detects it and "hands over" the phone identity to the new base station. A phone company has noticed that the majority of subscribers do not change their base station during their call, but the proportion of subscribers who change their base station is an upper truncated Poisson distribution with $\lambda = 0.04$, and truncation point 4. Find the expected percentage of subscribers who change their base station, and $\Pr[X \geq 2]$, where X denotes the number of hand overs.

8.81 If the CDF of a discrete random variable is

$$F(x) = \begin{cases} 0 & \text{if } x < 0 \\ 0.25 & \text{if } 0 \leq x < 2 \\ 0.50 & \text{if } 2 \leq x < 4 \\ 0.75 & \text{if } 4 \leq x < 6 \\ 1.0 & \text{if } x \geq 6 \end{cases}$$

find the PDF and the mean.

8.82 If the CDF of a continuous random variable is

$$F(x) = \begin{cases} 0 & \text{if } x < 2 \\ c\left(\dfrac{x^2}{2} - 2(x - 1)\right) & \text{if } 2 \leq x < 4 \\ c\left(\dfrac{-x^2}{2} + 2(3x - 7)\right) & \text{if } 4 \leq x < 5 \\ F(x) = 1 & \text{for } x \geq 5. \end{cases}$$

find the PDF and the mean.

8.83 If the CDF of a discrete random variable is

$$f(x) = \begin{cases} 0.25 & \text{if } x = -1 \\ 0.50 & \text{if } x = 0 \\ 0.25 & \text{if } x = +1 \end{cases}$$

how is the mean related to $P[X = 1]$.

8.84 If the CDF of a continuous random variable is

$$F(x) = \begin{cases} 0 & \text{if } x < 0 \\ c\dfrac{x^2}{2} & \text{if } 0 \le x < 1 \\ c(2x - 1 + x^2) & \text{if } 1 \le x < 2 \\ F(x) = 1 & \text{for } x \ge 2. \end{cases}$$

find the PDF and the mean.

8.85 If the PDF of a discrete random variable is

$$f(x) = \begin{cases} e^{-\lambda} p^x \displaystyle\sum_{j=1}^{x} \binom{x-1}{j-1} (\lambda q/p)^j /j! & \text{for } x = 1, 2, \ldots \\ e^{-\lambda} & \text{for } x = 0 \\ 0 & \text{elsewhere} \end{cases}$$

prove that the mean is λ/q. Find the PGF.

8.86 If the CDF of a continuous random variable is

$$F(\theta) = \begin{cases} 0 & \text{if } \theta < 0, \\ \tan(\theta)/2 & \text{if } 0 \le \theta \le \pi/4, \\ 1 - \tan(\pi/2 - \theta)/2 & \text{if } \pi/4 \le \theta \le \pi/2, \\ 1 & \text{elsewhere} \end{cases}$$

prove that the mean is $\pi/4$. Find the MGF.

8.87 Prove that the memory-less property of exponential distribution is equivalent to $G(u + v) = G(u) * G(v) \forall u, v > 0$, where $G(u) = \Pr[X > u]$

8.88 If both X and Y are independent gamma distributed, prove that (i) $E(Y|X) = cX + b$, (ii) $\text{Var}(Y|X) = b$, (iii) $E((Y - X)^2|X) = b$.

8.89 If $f(x; n, \mu) = \binom{n}{x} (1 - \mu/n)^n \mu^x / (n - \mu)^x$ where $x = 0, 1, \ldots, n$ prove that $E(X) = \mu$ and $\text{Var}(X) = \mu(1 - \mu/n)$.

8.90 If X, Y are independent normal random variables with the same variance, find $E[(X + Y)^4|(X - Y)]$

8.91 Suppose you toss a fair die once and note down the number N that shows up ($1 \leq N \leq 6$). You then toss a fair coin N times. Let X denote the number of heads that you get in N tosses of the coin. Find $E(X)$ and $V(X)$.

8.92 If X and Y are independent random variables, prove that $P(Y \leq X) = \int_{-\infty}^{\infty} F_Y(x) f_X(x) dx = 1 - \int_{-\infty}^{\infty} F_X(y) f_Y(y) dy$.

8.93 Prove that the population variance can be expressed as the values of CDF ($F_X(x)$) and its first two derivatives evaluated at $x = 1$.

8.94 Prove that the population variance can be expressed using the second derivative of $K_x(t)$ as $K_x''(x)$ evaluated at $x = 0$.

8.95 An electronic circuit has n^2 components that look identical. A technician has time to inspect just n of the components in any trip. What is the expected number of trips needed to inspect every component if the components are chosen arbitrarily in each repair trip, and inspected components are not marked.

8.96 A telephone carrier notices that the average duration of cell-phone calls among teenagers is distributed as a left-truncated

8.97 A site offers HTTP and FTP connections. The number of new customers who connect to HTTP server is Poisson distributed with $\lambda = 20$ for a time interval of 1 min. On the FTP server is Poisson distributed with $\lambda = 3$ for same time period. If both events are independent, what is the expected number of customers connecting to the site in 4 min?.

8.98 The number of cars that arrive at a gas station between 7 AM and 9 AM is Poisson distributed with mean 3 in 5 min. What is the expected number of minutes a person has to wait if there are no others in the queue? What is the expected number of cars that arrive in 27 s?

8.99 If X is CUNI(a, b) find the distribution and expected value of $Y = (2X - (a + b))/(b - a)$.

8.100 What is the expression to find $E[x^4]$ in terms of $x(x - 1)(x - 2)(x - 3)$ and lower order products?

exponential distribution with $\lambda = 1/2400$ s and truncation point 20 s. What is the expected percentage of phone calls that take more than 5 min? What is the variance of duration of all phone calls?

8.101 If X and Y are IID EXP(λ_i) with respective PDF $f(x)$ and $g(y)$ for $i = 1, 2$ prove that

$$\Pr(X < Y) = \int_0^\infty f(x)[1 - G(x)]dx = 1 - \int_0^\infty g(y)[1 - F(y)]dy = \lambda_1/(\lambda_1 + \lambda_2).$$

8.102 High-rise structures at earth-quake-prone areas are designed to withstand powerful earthquakes. From past data, it is found that the probability of an earthquake in a year is 0.091, and the probability of a building collapse after the earthquake is 0.004. The cost of constructing a high-rise building is C_0 and the cost of repair after damage is C_r. If a building portfolio comprises of n (> 2) buildings in a city neighborhood, find the expected value of the cost incurred in 10 years: (i) if no information on the earthquakes are available and (ii) if it is assumed that at least two earthquakes are likely to occur.

8.103 A POP (post-office protocol) based mail server sends each message and then waits for an ACK from the receiver. Only after the receipt of the ACK will the mail server send the next message in the queue. It is known that the delay in receipt of the ACK is exponentially distributed with mean 1/2 s. If three messages, each of size 1 K are to be sent, what is the expected number of seconds elapsed for successful transmission if the sending of each message itself takes half-second?.

8.104 An automated robot controlled inventory warehouse has racks of length 120 m on both sides of an alley. The robot is equally likely to break down anywhere on the stretch of 120 m. Where should a spare robot be located so that it can immediately take over the task of the broken down robot in minimal waste of time?

8.105 A computer virus can infect a laptop independently through (i) an email, (ii) an http, or (iii) a multimedia with respective Poisson probabilities 2%, 0.09%, and 5% in 1 day. If you use both email and multimedia connection for 30 h, what is the expected number of computer virus infections? If you use both http and multimedia for 20 h and email for 10 h, what is the expected number of infections?

8.106 If x_i – IDD beta -2 (a, b_i) find expected value of harmonic mean.

8.107 A software comprises of eight subsystems. Probability that the first five subsystems will throw a run-time exception (kind of error) in 8 h of use is POIS (0.03), and independently the last three subsystems is POIS (0.05). If the software is used for 80 h, (i) what is the expected number of exceptions? (ii) probability that no exceptions occurred.

8.108 An auto-emission test center has found that on an average one in eight automobiles fail in the emission test, and needs tune-up. The distribution of tune-up time in hours is EXP(2.5). If 100 vehicles are tested per month, find the expected number of hours spent on servicing of failed vehicles.

8.109 Suppose n letters are to be sent in n envelops. If the letters and envelops are shuffled and each letter is randomly assigned to an envelope, find the expected number of matches (letters that get into correct envelops).

8.110 A cereals manufacturer offers a promotional coupon with a new brand of cereal pack. Two types of coupons (that carry either 1 point or 2 points) are printed, and either of them is put in selected packs so that some packs do not contain a coupon. Probability that a customer will find a 1-point coupon is p, and a 2 points coupon is q. If a customer purchases n packs of the cereal, what is the expected number of points earned?

8.111 Prove that for the central chi-square distribution $E(1/X) = 1/(n-2)$.

8.112 If $X \sim \text{EXP}\ (\lambda)$ find $E[ce^x]$, where c is constant.

8.113 Prove that $K_X(0) = E(X)$

8.114 Prove $K_{aX+b}(t) = K_X(at) + bt$.

8.115 If $g(x)$ is a convex function, prove that $E[g(x)] \geq g(E[x])$ provided that $E[|g(x)|] < \infty$

8.116 Prove $\text{Cov}(X, Y) \leq [V(X)V(Y)]^{1/2}$ with equality when relationship is perfectly linear.

8.117 If X has Geometric distribution, find $E[(-1)^x]$.

8.118 If $X \sim \text{BINO}(n, p)$ prove that $E(n - X)^2 = nq[n - p(n-1)]$.

8.119 A consignment of 2^n missiles contain $\binom{n}{k}$ that have a range of $100 + k^2$ miles. If a group of m missiles are randomly picked up and fired, what is the expected miles covered from the firing point to hitting point?

8.120 If $X_1, X_2, \ldots, X_N$ are IID and $S_N = X_1 + X_2 + \cdots + X_N$, prove that $E(\sum_{i=1}^{N} X_i) = E(N)E(X)$ and $M_{S_N}(t) = \prod_{i=1}^{N} M_{X_i}(t)$.

8.121 If $X_1, X_2, \ldots, X_N$ are IID each with the same mean μ and same variance σ^2, find the second moment and variance of a random sum $S_N = X_1 + X_2 + \cdots + X_N$.

8.122 What is the limiting value of $\lim h \downarrow 0 (E[e^{hx} - 1])/h$.

8.123 An audio signal S is corrupted with background noise B. If S is uniformly distributed in the range $-c$ to $+c$, but the noise B is uniformly distributed in the range 0 to $2d$ where $d < c$, what is the expected value of signal plus noice? What is the covariance $\text{COV}(S, B)$ assuming that signal and noise are coming from independent sources?

8.124 Suppose two fair dice are tossed. Find the density function of $(X1, X2)$, where $X1$ and $X2$ are the scores that show up. Three random variables U,V,Y are defined as follows: $U = \min\{X1, X2\}$, the minimum score, $V = \max\{X1, X2\}$, the maximum score, and $Y = X1 + X2$, the sum of the scores. Find $E(U), E(V), E(Y)$, and $E(Y|X1)$.

9

GENERATING FUNCTIONS

After finishing the chapter, students will be able to

- Understand generating functions and their properties
- Comprehend generating functions and characteristic function
- Interpret moments and cumulants from generating functions
- Explore new type of generating functions for discrete CDF
- Apply the concepts to practical problems

9.1 TYPES OF GENERATING FUNCTIONS

Generating functions find a variety of applications in engineering and applied sciences. As the name implies, generating functions are used to generate different quantities with minimal work.

Definition 9.1 A generating function is a simple and concise expression in one or more dummy variables that captures the coefficients of a finite or infinite power series expansion, and generates a quantity of interest using calculus or algebraic operations, or simple substitutions.

Statistics for Scientists and Engineers, First Edition. Ramalingam Shanmugam and Rajan Chattamvelli.
© 2015 John Wiley & Sons, Inc. Published 2015 by John Wiley & Sons, Inc.

Depending on what we wish to generate, there are different generating functions. For example, moment generating function (MGF) generates moments of a population and probability generating function generates corresponding probabilities. These are specific to each distribution. An advantage is that if the MGF of an arbitrary random variable X is known, we can mathematically derive the MGF of any linear combination of the form $a * X + b$. This reasoning holds for other generating functions too.

Let $\{a_n\}$, $n = 0, 1, 2, \ldots, \infty$ be a sequence of bounded numbers. Then, the power series $f(x) = \sum_{n=0}^{\infty} a_n x^n$ is called the ordinary generating function (OGF) of $\{a_n\}$. Here, x is a dummy variable, n is the indexvar, and $a_n's$ are known constants. For different values of a_n, we get different OGFs. For example, if all $a_n = 1$, we get $f(x) = (1 - x)^{-1}$, and if $a_n = -1$ for n odd and $a_n = +1$ for n even, we get $(1 + x)^{-1}$. Similarly, if even coefficients $a_{2n} = +1$, and odd coefficients $a_{2n+1} = 0$, we get $\left(1 - x^2\right)^{-1}$. The function $g(x) = \sum_{n=0}^{\infty} a_n x^n / n!$ is called the exponential generating function (EGF), where the divisor of the nth term is $n!$. The generating functions used in Statistics can be finite or infinite, because they are defined on (sample spaces of) random variables. The above is a discrete generating function as it is defined for a discrete sequence. They may also be defined on continuous random variables as shown below.

■ EXAMPLE 9.1   *k*th derivative of EGF

Prove that $D^k g(x) = \sum_{n=0}^{\infty} a_{n+k} x^n / n!$

Solution 9.1 Consider $g(x) = a_0 + a_1 x / 1! + a_2 x^2 / 2! + a_3 x^3 / 3! + \cdots + a_k x^k / k! + \cdots$. Take the derivative with respect to x of both sides. As a_0 is a constant, its derivative is zero. Use derivative of $x^n = n * x^{n-1}$ for each term to get $g'(x) = a_1 + a_2 x / 1! + a_3 x^2 / 2! + \cdots + a_k x^{k-1} / (k - 1)! + \cdots$. Differentiate again (this time a_1 being a constant vanishes) to get $g''(x) = a_2 + a_3 x / 1! + a_4 x^2 / 2! + \cdots + a_k x^{k-2} / (k - 2)! + \cdots$. Repeat this process k times. All the terms whose coefficients are below a_k will vanish. What remains is $g^{(k)}(x) = a_k + a_{k+1} x / 1! + a_{k+2} x^2 / 2! + \cdots$. This can be expressed using the summation notation introduced in Chapter 1 as $g^{(k)}(x) = \sum_{n=k}^{\infty} a_n x^{n-k} / (n - k)!$. Using the change of indexvar introduced in Section 1.5 this can be written as $g^{(k)}(x) = \sum_{n=0}^{\infty} a_{n+k} x^n / n!$. Now if we put $x = 0$, all higher-order terms vanish except the constant a_k.   ■

■ EXAMPLE 9.2   OGF of $f(x)/(1 - x)$

If $f(x)$ is the OGF of the infinite sequence $a_0, a_1, \ldots, a_n, \ldots$, prove that $f(x)/(1 - x)$ is the OGF of the infinite sequence $a_0, a_0 + a_1, a_0 + a_1 + a_2, \ldots$

Solution 9.2 By definition $f(x) = a_0 + a_1 x + a_2 x^2 + \cdots + a_n x^n + \cdots$. Expand $(1 - x)^{-1}$ as a power series $1 + x + x^2 + \ldots$ and multiply with $f(x)$ to get the RHS as $g(x) = (1 - x)^{-1} * f(x) =$

$$\left(1 + x + x^2 + x^3 + \cdots\right)\left(a_0 + a_1 x + a_2 x^2 + \cdots + a_n x^n + \cdots\right) =$$

$$\left(\sum_{j=0}^{\infty} 1.x^j\right)\left(\sum_{k=0}^{\infty} a_k.x^k\right). \qquad (9.1)$$

Change the order of summation to get

$$\left(\sum_{k=0}^{\infty}\left(\sum_{j=0}^{k} a_j.1\right)x^k\right) = \sum_{k=0}^{\infty}\left(\sum_{j=0}^{k} a_j\right)x^k = a_0 + \left(a_0 + a_1\right)x$$

$$+ \left(a_0 + a_1 + a_2\right)x^2 + \cdots. \qquad (9.2)$$

This is the OGF of the given sequence. ∎

9.1.1 Generating Functions in Statistics

There are four popular generating functions used in statistics—namely (i) probability generating function (PGF), denoted by $P_x(t)$, (ii) MGF, denoted by $M_x(t)$, (iii) cumulant generating function (CGF), denoted by $K_x(t)$, and (iv) characteristic function, denoted by $\phi_x(t)$. In addition, there are still others to generate factorial moments (FMGF), inverse moments (IMGF), inverse factorial moments (IFMGF), absolute moments, as well as for odd moments and even moments separately. These are called "canonical functions" in some fields.

The PGF generates the probabilities of a random variable and is of type OGF. MGF (page 382) has further subdivisions as ordinary, and central mgf, factorial mgf, inverse mgf, inverse factorial mgf, CGF, and characteristic function (ChF). All of these can also be defined for *arbitrary origin*. The CGF is defined in terms of the MGF as $K_x(t) = \ln\left(M_x(t)\right)$, which when expanded as a polynomial in t gives the cumulants. As every distribution does not possess an MGF, the concept is extended to the complex domain by defining the ChF as $\phi_x(t) = E\left(e^{itx}\right)$. Note that the logarithm is to the base e (ln). If all of them exist for a distribution, then

$$P_x\left(e^t\right) = M_x(t) = e^{K_x(t)} = \phi_x(it). \qquad (9.3)$$

This can also be written in the alternate forms $P_x\left(e^{it}\right) = M_x(it) = e^{K_x(it)} = \phi_x(-t)$ or as $P_x(t) = M_x(\ln(t)) = e^{K_x(\ln(t))} = \phi_x(i \ln(t))$ (see Table 9.1).

9.2 PROBABILITY GENERATING FUNCTIONS (PGF)

The PGF of a random variable is used to generate probabilities. It is defined as

$$P_x(t) = E(t^x) = \sum_x t^x p(x) = p(0) + t * p(1) + t^2 * p(2) + \cdots + t^k * p(k) + \cdots,$$

$$(9.4)$$

TABLE 9.1 Summary Table of Generating Functions

Abbreviation	Symbol Used	Definition E = exp opr.	Generates What	How Obtained t = dummy-variable	Conditions
PGF	$P_X(t)$	$E(t^x)$	Probabilities	$p_k = \dfrac{\partial^k}{\partial t^k} P_X(t)\big|_{t=0}/k!$	Discrete
MGF	$M_X(t)$	$E(e^{tx})$	Moments	$\mu'_k = \dfrac{\partial^k}{\partial t^k} M_X(t)\big|_{t=0}$	Expectation exists
CMGF	$M_Z(t)$	$E(e^{t(x-\mu)})$	Central moments	$\mu_k = \dfrac{\partial^k}{\partial t^k} M_Z(t)\big|_{t=0}$	Expectation exists
ChF	$\phi_X(t)$	$E(e^{itx})$	Moments	$i^k \mu'_k = \dfrac{\partial^k}{\partial t^k} \phi_X(t)\big|_{t=0}$	Always exist
CGF	$K_X(t)$	$\log(E(e^{tx}))$	Cumulants	$\mu_k = \dfrac{\partial^k}{\partial t^k} K_X(t)\big|_{t=0}$	MGF exists
FMGF	$\Gamma_X(t)$	$E((1+t)^x)$	Factorial moments	$\mu_{(k)} = \dfrac{\partial^k}{\partial t^k} \Gamma_X(t)\big|_{t=0}$	Discrete

PGF is of type OGF. MGF and ChF are of type EGF. MGF need not always exist, but characteristic function always exists. Falling factorial moment is denoted as $\mu_{(k)} = E(x(x-1)(x-2)\cdots(x-k+1))$.

where the summation is over the range of X. This is a finite series for distributions with finite range. It may or may not possess a closed-form expression for other distributions. It converges for $|t| < 1$, and appropriate derivatives exist. Differentiating both sides of equation (9.4) k times with respect to t, we get $(\partial^k/\partial t^k) P_X(t) = k! p(k)$ +terms involving t. If we put $t = 0$, all higher-order terms that have "t" or higher powers vanish, giving $k!\, p(k)$, from which $p(k)$ is obtained as $(\partial^k/\partial t^k) P_X(0)/k!$. If the $P_X(t)$ involves powers or exponents, we take the log (with respect to e) of both sides and differentiate k times, and then use the following result on $P_X(t = 1)$ to simplify the differentiation.

■ **EXAMPLE 9.3 PGF special values $P_x(t = 0)$ and $P_x(t = 1)$**

Find $P_x(t = 0)$ and $P_x(t = 1)$ from the PGF of a discrete distribution.

Solution 9.3 As $\sum_k p(k)$, being the sum of the probabilities, is one, it follows trivially by putting $t = 1$ in equation (9.4) that is $P_x(t = 1) = 1$. Put $t = 0$ in equation (9.4) to get $P_x(t = 0) = p(0)$, the first probability.

Similarly, put $t = -1$ to get the RHS as $[p(0) + p(2) + \cdots +] - [p(1) + p[3] + p[5] + \cdots]$. ■

■ **EXAMPLE 9.4 PGF of Poisson distribution**

Find the PGF of a Poisson distribution, and obtain the difference between the sum of even and odd probabilities.

Solution 9.4 The PGF of a Poisson distribution is

$$P_x(t) = E(t^x) = \sum_{x=0}^{\infty} t^x e^{-\lambda} \lambda^x / x! = e^{-\lambda} \sum_{x=0}^{\infty} \frac{(\lambda t)^x}{x!} = e^{-\lambda} e^{t\lambda} = e^{-\lambda[1-t]}. \qquad (9.5)$$

Put $t = -1$ in equation (9.5) and use the above result to get the desired sum as $\exp(-\lambda[1 - (-1)]) = \exp(-2\lambda)$ (see Example 6.38 in Chapter 6, p. 233). ∎

◼ EXAMPLE 9.5 PGF of geometric distribution

Find the PGF of a geometric distribution, and obtain the difference between the sum of even and odd probabilities.

Solution 9.5 As the geometric distribution takes $x = 0, 1, 2, \ldots \infty$ values, we get the PGF as

$$P_x(t) = E(t^x) = \sum_{x=0}^{\infty} t^x q^x p = p \sum_{x=0}^{\infty} (qt)^x = p/(1 - qt). \tag{9.6}$$

In Chapter 6, page 221 we have evaluated $P[X \text{ is even}] = q^0 p + q^2 p + \cdots = p[1 + q^2 + q^4 + \cdots] = p/(1 - q^2) = 1/(1 + q)$, and $P[X \text{ is odd}] = q^1 p + q^3 p + \cdots = qp[1 + q^2 + q^4 + \cdots] = qp/(1 - q^2) = q/(1 + q)$. Using the above result, the difference between these must equal the value of $P_x(t = -1)$. Put $t = -1$ in equation (9.6) to get $p/(1 - q(-1)) = p/(1 + q)$, which is the same as $1/(1 + q) - q/(1 + q) = p/(1 + q)$. ∎

Closed-form expressions for $P_x(t)$ are available for most of the common discrete distributions. They are seldom used for continuous distributions because $\int t^x f(x)\, dx$ may not be convergent.

◼ EXAMPLE 9.6 PGF of BINO(n, p)

Find the PGF of BINO(n, p) and obtain the mean.

Solution 9.6 By definition

$$P_x(t) = E(t^x) = \sum_{x=0}^{n} \binom{n}{x} p^x q^{n-x} t^x = \sum_{x=0}^{n} \binom{n}{x} (pt)^x q^{n-x} = (q + pt)^n. \tag{9.7}$$

The coefficient of t^x gives the probability that the random variable takes the value x. To find the mean, we take the log of both sides. Then $\log(P_x(t)) = n*\log(q + pt)$. Differentiate both sides with respect to t to get $P'_x(t)/P_x(t) = n * p/(q + pt)$. Now put $t = 1$ and use $P_x(t = 1) = 1$ to get the RHS as $n*p/(q + p) = np$ as $q + p = 1$. ∎

Lemma 1 The PGF ($E(t^x)$) can be used to obtain the factorial moments using the relationship $\mu_{(r)} = P_x^{(r)}(1)$ (see Table 9.1).

9.2.1 Properties of PGF

1. $P^{(r)}(0)/r! = P[X = r]$.
   By restricting the argument of $P_x(t)$ to $|t| < 1$, it is easily seen that $P_x(t)$ is infinitely differentiable in t. Differentiating $P_x(t) = E(t^x)$ r times, we get

$$\frac{\partial^r}{\partial t^r} P_x(t) =$$

$$E[x(x-1)\cdots(x-r+1)\,t^{x-r}] = \sum_{x \geq r}[x(x-1)\cdots(x-r+1)\,t^{x-r}]f(x). \quad (9.8)$$

The first term in this sum is obviously $[r(r-1)\cdots(r-r+1)\,t^{r-r}]f(x=r) = [r!t^0]f(x) = r!f(x=r)$. By putting $t=0$, every term except the first vanishes, and the RHS becomes $r!f(x=r)$. Thus, $\frac{\partial^r}{\partial t^r}P_x(t=0) = r!f(x=r)$.

2. $P^{(r)}(1)/r! = E[X^{(r)}]$.
   By putting $t=1$ in equation 9.8, the RHS becomes $E[x(x-1)\cdots(x-r+1)]$, which is the rth factorial moment. Hence, some authors call this the factorial MGF (see Section 9.8, p. 391).

3. $\mu = E(X) = P'(1)$, and $\mu'_2 = E(X^2) = P'(1) + P''(1)$
   The first result follows directly from the above by putting $r=1$. As $X^2 = X(X-1) + X$, the second result also follows from it.

4. $V(X) = P'(1) + P''(1)[1 - P'(1)]$
   This result follows from the fact that $V(X) = E[X^2] - E[X]^2 = E[X(X-1)] + E[X] - E[X]^2$. Now use the above results.

5. $\int_t P(t)\,dt = E\left(\frac{1}{X+1}\right)$
   This is the first inverse moment, and holds for positive random variables.

6. $P_{cX}(t) = P_X(t^c)$
   This follows by writing t^{cX} as $(t^c)^X$.

7. $P_{X\pm c}(t) = t^{\pm c} * P_X(t)$
   This follows by writing $t^{X\pm c}$ as $(t^{\pm c})\,t^X$.

8. $P_X(t) = M_X(\ln(t))$
   From equation (9.3), we have $P_x(e^t) = M_x(t)$. Write $t' = e^t$ so that $t = \ln(t')$ to get the result.

9. $P_{(X\pm\mu)/\sigma}(t) = t^{\pm\mu/\sigma}P_X\left(t^{1/\sigma}\right)$
   This is called the change of origin and scale transformation of PGF. This follows by combining (6) and (7).

9.3 GENERATING FUNCTIONS FOR CDF (GFCDF)

As the PGF of a random variable generates probabilities, it can be used to generate the sum of left tail probabilities (CDF) as follows. We have seen in Example 9.2 that if $f(x)$ is the OGF of the sequence $a_0, a_1, \ldots, a_n, \ldots$, finite or infinite, then $f(x)/(1-x)$ is the OGF of the sequence $a_0, a_0 + a_1, a_0 + a_1 + a_2, \ldots$ By replacing a_i's by probabilities, we obtain a GF that generates the sum of probabilities as

$$G(x) = \sum_{k=0}^{\infty}\left(\sum_{j=0}^{k} p_k\right)x^k = p_0 + \left(p_0 + p_1\right)x + \left(p_0 + p_1 + p_2\right)x^2 + \cdots \quad (9.9)$$

This works only for discrete distributions.

◼ **EXAMPLE 9.7   GF for CDF of geometric distribution**

Obtain the GF for CDF of a geometric distribution.

Solution 9.7 We know that the PGF of geometric distribution is $p/(1 - qt)$ from which the GFCDF is obtained as $G(t) = p(1 - t)^{-1}/(1 - qt)$. Expand both $(1 - t)^{-1}$ and $(1 - qt)^{-1}$ as infinite series' and combine like powers to get

$$G(t) = p[1 + t(1 + q) + t^2(1 + q + q^2) + t^3(1 + q + q^2 + q^3) + \cdots]. \quad (9.10)$$

Write $(1 + q + q^2 + q^3 + \cdots + q^k)$ as $(1 - q^{k+1})/(1 - q)$ and cancel $(1 - q) = p$ with the numerator to get the generating function for geometric CDF as

$$G(t) = [1 + t(1 - q^2) + t^2(1 - q^3) + t^3(1 - q^4) + \cdots]. \quad (9.11)$$

◼

9.4   GENERATING FUNCTIONS FOR MEAN DEVIATION (GFMD)

The above result can be extended to obtain a GF for mean deviations of discrete distributions. We have seen in Chapter 6 that the MD of discrete distributions is given by

$$MD = 2 \sum_{x=ll}^{\mu-1} F(x). \quad (9.12)$$

where ll is the lower limit of the distribution, μ is the arithmetic mean, and $F(x)$ is the CDF. To obtain a GF for MD, first rewrite equation (9.9) as

$$G(t) = [1 + g_1 t + g_2 t^2 + g_3 t^3 + \cdots]. \quad (9.13)$$

where g_k denotes the sum of probabilities. Multiply both sides by $(1 - t)^{-1}$ and denote the LHS $(1 - t)^{-1} G(t)$ by $H(t)$ to get

$$H(t) = [1 + g_1 t + (g_1 + g_2) t^2 + (g_1 + g_2 + g_3) t^3 + \cdots]. \quad (9.14)$$

The above step is equivalent to applying the result in Example 9.2 in page 376 where $p_k = g_k$. As coefficients of $H(t)$ accumulate the sum of the CDF ("sum of the sum" of left tail probabilities), the MD is easily found as twice the coefficient of $t^{\mu-1}$ in $H(t)$. This can be stated as the following theorem.

Theorem 9.1 The MD of a discrete distribution is twice the coefficient of $t^{\mu-1}$ in the power series expansion of $(1 - t)^{-2} P_x(t)$, where μ is the mean (or the nearest integer to it) and $P_x(t)$ is the probability generating function.

In the above derivation, we have marked the probabilities as p_1, p_2, and so on. If they are denoted as p_0, p_1, p_2, and so on, we need to consider the sums $p_0 + p_1$ and so on. A similar result could be obtained using right tail probabilities.

◼ **EXAMPLE 9.8   MD of geometric distribution**

Find the MD of geometric distribution using Theorem 9.1.

Solution 9.8 We have seen in Example 9.7 in page 381 that the GF for CDF of a geometric distribution is

$$G(t) = [1 + t\left(1 - q^2\right) + t^2\left(1 - q^3\right) + t^3\left(1 - q^4\right) + \cdots].$$ (9.15)

Denote $(1 - q^{k+1})$ by g_k (note that there is no $(1 - q)$ term in equation (9.15) showing that the MD is zero when $q < p$ or equivalently $p > 1/2$) and obtain the GFMD with coefficients $h_k = \sum g_k = \sum\left(1 - q^{k+1}\right)$. As the mean of a geometric distribution is q/p, we can simply fetch the coefficient of $t^{\mu-1} = t^{[q/p-1]}$ in $H(t)$ and multiply by 2 to get the MD as $2\sum_{k=0}^{[q/p-1]}\left(1 - q^{k+1}\right)$, where $[q/p - 1]$ denotes the integer part. See Chapter 6, p. 217 for further simplifications.   ◼

9.5   MOMENT GENERATING FUNCTIONS (MGF)

The MGF of a random variable is used to generate the moments algebraically. Let X be a discrete random variable defined for all values of x. As e^{tx} has an infinite expansion in powers of x as $e^{tx} = 1 + (tx)/1! + (tx)^2/2! + \cdots + (tx)^n/n! + \cdots$, we multiply both the sides by $f(x)$, and take expectation on both the sides to get

$$M_x(t) = E\left(e^{tx}\right) = \begin{cases} \sum_x e^{tx}p(x) & \text{if } X \text{ is discrete;} \\ \int_{-\infty}^{\infty} e^{tx}f(x)\,dx & \text{if } X \text{ is continuous.} \end{cases}$$

In the discrete case, this becomes $M_x(t) = \sum_{x=0}^{\infty} e^{tx}f(x) = 1 + \sum_{x=0}^{\infty}(tx)/1!f(x) + \sum_{x=0}^{\infty}(tx)^2/2!f(x) + \cdots$. Replace each of the sums $\sum_{x=0}^{\infty} x^k f(x)$ by μ_k to obtain the following series (which is theoretically defined for all values, but depends on the distribution)

$$M_x(t) = 1 + \mu_1't/1! + \mu_2't^2/2! + \cdots + \mu_k't^k/k! + .$$ (9.16)

Analogous result holds for the continuous case by replacing summation by integration. By choosing $|t| < 1$, the above series can be made convergent for most random variables.

Theorem 9.2 The MGF (p. 377) and the PGF are connected as $M_X(t) = P_x\left(e^t\right)$, and $M_X(t = 0) = P_x\left(e^0\right) = P_x(1) = 1$.

Proof: This follows trivially by replacing t by e^t in equation (9.4). Note that it is also applicable to continuous random variables. Put $t = 0$ and use $e^0 = 1$ to get the second part.   ◼

■ EXAMPLE 9.9   MGF of binomial distribution from PGF

If the pgf of $BINO(n, p)$ is $(q + pt)^n$, obtain the MGF and derive the mean.

Solution 9.9 The mgf can be found from equation (9.7) by replacing t by e^t. This gives $M_X(t) = \left(q + pe^t\right)^n$. Take log to get $\log\left(M_X(t)\right) = n * \log\left(q + pe^t\right)$. Next differentiate as above: $M'_X(t)/M_X(t) = n * pe^t/\left(q + pe^t\right)$. Put $t = 0$ to get the mean as np. Take log again to get $\log\left(M'_X(t)\right) - \log\left(M_X(t)\right) = \log(np) + t - \log\left(q + pe^t\right)$. Differentiate again, and denote $M'_X(t)$ simply by M' and so on. This gives $M''/M' - M'/M = 1 - pe^t/\left(q + pe^t\right)$. Put $t = 0$ throughout and use $M'(0) = np$ and $M(0) = 1$ to get $M''(0)/np - np = 1 - p$ or equivalently $M''(0) = (q + np) * np$. Finally, use $\sigma^2 = M''_X(0) - [M'_X(0)]^2 = (q + np4) * np - (np)^2 = npq$.   ■

■ EXAMPLE 9.10   MGF of exponential distribution

Obtain the MGF of an exponential distribution.

Solution 9.10 Consider the PDF of an $EXP(\lambda)$ as $f(x; \lambda) = \lambda \exp(-\lambda x)$. By definition $M_x(t) = Ee^{tx} = \int_0^\infty e^{tx}\lambda \exp(-\lambda x)\,dx$. As λ is a constant, take it outside the integral, and combine the integrands to get $M_x(t) = \lambda \int_0^\infty \exp((t - \lambda)x)\,dx$. Write $(t - \lambda)x$ as $-(\lambda - t)x$ and integrate to get $M_x(t) = \lambda/(\lambda - t) = 1/(1 - t/\lambda)$.   ■

9.5.1   Properties of Moment Generating Functions

1. MGF of an origin changed variate can be found from MGF of original variable

$$M_{x \pm b}(t) = e^{\pm bt} * M_x(t). \tag{9.17}$$

This follows trivially by writing $E[e^{t[x \pm b]}]$ as $e^{\pm bt} * E[e^{tx}]$.

2. MGF of a scale changed variate can be found from MGF of original variable as

$$M_{c*x}(t) = M_x(c * t). \tag{9.18}$$

This follows trivially by writing $E[e^{t\,cx}]$ as $E[e^{(ct)*x}]$.

3. MGF of origin and scale changed variate can be found from MGF of original variable as

$$M_{c*x \mp b}(t) = e^{\pm bt} * M_x(c * t). \tag{9.19}$$

This follows by combining both the cases above.

Theorem 9.3 The MGF of a sum of independent random variables is the product of their MGFs. Symbolically $M_{X+Y}(t) = M_X(t) * M_Y(t)$.

Proof: We prove the result for the discrete case. $M_{X+Y}(t) = E\left(e^{t(x+y)}\right) = E\left(e^{tx}e^{ty}\right)$. If X and Y are independent, we write the RHS as $\sum_x e^{tx} f(x) * \sum_y e^{ty} f(y) = M_X(t) * M_Y(t)$. The proof for the continuous case follows similarly. This result can be extended to any number of pairwise independent random variables. ∎

If $X_1, X_2, \ldots, X_n$ are independent, and $Y = \sum_i X_i$ then $M_Y(t) = \prod_i M_{X_i}(t)$.

EXAMPLE 9.11   Moments from $M_x(t)$

Prove that $E(X) = \frac{\partial}{\partial t} M_x(t)\mid_{t=0}$ and $E\left(X^2\right) = \frac{\partial^2}{\partial t^2} M_x(t)\mid_{t=0}$.

Solution 9.11 We know that $M_x(t) = E\left(e^{tx}\right)$. Differentiating equation (9.16) with rspect to t gives $\frac{\partial}{\partial t} M_x(t) = \frac{\partial}{\partial t} E\left(e^{tx}\right) = E\left(\frac{\partial}{\partial t} e^{tx}\right) = E\left(xe^{tx}\right)$ because x is considered as a constant (and t is our variable). Putting $t = 0$ on the RHS we get the result, as $e^0 = 1$. Differentiating a second time, we get $\frac{\partial^2}{\partial t^2} M_x(t) = \frac{\partial}{\partial t} E\left(xe^{tx}\right) = E\left(x^2 e^{tx}\right)$. Putting $t = 0$ on the RHS, we get $M_x''(t = 0) = E\left(x^2\right)$. Repeated application of this operation allows us to find the kth moment as $M_x^{(k)}(t = 0) = E\left(x^k\right)$. This gives $\sigma^2 = M_x''(t = 0) - [M_x'(t = 0)]^2$. ∎

EXAMPLE 9.12   MGF of BINO(n, p)

Find the MGF of BINO(n, p), and obtain the first two moments.

Solution 9.12 $M_x(t) = E(e^{tx}) = \sum_{x=0}^{n} e^{tx} \binom{n}{x} p^x q^{n-x} = \sum_{x=0}^{n} \binom{n}{x} (pe^t)^x q^{n-x} = \left(pe^t + q\right)^n = \left(q + pe^t\right)^n$. Differentiating with respect to t gives $\frac{\partial}{\partial t} M_x(t) = \frac{\partial}{\partial t}\left(q + pe^t\right)^n = n\left(q + pe^t\right)^{n-1}.pe^t$ so that $M_x'(t = 0) = np(q + p) = np$. Differentiating one more time, we get $M_x''(t) = np\frac{\partial}{\partial t}[(q + pe^t)^{n-1}.e^t] = np[(q + pe^t)^{n-1} .e^t + e^t(n-1)\left(q + pe^t\right)^{n-2} pe^t]$. By putting $t = 0$, this becomes $np[1 + (n-1)p] = np + n(n-1)p^2$. This work can be greatly simplified by taking log, then differentiating and using $M_x(t = 0) = 1$. ∎

EXAMPLE 9.13   MGF of a Poisson distribution

Find the MGF for central moments of a Poisson distribution. Hence, show that $\mu_{r+1} = \lambda\left[\binom{r}{1}\mu_{r-1} + \binom{r}{2}\mu_{r-2} + \cdots \binom{r}{r}\mu_0\right]$.

Solution 9.13 First consider the ordinary MGF defined as $M_x(t) = E\left(e^{tx}\right) = \sum_{k=0}^{\infty} e^{tx} e^{-\lambda} \lambda^x /x! = e^{-\lambda}\sum_{k=0}^{\infty}\left(\lambda e^t\right)^x /x! = e^{-\lambda} e^{\lambda e^t} = e^{\lambda e^t - \lambda} = e^{\lambda(e^t - 1)}$. As the mean of a Poisson distribution is $\mu = \lambda$, we use the property 1 to get

$$M_{x-\lambda}(t) = e^{-\lambda t} M_x(t) = e^{-\lambda t} e^{\lambda(e^t-1)} = e^{\lambda(e^t-t-1)} = \sum_{j=0}^{\infty} \mu_j t^j / j!. \qquad (9.20)$$

Differentiate $\sum_{j=0}^{\infty} \mu_j t^j / j! = e^{\lambda(e^t-t-1)}$ by t to get

$$\sum_{j=0}^{\infty} \mu_j j t^{j-1} / j! = e^{\lambda(e^t-t-1)} \lambda \left(e^t - 1 \right) = \lambda \left(e^t - 1 \right) \sum_{j=0}^{\infty} \mu_j t^j / j!. \qquad (9.21)$$

Expand e^t as an infinite series. The RHS becomes $\lambda \sum_{k=1}^{\infty} t^k / k! \sum_{j=0}^{\infty} \mu_j t^j / j!$. This can be written as $\lambda \sum_{k=1}^{\infty} \sum_{j=0}^{\infty} \mu_j t^{j+k} / [j!k!]$. Equate coefficients of t^r on both sides to get $\mu_{r+1}/r! = \lambda \left[\frac{\mu_{r-1}}{1!(r-1)!} + \frac{\mu_{r-2}}{2!(r-2)!} + \cdots + \frac{\mu_0}{r!(r-r)!} \right]$. Cross-multiplying and identifying the binomial coefficients, this becomes

$$\mu_{r+1} = \lambda \left[\binom{r}{1} \mu_{r-1} + \binom{r}{2} \mu_{r-2} + \cdots + \binom{r}{r} \mu_0 \right]. \qquad (9.22)$$

∎

Corollary 1 Prove that if $E[|X|^k]$ exists and is finite, then $E[|X|^j]$ exists and is finite for each $j < k$.

Proof: We prove the result for the continuous case. The proof for discrete case follows easily by replacing integration by summation. As $E[|X|^k]$ exists, we have $\int_x |X|^k dF(x) < \infty$. Now consider an arbitrary $j < k$ for which

$$\int_{-\infty}^{\infty} |x|^j dF(x) = \int_{-1}^{+1} |x|^j dF(x) + \int_{|x|>1} |x|^j dF(x). \qquad (9.23)$$

As $j < k$, $|x|^j < |x|^k$ for $|x| > 1$. Hence integral (9.23) becomes

$$\int_{-\infty}^{\infty} |x|^j dF(x) < \int_{-1}^{+1} |x|^j dF(x) + \int_{|x|>1} |x|^k dF(x)$$

$$\leq \int_{-1}^{+1} dF(x) + \int_{|x|>1} |x|^k dF(x). \qquad (9.24)$$

The RHS of equation (9.24) is upper bounded by $1 + E[|X|^k]$, and is $< \infty$. This proves that the LHS exists for each j. ∎

◼ EXAMPLE 9.14   MGF of a gamma distribution

Find the MGF of a gamma distribution $f(x) = \theta^m e^{-\theta x} x^{m-1}/\Gamma(m)$, where $x \geq 0$ and $\theta > 0$, and obtain the first two moments.

Solution 9.14 $M_x(t) = E\left(e^{tx}\right) = \int_0^\infty e^{tx} \theta^m e^{-\theta x} x^{m-1}/\Gamma(m)\,dx$. Take the constants outside the integral to get $\theta^m/\Gamma(m) \int_0^\infty e^{-(\theta-t)x} x^{m-1} dx = [\theta/(\theta-t)]^m = (1-t/\theta)^{-m}$. Take log and differentiate with t, to get $M_x'(t)/M_x(t) = -m/(1-t/\theta)(-1/\theta)$ from which by putting $t=0$ we get the first moment as (m/θ). Taking the derivative again, we get $[M_x(t)M''_x(t) - (M_x'(t))^2]/[M_x(t)]^2 = (m/\theta)(1/\theta)/(1-t/\theta)^2$. Put $t=0$ and use $M_x(t=0)=1$ and $M_x'(t) = m/\theta$ to get $M''_x(t=0) = (m/\theta)^2 + (m/\theta^2)$, from which the variance is obtained using $\sigma^2 = M_x''(t=0) - [M_x'(t=0)]^2$ as m/θ^2.   ◼

9.6   CHARACTERISTIC FUNCTIONS (CHF)

The MGF of a distribution need not always exist. Those cases can be dealt with in the complex domain by finding the expected value of e^{itx}, where $i = \sqrt{-1}$, which always exist. Thus, the ChF of a random variable is defined as

$$\text{ChF} = E\left(e^{itx}\right) = \begin{cases} \sum_{x=-\infty}^{\infty} e^{itx} p_x & \text{if } X \text{ is discrete;} \\ \int_{x=-\infty}^{\infty} e^{itx} f(x)\,dx & \text{if } X \text{ is continuous.} \end{cases}$$

We have seen above that the ChF, if it exists, can generate the moments. Irrespective of whether the random variable is discrete or continuous, we could expand the ChF as a McClaurin series as

$$\phi_X(t) = \sum_{j=0}^{\infty} \mu_j'(it)^j/j! = \phi(0) + t\phi'(0) + t^2/2!\phi''(0) + \cdots. \tag{9.25}$$

which is convergent for an appropriate choice of t (which depends on the distribution). As $\phi_X(t)$ in the continuous case can be represented as $\phi_X(t) = \int_{-\infty}^{\infty} e^{itx} dF(x)$, successive derivatives with t gives $\int i^n x^n dF(x) = i^n \mu_n'$. Define $\delta^{(n)}(x)$ as the nth derivative of the delta function. Then, the PDF can be written as an infinite sum as

$$f(x) = \sum_{j=0}^{\infty} (-1)^j \mu_j' \delta^{(j)}(x)/j!. \tag{9.26}$$

See References 134 and 287 for further details.

◼ EXAMPLE 9.15   Characteristic function of the Cauchy distribution

Find the characteristic function of the Cauchy distribution.

Solution 9.15 We have $f(x) = [\pi(1+x^2)]^{-1}$, so that $\phi(it) = \frac{1}{\pi}\int_{-\infty}^{\infty}\frac{e^{it}}{1+x^2}dx = e^{-|t|}$ ∎

9.6.1 Properties of Characteristic Functions

Characteristic functions are Laplace transforms of the corresponding PDF. As all Laplace transforms have an inverse, we could invert it to get the PDF. Hence, there is a one-to-one correspondence between the ChF and PDF. This is especially useful for continuous distributions as shown below. There are many simple properties satisfied by the ChF.

1. $\overline{\phi(t)} = \phi(-t)$, $\phi(0) = 1$, and $|\phi(\pm t)| \leq 1$. In words, this means that the complex conjugate of the ChF is the same as that obtained by replacing t with -t in the ChF. The assertion $\phi(0) = 1$ follows easily because this makes e^{itx} to be 1.

2. $\phi_{ax+b}(t) = e^{ibt}\phi_x(at)$. This result is trivial as it follows directly from the definition.

3. If X and Y are independent, $\phi_{ax+by}(t) = \phi_x(at).\phi_y(bt)$. Putting $a = b = 1$, we get $\phi_{x+y}(t) = \phi_x(t).\phi_y(t)$ if X and Y are independent.

4. $\phi(t)$ is continuous in t, and convex for $t > 0$. This means that if t_1 and t_2 are two values of $t > 0$, then $\phi((t_1 + t_2)/2) \leq \frac{1}{2}[\phi(t_1) + \phi(t_2)]$.

5. $\partial^n \phi(t)/\partial t^n|_{t=0} = i^n E(X^n)$

▣ EXAMPLE 9.16 Symmetric random variables

Prove that the random variable X is symmetric about the origin if the chF $\phi(it)$ is real-valued for all t.

Solution 9.16 Assume that X is symmetric about the origin, so that $f(-x) = f(x)$. Then for a bounded and odd Borel function $g(x)$ we have $\int g(x)\,dF(x) = 0$. As $g(x)$ is odd, this is equivalent to $\int \sin(tx)\,dF(x) = 0$. Hence, $\phi(t) = E(e^{itx}) = E[\cos(tx)]$ is real. Also, as $\phi_{-X}(t) = \phi_X(-t) = \overline{\phi}_X(t) = \phi_X(\bar{t})$, $F_X(x)$ and $F_{-X}(x)$ are the same (Table 9.2). ∎

Theorem 9.4 The characteristic function uniquely determines a distribution. The inversion theorem provides a means to find the PDF from the characteristic function as $f(x) = 1/(2\pi)\int_{-\infty}^{\infty}\phi_x(it)e^{-itx}dt$.

9.6.1.1 Uniqueness Theorem Let the random variables X and Y have MGF $M_x(t)$ and $M_y(t)$, respectively. If $M_x(t) = M_y(t)\,\forall t$, then X and Y have the same probability distribution.

TABLE 9.2 Table of Characteristic Functions

Distribution	Density Function	Characteristic Function
Bernoulli	$p^x(1-p)^{1-x}$	$q + pe^{it}$
Binomial	$\binom{n}{x}p^x q^{n-x}$	$\left(q + pe^{it}\right)^n$
Negative binomial	$\binom{x+k-1}{x}p^k q^x$	$p^k\left(1-qe^{it}\right)^{-k}$
Poisson	$e^{-\lambda}\lambda^x/x!$	$\exp\left(\lambda\left(e^{it}-1\right)\right)$
Rectangular	$f(x) = \Pr[X = k] = 1/N$	$\left(1 - e^{itN}\right)/[N\left(e^{-it}-1\right)]$
Geometric	$q^x\mathrm{p}$	$p/\left(1-qe^{it}\right)$
Logarithmic	$q^x/[-x\log p]$	$\ln\left(1-qe^{it}\right)/\ln\left(1-q\right)$
Multinomial	$(n!/\prod_{i=1}^{k}x_i!)*\prod_{i=1}^{k}p_i^{x_i}$	$\left[\sum_{j=1}^{k}p_j e^{it_j}\right]^n$
Cont. uniform	$1/(b-a)\,\Delta a \le x \le b$	$\left(e^{ibt}-e^{iat}\right)/[(b-a)\,it]$
Exponential	$\lambda e^{-\lambda x}$	$\lambda/(\lambda - it)$
Gamma	$\lambda^m x^{m-1}e^{-\lambda x}/\Gamma(m)$	$(1 - it/\lambda)^{-m}$
Arcsine	$1/\pi\sqrt{(1-x^2)}$	$e^{-it/2}I_0(it/2)\,{}_1F_1\left(1/2, 1; it\right)$
Beta-I	$x^{a-1}(1-x)^{b-1}/B(a,b)$	${}_1F_1\left(a, a+b; it\right)$
Normal	$\dfrac{1}{\sigma\sqrt{2\pi}}e^{-\frac{1}{2}\left(\frac{x-\mu}{\sigma}\right)^2}$	$\exp\left(it\mu - \frac{1}{2}t^2\sigma^2\right)$
Cauchy	$1/[b\pi[1+(x-a)^2/b^2]]$	$\exp\left(ita - |t|b\right)$
IG	$\sqrt{\dfrac{\lambda}{2\pi x^3}}\exp\left\{-\dfrac{\lambda}{2\mu^2 x}(x-\mu)^2\right\}$	$\exp(\delta(1-\left(1-2\mu^2 it/\lambda\right)^{1/2}))$
Pareto	$ck^c x^{-(c+1)}$	$c(-ikt)^c\Gamma(-c, -ikt)$
Double expo.	$(1/2b)\exp(-|x-a|/b)$	$e^{iat}/\left(1+b^2 t^2\right)$
Chi-square	$x^{n/2-1}e^{-x/2}/[2^{n/2}\Gamma(n/2)]$	$(1-2it)^{-n/2}$
Student's T	$K\left(1+t^2/n\right)^{-(n+1)/2}$	$\exp\left(-|it\sqrt{n}|\right)S_n\left(|it\sqrt{n}|\right)$
F-distribution	$K\dfrac{x^{n/2-1}}{(m+nx)^{(m+n)/2}}$	$K\,{}_1F_1\left(m/2, 1-n/2; -nit/m\right)$
Z-distribution	$K\dfrac{e^{mz}}{(n+me^{2z})^{(m+n)/2}}$	$K\dfrac{e^{(mx)}}{\left(n+e^{(\lambda x)}\right)^{(m+n)/2}}$
Weibull	$ax^{a-1}e^{-x^a}$	$\sum_{k=0}^{\infty}(bit)^k\Gamma\left(1+k/a\right)/k!$
Rayleigh	$x/a^2\,e^{-x^2/(2a^2)}$	$1 - bt\exp\left(-\dfrac{b^2 t^2}{2}\right)\sqrt{\dfrac{\pi}{2}}[erfc\left(\dfrac{bt}{\sqrt{2}}\right)-1]$
Maxwell	$\sqrt{2/\pi}x^2 e^{-x^2/(2a)}/a^3$	$i\{\Phi(at)\sqrt{2/\pi} - \exp\left(-a^2 t^2/2\right)\left(a^2 t^2 - 1\right)\times\Phi(at)$

The constant $K = \Gamma((n+1)/2)/[\sqrt{n\pi}\,\Gamma(n/2)]$ for Student's T; $K = [\Gamma((m+n)/2)/\Gamma(n/2)]$ for F and Z distributions, MGFs are obtained by replacing it by t.

9.7   CUMULANT GENERATING FUNCTIONS (CGF)

The CGF is slightly easier to work with for exponential, normal, and Poisson distributions. It is defined in terms of the MGF as $K_x(t) = \ln\left(M_X(t)\right) = \sum_{j=1}^{\infty} k_j t^j / j!$, where k_j is the jth cumulant. This relationship shows that cumulants are polynomial functions of moments (low-order cumulants can also be exactly equal to corresponding moments). For example, for the general univariate normal distribution with mean $\mu_1 = \mu$ and variance $\mu_2 = \sigma^2$, the first and second cumulants are, respectively, $\kappa_1 = \mu$ and $\kappa_2 = \sigma^2$.

Theorem 9.5 $K_{aX+b}(t) = bt + K_x(at)$

Proof: $K_{aX+b}(t) = \log\left(M_{aX+b}(t)\right) = \log\left(e^{bt}M_X(at)\right) = bt + \log\left(M_X(at)\right) = bt + K_x(at)$ using $\log(ab) = \log(a) + \log(b)$, and $\log(e^x) = x$.   ∎

Corollary 2 CGF of a standardized variable can be expressed as $K_{(X-\mu)/\sigma}(t) = (-\mu/\sigma)t + K_x(t/\sigma)$.

Proof: This follows from the above theorem by setting $a = 1/\sigma$ and $b = -\mu/\sigma$.

   The cumulants can be obtained from moments and vice versa [225, 288]. This holds for cumulants about any origin (including zero) in terms of moments about the same origin.   ∎

9.7.1   Relations Among Moments and Cumulants

We have seen in Chapter 8, equation 8.19 (p. 357) that the central and raw moments are related as $\mu_k = E(X - \mu)^k = \sum_{j=0}^{k} \binom{k}{j} (-\mu)^{k-j} \mu'_j$. As the CGFs of some distributions are easier to work with, we can find cumulants and use the relationship with moments to obtain the desired moment.

Theorem 9.6 The rth cumulant can be obtained from the CGF as $\kappa_r = \frac{\partial^r K_x(t)}{\partial t^r}\big|_{t=0}$.

Proof: We have

$$K_x(t) = \sum_{r=0}^{\infty} \kappa_r t^r / r! = \kappa_0 + \kappa_1 t + \kappa_2 t^2 / 2! + \cdots . \qquad (9.27)$$

As done in the case of MGF, differentiate (9.27) k times and put $t = 0$ to get the kth cumulant. See References 134 and 289 for details.   ∎

📖   **EXAMPLE 9.17   Moments from cumulants**

   Prove that $\kappa_1 = \mu_1, \kappa_2 = \mu_2 = \sigma^2$, and $\kappa_3 = \mu_3 = E(X - \mu)^3$.

Solution 9.17 We know that $K_X(t) = \log\left(M_X(t)\right)$ or equivalently $M_X(t) = \exp\left(K_X(t)\right)$. We expand $M_X(t) = 1 + t/1!\mu_1 + t^2/2!\mu'_2 + t^3/3!\mu'_3 + \cdots$ and substitute for $K_X(t)$ also to get

$$\sum_{r=0}^{\infty} \mu'_r t^r / r! = \exp\left(\sum_{r=0}^{\infty} \kappa_r t^r / r!\right). \tag{9.28}$$

Differentiate n times and put $t = 0$ to get

$$\mu'_{n+1} = \sum_{j=0}^{n} \binom{n}{j} \mu'_{n-j} \kappa_{j+1}. \tag{9.29}$$

Put $n = 0,1$, and so on to get the desired result. There is another way to get the result for low order cumulants. Truncate $M_X(t)$ as $1 + t/1!\mu_1 + t^2/2!\mu'_2 + t^3/3!\mu'_3$. Expand the RHS using $\log(1+x) = x - x^2/2 + x^3/3 - x^4/4$, where $x = t/1!\mu_1 + t^2/2!\mu'_2 + t^3/3!\mu'_3$, and collect similar terms to get

$$K_X(t) = \mu_1 t + \left(\mu'_2 - \mu_1^2\right) t^2/2! + \left(\mu'_3 - 3\mu_1\mu'_2 + 2\mu_1^3\right) t^3/3! + \dots . \tag{9.30}$$

Compare the coefficients of $t^k/k!$ to get $\kappa_1 = \mu_1, \kappa_2 = \mu_2 - \mu_1^2 = \sigma^2$, and $\kappa_3 = \left(\mu'_3 - 3\mu_1\mu'_2 + 2\mu_1^3\right) = E(X - \mu)^3$.   ∎

Next, write $M_X(t)$ as $1 + [tx/1! + (tx)^2/2! + \cdots]$, expand $K_X(t) = \log\left(M_X(t)\right)$ as an infinite series to get

$$\sum_{r=0}^{\infty} \kappa_r t^r / r! = [tx/1! + (tx)^2/2! + \cdots] - [tx/1! + (tx)^2/2! + \cdots]^2/2! + \cdots . \tag{9.31}$$

Equate like coefficients of t to get

$$\kappa_{n+1} = \mu'_{n+1} - \sum_{j=0}^{n-1} \binom{n}{j} \mu'_{n-j} \kappa_{j+1}. \tag{9.32}$$

◼ **EXAMPLE 9.18   Moments of normal distribution from CGF**

Obtain the first three moments of normal distribution using CGF method.

Solution 9.18 We know that the MGF of $N\left(\mu, \sigma^2\right)$ is $\exp\left(t\mu + \frac{1}{2}t^2\sigma^2\right)$. Taking natural log, we get $K_x(t) = t\mu + \frac{1}{2}t^2\sigma^2$. Comparing the coefficients of $t/1!$ and $t^2/2!$, we get $\mu_1 = \mu$ and $\mu_2 = \sigma^2$. As the t^3 term is missing, $\mu_3 = 0$.   ∎

9.8   FACTORIAL MOMENT GENERATING FUNCTIONS (FMGF)

There are two types of factorial moments known as *falling factorial* and *raising factorial* moments. Among these, the falling factorial moments are more popular. The kth (falling) factorial moment of X is defined as $E[X(X-1)(X-2)\cdots(X-k+1)] = E[X!/(X-k)!]$, where k is an integer ≥ 1. It is easier to evaluate for those distributions that have an $x!$ or $\Gamma(x+1)$ in the denominator (e.g., binomial, negative binomial, hypergeometric, and Poisson distributions). The factorial moments and ordinary moments are related through the Stirling number of first kind as follows:–

$$X!/(X-r)! = \sum_{j=0}^{r} s(r,j) X^j \Rightarrow \mu'_{(r)} = \sum_{j=0}^{r} s(r,j)\mu'_j. \qquad (9.33)$$

A reverse relationship exists between the ordinary and factorial moments using the identity $X^r = \sum_{j=0}^{r} S(r,j) X!/(X-j)!$ as $\mu'_r = \sum_{j=0}^{r} S(r,j)\mu'_{(j)}$, where $S(r,j)$ is the Stirling number of second kind [134].

There are two ways to get (falling) factorial moments. The simplest way is by differentiating the PGF (see Section 9.2.1, p. 379). As $P_X(t) = E(t^x) = E\left(e^{x\log(t)}\right) = M_X(\log(t))$, we could differentiate it k times as in equation (9.8), page 380, and put $t = 1$ to get factorial moments.

We define it as $E[(1+t)^x]$, because if we expand it using binomial theorem, we get

$$E[(1+t)^x] = E[1 + tx + t^2 x(x-1)/2! + t^3 x(x-1)(x-2)/3! + \cdots]. \qquad (9.34)$$

By taking term-by-term expectations on the RHS, we get the factorial moments. The raising factorial moment is defined as $E[X(X+1)(X+2)\cdots(X+k-1)] = E[(X+k-1)!/(X-1)!]$. An analogous expression can also be obtained for raising factorials using the expansion $(1-t)^{-x} = \sum_{k=0}^{\infty} \binom{k+x-1}{k} t^k$. Taking term-by-term expectations as $E[(1-t)^{-x}] = E[1 + tx + t^2 x(x+1)/2! + t^3 x(x+1)(x+2)/3! + \cdots]$, we get raising factorial moments. We could also get raising factorial moments from PGF $P_x(t) = E(t^{-x})$. Differentiating it once gives $P'_x(t) = E\left(-xt^{-x-1}\right)$. From this, we get $P'_x(1) = E(-x)$. Differentiating it r times, we get $P_x^{(r)}(t) = E(-x(-x-1)(-x-2)\cdots(-x-r+1)t^{-x-r})$. Putting $t = 1$, this becomes

$$P_x^{(r)}(1) = (-1)^r E[x(x+1)\cdots E(x+r-1) = (-1)^r \mu^{(r)}. \qquad (9.35)$$

Replacing the summation by integration gives us the corresponding results for the continuous distributions.

To distinguish between the two, we will denote the falling factorial moment as $E\left(X_{(k)}\right)$ or $\mu_{(k)}$ and the raising factorial moment as $E\left(X^{(k)}\right)$ or $\mu^{(k)}$. Unless otherwise specified, factorial moment will mean falling factorial moment $\mu_{(k)}$.

▉ **EXAMPLE 9.19   Factorial moment of the Poisson distribution**

Find the kth factorial moment of the Poisson distribution, and obtain the first two moments.

Solution 9.19 By definition

$$\mu_{(k)} = \sum_{x=0}^{\infty} x(x-1)(x-2)\cdots(x-k+1)\, e^{-\lambda}\lambda^x/x! = e^{-\lambda}\lambda^k \sum_{x=k}^{\infty} \lambda^{x-k}/(x-k)!$$

$$= e^{-\lambda}\lambda^k \sum_{y=0}^{\infty} \lambda^y/y! = e^{-\lambda}\lambda^k e^{\lambda} = \lambda^k. \tag{9.36}$$

Alternatively, we could obtain the FMGF directly and get the desired moments. FMGF$_X(t) = E[(1+t)^x] = \sum_{x=0}^{\infty}(1+t)^x e^{-\lambda}\lambda^x/x! = e^{-\lambda}\sum_{x=0}^{\infty}[\lambda(1+t)]^x/x! = e^{-\lambda}e^{\lambda(1+t)} = e^{\lambda t}$. The kth factorial moment is obtained by differentiating this expression k times and putting $t = 0$. We know that the kth derivative of $e^{\lambda t}$ is $\lambda^k e^{\lambda t}$, from which the kth factorial moment is obtained as λ^k. Putting $k = 1$ and 2 gives the desired moments.   ▉

Corollary 3 The factorial moments of a Poisson distribution are related as $\mu_{(k)} = \lambda^r \mu_{(k-r)}$. In particular, $\mu_{(k)} = \lambda\mu_{(k-1)}$.

Proof: This follows easily because $\mu_{(k)}/\mu_{(k-r)} = \lambda^k/\lambda^{k-r} = \lambda^r$.   ▉

▉ **EXAMPLE 9.20   Factorial MGF of the geometric distribution**

Find the factorial MGF of the geometric distribution.

Solution 9.20 By definition FMGF$_X(t) =$

$$E[(1+t)^x] = \sum_{x=0}^{\infty}(1+t)^x q^x p = p\sum_{x=0}^{\infty}[q(1+t)]^x = p/[1-q(1+t)]. \tag{9.37}$$

As $1-q = p$, the denominator becomes $[1-q(1+t)] = p - qt$. Hence, FMGF$_X(t) = p/(p-qt)$. The kth factorial moment is obtained by differentiating this expression k times and putting $t = 0$. We know that the kth derivative of $1/(ax+b)$ is $a^r r!(-1)^r/(ax+b)^{r+1}$. Hence, kth derivative of $1/[1-q(1+t)]$ is $r!q^r/p^{r+1}$, as $q = 1-p$. This gives the kth factorial moment as $pr!q^r/p^{r+1} = r!(q/p)^r$.   ▉

9.9   CONDITIONAL MOMENT GENERATING FUNCTIONS (CMGF)

Consider an integer random variable that takes values ≥ 1. We define a sum of independent random variables as $S_N = \sum_{i=1}^{N} X_i$. For a fixed value of $N = n$, the distribution of the finite sum S_n can be obtained in closed form for many distributions when the variates are independent (see Table 7.47, Chapter 7). The conditional MGF can be expressed as

$$M_{X|Y}(t) = \int_x f(x|y) e^{tx} dx. \tag{9.38}$$

Replacing t by "it" gives the corresponding conditional characteristic function. If the variates are mutually independent, this becomes $M_{S|N}(t|N) = [M_X(t)]^N$.

9.10 CONVERGENCE OF GENERATING FUNCTIONS

Properties of generating functions are useful in deriving the distributions to which a sequence of generating functions converge. Let X_n be a sequence of random variables with ChF $\phi_X^n(t)$. If $\lim_{n\to\infty} \phi_X^n(t)$ converges to a unique limit, say $\phi_X(t)$ for all points in a neighborhood of $t = 0$, then that limit determines the unique CDF to which the distribution of X_n converge. Symbolically,

$$\lim_{n\to\infty} \phi_X^n(t) = \phi_X(t) \Rightarrow \lim_{n\to\infty} F_{X_i}(x) = F(x). \tag{9.39}$$

9.11 SUMMARY

This chapter introduced various generating functions encountered in statistics. These have wide applications in many other fields including astrophysics, fluid mechanics, spectroscopy, and various engineering fields. Examples are included to illustrate the use of various generating functions. The classical probability generating function of discrete distributions is extended to get a new generating function for the CDF. This is then used to derive the mean deviation by extracting just one coefficient.

See References 134, 284, 287 and 290 for further information.

EXERCISES

9.1 Mark as True or False

    a) PGF of discrete distribution can be obtained from its MGF

    b) PGFs of continuous distributions do not exist

    c) MGF is defined only for positive random variables

    d) Moments of order $k > n$ of a BINO(n,p) are all zeros

    e) All moments of F distribution are functions of numerator df

    f) All odd moments of Cauchy distribution are nonexistent

    g) All characteristic functions are periodic with period 2π

    h) ChF of a sum of random variables is always the product of the ChF.

9.2 Find the CDF and MGF of the distribution $f(x; a, b) = \frac{a}{b^a} x^{a-1} e^{-(x/b)^a}$, and obtain the first two moments.

9.3 Prove that $M_{ax}(t) = M_x(at)$ and $M_{ax+b}(t) = e^{bt} M_x(at)$. Deduce that $M_{(x-\mu)/\sigma}(t) = e^{-\mu t/\sigma} M_x(t/\sigma)$.

9.4 For Geometric distribution with PMF $f(x) = q^x p$, find $E(X)$ and $E(X^2)$ and deduce Var(X).

9.5 If $X_1, X_2, \ldots, X_n$ are IID and $S_n = X_1 + X_2 + \cdots + X_n$, prove that $M_{S_n}(t) = \prod_{i=1}^{n} M_{x_i}(t)$.

9.6 Find the MGF and derive the first two moments for the distribution $f(x) = \binom{x-1}{r-1} p^r q^{x-r}, x = r, r+1, \ldots$

9.7 What do you get as the result of $\frac{\partial^k}{\partial t^k} P_X(t)|_{t=1}$, where $P_X(t)$ is the probability generating function?

9.8 If $M_x(t) = 1/(b^2 - t^2)$ find the mean and variance.

9.9 If $X \sim$ NBIN(n, p), prove that the PGF is $P_x(t; n, p) = \left(\frac{p}{1-qt}\right)^n$.

9.10 Let $\{a_n\}$, $n = 0, 1, 2, \ldots, \infty$ be a sequence of bounded numbers with EGF $\sum_{n=0}^{\infty} a_n x^n / n!$. Find the sequence whose EGF is $e^x * \sum_{n=0}^{\infty} a_n x^n / n!$.

9.11 Find the PGF, mean, and variance of $f(x) = \left(1 - \frac{\mu}{n}\right)^n \binom{n}{x} \left(\frac{\mu}{n-\mu}\right)^x$, $x = 0, 1, \ldots n$. Derive the PGF of $n - X$ and obtain the GFCDF.

9.12 For the logarithmic series distribution $f(x) = c\theta^x / x$, where $0 < \theta < 1, x = 1, 2, \ldots$, and $c = -1/\log(1-\theta)$, find the MGF. Prove that the rth factorial moment $\mu_{(r)} = c(r-1)![\theta(1-\theta)]^r$.

9.13 What are the conditions for a function to be a moment generating function? Are these functions true MGF? (a) $e^{a(t-1)+b(t-1)^2}$, (b) $e^{a(t-1)/(1-bt)}$, (c) $e^{a(t-1)+b(t^2-1)}$, (d) $e^{|(t-1)/(t^2-1)|}$

9.14 Find the MGF and first three moments of the triangular distribution.

9.15 Find the ChF and first two moments of the Pareto distribution.

9.16 Prove that $K_{X+Y}(t) = K_X(t) + K_Y(t)$ if X and Y are independent.

9.17 When is $P_{X_1+X_2+\cdots+X_n}(t) = \prod_k P(X_i(t))$

9.18 Prove that the MGF of a truncated exponential distribution is $\frac{1}{\lambda-t}[\lambda - t\exp(-m(\lambda-t))]$ for $t \neq \lambda$.

9.19 The MGF of a random variable is $(e^{2t} - e^{-2t})/4t$. Find the mean, and the probability that $P[|x-1] < 0.5$.

9.20 Show that the MGF of CUNI $(-c, c)$ with PDF $f(x; c) = 1/(2c)$ for $-c < x < c$ is $\sinh(ct)/ct$. Prove that the even moments are given by $\mu_{2n} = c^{2n}/(2n+1)$.

9.21 Check whether the PGF exists for a discrete distribution with $f(x) = c/[x(x+1)]$, for $x = 1, 2, \ldots$ where c is constant. Obtain GFCDF and the MD.

9.22 If X and Y are related as $X = (2Y - (a+b))/(b-a)$, find the moments of X from those of Y. Find the relation among MGF of X and Y.

9.23 If X and Y are IID, and $Z = cX + (1-c)Y$, find the MGF of Z in terms of MGF of X and Y. Deduce the first two moments.

9.24 How are the PGF $(P_x(t))$, MGF $(M_x(t))$, ChF $(\phi_x(t))$, and KGF $(K_x(t))$ related.

9.25 If $X \sim \text{BINO}(n, p)$ find the PGF of the random variable $n - X$. What does $P_x(t = -1)$ give?

9.26 Obtain the CDF generating function for the Poisson distribution, and obtain the mean deviation.

9.27 Obtain the CDF generating function for the Logarithmic series distribution, and obtain the mean deviation.

9.28 If $X \sim N(\mu, \sigma^2)$, and $g(x)$ is a differentiable function of x with $E[X] < \infty$, then $E[g(x)(x - \mu)] = \sigma^2 E[g'(x)]$.

9.29 If $f(y|x) = \exp(-(y - x))$ for $y \geq x$, prove that $E(Y|X = x) = (1 + x)$.

9.30 If X is a positive random variable with PGF $P_x(t)$ find the PGF of the random variables $X - c$, and $|X|$.

9.31 Prove that for a non-negative random variable $E[X^2] = 2 \int_0^\infty x[1 - F_X(x)]dx$.

9.32 In a sequence of IID Bernoulli trials with probability of success p, suppose we count either the number of successes needed to get the first Failure, or the number of Failures needed to get the first Success. Find the PDF and the probability generating function.

9.33 The Maxwell distribution gives the velocity of a molecule at absolute temperature x as $f(X) = 4\pi x^2 (m/2\pi KT)^{3/2} \exp(-mx^2/2KT)$ where m = molecular weight, K = Boltzmann constant. Find the PGF and expected velocity at room temperature t_r.

9.34 If the PDF of a continuous random variable is $f(x) =$

$$
\begin{cases}
2(b + x)/[b(a + b)] \\
\qquad \text{if } -b \leq x < 0 \\
2(a - x)/[a(a + b)] \\
\qquad \text{if } 0 \leq x < a \\
0 \\
\qquad \text{elsewhere}
\end{cases}
$$

find the MGF and the mean.

9.35 If the PDF of a continuous random variable is $f(x) =$

$$
\begin{cases}
(a + x)/a^2 & \text{if } -a \leq x < 0 \\
(a - x)/a^2 & \text{if } 0 \leq x < a \\
0 & \text{elsewhere}
\end{cases}
$$

find the MGF and the mean.

9.36 What does $[K_X(t) + K_X(-t)]/2$ generate?

9.37 What does $[P_X(t) + P_X(-t)]/2$ generate?

9.38 If all moments of X exist, prove that $\phi(t) = \sum_{k=0}^{\infty} (it)^k/k! \, \mu'_k$

9.39 If $f(x) = (c + 1)x^c$ for $0 < x < 1$, find the PGF and $E[\ln(X)]$.

9.40 If $\phi(x)$ is a real-valued, monotonic function of a positive random variable X, prove that $E[\phi(x)] = \phi(0) + \int_0^\infty [1 - F(x)]\partial\phi(x)/\partial x$. Hence derive that $E[X^n] = n \int_0^\infty x^{n-1}[1 - F(x)]dx$.

9.41 If $Y \sim$ BETA$(a, b-a)$, prove that $M_x(t) = {}_1F_1(a, b; y) = K \int_0^1 e^{yt} t^{a-1}$ $(1-t)^{b-a-1} dt$ where $K = \frac{\Gamma(b)}{\Gamma(a)\Gamma(b-a)}$ and ${}_1F_1(a, b; y)$ is the confluent hypergeometric function. Hence, show that $E(Y) = a/b$.

9.42 Find the factorial MGF of hypergeometric and negative binomial distributions and show that the factorial moments are as given: (i) HYPG(N,n,p): $n_{(r)} N p_{(r)} / N_{(r)}$ if $f(x) = \binom{Np}{k} \binom{Nq}{n-k} / \binom{N}{n}$. (ii) NBIN$(n,p)$: $r(r+1)..(r+s)$ $(q/p)^s$.

9.43 Find the factorial MGF of binomial and geometric distributions and show that the factorial moments are as given: (i) BINO(n, p): $n_{(r)} p^r$. (ii) GEOM(p): $r! q^{r-1} / p^r$.

10

FUNCTIONS OF RANDOM VARIABLES

> **After finishing the chapter, students will be able to**
>
> - Understand distribution of functions of random variables
> - Distinguish linear and other transformations of random variables
> - Comprehend trigonometric transformations of random variables
> - Describe arbitrary transformations
> - Apply various transformations to practical problems

10.1   FUNCTIONS OF RANDOM VARIABLES

This chapter discusses the distribution of functions of a single random variable. There are many situations where simple functions of random variables have well-known distributions. One example is the relation between a standard normal and a chi-square distribution. As shown below, the square of a standard normal is chi-square distributed with one DoF. If there are several independent normal variates, the sum of the squares is also chi-square distributed with n DoF, where n is the number of variates.

Distribution of a function of random variable(s) has many applications in statistical inference. For instance, any general normal variate can be transformed into the standard normal ($N(0, 1)$) using a simple change of origin and scale transformation. The classical method known as the *method of distribution function* (MoDF) is useful when the CDF has closed form. The CDF of the transformed variable is easily

Statistics for Scientists and Engineers, First Edition. Ramalingam Shanmugam and Rajan Chattamvelli.
© 2015 John Wiley & Sons, Inc. Published 2015 by John Wiley & Sons, Inc.

TABLE 10.1   Summary Table of Transformation of Variates

Transformation $y = h(x)$	Transformed Density Function	Comments
$c * X + d$	$\frac{1}{|c|}f((y-d)/c)$	$c \neq 0$, not near 0
$|X|$	$g(y) = f(y) + f(-y)$	$g(y) = 2f(y)$ if Y symmetric
$|X - c|$	$g(y) = f(c+y) + f(c-y)$	$g(y) = 2f(c+y)$ if Y is symm.
x^2	$f(\sqrt{y})/2\sqrt{y}$	x positive
x^2	$[f(\sqrt{y})+f(-\sqrt{y})]/2\sqrt{y}$	$-\infty < x < \infty$
cx^2	$[f(\sqrt{y/c})+f(-\sqrt{y/c})]/2\sqrt{y/c}$	Any x
$1/[cx^2]$	$1/[2\sqrt{c}y^{3/2}]f(1/\sqrt{cy})$	
$\sqrt{x}$	$2yf(y^2)$	$yf(y^2)$ if Y is symmetric
$\sqrt{cx}$	$2(y/c)f(y^2/c)$	$2yf(y^2)$ if $c = 1$
$1/x$	$f(1/y)/y^2$	$x \neq 0$
x^α	$\frac{1}{\alpha}f(y^{1/\alpha})y^{1/\alpha - 1}$	$0 < \alpha < 1, x > 0$
x^n	$\frac{y^{1/n-1}}{n}[f(y^{1/n})+f(-y^{1/n})]$	For $y > 0$
e^x	$f(\ln(y))/y$	log to base e
e^{ax}	$f(\ln(y)/a)/ay$	log to base e
$\tan^{-1}(x)$	$f(\tan y)\sec^2(y)$	
$-2\ln(x)$	$\frac{1}{2}e^{-y/2}f(e^{-y/2})$	log base e
$-\ln(1-x)/\lambda$	$\lambda e^{-\lambda y}f(1 - e^{-\lambda y})$	log base e
$1/[1 + e^{-x}]$	$f(\ln(y/(1-y)))/[y(1-y)]$	Sigmoid function
$\ln\left(\dfrac{x}{1-x}\right)$	$f\left(\dfrac{e^y}{e^y+1}\right)\dfrac{e^y}{(1+e^y)^2}$	$0 < x < 1$
$\ln\left(\dfrac{1+x}{1-x}\right)$	$f\left(\dfrac{e^y-1}{e^y+1}\right)2\dfrac{e^y}{(1+e^y)^2}$	$0 < x < 1$
$\log_2(1+x)$	$\ln(2) 2^y f(2^y - 1)$	x is non-negative

The ranges are not shown above as it depends on the range of the original variate. Note that $\log x$ and $\sqrt{x}$ are concave functions.

obtained using MoDF when the transformation is strictly increasing or decreasing function, from which the PDF follows readily by differentiation (see table above) in the continuous case. This method can be used to find the PDF of the logistic, Gumbel, Pareto, and uniform distributions.

We come across functions of a variate in engineering applications as well. These are sometimes governed by physical laws such as that between resistance and current in a circuit. These are usually represented as exact or approximate functional relationships among dependent and independent variables. If the distribution of one of them is known, the behavior of the other can easily be modeled. As an example, consider the electrical resistance of semiconductors, which depends on the temperature in °K as

$$r(t) = (a/t^b)e^{c/t} \quad \text{where } t = \text{temperature in K°}, \quad a, b, c \text{ are constants.} \quad (10.1)$$

If the temperature variation is known, the resistance distribution can be derived for some values of the parameters. Similarly, the potential energy of weakly interacting dielectric gas subjected to an external electric field is modeled using an exponential law as $f(y) = \exp(-y/(KT))$, where y = potential energy, K = Boltzmann constant, and T = absolute temperature in °K.

There are many ways to derive such related distributions (see References 137, 225, 291–293. The most popular among them are the (i) CDF method (ii) MGF (or ChF) method (iii) trigonometric transformations (iv) geometric reasoning and (v) using Jacobians.

10.2 DISTRIBUTION OF TRANSLATIONS

These are obtained by a change of origin transformation $Y = X \pm c$. As the variate values are shifted either to the right (c is positive) or to the left (c is negative), the PDF remains the same, but the range is modified accordingly. A special translation is called reflection as $Y = -X + c$, where c is a location measure. If X is symmetric, $Y = -X + c$ results in the same distribution if c is the mean, median, or mode (all of which coincides for symmetric laws). This transformation is usually applied along with one of the following transformations such as change of scale or square transformation.

■ **EXAMPLE 10.1   Translations of CUNI(a, b) Distribution**

If X is CUNI(a, b), find the distribution of (i) $Y = X - a$ (ii) $Y = (X - (a + b)/2)$.

Solution 10.1 As both of these are change of origin transformations, the PDF remains the same as $f(y) = 1/(b - a)$. In case (i) the lower limit becomes 0 and upper limit becomes $(b - a)$, so that $f(y) = 1/(b - a), 0 < y < (b - a)$. In case (ii) the lower limit is $a - (a + b)/2 = (a - b)/2$, and the upper limit is $b - (a + b)/2 = (b - a)/2 = f(y) = 1/(b - a)$. This gives $f(y) = 1/(b - a), -(b - a)/2 < y < (b - a)/2$. As $(a + b)/2$ is the mean of a rectangular distribution, this transformation is a reflection.   ■

10.3 DISTRIBUTION OF CONSTANT MULTIPLES

First, consider the case of a discrete random variable X. If c is an integer, then $Y = c * X$ is a change of scale transformation that maps X values to positive or negative numbers. Hence, depending on the sign of c, Y could belong to the same family of distribution. As an example, if X is Poisson (λ) with mgf $\exp(\lambda(e^t - 1))$, Y has mgf $\exp(\lambda(e^{ct} - 1))$. This is the MGF of a Poisson distribution. If c is a fraction, the distribution is still well defined if we assume that X takes fractional values (but still it is discrete distribution due to the discontinuity).

■ **EXAMPLE 10.2   Constant Multiple of CUNI(a, b) Distribution**

If X is CUNI(a, b) find the distribution of $Y = (2X - (a + b))/(b - a)$.

Solution 10.2 Write $Y = (2X - (a + b))/(b - a)$ as $c * X + d$, where $c = 2/(b - a)$ and $d = -(a + b)/(b - a)$. Solve for X to get $x = (Y - d)/c$. Then use (10.3) to get $g(y) = f((y - d)/c)/c$. As X is CUNI(a, b), $f(x) = 1/(b - a), a < x < b$. As this does not involve x, putting $x = (y - d)/c$ has no effect. The range is modified as -1 and $+1$. Substitute the values of c and d to get $g(y) = (1/(b - a))/(2/(b - a)) = 1/2, -1 < y < 1$.   ■

10.4   METHOD OF DISTRIBUTION FUNCTIONS (MODF)

The MoDF is a simple and powerful method to find the PDF of a variety of continuous transformations. Consider the general transformation $Y = h(X)$. The MoDF works when (i) $h(x)$ is either an increasing or a decreasing function of x without discontinuities, (ii) the first derivative of $h(x)$ exists throughout the range of the variate, (iii) $h(x)$ is invertible (so that $x = h^{-1}(y)$, is uniquely solvable), and (iv) $F(x)$ is differentiable once. We illustrate the use of MoDF for various forms of $h(x)$ in their respective sections. Consider the transformation $Y = h(x) = c * X + b$, where X has a known distribution. It satisfies all the three conditions on $h(x)$. The CDF of Y is

$$G(y) = P(Y \leq y) = P(cX + b \leq y) = P(X \leq (y - b)/c) = F((y - b)/c). \quad (10.2)$$

Note that in equation (10.2), $G(.)$ is the CDF of Y, and $F(.)$ is the CDF of X. As the CDF of Y contains only y; and the constants (b, c), we have simply expressed y in terms of x. Differentiate with respect to y to get

$$g(y) = (\partial/\partial y) \ F((y - b)/c) = (1/c) * f((y - b)/c). \quad (10.3)$$

Here, we need to consider two cases as $c > 0$ or $c < 0$. The above result holds for $c > 0$. When $c < 0$ ($Y = h(X)$ is a decreasing function), we get

$$G_Y(y) = \Pr(cX + b \leq y) = \Pr(X \geq (y - b)/c) = 1 - F((y - b)/c). \quad (10.4)$$

Differentiation gives us $g(y) = (-\partial/\partial y)F((y - b)/c) = (-1/c) * f((y - b)/c)$. Combine both cases to get

$$g(y) = (\pm\partial/\partial y) \ F((y - b)/c) = (1/|c|) * f((y - b)/c), \quad (10.5)$$

where the vertical line means "absolute value," which absorbs the $\pm$ sign. Thus, the PDF of Y is easily obtained from that of X. The only assumption we have made is that $F(y)$ is differentiable, as the other conditions are satisfied by the linear transformation. The constant c can be any nonzero real number. Depending on whether $|c| < 1$ or > 1, the new range is either expanded or contracted.

10.4.1   Distribution of Absolute Value ($|X|$) Using MoDF

The MoDF can easily be adapted to find the distribution of $Y = |X|$. This is meaningful only when X takes both positive and negative values (if X assumes only negative values, then $|x| = -x$). As above, let $G(y)$ be the CDF of Y. Then,

$$G(y) = P(Y \le y) = P(|X| \le y) = P(-y \le X \le y) = F(y) - F(-y). \qquad (10.6)$$

Differentiate both sides to get $g(y) = f(y) + f(-y)$. In the particular case when Y is symmetric, $f(y) = f(-y)$ so that $g(y) = 2 f(y)$.

■ **EXAMPLE 10.3   Absolute value of CUNI$(-\pi/2, \pi/2)$**

If X is CUNI$(-\pi/2, \pi/2)$ variate, find the distribution of $Y = |X|$.

Solution 10.3 As the CUNI$(-\pi/2, \pi/2)$ distribution exhibits special symmetry around zero point, $f(-x) = f(x)$. We know $f(x) = 1/\pi$ for $-\pi/2 \le x \le \pi/2$. Thus, using equation (10.8) the PDF of $Y = |X|$ is $f(y) = 2 * f(x) = 2/\pi$ for $0 \le y \le \pi/2$. ■

■ **EXAMPLE 10.4   Distribution of Absolute value of Cauchy Variate**

If X is a standard Cauchy variate, find the distribution of $Y = |X|$.

Solution 10.4 We know that $f(x) = 1/[\pi(1 + x^2)]$, which is symmetric in x. Using equation (10.8), the distribution of Y is $2 * f(x) = 2/[\pi(1 + x^2)]$ for $0 \le x \le \infty$. ■

10.4.2   Distribution of $F(x)$ and $F^{-1}(x)$ Using MoDF

Let $F(x)$ be the CDF and $F^{-1}(x)$ be the inverse CDF of a continuous random variable. Obviously, the minimum value that $F(x)$ can take is 0, and the maximum value is 1. The following section derives the distribution of $F(x)$, irrespective of the range.

10.4.2.1   Distribution of $F(x)$ If X is a continuous variate, $U = F(x)$ is uniformly distributed in $[0, 1]$. Consider

$$F(u) = P(U \le u) = P(F(x) \le u) = P(x \le F^{-1}(u)) = F[F^{-1}(u)] = u. \qquad (10.7)$$

We have seen in Chapter 7 that the CDF of a rectangular distribution is $(x - a)/(b - a)$. Put $a = 0, b = 1$ to get $F(x) = x$. Equation (10.7) then shows that $F(U)$ has a unit rectangular distribution.

10.4.2.2 Distribution of $F^{-1}(x)$ Distribution of $Y = F^{-1}(x)$ is well tractable in the continuous case when x is defined on unit interval. If we define $F^{-1}(x)$ as the minimum value of y satisfying $F(y) \geq x$, we could use the MoDF to find the distribution in certain cases. Obviously, $F^{-1}(x)$ is nondecreasing and satisfy (i) $F^{-1}(F(x)) \leq x$, for $-\infty < x < \infty$ and (ii) $F(F^{-1}(y)) \geq y$ for $0 < y < 1$. Let $Y = F^{-1}(x)$. If there are no discontinuities,

$$G(y) = P(Y \leq y) = P(F^{-1}(x) \leq y) = P(x \leq F(y)) = F[F(y)]. \tag{10.8}$$

where we have used $F(F^{-1}(x)) = x$ (strictly). In the particular case, when y is CUNI[0,1], we have $F(y) = y$. Substitute in the last term (in square bracket) of equation (10.8) to get the CDF of $F^{-1}(x)$ as $G(y) = F(F(y)) = F(y) = y$ showing that Y is uniformly distributed with unit range. Next, consider the general case. As the derivative of $F(F(y))$ is unambiguously determined when the argument (of outer $F()$) ranges over the unit interval, we could differentiate both sides of equation (10.8) to get

$$g(y) = (\partial/\partial y) \, F[F(y)] = (\partial/\partial F(y)) \, F[F(y)] * \partial(F(y)/\partial y) = f[F(y)] * f(y) \tag{10.9}$$

where we have used the function-of-a-function rule of differentiation because inner $F(y)$, being the CDF, satisfies $0 \leq F(y) \leq 1$. Use $F(y) - F(y - 1) = f(y)$ in the discrete case. An application of the above result is to find the mean deviation of continuous distributions given below.

Theorem 10.1 If $f(F^{-1}(x))$ of a continuous distribution has closed form, and is integrable in the proper range, the mean deviation is given by

$$MD = 2 \int_{t=0}^{F(\mu)} t \, dt / f(F^{-1}(t)). \tag{10.10}$$

Proof: The proof follows easily by putting $y = F^{-1}(x)$ and using properties of distribution functions in the above result. This is illustrated below.   ∎

EXAMPLE 10.5   Mean deviation of Exponential Distribution

If X is EXP(λ) find the mean deviation.

Solution 10.5 Consider the exponential distribution $f(x) = \lambda e^{-\lambda x}$ with mean $\mu = 1/\lambda$ and CDF $1 - e^{-\lambda x}$. Put $x = 1/\lambda$ in the CDF to get $F(\mu) = 1 - e^{-\lambda(1/\lambda)} = 1 - e^{-1} = (e - 1)/e$. The inverse CDF is $F^{-1}(x) = -(\log (1 - x))/\lambda$, where log is to the base e (i.e., $\ln(1 - x)$). Put the values in equation (10.10) to get the simple integral

$$MD = 2 \int_{t=0}^{(e-1)/e} t dt / \lambda(1 - t). \tag{10.11}$$

where we have used

$$f(F^{-1}(x)) = \lambda e^{(-\lambda)*(-\ln\ (1-x)/\lambda)} = \lambda e^{\ln\ (1-x)} = \lambda(1-x).$$

Write t in the numerator as $1 - (1 - t)$ and split the integral into two. They evaluate to $(1/e - 1)/\lambda$ and $1/\lambda$. Adding them together gives the mean deviation as $1/(e\lambda)$.   ∎

◨ EXAMPLE 10.6   Inverse of CUNI(a, b) Distribution Function

If X is CUNI(a, b) find the distribution of $F^{-1}(x)$.

Solution 10.6 We know that the CDF of CUNI(a, b) is $F(x) = (x - a)/(b - a)$. Using equation (10.9), the PDF of $Y = F^{-1}(x)$ is $f((y - a)/(b - a)) * f(y)$. As $f(y) = 1/(b - a)$ irrespective of the value of y, and $f((y - a)/(b - a))$ has unit range, we get the PDF of Y as $g(y) = 1/(b - a), a \le y \le b$.   ∎

◨ EXAMPLE 10.7   Mean deviation of Uniform Distribution

If X is CUNI(a, b), find the mean deviation.

Solution 10.7 As the CDF of CUNI(a, b) is $F(x) = (x - a)/(b - a), F(\mu) = F((a + b)/2) = 1/2$ (this also follows trivially from the fact that CUNI(a,b) has special symmetry so that the area up to the mean is 1/2). As the density is constant throughout the range, $f(F^{-1}(t)) = 1/(b - a)$ always. Substitute these values in equation (10.10) to get the MD as

$$\text{MD} = 2\int_{t=0}^{1/2} t\,dt/(1/(b-a)) = 2(b-a)\ t^2/2\big|_0^{1/2} = (b-a)/4. \qquad (10.12)$$

This tallies with the result given in page 264.   ∎

10.5   CHANGE OF VARIABLE TECHNIQUE

The Change of Variable Technique (CoV-T) (also called Transformation of Variable Technique (ToV-T)) is a useful method to find distributions of simple continuous differentiable functions of real-valued random variables. It works on the principle that the average value of an integral $f_{\text{avg}}(x) = 1/(b - a) \int_a^b dF(x)$ can be equalized under an arbitrary continuous transformation $y = h(x)$ by the integral $g_{\text{avg}}(y) = 1/(h(b) - h(a)) \int_{h(a)}^{h(b)} dG(y)$ provided that $h(b) \rightarrow h(a)$ as $b \rightarrow a$. Equating the above gives

$$\int_a^b dF(x) = (b - a)/[(h(b) - h(a))] \int_{h(a)}^{h(b)} dG(y). \qquad (10.13)$$

Now consider $\lim_{b \to a}[(h(b) - h(a))/(b - a)]$, which is the limiting value of the derivative $h(x)$ at $x = a$. If this derivative of $h'(x)$ exists for each point x in the interval (a, b), then the RHS will be finite. When $b \to a$ from above, the LHS integral approaches $f(x)$ and RHS integral approaches $g(y)/h'(y)$. This allows us to equate the width of an infinitesimal strip under the function $f(x)$ as $f(x)dx = g(y)dy$ for all points x in (a, b). Because $y = h(x)$ is invertible, we get $g(y) = f(x)|\partial h(y)/\partial y| = |1/J|f(h^{-1}(y))$, where J is called the Jacobian of the transformation. The only conditions in this transformation are that the mapping is once differentiable (i.e., $h'(x)$ exists) and it is invertible (x can be expressed in terms of y). This can also be proved using the CDF method as follows.

Theorem 10.2 The CDF of one-variable transformation is given by $G(y) = F(u^{-1}(y))$ if $u(x)$ is strictly increasing and $1 - F(u^{-1}(y))$ if $u(x)$ is strictly decreasing. The PDF in both cases is $g(y) = f(u^{-1}(y))|\frac{\partial x}{\partial y}|$.

Proof: If $u(x)$ is a strictly increasing function and invertible,

$$G(y) = \Pr[Y \le y] = \Pr[u(x) \le y] = \Pr[x \le u^{-1}(y)] = F(u^{-1}(y)). \qquad (10.14)$$

As the derivative is positive, the transformation results in $g(y)\partial y = f(x)\partial x$, so that $g(y) = f(u^{-1}(y))|\frac{\partial x}{\partial y}|$. If $u(x)$ is a strictly decreasing function, the derivative is negative so that the transformation results in $G(y) = 1 - F(u^{-1}(y))$ and $g(y) = -f(u^{-1}(y))|\frac{\partial x}{\partial y}|$. This is the reason why we take the absolute value of the Jacobian. These results can easily be generalized to n-dimensions, as shown in Chapter 11.   ∎

The standardization transformation $Z = (X - \mu)/\sigma$, where μ is the mean and σ is the standard deviation, is the simplest and the most frequent CoVT. The Jacobian in this case is $|\partial z/\partial x| = 1/\sigma$. When applied to an arbitrary normal distribution, this results in a standard normal distribution, which is extensively tabulated.

10.5.1   Linear Transformations

Any random variable X can be transformed linearly using $y = cx + b$, where $c \ne 0$. As they are linearly related, we could directly invert it to get $x = (y - b)/c$ and $|J| = 1/|c|$. Let $g_Y(y)$ denote the PDF of Y, and $f_X(x)$ denote the PDF of X. Then,

$$g(y) = f(x)|\partial x/\partial y| = |J|f(h^{-1}(y)) = (1/|c|)f((y - b)/c), \qquad (10.15)$$

which is the same result obtained by the MoDF technique. When the variate is discrete, we simply ignore the $1/|c|$ multiplier. In general, if the transformation is $Y = h(X)$, the PDF of Y is $g(y) = f(h^{-1}(y))$ if X is discrete and $g(y) = f(h^{-1}(y))|\partial h^{-1}(y)/\partial y|$ if X is continuous.

◼ **EXAMPLE 10.8** **Linear functions of binomial distribution**

If $X \sim \text{BINO}(n, p)$ find distributions of $Y = n - X$, and find $E(Y), V(Y)$.

Solution 10.8 As $Y = n - X$ takes integer values in reverse, it has the same distribution. The PMF is $f(y) = \binom{n}{(n-y)} p^{n-y} q^y$. Using $\binom{n}{(n-y)} = \binom{n}{y}$, this can also be written as $f(y) = \binom{n}{y} q^y p^{n-y}$, $y = 0, 1, .., n$; which is the PMF of a $\text{BINO}(n, q)$. Hence, $E(Y) = nq, V(Y) = npq$. ◼

10.6 DISTRIBUTION OF SQUARES

Distribution of squares is especially important in statistical inference and analysis of variance. This is because we encounter sums of squares or functions thereof. For instance, the ANOVA procedure is dependent on decomposing the total sum of squares as between treatment and within sums of squares. Similarly, confidence intervals for variances are constructed using the distribution of sample variance, and testing of regression coefficients in multiple linear regression (MLR) uses the ratio of sums of squares. All of these require the distribution of appropriate sums of squares under normality assumption. In such cases we need to find the distribution of sums of independent normal random variates. Although the distribution of squares of other random variables is seldom used in practice, they do have great theoretical significance. A special case of the above is the relation between Student's T and Snedecor's F distributions. If $T \sim T_n$, then the PDF of T_n^2 has an F distribution with 1 and n DoF. By definition, $T = Z / \sqrt{\chi_n^2/n}$ so that $T_n^2 = Z^2/(\chi_n^2/n)$. As the numerator and denominator are independent, both of them are chi-squared distributed so that their ratio has an F distribution.

There exist many methods to derive the distribution of squares. Let $Y = X^2$ so that $dy/dx = 2x$ and $dx/dy = 1/(2\sqrt{y})$. If X takes positive values, the PDF of Y is obtained as

$$g(y) = f(x)|\partial x/\partial y| = f(\sqrt{y})/(2\sqrt{y}). \tag{10.16}$$

The corresponding relationship between distribution functions for strictly increasing functions is

$$G(y) = \Pr(X^2 \leq y) = \Pr[-\sqrt{y} \leq X \leq \sqrt{y}] = F(\sqrt{y}) - F(-\sqrt{y}). \tag{10.17}$$

As (10.17) is valid for any $x(-\infty < x < \infty)$, we differentiate it with respect to y to get the PDF as $g(y) = (f(\sqrt{y}) + f(-\sqrt{y}))/2\sqrt{y}$ (see Table 10.1).

◼ **EXAMPLE 10.9** **Distribution of the square of a T variate**

If X is $T(n)$, find the distribution of X^2/n.

Solution 10.9 The PDF of Student's T distribution is given in Section 7.18 (p. 7–85) as

$$f(t) = K(1 + t^2/n)^{-(n+1)/2}. \tag{10.18}$$

where $K = \Gamma((n + 1)/2)/[\sqrt{n\pi} \ \Gamma(n/2)]$. Using equation (10.17), the PDF of $Y = T^2/n$ is obtained as $G(y) = \Pr(Y \leq y) =$

$$\Pr(T^2/n \leq y) = \Pr(-\sqrt{ny} \leq T \leq \sqrt{ny}) = F(\sqrt{ny}) - F(-\sqrt{ny}). \tag{10.19}$$

Differentiate (10.19) with respect to y to get the PDF of Y as

$$\sqrt{n}[f(\sqrt{ny}) + f(-\sqrt{ny})](1/2\sqrt{y}). \tag{10.20}$$

As the T distribution is symmetric, $f(\sqrt{y}) = f(-\sqrt{y})$. Substitute in equation (10.18), and cancel out $\sqrt{n}$ to get the desired PDF as

$$g(y, n) = \Gamma((n + 1)/2)/[\sqrt{\pi} \ \Gamma(n/2)](1 + y)^{-(n+1)/2}/\sqrt{y}. \tag{10.21}$$

Write $\sqrt{y}$ in the denominator as $y^{1/2-1}$ and take numerator expression to the denominator. Then this is found to be a BETA-II distribution

$$(1/B(1/2, n/2))y^{1/2-1}/(1 + y)^{(n+1)/2}. \tag{10.22}$$

where $B(1/2, n/2) = \Gamma((n + 1)/2)/[\sqrt{\pi} \ \Gamma(n/2)]$ is the CBF.   ∎

10.7   DISTRIBUTION OF SQUARE-ROOTS

As the square root of a negative number is imaginary, this transformation is defined only for random variables that take positive values. It makes sense for continuous variates than discrete ones. Let $Y = \sqrt{X}$, which gives $X = Y^2$ and $dx/dy = 2y$. The straightforward way to find the PDF of Y is to use

$$g(y) = f(x)|\partial x/\partial y| = 2yf(y^2). \tag{10.23}$$

If the resulting distribution of Y is symmetric, we need to divide the final density by 2 to get the correct PDF, because both $(-y)^2$ and $(+y)^2$ map to x. Symbolically, $g(y) = yf(y^2)$ if Y is symmetric. This is summarized as

$$g(y) = f(x)|\partial x/\partial y| = \begin{cases} 2yf(y^2) & \text{if } Y \text{ is asymmetric;} \\ yf(y^2) & \text{if } Y \text{ is symmetric.} \end{cases}$$

■ **EXAMPLE 10.10   Distribution of the square-root of a χ^2 variate**

If X is χ_n^2, find the distribution of $\sqrt{X}$.

Solution 10.10 The PDF of χ_n^2 variate is $f(x) = e^{-x/2}x^{n/2-1}/[2^{n/2}\Gamma(n/2)]$, where n is the DoF ≥ 1. Put $Y = \sqrt{X}$ and use equation (10.23) to get

$$f(y) = 2y * e^{-y^2/2}(y^2)^{n/2-1}/[2^{n/2}\Gamma(n/2)]. \qquad (10.24)$$

As $y * (y^2)^{n/2-1} = y^{n-1}$ the PDF becomes $f(y) = e^{-y^2/2}y^{n-1}/[2^{n/2-1}\Gamma(n/2)]$. This is the chi-distribution, or the standard form of Rayleigh distribution (p. 7–107).   ■

■ **EXAMPLE 10.11   Distribution of the square-root of an F variate**

If X is $F(1, n)$, find the distribution of $\sqrt{X}$.

Solution 10.11 The PDF of F distribution was given in Chapter 7 (p. 316) as

$$f(x; m, n) = \frac{\Gamma((m+n)/2)m^{m/2}n^{n/2}}{\Gamma(m/2)\Gamma(n/2)} \frac{x^{m/2-1}}{(n+mx)^{(m+n)/2}}, \qquad 0 < x < \infty. \qquad (10.25)$$

Put $m = 1$ to get the PDF of $F(1, n)$ as

$$f(x; n) = \frac{\Gamma((1+n)/2)n^{n/2}}{\Gamma(1/2)\Gamma(n/2)} \frac{x^{1/2-1}}{(n+x)^{(1+n)/2}}, \qquad 0 < x < \infty. \qquad (10.26)$$

Now use equation (10.23) to get

$$g(y) = 2yf(y^2)$$
$$= 2y\Gamma((1+n)/2)n^{n/2}/[\Gamma(1/2)\Gamma(n/2)](y^2)^{1/2-1}/(n+y^2)^{(1+n)/2}. \qquad (10.27)$$

The y cancels out with $(y^2)^{1/2-1} = 1/y$. Take n outside from the bracket in the denominator and cancel out with $n^{n/2}$ in the numerator to get a $\sqrt{n}$ in the denominator. As the PDF now involves only powers of y^2, it is symmetric. Hence, we need to divide the resulting PDF by 2 to get the correct PDF as

$$g(y) = (1 + y^2/n)^{-(n+1)/2}/[\sqrt{n}B(1/2, n/2)]. \qquad (10.28)$$

which is the Student's $T(n)$ distribution.   ■

10.8   DISTRIBUTION OF RECIPROCALS

Distribution of reciprocals is defined only in some particular cases. If the value of a random variable X at $x = 0$ is nonzero, the random variable $Y = 1/X$ is well defined. It can also be used (along with Section 10.5.1 discussed in p. 404) to find the distribution of $(X - 1)/X = 1 - 1/X$ and $(1 - X)/X = 1/X - 1$. The straightforward way to find the PDF of Y is to use

$$g(y) = f(x)|\partial x/\partial y| = f(1/y)/y^2. \tag{10.29}$$

■ **EXAMPLE 10.12   Distribution of the reciprocal of a Cauchy variate**

If X is Cauchy distributed, find the distribution of $Y = 1/X$.

Solution 10.12 We have seen in Chapter 7 that $f(x) = \frac{1}{\pi} \frac{1}{1+x^2}$, $-\infty < x < \infty$. As $f(x = 0)$ is $1/\pi$, distribution of the reciprocal is well-defined. Using equation (10.29) the PDF becomes $f(y) = \frac{1}{\pi} \frac{1}{1 + (1/y)^2}/y^2$. The y^2 cancels out from the numerator and denominator, giving $f(y) = \frac{1}{\pi} \frac{1}{1 + y^2}$, which is Cauchy distributed.   ■

■ **EXAMPLE 10.13   Reciprocal of a unit rectangular variate**

If X is $U(0, 1)$ distributed, find the distribution of $Y = 1/X$.

Solution 10.13 Let $G(y)$ be the CDF of Y. Then,

$$G(y) = \Pr[Y \le y] = \Pr[1/X \le y] = \Pr[X \ge 1/y] = 1 - 1/y. \tag{10.30}$$

Differentiate with respect to y to get the PDF as $g(y) = 1/y^2$, for $y \ge 1$.   ■

10.9   DISTRIBUTION OF MINIMUM AND MAXIMUM

The distribution of minimum and maximum (called extremes) finds applications in many fields. For example, distribution of maximum is used in life sciences to model the survival time of species, produce, machines, and various products. It is also used in reliability theory to model the life of equipments and parts, various devices, and consumer items (such as light bulbs and computer chips). The study of extremes is called order statistics. Let $X_1, X_2, \ldots, X_n$ be a random sample from an arbitrary distribution with PDF $f(y)$ and CDF $F(y)$. Let $Y_1 = \min(X_1, X_2, \ldots, X_n)$. Then,

$$1 - F_{Y_1}(y) = P(Y_1 > y) = P(X_1 > y) * \cdots * P(X_n > y) = [1 - F(y)]^n. \tag{10.31}$$

using independence. Differentiate the above to get the PDF of Y_1 as

$$f(y_1) = n(1 - F(y))^{n-1} f(y). \tag{10.32}$$

Similarly, the PDF of y_n using CDF method is $f(y_n) = nf(y)[F(y)]^{n-1}$.

10.10   DISTRIBUTION OF TRIGONOMETRIC FUNCTIONS

Trigonometric functions of some random variables are easy to work with. One example is the Cauchy distribution. If X has a standard Cauchy distribution, then $\cos(X)$ has the same distribution. Trigonometric functions are also utilized to derive some distributions using geometric concepts. One example is the correlation coefficient. The cosine of the angle between two normalized vectors in n-dimensional Euclidean space is called the correlation coefficient.

◼ **EXAMPLE 10.14   Distribution of $U = \tan(X)$**

If X has a $U(0, 1)$ distribution, find the distribution of $U = \tan(X)$.

Solution 10.14   The PDF of X is $f(x) = 1, 0 < x < 1$. The inverse transformation is $X = \tan^{-1}(U)$. This gives $|\partial x/\partial u| = 1/(1 + u^2)$. The range of U is modified as $\tan(0) = 0$ to $\tan(1) = \pi/4$. Hence, the distribution of U is $f(u) = 1/(1 + u^2)$ for $0 < u < \pi/4$.   ◼

◼ **EXAMPLE 10.15   Distribution of $U = \sin(X)$**

If X has a $CUNI[-\pi/2, \pi/2]$ distribution, find the distribution of $U = \sin(X)$.

Solution 10.15   The inverse transformation is $x = \sin^{-1}(u)$ so that $|\partial x/\partial u| = 1/\sqrt{1 - u^2}$.   When $x = -\pi/2, u = \sin(-\pi/2) = -\sin(\pi/2) = -1$.   When $x = \pi/2, u = \sin(\pi/2) = 1$. As the PDF of X is $1/\pi$, the PDF of U is $f(u) = (1/\pi)1/\sqrt{1 - u^2}$, for $-1 < u < +1$.   ◼

10.11   DISTRIBUTION OF TRANSCENDENTAL FUNCTIONS

Distributions of transcendental functions are quite useful in engineering and related fields. Several laws and principles in engineering and physical sciences are modeled as mathematical equations called functionals involving the unknown variables and known constants. In most applications, one variable (called dependent variable) is modeled as a function of two or more other variables (called independent variables), which can include time. If the number of variables involved is two, the CoVT technique is useful to derive the distribution of one, using the distribution of the other. For example, in data compression and telecommunications, the speech amplitude is modeled using the Laplacian law $f(x) = (1/\sigma\sqrt{2})\exp(-\sqrt{2}|x|/\sigma)$, and compressed using the μ−law as $A * \mathrm{sign}(x)[\ln(1 + \mu|x|)/\ln(1 + \mu)]$, where "$A$" is the peak input magnitude and μ is the compression constant (typically set to high, say 255). Similarly, in wireless communication of fading channels, if X is $N(\mu, \sigma^2)$ distributed, $Y = e^X$ has a lognormal distribution. This has applications in mining engineering, reliability, and spectroscopy among many other fields (Chapter 7, p. 297).

▐ EXAMPLE 10.16   Logarithmic transformation of CUNI Distribution

If X is CUNI$[a, b]$ distributed (Chapter 7, p. 261), find the distribution of $Y = -\log(X - a)/(b - a)$.

Solution 10.16 The CDF of Y is $F(y) = P[Y \leq y] = P[(X - a)/(b - a) \geq e^{-y}] = P[X \geq a + (b - a)e^{-y}]$. As the CDF of CUNI$(a, b)$ is $(x - a)/(b - a)$ this becomes $1 - P[X \leq a + (b - a)e^{-y}] = 1 - e^{-y}$. From this the PDF is obtained by differentiation as $f(y) = e^{-y}$. Hence, Y is EXP(1).   ▪

▐ EXAMPLE 10.17   Transformation of Arc-Sine Distribution

If X is Arc-Sine distributed (Chapter 7, p. 279), find the distribution of $Y = -\log(X)$.

Solution 10.17 Let $F(y)$ be the CDF of Y. Then, $F(y) = P[Y \leq y] = P[X \geq e^{-y}] = 1 - (2/\pi)\sin^{-1}(\sqrt{e^{-y}})$. Differentiate with respect to y to get the PDF as $f(y) = (2/\pi)e^{-y/2}/[2(\sqrt{1 - e^{-y}})]$. The 2 cancels out from numerator and denominator giving

$$f(y) = (1/\pi)e^{-y/2}/\sqrt{1 - e^{-y}}, \quad 0 \leq y < \infty. \tag{10.33}$$

▪

10.11.1   Distribution of Sums

If X and Y are independent random variables, we may need to find the distribution of $U = X + Y$. This is easy to find using convolution of integrals when both X and Y are continuous. Let $F_U(u)$ be the CDF of U. Then,

$$F_U(u) = \Pr[U \leq u] = \int\int_{x+y\leq u} f(x)g(y)dxdy. \tag{10.34}$$

due to the assumption of independence. As $x + y \leq u$ represents the region below a straight line, the limits can be adjusted as $\int_{x=-\infty}^{\infty}\int_{y=-\infty}^{u-x} f(x)g(y)dxdy$. This can be factored as $\int_{x=-\infty}^{\infty}\left(\int_{y=-\infty}^{u-x} g(y)dy\right)f(x)dx$. Denote the inner integral by $G_Y(u - x)$. The above then becomes $\int_{x=-\infty}^{\infty} G_Y(u - x)f(x)dx$. Owing to the symmetry of X and Y, we can rearrange the integrals to get a similar expression as $\int_{x=-\infty}^{\infty} F_X(u-y)g(y)dy$. Differentiation with respect to u gives the PDF of $X + Y$ as $g_U(u) =$

$$\frac{\partial}{\partial u} \int_{x=-\infty}^{\infty} G_Y(u-x)f(x)dx = \int_{-\infty}^{\infty} g(u-x)f(x)dx = \int_{-\infty}^{\infty} f(u-y)g(y)dy. \quad (10.35)$$

This is called the convolution of X and Y. The MGF is related as $M_{X+Y}(t) = M_X(t) * M_Y(t)$ and the CGF is related as $K_{X+Y}(t) = K_X(t) + K_Y(t)$ if X and Y are independent. This method can easily be extended to find the distribution of the difference of independent random variables.

EXAMPLE 10.18   Sum of Poisson Variates

If X and Y are Poisson with parameters λ_1 and λ_2, find the PMF of $X + Y$.

Solution 10.18 We could easily get the required PDF using the MGF technique. However, we proceed as follows to clarify the change of variable technique. The joint PMF of X and Y is $f(x, y) = e^{-\lambda_1}\lambda_1^x/x! e^{-\lambda_2}\lambda_2^y/y!$. Let $U = X + Y, Y = V$. The inverse mapping is $V = Y, X = U - V$. Thus, $f_{U,V}(u, v) = e^{-\lambda_1}\lambda_1^{u-v}/(u-v)! e^{-\lambda_2}\lambda_2^v/v!$, where $v = 0,1,2..$ and $u = v, v+1,...$ The PMF of U is obtained by summing over the entire range of V. As U has V as the lower bound, we need sum over v from 0 to u. Hence $f_U(u) = e^{-(\lambda_1+\lambda_2)} \sum_{v=0}^{u} \lambda_1^{u-v}\lambda_2^v/[v!(u-v)!]$. Multiply and divide by $u!$ and take it outside the summation in the denominator. This gives $f_U(u) =$

$$\frac{e^{-(\lambda_1+\lambda_2)}}{u!} \sum_{v=0}^{u} \binom{u}{v}\lambda_1^{u-v}\lambda_2^v = \frac{e^{-(\lambda_1+\lambda_2)}}{u!}(\lambda_1 + \lambda_2)^u, u = 0, 1, 2 \ldots . \quad (10.36)$$

■

EXAMPLE 10.19   Sum of Exponential Variates

If X and Y are independent exponential variates with the same shape parameter, find the distribution of $U = X + Y$.

Solution 10.19 We could use equation (10.35) to get the PDF. It is much easier to use the MGF. We know that the MGF of X is $(1 - t/\lambda)^{-1}$. As X and Y are independent, $M_{X+Y}(t) = M_X(t) * M_Y(t) = (1 - t/\lambda)^{-2}$, which is the MGF of gamma$(2, \lambda)$.

■

EXAMPLE 10.20   PMF of sum X+Y

The joint PMF of X and Y is given in Table 10.2. Find the pmf of $u = X + Y$.

TABLE 10.2   Joint Distribution

X	Y	$f(x, y)$	Total
1	1	1/12	1/12
1	2	2/12	3/12
1	3	3/12	6/12
2	1	3/12	9/12
2	2	2/12	11/12
2	3	1/12	12/12

TABLE 10.3   Distribution of $X + Y$

$X + Y$	$f(x, y)$
2	1/12
3	5/12
4	5/12
5	1/12
Total	1.0

Solution 10.20 Clearly, $X + Y$ takes values in the range [2,5]. $P[X + Y = 2] = P[X = 1, Y = 1] = 1/12$. $P[X + Y = 3] = P[X = 1, Y = 2] + P[X = 2, Y = 1] = 2/12 + 3/12 = 5/12$.   Similarly, $P[X + Y = 4] = P[X = 1, Y = 3] + P[X = 2, Y = 2] + P[X = 3, Y = 1] = 2/12 + 3/12 = 5/12$, and so on. The results are given in Tables 10.2 and 10.3.   ∎

10.11.2   Distribution of Arbitrary Functions

Consider the transformation $y = g(x)$, where $g(x)$ is a one–one mapping that is invertible. This means that x can be expressed in terms of Y as $x = g^{-1}(y)$. Then, the PDF of Y can be represented in the continuous case as follows. First express the CDF of Y in terms of the CDF of X as

$$F_Y(y) = P(Y \leq y) = P(g(x) \leq y) = P(x \leq g^{-1}(y)) = F_X(g^{-1}(y)). \qquad (10.37)$$

Differentiate both sides to get the PDF as

$$f_Y(y) = f_X(g^{-1}(y))\partial g^{-1}(y)/\partial y. \qquad (10.38)$$

and in the discrete case $Y = r(X)$ as

$$f_Y(y) = Pr_X(r(x) = y) = \sum_{x:r(x)=y} Pr_X(X = x). \qquad (10.39)$$

📖 **EXAMPLE 10.21   Distribution of Integer Part**

If X has Cauchy distribution, find the distribution of the integer part $Y = \lfloor X \rfloor$.

Solution 10.21 We have $f(x) = 1/[\pi(1+x^2)]$ for $-\infty < x < \infty$. The random variable Y takes integer values on the entire real line including $y = \mp\infty$ (see discussion in Section 7.2.1, p. 260). Specifically,

$$\Pr[Y = y] = \Pr[y \le X < y+1] = \int_y^{y+1} 1/[\pi(1+x^2)]dx = (1/\pi)\tan^{-1}(x)|_y^{y+1}$$

$$= (1/\pi)[\tan^{-1}(y+1) - \tan^{-1}(y)].$$

A similar expression could be obtained for $x < 0$ as $\Pr[Y = y] = \Pr[y-1 \le X \le y]$. As alternate terms in above equation (10.40) cancel out, this form is useful to compute the CDF, and to prove that the probabilities add up to 1. For example, sum from $-\infty$ to ∞ to get $\tan(\infty) - \tan(-\infty) = \pi/2 - (-\pi/2) = \pi$, which cancels with π in the denominator. Now use the identity $\tan^{-1}(x) - \tan^{-1}(y) = \tan^{-1}((x-y)/(1+xy))$ to get

$$\Pr[Y = y] = (1/\pi)\tan^{-1}(y+1-y)/[1+y(y+1)]$$

$$= (1/\pi)\tan^{-1}(1/[1+y(y+1)]), \qquad (10.40)$$

which is the desired form. As the terms do not cancel, this form is not useful to compute the CDF.   ∎

This example has an enormous use. It shows another way to specify a PMF as the difference of two functions (say $[\tan^{-1}(x+1) - \tan^{-1}(x)]$ or $[\exp(-\lambda x) - \exp(-\lambda(x+1))]$) that simply cancels out when summed over the proper range of x, leaving behind only two extreme terms whose difference appears as the normalizing constant in the denominator. This allows us to define a variety of new discrete distributions. As the PMF should be positive, we should form the difference separately for positive and negative values, to make it non-negative. The shape of the distribution (whether it is unimodal and tails off to the extremes, or it is U-shaped or J-shaped, etc.) must be known to form the PMF.

📖 **EXAMPLE 10.22   Distribution of Fractional Part**

If X has an exponential distribution, find the distribution of the fractional part $Y = X - \lfloor X \rfloor$.

Solution 10.22 It was shown in Chapter 6 (p. 218) that if X has an Exponential distribution, the distribution of $Y = \lfloor X \rfloor$ is GEO$(1 - \exp(-\lambda))$. The possible values of $Y = X - \lfloor X \rfloor$ are $0 \le y \le 1$. If we assume that the integer and fractional parts are independent, Y is the difference between an exponential and geometric random variables. This is of mixed type (as geometric distribution is discrete), where the continuous distribution dominates. This means that Y has a continuous distribution. Using the MoDF it is easy to show that Y is distributed as $f(y) = \lambda \exp(-\lambda y)/[1 - \exp(-\lambda)]$, for $0 \leq y \leq 1$.   ∎

10.11.3   Distribution of Logarithms

The logarithmic transformation can be applied to any random variable that takes non-negative values. As $\log(0) = -\infty$, this transforms $(0, \infty)$ to the new range $(-\infty, \infty)$. As the $\log()$ is a real function of its argument, this transformation is applied to continuous random variables. Unless otherwise stated, the base of the logarithm is assumed to be e. A special transformation encountered in communication theory is $Y = \log_2(1 + X)$. Using the method described above, the PDF of Y is given by $g_Y(y) = \ln(2) f(2^y - 1)2^y$.

10.11.4   Special Functions

There exist many symmetric and skew symmetric functions that possess interesting properties. These can be in single or multiple variables. Examples are $x/(1 - x), (1 + x)/(1 - x)$, and $e^x/(1 + e^x)$. Consider $Y = X/\sqrt{1 + X^2}$. Square both sides to get $Y^2 = X^2/(1 + X^2)$, from which $x = y/\sqrt{1 - y^2}$. Differentiate with respect to y to get $|J| = |\partial x/\partial y| = 1/(1 - y^2)^{3/2}$. From this, the PDF of Y is easy to obtain as

$$g(y) = f(y/\sqrt{1 - y^2})/(1 - y^2)^{3/2}. \qquad (10.41)$$

If X is standard Cauchy distributed, then $Y = X/\sqrt{1 + X^2}$ has PDF

$$g(y) = (1/\pi)(1 - y^2)/(1 - y^2)^{3/2} = (1/\pi)\, 1/(1 - y^2)^{1/2}, \quad -1 \leq y \leq +1. \quad (10.42)$$

◼ EXAMPLE 10.23   Distribution of Ratio of sums

Let X_i's be IID EXP$(\lambda\theta)$ for $i = 1, 2, .., m$. Let Y_j's be IID EXP(θ) for $j = 1, 2, .., n$. If X_i's and Y_j's are pair-wise independent, find the distribution of the ratio $W = U/V = \sum_{i=1}^{m} X_i / \sum_{j=1}^{n} Y_j$.

Solution 10.23 As X_i's are IID, the joint PDF is the product of individual PDFs. We first use the MGF technique to find the distribution of numerator and denominator. The MGF of EXP$(\lambda\theta)$ is $M_x(t) = 1/[1 - \lambda\theta t]$. As the X_i's are IID, $M_u(t) = 1/[1 - \lambda\theta t]^m$. Similarly, $M_v(t) = 1/[1 - \theta t]^n$. These are the MGFs of gamma distributions. Hence, W is the ratio of two independent gamma variates. See References 134, 138, 285 and 294 for further properties.   ∎

10.12 TRANSFORMATIONS OF NORMAL VARIATES

Most of the transformations discussed above are applicable to the normal variate. As this has many practical applications, this section briefly discusses some of them.

10.12.1 Linear Combination of Normal Variates

Linear combination of any number of independent normal variates is normally distributed. This can be proved using induction. A simpler method is to use the MGF technique. We saw in Chapter 9 that the ChF of $N(\mu, \sigma^2)$ is $\exp(it\mu - \frac{1}{2}t^2\sigma^2)$. If $X_1, X_2, \ldots, X_n$ are independent normal random variables $N(\mu_i, \sigma_i^2)$, the ChF of $Y = X_1 + X_2 + \cdots + X_n$ is

$$\phi_y(t) = \prod_{i=1}^{n} \phi_{x_i}(t) = \exp\left(it \sum_i \mu_i - \frac{1}{2}t^2 \sum_i \sigma_i^2 \right). \tag{10.43}$$

As this is the ChF of a normal variate with mean $\sum_i \mu_i$ and variance $\sum_i \sigma_i^2$, it follows that Y is normally distributed. In the particular case, when each of the μ_i's and σ_i^2 are equal, we have $N(n\mu, n\sigma^2)$. If $Y = c_1X_1 + c_2X_2 + \cdots + c_nX_n$, the ChF of Y is

$$\phi_y(t) = \prod_{i=1}^{n} \phi_{c_i x_i}(t) = \prod_{i=1}^{n} \phi_{x_i}(c_i t) = \exp\left(it \sum_i c_i \mu_i - \frac{1}{2}t^2 \sum_i c_i^2 \sigma_i^2 \right). \tag{10.44}$$

This shows that $\sum_i c_i Y_i$ is normal with mean $\mu = \sum_i c_i \mu_i$, and variance $\sigma^2 = \sum_i c_i^2 \sigma_i^2$.

Corollary 1 $cX + b$ is normal with mean $\mu' = c * \mu + b$ and variance $\sigma'^2 = c^2\sigma^2$.

10.12.2 Square of Normal Variates

The square of a normal variate is χ_1^2 distributed. In general, the sum of squares of any number of independent normal variates is χ_n^2 distributed. Similarly, the ratio of independent normal variates is Cauchy distributed. This is proved in Chapter 11.

■ **EXAMPLE 10.24   Distribution of the square of a normal variate**

If X is $N(0, 1)$, find the distribution of X^2.

Solution 10.24 Applying the result, we get $G(y) =$

$$\Pr(Y \leq y) = \Pr(X^2 \leq y) = \Pr(-\sqrt{y} \leq X \leq \sqrt{y})$$

$$= \int_{-\sqrt{y}}^{\sqrt{y}} (1/\sqrt{2\pi})e^{-z^2/2}dz. \tag{10.45}$$

Differentiate with respect to y to get

$$g(y) = (1/\sqrt{2\pi})\frac{\partial}{\partial y}\int_{-\sqrt{y}}^{\sqrt{y}} e^{-z^2/2}dz. \qquad (10.46)$$

Using Leibnitz theorem this reduces to $(1/\sqrt{2\pi})(e^{-y/2} * 1/(2\sqrt{y}) - e^{-y/2} * (-1)/(2\sqrt{y})) = (1/\sqrt{2\pi})y^{1/2-1}e^{-y/2}$. This is the PDF of a chi-square variate with 1 DoF. Alternatively, write (10.46) as $\frac{\partial}{\partial y}[\Phi(\sqrt{y}) - \Phi(-\sqrt{y})]$ and proceed as above.

■

10.13 SUMMARY

This chapter derives and explains the formulas for the probability distribution of a sum, difference, product, and ratio of two independent random variables. The distribution of squares, square-roots, of univariate and other transformations of two or more random variables are derived and illustrated. Distribution of integer and fractional parts of some continuous random variables are discussed, as also the distribution of CDF ($F(x)$) and its inverse ($F^{-1}(x)$). These known results are used to derive an expression for the mean deviation of some continuous distributions as twice the simple integral of $t/f(F^{-1}(t))$ from zero to $F(\mu)$, where μ is the mean and $F(x)$ is the CDF ($F(\mu) = 1/2$ for symmetric laws). Several examples are included to understand the need and usefulness of the transformations. Advanced treatment can be found in References 134, 138, 225 and 295 and engineering applications in References 285 and 291.

EXERCISES

10.1 Mark as True or False

   a) The CoVT is applicable to both discrete and continuous distributions.
   b) If the range of a random variable X includes the origin, we cannot use the transformation $Y = 1/X$
   c) A translation $Y = X \pm c$ is not applicable to discrete variates
   d) The distribution of the square of a normal variate is Student's T
   e) The reciprocal of a Cauchy variate is Cauchy distributed.

10.2 If X is $N(\mu, \sigma^2)$, find distribution of (i) $Y = c * X + d$, (ii) $Y = |X - \mu|$.

10.3 If X is EXP(λ), find the distribution of $Y = c * X$.

10.4 If $f(x) = ke^{-kx}, x \geq 0$, find the distribution of $Y = \sqrt{x}$.

10.5 If X is BINO(n, p) find distribution of (i) $Y = 1 - X/n$, (ii) $Y = n - X$.

10.6 If X is GEOM(p), find the distribution of $Z = \sum_{i=1}^{n} X_i$, where each X_i is independent.

10.7 If F and G are two CDF's symmetric around zero, with unit second moment, prove that $H = c * F + (1 - c) * G$ is identically distributed.

10.8 If X is DUNI(k) with $f(x) = 1/k, x = 1, 2, .., k$, find distribution of $Y = X + b$

10.9 If X is BETA-I(p, q), find the distribution of $(1 - X)/X$.

10.10 If X is CUNI(0, 1) prove that $Y = -\ln(X)$ is standard exponential. What is the distribution when the logarithm is not to the base e.

10.11 If $h(x)$ is a monotonic function, prove that the CDF of $Y = h(x)$ can be expressed as $F_Y(y) = F_X(h^{-1}(y))$ if x is increasing, and $F_Y(y) = 1 - F_X(h^{-1}(y))$ otherwise.

10.12 If X is CUNI(0, 1), find the distribution of $Y = 1 - e^{-x}$

10.13 If X is CUNI[0,1], find the distribution of $U = c * \tan(1/X)$

10.14 If X_1, X_2 are IID CUNI(0, 1), find the distribution of $Y_1 = \sqrt{-2\log_e(x_1)} \cos(2\pi x_2)$ and $Y_2 = \sqrt{-2\log_e(x_1)} \sin(2\pi x_2)$ (the inverse transformation being $X_1 = \exp\left[-\frac{1}{2}(y_1^2 + y_2^2)\right]$ and $x_2 = \frac{1}{2\pi} \arctan(y_2/y_1)$).

10.15 Prove that the sum of independent exponential random variables has a gamma distribution.

10.16 If conditional distribution of X is BINO(n, p), where p is distributed as BETA(p, q), find the mean and variance of X.

10.17 If X and Y are exponentially distributed, find the distribution of $X + Y$ and $X - Y$.

10.18 If X is distributed as GAMMA ($\alpha, 1$), find the distribution of $Y = \log(X/\alpha)$.

10.19 If X is CUNI(a, b), find the PDF of $Y = -\log(X)$. If $a = -b$, find the distribution of $Y = X^2$.

10.20 If $X \sim \chi_n$ (i.e., $X \sim \sqrt{\chi_n^2}$) and $Y \sim \text{BETA}(\frac{n-1}{2}, \frac{n-1}{2})$ is independent of X, prove that $(2Y - 1) X \sim N(0, 1)$.

10.21 If X is BETA-I(p, q) distributed, show that the variate $Y = \ln(X/(1 - X))$ is generalized logistic(p, q).

10.22 Find the distribution of absolute value of a general normal $N(\mu, \sigma^2)$, and its mean.

10.23 If $X \sim$ Weibull(a, b), find the distribution of $Y = \log(X/b)$.

10.24 If X is CUNI($-(a + b)/2$, $(a + b)/2$) variate, find the distribution of $Y = |X|$.

10.25 If $X \sim$ standard Weibull distribution with PDF $f(x) = bx^{b-1}e^{-x^b}$ find the distribution of $Y = X^{(b)}$.

10.26 If U is CUNI(0, 1) find the distribution of (i) $|U - \frac{1}{2}|$, (ii) $1/(1 + U)$.

10.27 If $f(x) = (c - 1)/(1 + x)^c$ for $0 < x < \infty$, find the distribution of $Y = 1/(1 + x)^c$ and find the mean and variance.

10.28 If $Y = g(x)$ is an arbitrary function of a discrete random variable, prove that the PMF of Y is $f(y) = \sum_{x \in g^{-1}(y)} f(x)$.

11

JOINT DISTRIBUTIONS

After finishing the chapter, students will be able to

- Distinguish joint and conditional distributions
- Find distribution of functions of a random variable
- Understand linear transformations of random variables
- Comprehend Jacobian of transformations
- Describe arbitrary transformations
- Apply Polar Transformations
- Utilize "do-little" technique to quickly find Jacobians

11.1   JOINT AND CONDITIONAL DISTRIBUTIONS

Definition 11.1 Joint distribution is the distribution of two or more (dependent or independent) random variables.

Usually, the variables involved are either all discrete or all continuous. Symbolically, it is represented as $p(x, y) = p(X = x \text{ and } Y = y)$ (such that $\sum_{(x,y) \in A} p(x, y) = 1$ for discrete case and $(\int\int_A p(x, y) dx dy = 1$ for the continuous case).

Statistics for Scientists and Engineers, First Edition. Ramalingam Shanmugam and Rajan Chattamvelli.
© 2015 John Wiley & Sons, Inc. Published 2015 by John Wiley & Sons, Inc.

11.1.1   Marginal Distributions

Definition 11.2 Marginal distributions are distributions of individual variates.

Marginal PDF's can be obtained from joint PDF's by summation (in discrete case) or integration (in continuous case) as follows:

$$f(x) = \sum_{y=-\infty}^{\infty} f(x, y) \quad \text{and} \quad f(y) = \sum_{x=-\infty}^{\infty} f(x, y) \quad \text{(discrete)}, \tag{11.1}$$

$$f(x) = \int_{y=-\infty}^{\infty} f(x, y) dy \quad \text{and} \quad f(y) = \int_{x=-\infty}^{\infty} f(x, y) dx \quad \text{(continuous)}, \tag{11.2}$$

where the summation or integration is carried out only throughout the range of proper variate. Extension to more than two variates is straightforward. Joint PDF is the product of constituent marginal PDFs when the variables are independent.

$$f(x, y) = f(x) * f(y) \quad \text{and} \quad F(x, y) = F(x) * F(y). \tag{11.3}$$

This has important applications in obtaining likelihoods, finding estimators, and so on.

■ **EXAMPLE 11.1   Find marginal distribution**

If the joint PDF of X and Y is given by $f(x, y) = Kx(1 + y), \{x = 1, 2\}, \{y = 1, 2, 3\}$ find the marginal PMF of x and y.

Solution 11.1 Both X and Y are discrete in this example. As the total probability is unity, we have $K[2 + 3 + 4 + 4 + 6 + 8] = 1$, giving $K = 1/27$.   ■

To obtain the marginal distribution of X, we sum out Y over its entire range. Hence, $f(x) = (x/27)[9] = x/3, \{x = 1, 2\}$. Similarly, $f(y) = ((1 + y)/27)[3] = (1 + y)/9, \{y = 1, 2, 3\}$.

■ **EXAMPLE 11.2   Find marginal distribution**

A radioactive source is emitting α-particles intermittently in different directions. The number of particles emitted in a fixed time interval is Poisson(λ). A particle recorder is placed at a point in direct line-of-sight. It has probability p of recording any particle coming toward it. Find the PMF of the number of particles recorded.

Solution 11.2 Let X be the number of particles emitted and Y be the number of particles recorded. Then, we are given that

$$p(x) = \exp(-\lambda)\lambda^x/x! \text{ and } p(y|x) = \binom{x}{y} p^y(1 - p)^{x-y}. \tag{11.4}$$

As these two sources are independent, $f(x, y)$ is the product of the individual PDFs. From this the marginal distribution of Y is obtained using equation (11.1) as

$$f(y) = \sum_x f(x, y) = \exp(-\lambda)p^y/y! \sum_{x=y}^{\infty} \lambda^x q^{x-y}/(x-y)! \qquad (11.5)$$

Put $t = x - y$ in equation (11.5) so that t varies between 0 and ∞,

$$f(y) = \exp(-\lambda)(\lambda p)^y/y! \sum_{t=0}^{\infty} (\lambda q)^t/t! = \exp(-\lambda)(\lambda p)^y/y! * \exp(\lambda q). \qquad (11.6)$$

This simplifies to $\exp(-\lambda p)(\lambda p)^y/y!$, for $y = 0, 1, \ldots$, as $q = 1 - p$. ■

11.1.2   Conditional Distributions

Conditional distributions are obtained from joint distributions by conditioning on one or more variables. Conditional PDF's can be expressed in terms of joint PDF's using laws of conditional probabilities.

$$f(x|y) = f(x, y)/f(y) \quad \text{and} \quad f(y|x) = f(x, y)/f(x). \qquad (11.7)$$

It is easy to see that multiple conditional distributions exist by conditioning y at different levels.

▣ EXAMPLE 11.3   Gamma distributed Poisson parameter

Assume that the number of accidents follows a Poisson law with parameter λ. If λ itself is distributed according to the gamma law, prove that the unconditional distribution is negative binomial distributed.

Solution 11.3 Let $f(x, \lambda)$ represent the PDF of Poisson distribution and $f(\lambda, m, p)$ denote the gamma PMF. Owing to independence, the joint distribution is the product of the marginals and the unconditional distribution of x is obtained by integrating out λ as

$$f(x, m, p) = \int_{\lambda=0}^{\infty} [e^{-\lambda} \lambda^x]/x! * [m^p/\Gamma(p)]e^{-m\lambda} \lambda^{p-1} d\lambda. \qquad (11.8)$$

Take constants independent of λ outside the integral to get

$$f(x, m, p) = m^p/[\Gamma(p) \, x!] \int_{\lambda=0}^{\infty} e^{-\lambda(m+1)} \lambda^{x+p-1} d\lambda. \qquad (11.9)$$

The integral in equation (11.9) is easily seen to be the gamma integral, so that it becomes

$$m^p/[\Gamma(p)x!]\Gamma(x+p)/[(1+m)^{x+p}] = \binom{x+p-1}{x}(m/(1+m))^p(1/(1+m))^x. \tag{11.10}$$

where the last expression is obtained by writing $(1+m)^{x+p} = (1+m)^x * (1+m)^p$ and expanding the gamma functions as $\Gamma(x+p) = (x+p-1)!$ and $\Gamma(p) = (p-1)!$. This is a negative binomial distribution with $p = 1/(1+m)$.

∎

▣ EXAMPLE 11.4 Find conditional distribution from trinomial

Let X and Y be jointly distributed as trinomial with pmf

$$f(x,y,n,p) = n!/[x!y!(n-x-y)!]p_1^x p_2^y(1-p_1-p_2)^{n-x-y}, x+y \leq n. \tag{11.11}$$

Find the conditional distribution of (i) $Y|X=x$ and (ii) $X|X+Y=n$. Obtain $E(Y|x)$ and $E(X|X+Y=n)$.

Solution 11.4 We get the marginal PMF of y (resp x) by summing over x (resp y). Multiply and divide the RHS by $(n-y)!$ and sum over x to get

$$f(y) = \sum_x \frac{n!(n-y)!}{x!y!(n-y)!(n-x-y)!}p_1^x p_2^y(1-p_1-p_2)^{n-x-y}$$

$$= \frac{n!}{y!(n-y)!}p_2^y \sum_x \frac{(n-y)!}{x!(n-x-y)!}p_1^x(1-p_1-p_2)^{n-x-y}$$

$$= \binom{n}{y}p_2^y \sum_x \binom{n-y}{x}p_1^x(p_3)^{n-x-y}. \tag{11.12}$$

where $p_3 = 1-p_1-p_2$. Expression inside the summation is simply the successive terms of the binomial expansion of $(p_1+p_3)^{n-y}$. However, $(p_1+p_3) = p_1 + (1-p_1-p_2) = (1-p_2)$. Substitute in the above to get $f(y) = \binom{n}{y}p_2^y(1-p_2)^{n-y}$, which is BINO$(n,p_2)$. Similarly, $X \sim$ BINO(n, p_1). The PMF of $Y|x$ is

$$\frac{f(x,y)}{f(x)} = \frac{(n-x)!}{y!(n-x-y)!}\left(\frac{p_2}{1-p_1}\right)^y\left(\frac{1-p_1-p_2}{1-p_1}\right)^{n-x-y}, \tag{11.13}$$

where $y = 0,1,\ldots,n-x$. This is the pmf of BINO$(n-x, p_2/(1-p_1))$. Hence, $E(Y|x) = (n-x)p_2/(1-p_1)$.

(ii) $X+Y$ is clearly distributed as a BINO(n, p_1+p_2), so that

$$P(X+Y=n) = \binom{n}{n}(p_1+p_2)^n(1-p_1-p_2)^{n-n} = (p_1+p_2)^n. \tag{11.14}$$

The PDF of $Y|X + Y = n$ is thus $f(x, n - x)/P[X + Y = n] = \frac{n!}{x!(n-x)!(n-x-(n-x))!}$ $p_1^x p_2^{(n-x)}(1 - p_1 - p_2)^{n-x-(n-x)}/(p_1 + p_2)^n = \frac{n!}{x!(n-x)!}p_1^x p_2^{(n-x)}/(p_1 + p_2)^n$. Splitting $(p_1 + p_2)^n$ as $(p_1 + p_2)^x(p_1 + p_2)^{n-x}$, the above reduces to a BINO$(n, p_1/(p_1 + p_2))$. From this, we get $E(X|X + Y = n) = np_1/(p_1 + p_2)$. ∎

11.2 JACOBIAN OF TRANSFORMATIONS

The Jacobian is a useful concept in various fields of applied sciences, including vector calculus, differential equations, atmospheric sciences, astronomy, and statistics, to name a few. The Jacobian determinant measures the stretching effect of a mapping or transformation as explained later. Carl Gustav Jacobi (1804–1851), whose work originated in mathematical physics, invented it in 1841. It could mean either the Jacobian matrix or its determinant (if the matrix is square).

The Jacobian matrix could be rectangular when a mapping is induced from the Euclidean space $\mathbb{R}^n \to \mathbb{R}^m$, where $m < n$. This matrix contains the partial derivatives of the output variables with respect to the input variables in modeling problems that involve many input and output variables (which need not tally in number). In other words, the Jacobian relates infinitesimal areas in the input space to infinitesimal areas in the output space of the same dimensionality (areas in 2D, volume elements in $\geq$ 3D). By analyzing the rows of the Jacobian matrix, we can study the impact or sensitivity on output variables due to a selected subset of input variables (by keeping the other variables at fixed levels).

As the determinant of a square matrix exists only if the matrix is of full-rank, there are some regularity conditions to be satisfied by the transformations. We assume that there are m real-valued functions $y_1 = f_1(x_1, x_2, \ldots, x_n), y_2 = f_2(x_1, x_2, \ldots, x_n), \ldots, y_m = f_m(x_1, x_2, \ldots, x_n)$. Then the Jacobian matrix comprises of all first-order partial derivatives of mapping functions:

$$J = \frac{\partial(x_1, x_2, \ldots, x_n)}{\partial(y_1, y_2, \ldots, y_m)} = \begin{bmatrix} \dfrac{\partial x_1}{\partial y_1} & \dfrac{\partial x_2}{\partial y_1} & \cdots & \dfrac{\partial x_n}{\partial y_1} \\ \dfrac{\partial x_1}{\partial y_2} & \dfrac{\partial x_2}{\partial y_2} & \cdots & \dfrac{\partial x_n}{\partial y_2} \\ \vdots & \vdots & \cdots & \vdots \\ \dfrac{\partial x_1}{\partial y_m} & \dfrac{\partial x_2}{\partial y_m} & \cdots & \dfrac{\partial x_n}{\partial y_m} \end{bmatrix}_{(m \times n)}.$$

The (i,j)th entry of the above matrix affirms that a small change dx_i in the original variate x should contract to $(\partial y_i/\partial x)dx_i$ in the transformed space. When $m = n$, the transformation is concisely expressed as the determinant of above matrix. The determinant of a square matrix $|J|$ is the same as the determinant of its transpose matrix $|J'|$. This means that the variable order is unimportant in statistical applications. A geometric interpretation of Jacobians is that it represents the best linear approximation to mapped domain at a general point using a tangent plane in the transformed space. Thus, we get equivalent density contractions of space in transformed domain by multiplying the original function by the Jacobian (which acts as a magnification or contraction factor). For example,

$$\int\int_{R^2} f(x, y)dxdy = \int\int_{R'_2} f(g(u, v), h(u, v)) \left| \frac{\partial(x, y)}{\partial(u, v)} \right| dudv. \qquad (11.15)$$

Jacobian determinant is used to obtain the distribution of one-to-one (bijective) invertible functions of continuous random variables in statistics (the one-to-one condition can be relaxed in certain situations). Derivation is considerably simplified when the original variates are either statistically independent, or are identically distributed. All such transformation functions should be at least once differentiable. We denote the Jacobian determinant simply as $|J|$ (instead of $\|J\|$), where the vertical line has double meaning—it denotes the *absolute value* of the *determinant* (the double vertical bar | has various meanings in different fields—it denotes the absolute value of the argument in algebra, determinant in matrices, norm of a vector or a matrix in geometry, cardinality of a set or a set expression (such as A∩B) in set-theory and probability theory, and so on. In some of the discussions below, the $|J|$ denotes only the determinant *without* absolute value. See Example 11.8 (p. 431), Table 11.1 (p. 425), etc.).

11.2.1 Functions of Several Variables

Distribution of a function of random variable(s) has many applications in engineering and applied sciences. These are easily obtained when the variates are independent. It is fairly easy to obtain the joint distribution of identically distributed variables using a correct set of transformations. In majority of problems of this type, we have to employ one of the transformations summarized in Table 11.1. For functions of two variables, we have to choose a convenient auxiliary function such that the Jacobian is nonzero, and auxiliary variable is easy to integrate out.

The 2D Jacobian works on the principle that the average value of a double integral $f_{avg}(x, y) = 1/[(b - a) * (d - c)] \int_a^b \int_c^d dF(x, y)$ can be equalized under an arbitrary transformation $u = h(x, y); v = g(x, y)$ as done in the univariate case. If the derivatives of $h(x, y)$ and $g(x, y)$ exist for each point in the rectangular region $[a, b]x[c, d]$, then the above limit will be finite. This allows us to equate the width of an infinitesimal strip under the surface $f(x, y)$ as $f(x, y)dxdy = \gamma(u, v)dudv$ for all points within the region. From this, we get $\gamma(u, v) = f(x, y)|J| = |J|f(h^{-1}(u, v), g^{-1}(u, v))$, where J is the Jacobian of the transformation. The only conditions in this transformation are that the mapping is once differentiable (i.e., $h^{-1}(u, v), g^{-1}(u, v)$) exists), and it is invertible (x, y can be expressed in terms of u, v).

11.2.2 Arbitrary Transformations

The above transformation can be applied to arbitrary continuously differentiable, and invertible functions in higher dimensions (as bivariate, trivariate, and multivariate transformations). Let X, Y be jointly distributed according to some PDF $f(x, y)$. Consider arbitrary continuously differentiable, and invertible functions of the form

TABLE 11.1   Common Transformation of Two Variables

Transformation	Inverse Transformation	Jacobian $|J|$
$u = x + y, v = x - y$	$x = (u + v)/2, y = (u - v)/2$	$-1/2$
$u = x \pm y, v = y$	$x = u \pm v, y = v$	1
$u = x + y, v = x/y$	$x = uv/(1 + v), y = u/(1 + v)$	$-u/(1 + v)^2$
$u = x + y, v = x/(x + y)$	$x = uv, y = u(1 - v)$	$-u$
$u = ax + by, v = cx + dy$	$x = (du - bv)/D,$ $y = (av - cu)/D$	$1/D, D = (ad - bc) \neq 0$
$u = x/y, v = xy$	$x = \sqrt{uv}, y = \sqrt{v/u}$	$1/(2u)$
$u = mx/ny, v = y$	$x = nuv/m, y = v$	nv/m
$u = xy, v = y$	$x = u/v, y = v$	$1/v$
$u = \sqrt{x}, v = y$	$x = u^2, y = v$	$2u$
$u = \sqrt{x + y}, v = \sqrt{y}$	$x = u^2 - v^2, y = v^2$	$4uv$
$u = \sqrt{x^2 + y^2}, v = y$	$x = \pm\sqrt{u^2 - v^2}, y = v$	$u/\sqrt{u^2 - v^2}$
$u = \sqrt{x^2 + y^2}, v = x/\sqrt{x^2 + y^2}$	$x = uv, y = u\sqrt{1 - v^2}$	$-u/\sqrt{1 - v^2}$
$u = \sqrt{nx}/\sqrt{y}, v = \sqrt{y}$	$x = uv/\sqrt{n}, y = v^2$	$2v^2/\sqrt{n}$
$u = x/y, v = xy/\sqrt{x^2 + y^2}$	$x = v\sqrt{u^2 + 1}, y = v\sqrt{u^2 + 1}/u$	$v(1 + 1/u^2)$
$x = r\cos(\theta), y = r\sin e(\theta)$	$r = \sqrt{x^2 + y^2}, \theta = \tan^{-1}(y/x)$	$-r^2\sin(\phi)$
$x = r\cos(\theta)\sin(\phi),$ $y = r\sin(\theta)\sin(\phi),$ $z = r\cos(\phi)$	$r = \sqrt{x^2 + y^2 + z^2},$ $\theta = \tan^{-1}(y/x)$	r
$x = r\cos^2(\theta), y = r\sin^2(\theta)$	$r = x + y, \theta = \tan^{-1}(\sqrt{y/x})$	$r\ \sin(2\theta)$

The order of the variables can be exchanged to get identical results, but the sign of Jacobian could differ. For example, if $u = xy, v = x, |J| = -1/v$. In the last case, as $x + y = r$, we can also write $\theta = \frac{1}{2}\cos^{-1}((x - y)/(x + y))$. The sign of $|J|$ is ignored in statistical applications.

$u = g(x, y)$ and $v = h(x, y)$. If the mapping from (x, y) to (u, v) is one-to-one, it is invertible. We can express x and y in terms of u and v (say $x = G(u, v), y = H(u, v)$). The differential relation $f(x, y)dxdy = f(u, v)dudv$ translates into

$$f(u, v) = f(x, y)\begin{vmatrix} \dfrac{\partial x}{\partial u} & \dfrac{\partial x}{\partial v} \\ \dfrac{\partial y}{\partial u} & \dfrac{\partial y}{\partial v} \end{vmatrix} = f(G(u, v), H(u, v))\begin{vmatrix} \dfrac{\partial x}{\partial u} & \dfrac{\partial x}{\partial v} \\ \dfrac{\partial y}{\partial u} & \dfrac{\partial y}{\partial v} \end{vmatrix}.$$

Here, either u or v is the required transformation, and the other is called the *auxiliary* function. The choice of the auxiliary function is quite often arbitrary. It can be as simple as one of the original variables, provided that the inverse transformation is easy to find. It can be polar or trigonometric transformation when expressions such as $\sqrt{x^2 + y^2}$ or $x^2 \pm y^2$ are present. Bivariate linear transformation is a special case of the above in which the dependency is $u = c_1x + c_2y$ and $v = c_3x + c_4y$, where $c_i's$ are arbitrary constants. The Jacobian of the transformation considerably simplifies in this case as

$$|J| = \left|\frac{\partial(x, y)}{\partial(u, v)}\right| = \begin{vmatrix} c_1 & c_2 \\ c_3 & c_4 \end{vmatrix}$$

A challenge in this type of transformations is the range mapping (see Figure 11.4). We could visualize the transformed mapping easily in the bivariate case, but it is not easy in higher dimensions.

Finding the determinant of a transformation involves much work in some cases. This can be reduced by the *do-little* method. Consider the transformation $u = g(x, y)$ and $v = h(x, y)$. If $v = y$ (or analogously $u = x$), we can reduce the work as follows. Simply substitute for $y = v$ in u to get $u = g(x, v)$. Next, find the derivative $\partial u/\partial x = \partial g(x, v)/\partial x$. Substitute for any x and take the reciprocal to get the Jacobian. Alternatively, express $x = G(u, v)$ and find $J = \partial x/\partial u = \partial G(u, v)/\partial u$. As examples, consider the transformation $u = x + y, v = y$. Put $y = v$ to get $u = x + v$, and $1/J = \partial u/\partial x = 1$; as v is constant. Alternatively, solve for x to get $x = u - y = u - v$. Then, $J = \partial x/\partial u = (\partial/\partial u)(u - v) = 1$. Similarly, if $u = xy, v = y$, then $x = u/v$, and $J = \partial x/\partial u = 1/v$; and for $u = x/y, v = y$ we have $x = uv$ and $J = \partial x/\partial u = v$. As another example, if $u = x/(x + y)$ and $v = y, x = uv/(1 - u)$, and $J = \partial x/\partial u = v/(1 - u)^2$. This works even for constant multiples. If $u = kx/y, v = y$; we have $x = uv/k$ and $J = v/k$. This idea can be extended to those cases where one of the input variables is a function of just one output variable in the two variables case. Consider $x = u^2 - v^2$ and $y = v^2$ (independent of u) so that $J = 2u*2v = 4uv$ (see Table 11.1).

▊ EXAMPLE 11.5   Distribution of sum of rectangular variates

An electronic circuit consists of two independent identical transistors connected in parallel. Let X and Y be the lifetimes of them, distributed as CUNI$(0, b)$ with PDF $f(x) = 1/b, 0 < x < b$. Find the distribution of (i) $Z = X + Y$ (ii) $U = XY$.

Solution 11.5 Consider the transformation $Z = X + Y, W = Y$. The inverse transformation is $Y = W, X = Z - W$. The absolute value of the Jacobian is

$$|J| = \begin{vmatrix} -1 & 1 \\ 1 & 0 \end{vmatrix} = |-1| = 1$$

(here the first vertical bar denotes determinant and second one denotes absolute value). As X and Y are IID, the joint PDF is $f(x, y) = 1/b^2$. The joint PDF of W and Z is $f(w, z) = 1/b^2|J| = 1/b^2$. The range for Z is $[0, +2b]$, and for W is $[0, b]$. As $0 < x < b$, we need to impose the condition $0 < z - w < b$. This in turn results in two regions of integration as shown in Figure 11.1. For $0 < Z < b$, w varies between 0 and z so that $f(z) = \int_w f(w, z)dw = \int_{w=0}^{z} dw/b^2 = z/b^2$. For $b < z < 2b$, w varies between $z - b$ and b, so that $f(z) = \int_w f(w, z)dw =$

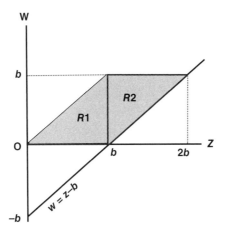

Figure 11.1   Region of integration (i).

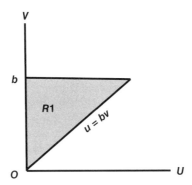

Figure 11.2   Region of integration (ii).

$\int_{w=z-b}^{b} dw/b^2 = (2b - z)/b^2$. Combining both the cases, we can write the PDF as $f(z) = (b - |b - z|)/b^2$ for $0 < z < 2b$, because when $z < b$, $|b - z| = b - z$; and for $z \geq b$, $|b - z| = z - b$. In the second case, we put $U = XY$ and $V = Y$ so that $X = U/V$, and $J = 1/v$. The range for U is $[0, b^2]$, and for V is $[0, b]$. As $0 < x < b$, we need to impose the condition $0 < U/V < b$, or equivalently $u < bv$. The region of integration is shown in Figure 11.2. The joint PDF of u and v is $f(u, v) = 1/(vb^2)$, $0 < u < bv < b^2$. The PDF of U is obtained as $f(u) = \int_{u/b}^{b} 1/(vb^2)dv = (1/b^2)(\ln(b) - \ln(u/b))$, $0 < u < b^2$. Using $\log(x/y) = \log(x) - \log(y)$, this can be simplified as $(1/b^2)[2\ln(b) - \ln(u)]$. ∎

11.2.3   Image Jacobian Matrices

Image Jacobian matrices used in robotics, unmanned aerial vehicles (UAVs), image and video compression, medical imaging, and so on are often rectangular. In image compression and video processing applications, we look for an unknown displacement vector (or matrix) to minimize two successive time frames (or subframes of appropriate sizes) so as to align successive images with minimal loss of information. Sparse residual distortions indicate almost still image frames. In this case, the Jacobian matrix becomes

$$1/J = \frac{\partial(y_1 - u_{y_1}, \ldots, y_m - u_{y_m})}{\partial(x_1, x_2, \ldots, x_n)} = \begin{bmatrix} \dfrac{\partial y_1 - u_{y_1}}{\partial x_1} & \dfrac{\partial y_1 - u_{y_1}}{\partial x_2} & \cdots & \dfrac{\partial y_1 - u_{y_1}}{\partial x_n} \\ \dfrac{\partial y_2 - u_{y_2}}{\partial x_1} & \dfrac{\partial y_2 - u_{y_2}}{\partial x_2} & \cdots & \dfrac{\partial y_2 - u_{y_2}}{\partial x_n} \\ \vdots & \vdots & \cdots & \vdots \\ \dfrac{\partial y_m - u_{y_m}}{\partial x_1} & \dfrac{\partial y_m - u_{y_m}}{\partial x_2} & \cdots & \dfrac{\partial y_m - u_{y_m}}{\partial x_n} \end{bmatrix},$$

with the determinant sign (when the matrix is square, i.e., $m = n$) indicating volumetric expansion ($|J| > 1$), shrinkage ($|J| < 1$), or steadiness ($|J| = 1$).

The Jacobian determinant is a function of the variates (or a constant) when applied to variate transformations in statistics (Table 11.1). This means that the determinant can be made nonzero almost always. As the first derivative of linear functions is a constant, $|J|$ is a scalar constant for linear transformations (If $Y = AX$, then $|J| = |A|$ for multivariate transformation). The range of the transformed variates should be adjusted to account for this fact. As the Jacobian determinant is used as a multiplier, we take the absolute value of Jacobian in statistical applications (the sign of a determinant depends on the order of the columns (variables) in the corresponding matrix). We can do better without the Jacobian method for simple transformation of variates such as translations ($u = x + c, v = y + d$), and scaling ($u = c * x, v = d * y$) (using the CDF or MGF methods). The power of the Jacobian method becomes apparent when variable interactions are present. See References 137, 293 and 296 for further examples.

■ EXAMPLE 11.6   Functions of exponential distribution

Let X_i's be IID EXP(λ) with PDF $f(x_i) = \frac{1}{\lambda}e^{-x_i/\lambda}$. Define new variates Y_i's as $Y_1 = X_1/(X_1 + X_2 + \cdots + X_n), Y_2 = (X_1 + X_2)/(X_1 + X_2 + \cdots + X_n)$, etc $Y_k = (X_1 + X_2 + \cdots + X_k)/(X_1 + X_2 + \cdots + X_n)$, and $Y_n = (X_1 + X_2 + \cdots + X_n)$. Prove that the joint distribution of $(Y_1, Y_2, \ldots, Y_n)$ depends on y_n and y_{n-1} only.

Solution 11.6 As X_i's are IID, the joint PDF is the product of individual PDFs. Thus $f(x_1, x_2, \ldots, x_n) = \frac{1}{\lambda^n}e^{-\sum_{i=1}^{n} x_i/\lambda}$. The inverse mapping is $x_1 = y_1 y_n, x_2 = y_n(y_2 - y_1), x_3 = y_n(y_3 - y_2), x_k = y_n(y_k - y_{k-1}) \ldots, x_n = y_n(1 - y_{n-1})$. The Jacobian is

$$|J| = \left| \frac{\partial(x_1, x_2, \ldots, x_n)}{\partial(y_1, y_2, \ldots, y_n)} \right|$$

$$= \begin{vmatrix} y_n & 0 & 0 & 0 & \cdots & 0 & y_1 \\ -y_n & y_n & 0 & 0 & \cdots & 0 & y_2 - y_1 \\ 0 & -y_n & y_n & 0 & \cdots & 0 & y_3 - y_2 \\ 0 & 0 & -y_n & y_n & \cdots & 0 & y_4 - y_3 \\ \vdots & \vdots & \vdots & \vdots & \cdots & 0 & y_k - y_{k-1} \\ 0 & 0 & 0 & 0 & \cdots & -y_n & (1 - y_{n-1}) \end{vmatrix}.$$

∎

To evaluate this determinant, we apply the row transformations $R'_2 = R_2 + R_1$, $R'_3 = R_3 + R'_2, \cdots, R'_{n-1} = R_{n-1} + R'_{n-2}$, and keep the nth row intact. The determinant reduces to

$$\begin{vmatrix} y_n & 0 & 0 & 0 & \cdots & 0 & y_1 \\ 0 & y_n & 0 & 0 & \cdots & 0 & y_2 \\ 0 & 0 & y_n & 0 & \cdots & 0 & y_3 \\ 0 & 0 & 0 & y_n & \cdots & 0 & y_4 \\ \vdots & \vdots & \vdots & \vdots & \cdots & 0 & y_k \\ 0 & 0 & 0 & 0 & \cdots & -y_n & (1 - y_{n-1}) \end{vmatrix}.$$

By expanding this determinant along the first column, we get $|J| = y_n^n(1 - y_{n-1})$. Thus, $f(y_1, y_2, \ldots, y_n) = \frac{1}{\lambda^n} e^{-y_n/\lambda} y_n^n (1 - y_{n-1})$.

This technique is applicable to discrete random variables as well. Let $u = g(x, y)$ and $v = h(x, y)$ be the bivariate mapping as before. Find the inverse transformation (express x and y as functions of u and v, say $x = f_1(u, v)$ and $y = f_2(u, v)$. Then the joint pmf of u and v is $p_{UV}(u, v) = p_{XY}(f_1(u, v), f_2(u, v))$.

11.2.4 Distribution of Products and Ratios

The distribution of products and ratios of independent random variables are of interest in some applications. These can be obtained by the Jacobian technique when the variables involved are continuous and independent. Make the transformation $U = X/Y$ and $V = XY$. Then, $x = \sqrt{uv}$, and $y = \sqrt{v/u}$, so that the Jacobian is $1/(2u)$. The joint PDF is the product of the marginal PDFs (due to independence assumption). From this, the PDF of either of them can be obtained by integrating out the other. An alternate and simple method exists using the MoDF discussed in the last chapter.

Let $F(z)$ be the CDF of the product. By definition,

$$F(z) = P[Z \le z] = \int \int_{xy \le z} f(x, y) dx dy. \tag{11.16}$$

As $xy = c$ represents a parabolic curve, we split the range of integration of y from $(-\infty, z/x]$ and from $[z/x, \infty)$ to get

$$F(z) = \int_{-\infty}^{0} \left[\int_{z/x}^{\infty} f(x, y) dy \right] dx + \int_{0}^{\infty} \left[\int_{-\infty}^{z/x} f(x, y) dy \right] dx. \tag{11.17}$$

Using the transformation $U = XY$ this becomes

$$F(z) = \int_{-\infty}^{0} \left[\int_{-\infty}^{0} f(x, u/x) du/x \right] + \int_{0}^{\infty} \left[\int_{-\infty}^{z} f(x, u/x) du/x \right] dx. \qquad (11.18)$$

This, on rearrangement, becomes

$$F(z) = \int_{-\infty}^{z} \left[\int_{-\infty}^{\infty} (1/|x|) f(x, u/x) dx \right] du. \qquad (11.19)$$

Differentiate with respect to z to get the PDF as

$$f(z) = \int_{-\infty}^{\infty} (1/|x|) f(x, z/x) dx. \qquad (11.20)$$

It is shown below that if X and Y are IID normal variates, the ratio X/Y has a Cauchy distribution. Analogously $U = X/Y$ has PDF

$$f(u) = \int_{-\infty}^{\infty} |x| f(x, ux) dx. \qquad (11.21)$$

■ **EXAMPLE 11.7   Ratio of uniform distributions**

If X and Y are CUNI$(0, b)$ distributed, find the distribution of $U = X/Y$.

Solution 11.7 Let $U = X/Y$, $V = Y$ so that the inverse mapping is $Y = V, X = UV$. The Jacobian is $|J| = v$. The joint PDF is $f(x, y) = 1/b^2$. Hence, $f(u, v) = v/b^2$. The PDF of u is obtained by integrating out v. A plot of the mapping is shown in Figure 11.3. The region of interest is a rectangle of sides $1 \times b$ at the left, and a curve $uv = b$ to its right. Integrating out v, we obtain $f(u) = \int_{0}^{b} \frac{v}{b^2} dv$ for $0 < u \le 1$, and $f(u) = \int_{0}^{b/u} v/b^2 dv = 1/(2u^2)$ for $1 < u < \infty$.

$$f(u) = \begin{cases} 1/2 \text{ for } & 0 < u < 1; \\ 1/(2u^2) & 1 < u < \infty \end{cases}$$

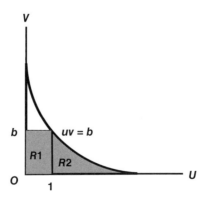

Figure 11.3   Region of integration for X/Y.

See Reference 294 for the distribution of ratios of exponential variates, and Reference 297 for ratios of gamma variates. ∎

EXAMPLE 11.8 Sum and ratio of gamma distribution

If X and Y are IID GAMM(α, β_i), find the distribution of (i) $X + Y$, (ii) X/Y.

Solution 11.8 Let $U = X + Y$ and $V = X/Y$. Solving for X and Y in terms of U and V, we get $x = \frac{uv}{1+v}$, and $y = \frac{u}{1+v}$. The Jacobian of the transformation is

$$
|J| = \begin{vmatrix} \dfrac{v}{1+v} & \dfrac{u}{(1+v)^2} \\ \dfrac{1}{1+v} & -\dfrac{u}{(1+v)^2} \end{vmatrix} = \frac{-u}{(1+v)^2}.
$$

The joint PDF of X and Y is $f(x,y) = \dfrac{\alpha^{\beta_1+\beta_2}}{\Gamma(\beta_1)\Gamma(\beta_2)} e^{-\alpha(x+y)} x^{\beta_1-1} y^{\beta_2-1}$. Multiply by the Jacobian, and substitute for x, y to get

$$
f(u,v) = \frac{\alpha^{\beta_1+\beta_2}}{\Gamma(\beta_1)\Gamma(\beta_2)} e^{-\alpha u} (uv/(1+v))^{\beta_1-1} (u/(1+v))^{\beta_2-1} \frac{u}{(1+v)^2}. \tag{11.22}
$$

The PDF of u is obtained by integrating out v as

$$
f(u) = \frac{\alpha^{\beta_1+\beta_2}}{\Gamma(\beta_1)\Gamma(\beta_2)} e^{-\alpha u} u^{\beta_1+\beta_2-1} \int_0^\infty v^{\beta_1-1}/(1+v)^{\beta_1+\beta_2} dv. \tag{11.23}
$$

Put $1/(1+v) = t$ so that $v = (1-t)/t$, and $dv = -1/t^2 dt$. This gives us

$$
f(u) = \frac{\alpha^{\beta_1+\beta_2}}{\Gamma(\beta_1)\Gamma(\beta_2)} e^{-\alpha u} u^{\beta_1+\beta_2-1} \int_0^1 t^{\beta_2-1}(1-t)^{\beta_1-1} dt
$$

$$
= \frac{\alpha^{\beta_1+\beta_2}}{\Gamma(\beta_1)\Gamma(\beta_2)} e^{-\alpha u} u^{\beta_1+\beta_2-1} B(\beta_1, \beta_2). \tag{11.24}
$$

This simplifies to

$$
f(u) = \frac{\alpha^{\beta_1+\beta_2}}{\Gamma(\beta_1+\beta_2)} e^{-\alpha u} u^{\beta_1+\beta_2-1}, \tag{11.25}
$$

which is GAMMA(β_1, β_2). ∎

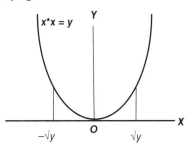

Figure 11.4 Region of integration $Y = X^2$

The PDF of v is found by integrating out u as

$$f(v) = \frac{\alpha^{\beta_1+\beta_2}}{\Gamma(\beta_1)\Gamma(\beta_2)} \frac{v^{\beta_1-1}}{(1+v)^{\beta_1+\beta_2}} \int_{u=0}^{\infty} u^{\beta_1+\beta_2-1} e^{-\alpha u} du. \qquad (11.26)$$

This simplifies to $f(v) = \frac{\Gamma(\beta_1+\beta_2)}{\Gamma(\beta_1)\Gamma(\beta_2)} \frac{v^{\beta_1-1}}{(1+v)^{\beta_1+\beta_2}}$, which is BETA2$(\beta_1, \beta_2)$ (also called Pearson type VI distribution).

◼ EXAMPLE 11.9   Ratio of pairwise independent distributions

Let X_i's be IID EXP$(\lambda\theta)$ for $i = 1, 2, .., m$. Let Y_j's be IID EXP(θ) for $j = 1, 2, .., n$. If X_i's and Y_j's are pair-wise independent, find the distribution of $W = U/V = \sum_{i=1}^{m} X_i / \sum_{j=1}^{n} Y_j$.

Solution 11.9 As X_i's are IID, the joint PDF is the product of individual PDFs. We first use the MGF technique to find the distribution of numerator and denominator. The MGF of EXP$(\lambda\theta)$ is $M_x(t) = 1/[1 - \lambda\theta t]$. As the X_i's are IID, $M_u(t) = 1/[1 - \lambda\theta t]^m$. Similarly, $M_v(t) = 1/[1 - \theta t]^n$. These are the MGFs of gamma distributions. Hence, W is the ratio of two independent gamma variates, whose distribution is found in Example 11.8.   ◼

◼ EXAMPLE 11.10   From gamma to beta

If X and Y are IID GAMM(α, β_i), prove that $X/(X + Y)$ is BETA1 distributed.

Solution 11.10 We find the distribution of $U = X + Y$ and $V = X/(X + Y)$. The joint PDF is

$$f(x, y) = \frac{\alpha^{\beta_1+\beta_2}}{\Gamma(\beta_1)\Gamma(\beta_2)]} x^{\beta_1-1} y^{\beta_2-1} e^{-\alpha(x+y)}. \qquad (11.27)$$

The inverse mapping is $x = uv, y = u(1 - v)$, so that the Jacobian is u. The joint PDF of u and v is

$$f(u, v) = \frac{\alpha^{\beta_1+\beta_2}}{\Gamma(\beta_1)\Gamma(\beta_2)} (uv)^{\beta_1-1}(u - uv)^{\beta_2-1} e^{-\alpha u} u, \qquad (11.28)$$

$0 < u < 1, 0 < v < \infty$. Combining common terms this becomes

$$\frac{\alpha^{\beta_1+\beta_2}}{\Gamma(\beta_1)\Gamma(\beta_2)} e^{-\alpha u} u^{\beta_1+\beta_2-1} v^{\beta_1-1}(1 - v)^{\beta_2-1}. \qquad (11.29)$$

Integrating out u, it is easy to show that v has a BETA1(β_1, β_2) distribution.   ◼

EXAMPLE 11.11 Ratio of independent normal distributions

If X,Y are IID $N(0, \sigma_i^2)$, find the distribution of $U = X/Y$, and $V = XY/\sqrt{X^2 + Y^2}$.

Solution 11.11 Here, both U and V have range $-\infty$ to ∞. From $U = X/Y$, we get $x = uy$. Substituting for x in V, we get $V = (uy)y/\sqrt{(uy)^2 + y^2}$. Taking the y^2 within the square-root in the denominator outside, and canceling it out with the y in the numerator, this becomes $v = uy/\sqrt{u^2 + 1}$ so that $y = v\sqrt{u^2 + 1}/u$, and $x = v\sqrt{u^2 + 1}$. The Jacobian is easily obtained as

$$J = \begin{vmatrix} \frac{u}{v}\sqrt{u^2 + 1} & \sqrt{u^2 + 1} \\ \frac{-v}{u^2\sqrt{u^2 + 1}} & \sqrt{u^2 + 1}/u \end{vmatrix} = v(1 + 1/u^2).$$

We first find the distribution of U. The joint PDF of X and Y is

$$f(x, y) = \frac{1}{2\pi\sigma_1\sigma_2} e^{-\frac{1}{2}(x^2/\sigma_1^2 + y^2/\sigma_2^2)}.\tag{11.30}$$

Substituting the values of x and y, $(x^2/\sigma_1^2 + y^2/\sigma_2^2) = v^2(1 + u^2)\left(\frac{1}{\sigma_1^2} + \frac{1}{\sigma_2^2 u^2}\right)$. Write $A = (1 + u^2)(\frac{1}{\sigma_1^2} + \frac{1}{\sigma_2^2 u^2})$, which is independent of v. Then, $f(u, v) = \frac{1}{2\pi\sigma_1\sigma_2} e^{-\frac{1}{2}Av^2}$. Multiply by the Jacobian, and integrate out v, to get the PDF of u as

$$f(u) = \frac{1}{2\pi\sigma_1\sigma_2}(1 + 1/u^2) \int_{-\infty}^{\infty} v e^{-\frac{1}{2}Av^2} dv.\tag{11.31}$$

Put $v^2/2 = t$ in equation (11.31), so that $vdv = dt$, to get

$$f(u) = \frac{1}{2\pi\sigma_1\sigma_2}(1 + 1/u^2) \int_0^{\infty} e^{-At} dt = \frac{1}{2\pi\sigma_1\sigma_2}(1 + 1/u^2)[0 + 1/A].\tag{11.32}$$

Substituting $A = (1 + u^2)(\frac{1}{\sigma_1^2} + \frac{1}{\sigma_2^2 u^2})$ this becomes

$$f(u) = \frac{1}{2\pi\sigma_1\sigma_2}(1 + 1/u^2)\frac{1}{(1 + u^2)}\left[1/\left(\frac{1}{\sigma_1^2} + \frac{1}{\sigma_2^2 u^2}\right)\right].\tag{11.33}$$

This, after simplification, becomes

$$f(u) = \frac{\sigma_1\sigma_2}{2\pi} \frac{1}{(\sigma_1^2 + u^2\sigma_2^2)}, \quad -\infty < u < \infty.\tag{11.34}$$

This is the PDF of a scaled Cauchy distribution with scaling factor σ_2/σ_1. When $\sigma_1 = \sigma_2$, equation (11.34) reduces to the standard Cauchy distribution. Thus, the ratio of two independent normals (with same variance) is Cauchy distributed. This result can be used to characterize the normal law as follows: ■

Remark 1 If X and Y are independent random variables with the same variance, whose ratio is Cauchy distributed, then X and Y are $N(0, \sigma^2)$ distributed.

We have assumed the independence of the normal variates in the above derivation. If they are dependent with correlation ρ, the ratio is no longer Cauchy distributed. To find the PDF of V, integrate out u from $-\infty$ to ∞. Write the exponent as

$$v^2(1 + u^2)\left(\frac{1}{\sigma_1^2} + \frac{1}{\sigma_2^2 u^2}\right) = v^2[(1/\sigma_1^2 + 1/\sigma_2^2) + (u^2/\sigma_1^2 + 1/(\sigma_2^2 u^2))].$$

Multiply by the Jacobian, and integrate out u, to get the PDF of v as

$$f(v) = \frac{1}{2\pi\sigma_1\sigma_2} v e^{-\frac{1}{2}Av^2} \int_{-\infty}^{\infty} e^{-\frac{1}{2}v^2\left(\frac{u^2}{\sigma_1^2} + \frac{1}{\sigma_2^2 u^2}\right)}(1 + 1/u^2)du, \qquad (11.35)$$

where $A = (1/\sigma_1^2 + 1/\sigma_2^2)$. Split the integral

$$\int_{-\infty}^{\infty} e^{-\frac{1}{2}v^2\left(\frac{u^2}{\sigma_1^2} + \frac{1}{\sigma_2^2 u^2}\right)}(1 + 1/u^2)du = \int_{-\infty}^{\infty} e^{-\frac{1}{2}v^2\left(\frac{u^2}{\sigma_1^2} + \frac{1}{\sigma_2^2 u^2}\right)} du$$

$$+ \int_{-\infty}^{\infty} e^{-\frac{1}{2}v^2\left(\frac{u^2}{\sigma_1^2} + \frac{1}{\sigma_2^2 u^2}\right)}(1/u^2)du = I_1 + I_2 (\text{say}). \qquad (11.36)$$

To evaluate I_1, we use the formula

$$\int_{-\infty}^{\infty} e^{-ax^2 - b/x^2} dx = \sqrt{\frac{\pi}{a}} e^{-2\sqrt{ab}} \qquad (11.37)$$

(see equation 7.4.3 in page 302 of [170] or [298], where $a = v^2/(2\sigma_1^2)$, and $b = v^2/(2\sigma_2^2)$, so that $\sqrt{ab} = v^2/(2\sigma_1\sigma_2)$. Thus, $I_1 = \frac{\sigma_1}{v}\sqrt{2\pi} e^{-\frac{v^2}{\sigma_1\sigma_2}}$. To evaluate I_2, we make a simple substitution $t = 1/u$, so that $dt = -du/u^2$. Exponent of the integrand is $\left(\frac{u^2}{\sigma_1^2} + \frac{1}{\sigma_2^2 u^2}\right) = \left(\frac{1}{\sigma_1^2 t^2} + \frac{t^2}{\sigma_2^2}\right)$. Hence, $I_2 = I_1$ (with σ_1 and σ_2 swapped) $= \frac{\sigma_2}{v}\sqrt{2\pi} e^{-\frac{v^2}{\sigma_1\sigma_2}}$. Substitute these values in equation (11.35) to get the PDF of v as

$$f(v) = \frac{1}{2\pi\sigma_1\sigma_2} v e^{-\frac{1}{2}Av^2} \left[\frac{\sigma_1}{v}\sqrt{2\pi} e^{-\frac{v^2}{\sigma_1\sigma_2}} + \frac{\sigma_2}{v}\sqrt{2\pi} e^{-\frac{v^2}{\sigma_1\sigma_2}}\right]. \qquad (11.38)$$

Canceling out common factors (v, $\sqrt{2\pi}$ from numerator and denominator) and taking the $\sigma_1\sigma_2$ in the denominator into the brackets, we simplify this to the form

$$\frac{1}{\sqrt{2\pi}} e^{-\frac{v^2}{2}\left[A + \frac{2}{\sigma_1\sigma_2}\right]}(1/\sigma_1 + 1/\sigma_2). \tag{11.39}$$

Substitute $A = (1/\sigma_1^2 + 1/\sigma_2^2)$, and note that $(1/\sigma_1^2 + 1/\sigma_2^2 + \frac{2}{\sigma_1\sigma_2}) = (1/\sigma_1 + 1/\sigma_2)^2$, we find that this is the PDF of a normal distribution with variance $(\sigma_1\sigma_2/[\sigma_1 + \sigma_2])^2$.

Remark 2 If X and Y are independent continuous random variables with variances σ_1^2 and σ_2^2 such that $XY/\sqrt{X^2 + Y^2}$ is $N(0, (\sigma_1\sigma_2/[\sigma_1 + \sigma_2])^2)$ distributed, then X and Y are $N(0, \sigma_i^2)$ distributed.

11.3   POLAR TRANSFORMATIONS

Polar transformation finds applications in integral calculus, differential equations, statistics, and image-based computing (log-polar transformations), among many other fields. It is so-called because the Cartesian points that use horizontal and vertical coordinate axes are transformed into polar coordinates that use radius and angle with respect to a fixed set of coordinate axes. The most popular polar transformations are discussed in the following section. Advanced treatment on this topic can be found in References 296, 299 and 300, and so on.

11.3.1   Plane Polar Transformations (PPT)

The name comes from the fact that it is applied in 2D for Cartesian to polar mapping. Let (x, y) represent the Cartesian coordinates and (r, θ) denote the corresponding polar coordinates. Then, the mapping is defined by the relation $x = r\cos(\theta)$ and $y = r\sin(\theta)$. The inverse relation is $r = \sqrt{x^2 + y^2}, \theta = \tan^{-1}(y/x)$. The Jacobian of the transformation is

$$J = \begin{vmatrix} \dfrac{\partial x}{\partial r} & \dfrac{\partial x}{\partial \theta} \\[2mm] \dfrac{\partial y}{\partial r} & \dfrac{\partial y}{\partial \theta} \end{vmatrix} = \begin{vmatrix} \cos(\theta) & -r\sin(\theta) \\ \sin(\theta) & r\cos(\theta) \end{vmatrix} = r.$$

As $x^2 + y^2 = r^2$, this transformation is especially useful in statistics when the PDF contains functions of the form x^2 or $1 \pm x^2$ (in the univariate case) and $\sum_i x_i^2$ or $1 \pm \sum_i x_i^2$ (in the multivariate case).

EXAMPLE 11.12   Find distribution of $\sqrt{X^2 + Y^2}$

If X and Y are IID standard normal $N(0, \sigma^2)$, find the PDF of $\sqrt{X^2 + Y^2}$.

Solution 11.12 As they are independent, the joint PDF is

$$f(x, y) = \frac{1}{\sigma\sqrt{2\pi}}e^{-x^2/2\sigma^2}\frac{1}{\sigma\sqrt{2\pi}}e^{-y^2/2\sigma^2} = \frac{1}{2\pi\sigma^2}e^{-(x^2+y^2)/2\sigma^2}. \tag{11.40}$$

Make the transformation $x = r\cos(\theta)$ and $y = r\sin(\theta)$, so that $x^2 + y^2 = r^2$ and $\theta = \tan^{-1}(y/x)$. The Jacobian is r. Hence, the joint PDF of r and θ is $f(r, \theta) = \frac{1}{2\pi\sigma^2}re^{-r^2/2\sigma^2}$. The density of r is found by integrating out θ as

$$f(r) = \int_{\theta=0}^{2\pi} \frac{1}{2\pi\sigma^2}re^{-r^2/2\sigma^2}\,d\theta = \frac{1}{\sigma^2}re^{-r^2/2\sigma^2}, \tag{11.41}$$

which is the Rayleigh distribution. As the joint PDF of r and θ is independent of θ, this is an indication that θ is uniformly distributed.   ∎

■ **EXAMPLE 11.13** χ_n^2 **to Student's T distribution**

If X and Y are IID χ_n^2 distributions, prove that $Z = \frac{\sqrt{n}}{2}\frac{X-Y}{\sqrt{XY}}$ has a Student's T distribution.

Solution 11.13 Consider the polar transformation $x = r\cos^2(\theta)$ and $y = r\sin^2(\theta)$. Then, $x + y = r(\sin^2(\theta) + \cos^2(\theta)) = r$, $x - y = r(\cos^2(\theta) - \sin^2(\theta)) = r\cos(2\theta)$, $\sqrt{xy} = r\sin(\theta)\cos(\theta) = \frac{r}{2}\sin(2\theta)$, so that $xy = \frac{r^2}{4}\sin^2(2\theta)$. The Jacobian of the transformation is

$$J = \begin{vmatrix} \dfrac{\partial x}{\partial r} & \dfrac{\partial x}{\partial \theta} \\[2mm] \dfrac{\partial y}{\partial r} & \dfrac{\partial y}{\partial \theta} \end{vmatrix} = \begin{vmatrix} \cos^2(\theta) & -r\sin(2\theta) \\[1mm] \sin^2(\theta) & r\sin(2\theta) \end{vmatrix} = r\,\sin(2\theta).$$

Substituting the values of X and Y in $Z = \frac{\sqrt{n}}{2}\frac{X-Y}{\sqrt{XY}}$, we get $Z = \frac{\sqrt{n}}{2}2\cot(2\theta) = \sqrt{n}\cot(2\theta)$. The joint PDF of X and Y is

$$f(x, y) = \frac{1}{2^n\Gamma(n/2)^2}e^{-(x+y)/2}(xy)^{\frac{n}{2}-1}. \tag{11.42}$$

Multiply by the Jacobian and substitute the values of x and y to get

$$f(r, \theta) = \frac{1}{2^n\Gamma(n/2)^2}e^{-r/2}\left(\frac{r^2}{4}\sin^2(2\theta)\right)^{\frac{n}{2}-1}r\,\sin(2\theta). \tag{11.43}$$

This, after simplification, reduces to

$$f(r, \theta) = \frac{r^{n-1}e^{-r/2}}{2^{2n-2}\Gamma(n/2)^2} \sin(2\theta)^{n-1}. \tag{11.44}$$

Distribution of θ is obtained by integrating out r. Thus,

$$f(\theta) = \frac{\sin(2\theta)^{n-1}}{2^{2n-2}(\Gamma(n/2))^2} \int_{r=0}^{\infty} r^{n-1}e^{-r/2}dr$$

$$= \frac{\sin(2\theta)^{n-1}}{2^{2n-2}(\Gamma(n/2))^2} 2^n\Gamma(n) = \frac{\Gamma(n)}{2^{n-2}(\Gamma(n/2))^2} \sin(2\theta)^{n-1}. \tag{11.45}$$

Consider the transformation $t = \sqrt{n}\cot(2\theta)$, so that $\frac{\partial\theta}{\partial t} = 1/[2\sqrt{n}\csc^2(2\theta)] = 1/[2\sqrt{n}(1 + \cot^2(2\theta)]$. Writing $\sin(2\theta) = 1/\sqrt{1 + \cot^2(2\theta)}$, and multiplying by the Jacobian, we get

$$f(t) = \frac{\Gamma(n)}{\sqrt{n}2^{n-1}\Gamma(n/2)^2} \frac{1}{(1 + t^2/n)^{(n+1)/2}}. \tag{11.46}$$

Multiply the numerator and denominator by $\Gamma((n+1)/2)\Gamma(1/2)$ and use the formula $\Gamma(n)\Gamma(1/2) = 2^{n-1}\Gamma(n/2)\Gamma(\frac{n+1}{2})$ to get the constant multiplier in the form $1/[\sqrt{n}B(\frac{1}{2}, \frac{n}{2})]$ (where $\Gamma(1/2) = \sqrt{\pi}$). This is the Student T distribution $T(n)$. See Cacoullos [254] for a CDF derivation of this and related results. ∎

■ **EXAMPLE 11.14   Functions of Weibull distributions**

If X, Y are IID Weibull(2, b) with PDF $f(x, b) = \frac{2}{b^2}xe^{-x^2/b^2}$, for $x \geq 0$, show that $Z = XY/(X^2 + Y^2)$ has PDF $f(z) = 2z/\sqrt{1 - 4z^2}$, which is independent of b.

Solution 11.14 Put $x = r\cos(\theta)$ and $y = r\sin(\theta)$, so that $x^2 + y^2 = r^2$ and $\theta = \tan^{-1}(y/x)$. The Jacobian of the transformation is r. Hence, the joint PDF of r and θ is $f(r, \theta) = \frac{4r^3}{b^4}\sin(\theta)\cos(\theta)e^{-r^2/b^2}$. Using $2\sin(\theta)\cos(\theta) = \sin(2\theta)$ this becomes $f(r, \theta) = \frac{2r^3}{b^4}\sin(2\theta)e^{-r^2/b^2}$. The PDF of θ is obtained by integrating out r as

$$f(\theta) = \frac{2\sin(2\theta)}{b^4} \int_0^{\infty} r^3 e^{-r^2/b^2} dr. \tag{11.47}$$

Put $r^2/b^2 = t$, so that $rdr = (b^2/2)dt$. Then, $f(\theta) = \frac{2\sin(2\theta)}{b^4}\frac{b^4}{2}\int_0^{\infty} te^{-t}dt = \sin(2\theta)$. Putting the values of x and y in Z gives $Z = \frac{1}{2}\sin(2\theta)$ so that $\frac{dz}{d\theta} = \cos(2\theta) = \sqrt{1 - 4z^2}$. Hence, $f(z) = 2z/\sqrt{1 - 4z^2}, -\frac{1}{2} \leq z \leq \frac{1}{2}$. ■

11.3.2 Cylindrical Polar Transformations (CPT)

This transformation is a simple extension of the above to 3D. The mapping is defined by the relations $x = r\ \cos(\theta), y = r\ \sin(\theta)$, and $z = z$. The Jacobian of this transformation also is r. This defines a cylinder of base r in the polar coordinates. Inverse mapping is easily obtained as $r = \sqrt{x^2 + y^2}, \theta = \tan^{-1}(y/x)$.

11.3.3 Spherical Polar Transformations (SPT)

This is a more general form of the above PPT to 3D. The mapping is defined by the relations $x = r\ \cos(\theta)\cos(\phi), y = r\ \sin(\theta)\cos(\phi)$, and $z = r\ \sin(\phi)$, so that $x^2 + y^2 + z^2 = r^2\ \cos^2(\phi)[\cos^2(\theta) + \sin^2(\theta)] + r^2\ \sin^2(\phi) = r^2$. The inverse mapping is defined as $r = \sqrt{x^2 + y^2 + z^2}, \theta = \tan^{-1}(y/x)$, and $\phi = \sin^{-1}(z/r)$. The Jacobian of this transformation is $r^2\cos^2(\phi)$. An equivalent mapping is defined by the relations $x = r\ \cos(\theta)\sin(\phi), y = r\ \sin(\theta)\sin(\phi)$, and $z = r\ \cos(\phi)$. The inverse mapping is given by $r = \sqrt{x^2 + y^2 + z^2}, \theta = \tan^{-1}(y/x)$, and $\phi = \cos^{-1}(z/\sqrt{x^2 + y^2 + z^2})$, or $\phi = \tan^{-1}(\sqrt{x^2 + y^2}/z)$. The Jacobian of this transformation is $-r^2\ \sin(\phi)$ (so that $dxdydz = r^2\ \sin(\phi)drd\theta d\phi$). The name *spherical transformation* comes from the fact that its domain is of the form $x^2 + y^2 + z^2$ or arithmetic functions of it.

11.3.4 Other Methods

The SPT can be generalized to n-dimensions in multiple ways. One simple way is to use the Helmert transformation $x_1 = r\ \cos(\theta_1), x_2 = r\ \sin(\theta_1)\cos(\theta_2), x_3 = r\ \sin(\theta_1)\sin(\theta_2)\cos(\theta_3), \cdots , x_{n-1} = r\sin(\theta_1)\sin(\theta_2) \ldots \sin(\theta_{n-2})\cos(\theta_{n-3})$, and $x_n = r\ \sin(\theta_1)\sin(\theta_2) \ldots \sin(\theta_{n-1})$. The Jacobian is given by $|J| = r^{n-1}\sin^{n-2}(\theta_1)\sin^{n-3}(\theta_2)\sin^{n-4}(\theta_3)\cdots \sin(\theta_{n-2})$. Squaring and adding each term, we get $r^2 = x_1^2 + x_2^2 + \cdots + x_n^2$.

Toroidal polar transformation (TPT) is an extension of SPT (see Table 11.2), defined as

$$x = (r\ \cos(\theta) + R)\cos(\phi), y = (r\ \cos(\theta) + R)\sin(\phi), \text{ and } z = r\ \sin(\theta), \quad (11.48)$$

so that $x^2 + y^2 + z^2 = (r\cos(\theta) + R)^2\ [\cos^2(\phi) + \sin^2(\phi)] + r^2\ \sin^2(\theta) = r^2 + R^2 + 2rR\cos(\theta)$. The inverse mapping is defined as

$$r = \{[(x^2 + y^2)^{1/2} - R]^2 + z^2\}^{1/2}, \phi = \tan^{-1}(y/x), \text{ and } \theta = \sin^{-1}(z/r). \quad (11.49)$$

The Jacobian is $1/J=$

$$\begin{vmatrix} \cos(\theta)\cos(\phi) & \cos(\theta)\sin(\phi) & \sin(\theta) \\ -B * \sin(\phi) & B * \cos(\phi) & 0 \\ -r\cos(\phi)\sin(\theta) & -r\sin(\phi)\sin(\theta) & r\cos(\theta) \end{vmatrix}.$$

TABLE 11.2 Common Polar Transformation of Three Variables

Name	Transformation	Jacobian $|J|$
Cylindrical	$x = r\,\cos(\theta), y = r\,\sin(\theta)$, and $z = z$	$-1/2$
Spherical	$x = r\,\cos(\theta)\cos(\phi),$ $y = r\,\sin(\theta)\cos(\phi), z = r\,\sin(\phi)$	$r^2\cos^2(\phi)$
Spherical	$x = r\,\cos(\theta)\sin(\phi),$ $y = r\,\sin(\theta)\sin(\phi), z = r\,\cos(\phi)$	$-r^2\,\sin(\phi)$
Toroidal	$x = (r\,\cos(\theta) + R)\cos(\phi),$ $y = (r\,\cos(\theta) + R)\sin(\phi), z = r\,\sin(\theta)$	$(r\,\cos(\theta) + R)$
Toroidal	$x = r * \cos(\theta)\cos(\phi), y = Cr * \sin(\theta),$ $z = Dr * \sin(\phi)$	$r^2\,(m^2\cos^2(\phi) + n^2\cos^2(\theta))/[C * D].$

The sign of $|J|$ is ignored in statistical applications. The inverses of the transformations appear in the respective sections. Last row has $C = \sqrt{1 - m^2\sin^2(\phi)}, D = \sqrt{1 - n^2\sin^2(\theta)}$

where $B = (r\,\cos(\theta) + R)$. To evaluate this determinant, take out B from second row, r from third row, multiply new first column by $\cos(\phi)$, new second column by $\sin(\phi)$, and add new second column to the first column ($C_1 = C_1 + C_2$). The $(2, 2)$th element also becomes zero. Then, expand the determinant along second row to get the Jacobian as $B = (r\,\cos(\theta) + R)$.

Another general transformation is given by $x = r\cos(\theta)\cos(\phi), y = r\sin(\theta)$ $\sqrt{1 - m^2\sin^2(\phi)}, z = r\sin(\phi)\sqrt{1 - n^2\sin^2(\theta)}$, where $m^2 + n^2 = 1$. Squaring and adding gives us $x^2 + y^2 + z^2 = r^2$. The Jacobian in this case is $1/J=$

$$\begin{vmatrix} \cos(\theta)\cos(\phi) & -r\sin(\theta)\cos(\phi) & -r\cos(\theta)\sin(\phi) \\ \sin(\theta)\sqrt{1 - m^2\sin^2(\phi)} & r\cos(\theta)\sqrt{1 - m^2\sin^2(\phi)} & \dfrac{-m^2r\sin(\theta)\sin(\phi)\cos(\phi)}{\sqrt{1-m^2\sin^2(\phi)}} \\ \sin(\phi)\sqrt{1 - n^2\sin^2(\theta)} & \dfrac{-n^2r\sin(\theta)\sin(\phi)\cos(\theta)}{\sqrt{1-n^2\sin^2(\theta)}} & r\cos(\phi)\sqrt{1 - n^2\sin^2(\theta)} \end{vmatrix}.$$

To evaluate this determinant, take r as a common factor from second and third columns, multiply first column by $\sin(\theta)$ and second column by $\cos(\theta)$, and apply $C_1 = C_1 + C_2$ (i.e., add new second column to new first column). The first element at $(1,1)$ reduces to zero, so that the determinant becomes that of two 2×2 matrices. Using the relationship $m^2 + n^2 = 1$, this is easily seen to be $r^2\,(m^2\cos^2(\phi) + n^2\cos^2(\theta))/[\sqrt{1 - m^2\sin^2(\phi)}\sqrt{1 - n^2\sin^2(\theta)}]$ (see Table 11.2).

There are many other ways to find the distribution of transformed variables. One possibility is to use the characteristic function (if it is easily invertible) of the original variates. Let $U = g(x_1, x_2, ..., x_n)$ be the transformation required. If $\phi_Z(t) = E(e^{itg(x)}) = \int \int e^{itg(x)} f(x_1, x_2, ..., x_n)dx_1..dx_n$ is easy to evaluate, we could simply use inversion theorem to get the PDF of U.

11.4   SUMMARY

This chapter discusses the methodology to obtain the marginal, joint, and conditional probability distributions for both the discrete (i.e., count) distributions and continuous distributions. The concept and tools for the Jacobian to derive the joint probability distribution of functions of continuous random variables are introduced and illustrated in this chapter. Distribution of functions of two or more variates has received much attention in the literature. Most of the research in this field uses the normal [218] and exponential [294] distributions

See Shepp [217], Quine [218], Baringhaus et al. [301], and Jones [302] for alternative derivations of the result in Example 11.0 (p. 433). Bansal et al. [303] uses the uniqueness of moments to prove that the distribution of $2XY/\sqrt{X^2 + Y^2}$ is identical to that of X and Y.

EXERCISES

11.1 Mark as True or False

    a) The Jacobian method is applicable to both discrete and continuous variate transformations.

    b) If the range of a random variable X includes the origin, we cannot use the transformation $Y = 1/X$.

    c) Marginal distributions can be obtained from joint distributions.

    d) Marginal distributions determine joint distributions only when variates are independent.

    e) Joint PDF of random variables can be obtained uniquely from joint CDF.

11.2 If $f(x, y) = c(x + y)$ is the joint PDF of two discrete random variables $(x = 1, 2, 3; y = 1, 2, 3, 4)$, find the constant c and hence obtain the conditional distribution of Y given $X = k$, and the distribution of X^2.

11.3 What is the Jacobian of the transformation $u = x(1 - y), v = xy$. Use it to find the distribution of u and v when x and y are (i) CUNI(0, 1) and (ii) BETA(0, 1).

11.4 Find the unknown K in the following joint PDFs: (a) $f(x, y) = Kx^2y, 0 < x < 1, 0 < y < 2$ (b)

$f(x, y, z) = K(x + 2y + 3z)$ for $0 < x < 1, 0 < y < 2, 0 < z < 1$.

11.5 What is the Jacobian of the transformation $u = aX + bY, v = cX - dY$. If X and Y are triangular, find the distribution of U and V.

11.6 Find the Jacobian of the *rotation* transformation $u = x\cos(\theta) - y\sin(\theta)$ and $v = x\sin(\theta) + y\cos(\theta)$.

11.7 If $f(x, y) = K * x^3 y^2 e^{-(x+(y/2))}$ for $x, y > 0$, find constant K. Are X and Y independent?

11.8 What is the Jacobian of the transformation $x = r\cosh(\theta)\cosh(\phi), y = r\sinh(\theta)\cosh(\phi), z = r\sinh(\theta)$?

11.9 If $f(x, y) = K(x + y)$ for $0 < x < 1, 0 < y < 1$ find constant K and obtain the marginal distributions.

11.10 Find the inverse mapping and Jacobian for the transformation $U = 2xy/(x^2 + y^2)$, $v = x^2 + y^2 + z^2$, $w = (x^2 + y^2)/z^2$ (hint: use spherical polar transform).

11.11 Find the inverse mapping and Jacobian for the transformation $U = x + y$, $v = x^2 - y^2$ if x and y are IID chi-square variates.

11.12 Find the Jacobian of the *shear* transformation (parallel to the Y-axis) $u = ax + y$, $v = y$. Use it to find the distribution of u and v

when x and y are (i) CUNI(0, 1), (ii) GAMM(m_i, p), $i = 1, 2$, and (iii) χ_n^2.

11.13 If $f(x, y) = \Gamma(m)/[\Gamma(a)\Gamma(b)\Gamma(m-a-b-1)]x^{a-1}y^{b-1}(1 - x - y)^{m-a-b-1}$, prove that the marginals and conditional distributions of $X|Y$, $Y|X$ are all beta distributed when the variables are independent.

11.14 Find the Jacobian of the *rotation* transformation $u = x\cos(\theta) - y\sin(\theta)$, and $v = x\sin(\theta) + y\cos(\theta)$.

11.15 If X is BETA-I(p, q) and Y is independent BETA-I($p + q, r$) find the distribution of $X/(X + Y)$.

11.16 If X_1, X_2 are IID N(0,1), find the distribution of $Y_1 = \sqrt{-2\log_e(x_1)} \cos(2\pi x_2)$ and $Y_2 = \sqrt{-2\log_e(x_1)} \sin(2\pi x_2)$ (the inverse being $X_1 = \exp\left[-\frac{1}{2}(y_1^2 + y_2^2)\right]$ and $x_2 = \frac{1}{2\pi}\arctan(y_2/y_1)$).

11.17 If X and Y are independent Cauchy distributed, prove that (i) $Z = (X + Y)/2$ and (ii) $(X - Y)/(1 + XY)$ are also Cauchy distributed.

11.18 If X and Y are independent Gamma distributed, prove that $Z = X/(X + Y)$ is Type I beta distributed.

11.19 Find the Jacobian of the transformation $U_1 = \frac{X_1}{X_n}, U_2 = \frac{X_2}{X_n}, \ldots, U_{n-1} = \frac{X_{n-1}}{X_n}$, and $X_1^2 + X_2^2 + \cdots + X_n^2 = 1$. If X_i's are IID N(0,1), find the joint distribution of U_i's.

11.20 Suppose that $X \sim$ BINO(n, θ), where $\theta \sim$ BETA(a, b). Find the unconditional distribution of X, conditional distribution of $\theta|X$ and

prove that $E(\theta|X) = (a + X)/(a + b + n)$.

11.21 Find the inverse mapping and Jacobian for the transformation $x = r\sin(\theta)\sin(\phi), y = r\sin(\theta)\cos(\phi), z = r\cos(\theta)\sin(\xi), w = r\cos(\theta)\cos(\xi)$.

11.22 Suppose two fair dice are tossed. Find the density function of $(X1, X2)$ where $X1$ and $X2$ are the scores that show up.

11.23 If X, Y are IID CUNI(a, b), find the PDF of $U = -\log(X/Y)$. If $a = -b$, find the distribution of $V = X^2$.

11.24 If $f(x, y) = Ke^{-(aX+bY)}$, find K and obtain the PDF of X/Y and $X - Y$.

11.25 If $X \sim \chi_n$ (i.e., $X \sim \sqrt{\chi_n^2}$) and $Y \sim \text{BETA}(\frac{n-1}{2}, \frac{n-1}{2})$ is independent of X, prove that $(2Y - 1)X \sim N(0, 1)$.

11.26 If $X \sim \chi_m^2$, and $Y \sim \chi_n^2$ prove that $X/(X + Y) \sim \text{BETA}(m/2, n/2)$, and $X/Y \sim \text{BETA2}(m/2, n/2)$.

11.27 If $X \sim \chi_m^2$, $Y \sim \chi_n^2$ and $Z \sim \chi_p^2$ Find distribution of $u = x/y, v = (x + y)/z$ and $w = x + y + z$.

11.28 If $X \sim \chi_{2m+2n}^2$ and $Y \sim \text{BETA1}(p, q)$ be independent, find distribution of XY and $X(1 - Y)$.

11.29 If X and Y are Gamma distributed with parameters (p,m) and (q,m) find the distribution of X/Y and $X/(X + Y)$.

11.30 If X, Y, Z are CUNI$(-1,+1)$, find the PDF of (i) XY, (ii) XY/Z, (iii) $(X + Y)/(X - Y)$, (iv) $(X + Y - Z)/(Y + Z - X)$

11.31 If $U = \frac{XY}{Z}, V = \frac{YZ}{X}$, and $W = \frac{ZX}{Y}$, prove that the Jacobian is a constant. Find the distribution of U when X, Y, Z are (i) GAMM(m, p) and (ii) CUNI$(0, b)$.

11.32 If $X_1, X_2, ..., X_n$ are IID $\Gamma(m, p)$ prove that $Y_n = \min(X_1, X_2, ..., X_n)$ is distributed as GAMMA(mn, p).

11.33 Express the Cartesian coordinates in terms of cylindrical and spherical polar coordinates.

11.34 If X and Y are exponentially distributed, find the distribution of $X + Y$ and $X - Y$.

11.35 If X_1, X_2, X_3 are IID GAMMA(m_k, p) for $k = 1,2,3$ prove that the joint distribution of $Y_1 = X_1/X_3$ and $Y_2 = X_2/X_3$ is bivariate BETA-II(m_k) with PDF

$$f(y_1, y_2) = \frac{\Gamma(m_1 + m_2 + m_3)}{\Gamma(m_1)\Gamma(m_2)\Gamma(m_3)} \frac{y_1^{m_1-1} y_2^{m_2-1}}{(1 + y_1 + y_2)^{m_1 + m_2 + m_3}}, y_1, y_2 > 0.$$

11.36 Find the Jacobian of the transformation $x_1 = r \sin(\theta_1) \sin(\theta_2) \cdots \sin(\theta_{n-2}) \sin(\theta_{n-1}), x_2 = r \cos(\theta_1) \sin(\theta_2) \sin(\theta_3) \cdots \sin(\theta_{n-2}) \sin(\theta_{n-1}), x_3 = r \cos(\theta_2) \sin(\theta_3) \cdots \sin(\theta_{n-2}) \sin(\theta_{n-1}), x_{n-1} = r \cos(\theta_{n-2}) \sin(\theta_{n-1}),$ and $x_n = r \cos(\theta_{n-1})$. What is the inverse mapping?.

11.37 If U,V are IID CUNI$(0, b)$ find the distribution of (i) $U + V$, (ii) $|U - V|$

11.38 If X and Y have joint PDF $f(x, y) = \exp(-x - y), x, y \geq 0$, find the distribution of X/Y and $X + Y$ assuming independence.

11.39 Find the Jacobian of the transformation $y_1 = \sum_{i=1}^{n} X_i / \sqrt{n}, Y_2 = (X_1 - X_2)/\sqrt{2}, Y_i =$

$(X_1 + X_2 + \cdots + X_{i-1} - (i - 1) X_n)/\sqrt{n(n - 1)}$ for $i = 3, 4, ..n$.

11.40 If Y has a chi-distribution with m DoF and Z is BETA-I$((m - 1)/2, (m - 1)/2)$ is independent of Y, then $(2Z - 1)Y$ is standard normal.

11.41 Find the inverse mapping and Jacobian for the transformation $u = x/(x + y), v = y/(x + y + z), w = x + y + z$.

REFERENCES

[1] Stevens SS. On the theory of scales of measurement. Science 1946; 103: 677–680.

[2] Chattamvelli R. *Data Mining Algorithms*. Oxford: Alpha Science; 2011.

[3] Yilmaz E, Aslam JA, Robertson S. A new rank correlation coefficient for information retrieval. SIGIR '08: Proceedings of the 31st Annual International ACM SIGIR Conference on Research and Development in Information Retrieval; Singapore; 2008.

[4] Chattamvelli R. A note on the noncentral beta distribution function. Am Stat 1995; 49: 231–234.

[5] Chattamvelli R. On the doubly noncentral F distribution. Comput Stat Data Anal 1996; 20: 481–489.

[6] Chattamvelli R, Shanmugam R. Efficient computation of the noncentral χ^2 distribution. Commun Stat Simul Comput 1995; 24(3): 675–689.

[7] Chattamvelli R, Shanmugam R. Computing the noncentral beta distribution function, algorithm AS310. Appl Stat 1998; 41: 146–156.

[8] Knuth DE. *Fundamental Algorithms*. Volume 1. 3rd ed. Reading (MA): Addison-Wesley; 1997. xx+650 pp. ISBN 0-201-89683-4 Pearson Education Inc.; 1997.

[9] Jones MC. On some expressions for variance, covariance, skewness and L-moments. J Stat Plann Inference 2004; 126: 97–106.

[10] Holte RC. Very simple classification rules perform well on most commonly used datasets. Mach Learn 1993; 11: 63–90.

[11] Kerber R. Chimerge: discretization of numeric attributes. Proceedings of 10th National Joint Conference on AI. San Jose (CA): MIT Press; 1992. p 123–128.

[12] Cho DW, Im KS. A test of normality using Geary's skewness and kurtosis statistics. Econ Lett 2002; 53: 247–251.

[13] Jarque CM, Bera AK. Efficient tests for normality, homoscedasticity, and serial independence of regression residuals. Econ Lett 1980; 6(3): 255–259.

[14] Lin CC, Mudholkar GS. A simple test for normality against asymmetric alternatives. Biometrika 1980; 67: 455–461.

[15] Zeng A, Gao Q-G, Pan D. A Global unsupervised data discretization algorithm based on collective correlation coefficient. IEA/AIE; Syracuse University, New York; 2011. p 146–155.

[16] Lustgarten JL, Visweswaran S, Gopalakrishnan V, Cooper GF. Application of an efficient Bayesian discretization method to biomedical data. BMC Bioinformatics (BMCBI) 2011; 12: 309–311.

[17] Battiato S, Cantone D, Catalano D, Cincotti G, Hofri M. An efficient algorithm for the approximate median selection problem. 4th Italian Conference, CIAC, Rome. Volume 1767, Lecture Notes in CS. Rome, Italy: Springer; 2000. p 226–240.

[18] Groeneveld RA, Meeden G. The mode, median and mean inequality. Am Stat 1977; 31(3): 120–121.

[19] Hippel V, Paul T. Mean median, skew: correcting a textbook rule. J Stat Educ 2005; 13(2): 95.

[20] Mercer PR. Refined arithmetic, geometric and harmonic mean inequalities. Rocky Mt J Math 2003; 33(4): 1459–1464.

[21] Weisberg HF. *Central Tendency and Variablity*. Newbury Park (CA): Sage Publishers; 1992.

[22] Chattamvelli R. *Statistical Algorithms*. Oxford: Alpha Science; 2012.

[23] Cormen TH, Leiserson CE, Rivest RL, Stein C. *Introduction to Algorithms*. Cambridge (MA): MIT Press; 2009.

[24] Fabian Z. New measures of central tendency and variability of continuous distributions. Commun Stat Theory Methods 2008; 37(2): 159–174.

[25] Tsimashenka I, Knottenbelt W, Harrison P. Controlling variability in split-merge systems. *Analytical and Stochastic Modeling Techniques and Applications*. Lecture Notes in Computer Science 7314. New York: Springer; 2012.

[26] Cook D, Swayne DF. *Interactive and Dynamic Graphics for Data Analysis*. New York: Springer; 2007.

[27] Oja H. Descriptive statistics for multivariate distributions. Stat Probab Lett 1983; 33: 327–333.

[28] Seier E, Bonett DG. A Polyplot for visualizing location, spread, skewness, and kurtosis. Am Stat 2011; 65(4): 258–261.

[29] Yatracos YG. Variance and clustering. Proc Am Math Soc 1998; 126: 1177–1179.

[30] Boos DD, Brownie C. Comparing variances and other measures of dispersion. Stat Sci 2004; 19(4): 571–578.

[31] Chan TF, Golub GH, Leveque R. Algorithms for computing the sample variance: analysis and recommendations. Am Stat 1983; 37(3): 242–247.

[32] Forkman J, Verrill S. The distribution of McKay's approximation for the coefficient of variation. Stat Probab Lett 2008; 78(1): 10–14.

[33] Croucher JS. An upper bound on the value of the standard deviation. Teach Stat 2004; 26(2): 54–55.

[34] Mcleod AJ, Henderson GR. Bounds for the sample standard deviation. Teach Stat 1984; 6(3): 72–76.

[35] Thomson GW. Bounds for the ratio of range to standard deviation. Biometrika 1955; 42: 268–269.

[36] Shiffler RE, Harsha PD. Upper and lower bounds for the sample standard deviation. Teach Stat 1980; 2(3): 84–86.

[37] Joarder AH, Latif RM. Standard deviation for small samples. Technical report series, TR329. Dhahran, Saudi Arabia: King Fahd University of Petroleum and Minerals; 2005.

[38] Glasser GJ. Variance formulas for the mean difference and coefficient of concentration. J Am Stat Assoc 1962; 57: 648–654.

[39] Joarder AH. On some representations of sample variance. Int J Math Educ Sci Technol 2002; 33(5): 772–784.

[40] Heffernan P. New measures of spread and a simpler formula for the normal distribution. Am Stat 1988; 42(2): 100–102.

[41] Zhang Y, Wu H, Cheng L. Some new deformation formulas about variance and covariance. Proceedings of 4th International Conference on Modelling, Identification and Control; Wuhan, China; 2012. p 987–992.

[42] Domingo-Ferrer J, Solanas A. A measure of variance for hierarchical nominal attributes. Inf Sci 2008; 178(24): 4644–4655.

[43] Mandrekar JN, Mandrekar SJ, Cha SS. *Evaluating Methods of Symmetry*. Rochester (MN): Mayo Clinic; 2004.

[44] Petitjean M. Chirality and symmetry measures: a transdisciplinary review. Entropy 2003; 5(3): 271–312.

[45] Székely GJ, Móri TF. A characteristic measure of asymmetry and its application for testing diagonal symmetry. Commun Stat Theory Methods 2001; 30(8,9): 1633–1639.

[46] Shanmugam R. Is Poisson dispersion diluted or over-saturated? An index is created to answer. Am J Biostat 2011; 2(2): 56–60.

[47] Hildebrand DK. Kurtosis measures modality? Am Stat 1971; 25: 42–43.

[48] Holgersson HET. A modified skewness measure for testing asymmetry. Commun Stat Simul Comput 2010; 39(2): 335–346.

[49] Tabachnick BG, Fidell LS. *Using Multivariate Statistics*. 3rd ed. New York: Harper-Collins; 1996.

[50] Royston P. Which measures of skewness and kurtosis are best? Stat Med 1992; 11: 333–343.

[51] Pearson K. Skew variation, a rejoinder. Biometrika 1905; 4: 169–212.

[52] Balanda KP, MacGillivray HL. Kurtosis and spread. Can J Stat 1990; 18: 17–30.

[53] Hosking JMR. Moments or L-moments? An example comparing two measures of distributional shape. Am Stat 1992; 46: 186–189.

[54] Harremoës P. *An Interpretation of Squared Skewness Minus Excess Kurtosis*. New York: Elsevier Science; 2002.

[55] MacGillivray HL. Skewness and asymmetry: measures and orderings. Ann Stat 1986; 14: 994–1011.

[56] Bowley AL. *Elements of Statistics*. New York: Scribner; 1920.

[57] MacGillivray HL. Shape properties of the g-, h-, and Johnson families. Commun Stat Theory Methods 1992; 21: 1233–1250.

[58] Liu W, Chawla S. A quadratic-mean based supervised learning model for managing data skewness. SDM; Mesa (AZ); 2011. p 188–198.

[59] van Zwet WR. *Convex Transformations of Random Variables*. The Netherlands: Mathematisch Centrum Amsterdam; 2005.

[60] Johnson NL, Kotz S, Balakrishnan N. *Continuous Univariate Distributions*. Volume 2. New York: John Wiley and Sons; 1998.

[61] Seier E, Bonett D. Two families of kurtosis measures. Metrika 2003; 58: 59–70.

[62] Mudholkar GS, Natarajan R. The inverse Gaussian models: analogues of symmetry, skewness and kurtosis. Ann Inst Stat Math 2002; 54(1): 138–154.

[63] Brys G, Hubert M, Struyf A. A comparison of some new measures of skewness. *International Conference on Robust Statistics*. Heidelberg: Physica-Verlag; 2011. p 97–112.

[64] Wang J. On nonparametric multivariate scale, kurtosis and tail-weight measures [PhD thesis]. Department of Mathematical Sciences, University of Texas at Dallas; 2003.

[65] Chissom BS. Interpretation of the kurtosis statistic. Am Stat 1970; 24(4): 19–22.

[66] Darlington RB. Is kurtosis really "peakedness"? Am Stat 1970; 24(2): 19–22.

[67] Fung T, Seneta E. Tailweight quantiles and kurtosis: a study of competing distributions. Oper Res Lett 2007; 35(4): 448–454.

[68] Horn PS. A measure for peakedness. Am Stat 1983; 37(1): 55–56.

[69] Moors JJA. The meaning of kurtosis. Am Stat 1986; 40: 283–284.

[70] Oja H. On location, scale, skewness and kurtosis of univariate distributions. Scand J Stat 1981; 8: 154–168.

[71] Stavig GR. A robust measure of kurtosis. Percept Mot Skills 1982; 55: 666.

[72] Groeneveld RA, Meeden G. Measuring skewness and kurtosis. Statistician 1984; 38: 391–399.

[73] Groeneveld RA. A class of quantile measures for kurtosis. Am Stat 1998; 52(3): 325–329.

[74] Ruppert D. What is kurtosis? An influence function approach. Am Stat 1987; 41(1): 1–5.

[75] Finucan HM. A note on kurtosis. J R Stat Soc Ser B 1964; 26: 111–112.

[76] MacGillivray HL, Balanda KP. The relationships between skewness and kurtosis. Aust J Stat 1988; 30(3): 319–337.

[77] Balanda KP, MacGillivray HL. Kurtosis: a critical review. Am Stat 1988; 42: 111–119.

[78] DeCarlo LT. On the meaning and use of kurtosis. Psychol Methods 1997; 2(3): 292–307.

[79] Pearson K. Mathematical contributions to the theory of evolution,-xix; second supplement to a memoir on skew variation. Philos Trans R Soc Lond A 1916; 216: 429–457.

[80] Klaassen CAJ, Mokveld PJ, van Es B. Squared skewness minus kurtosis bounded by 186/125 for unimodal distributions. Stat Probab Lett 2000; 50(2): 131–135.

[81] Rohatgi VK, Székely GJ. Sharp inequalities between skewness and kurtosis. Stat Probab Lett 1989; 8(4): 297–299.

[82] Gupta AK, Móri TF, Székely GJ. Testing for Poissonity-normality vs. other infinite divisibility. Stat Probab Lett 1994; 19: 245–248.

[83] Teuscher F, Guiard V. Sharp inequalities between skewness and kurtosis for unimodal distributions. Stat Probab Lett 1995; 22(3): 257–260.

[84] Tracy RL, Doane DP. Using the studentized range to assess kurtosis. J Appl Stat 2005; 32(3): 271–280.

[85] Fiori AM. Measuring kurtosis by right and left inequality orders. Commun Stat Theory Methods 2008; 37(17): 2665–2680.

[86] Jones MC, Rosco JF, Pewsey A. Skewness-invariant measures of kurtosis. Am Stat 2011; 65(2): 89–95.

[87] Vrabie V, Granjon P, Serviere C. Spectral kurtosis: from definition to application. Workshop on Nonlinear Signal and Image Processing, NSIP03; Grado-Trieste, Italy; 2003.

[88] Rao VB. Kurtosis as a metric in the assessment of gear damage. Shock Vib Dig 1999; 31(6): 443–448.

[89] Vrabie V, Granjon P, Maroni C-S, Leprettre B. Application of spectral kurtosis to bearing fault detection in induction motors. 5th International Conference on Acoustical and Vibratory Surveillance Methods and Diagnostic Techniques (Surveillance5); 2004. Senlis. p 1–10.

[90] Pakrashi V, Basu B, O'Connor A. A structural damage detector and calibrator using a wavelet kurtosis technique. Eng Struct 2007; 29: 2097–2108.

[91] Randall RB. Applications of spectral kurtosis in machine diagnostics and prognostics. Key Eng Mater Damage Assess Struct 2005; 6: 21–36.

[92] Antoni J. The spectral kurtosis of nonstationary signals: formalization, some properties, and application. 12th European Signal Processing Conference; Vienna, Austria; 2004.

[93] Combet F, Gelman L. Optimal filtering of gear signals for early damage detection based on the spectral kurtosis. *Mech Syst Signal Process* 2009; 23 (3) 652–668.

[94] Otonnello C, Pagnan S. Modified frequency domain kurtosis for signal processing. Electron Lett 1994; 30(14): 1117–1118.

[95] Cocconcelli M, Zimrov R, Rubini R, Bartelmus W. Kurtosis over energy distribution approach for STFT enhancement in ball bearing diagnosis. Condition Monitoring of Machinery in Non-Stationary Operations -Part 1; Berlin and Heidelberg: Springer; 2012. p 51–59.

[96] Liu J, Ghafari S, Wang W, Golnaraghi F, Ismail F. Bearing fault diagnostics based on reconstructed features. IEEE Industry Applications Society Annual Meeting, IAS '08; Edmonton, Albeerta, Canada: 2008. p 1–7.

[97] Millioz FE, Martin N. Circularity of the STFT and Spectral Kurtosis for time-frequency segmentation in Gaussian environment. IEEE Trans Signal Process 2011; 59(2): 515–524.

[98] Rajan J, Poot D, Juntu J, Sijbers J. Noise measurement from magnitude MRI using local estimates of variance and skewness. Phys Med Biol 2010; 55(22): 6973–6975.

[99] Mardia KV. Measures of multivariate skewness and kurtosis with applications. Biometrika 1970; 57: 519–530.

[100] Mardia KV, Zemroch PJ. Measures of multivariate skewness and kurtosis. Appl Stat 1975; 24(2): 262–265.

[101] Wang J, Serfling R. Nonparametric multivariate kurtosis and tailweight measures. J Nonparametr Stat 2005; 17: 441–456.

[102] Castella M, Moreau E. A new method for kurtosis maximization and source separation. ICASSP; Dallas (TX); 2010. p 2670–2673.

[103] Bowman KO, Shenton LR. Omnibus test contours for departures from normality based on $\sqrt{b_1}$ and b_2. Biometrika 1975; 62: 243–250.

[104] Pearson K. Contributions to the mathematical theory of evolution,-II. Skew variation in homogeneous material. Philos Trans R Soc Lond A 1895; 186: 343–414.

[105] Abbasi, N. Asymptotic distribution of coefficients of skewness and kurtosis. J Math Stat Sci Publ 2009; 5(4): 365–368.

[106] Asmussen S, Rydén T. A note on skewness in regenerative simulation. Commun Stat Simul Comput 2011; 40(1): 45–57.

[107] Ignaccolo M, Michele CD. Skewness as a measure of instantaneous renormalized drop diameter distributions. Hydrol Earth Sci 2012; 16(2): 319–327.

[108] Bhattacharyya RE, Kar S, Majumder DD. Fuzzy mean-variance-skewness portfolio selection models by interval analysis. Comput Math Appl 2011; 61(1): 126–137.

[109] Kotz S, Seier E. Visualizing peak and tails to introduce kurtosis. Am Stat 2008; 62(4): 348–354.

[110] Cheng PC-H, Pitt NG. Diagrams for difficult problems in probability. Math Gaz 2003; 87(508): 86–97.

[111] de Montmort PR. *Essay d'Analyse sur des Jeux de Hazard*. Paris: Quillau; 1708.

[112] Hacking I. *The Emergence of Probability: A Philosophical Study of Early Ideas about Probability, Induction and Statistical Inference*. Cambridge: Cambridge University Press; 2006.

[113] Feller W. *Introduction to Probability Theory and its Applications*. New York: John Wiley and Sons; 1968.

[114] Port SC. *Theoretical Probability for Applications*. New York: John Wiley and Sons; 1994.

[115] Johnson NL, Kotz S. *Urn Models and their Application*. New York: John Wiley and Sons; 1977.

[116] Devore JL. *Probability and Statistics for Engineering and the Sciences*. 4th ed. Belmont (CA): Wadsworth; 1991.

[117] Hájek A. In: Edward NZ, editor. *Interpretations of Probability*. Winter 2012 ed. Stanford (CA): Stanford Encyclopedia of Philosophy; 2012.

[118] Ionut F, Tudor CA. *Handbook of Probability*. New York: John Wiley and Sons; 2013.

[119] Spiegel MR, Stephens LJ. *Theory and Problems of Statistics*. 3rd ed. New York: McGraw-Hill; 1999.

[120] Stroock D. *Probability Theory*. Cambridge: Cambridge University Press; 2011.

[121] Balakrishnan N, Nevzorov VB. *A Primer on Statistical Distributions*. New York: John Wiley and Sons; 2003.

[122] Evans M, Hastings N, Peacock B. *Statistical Distributions*. New York: John Wiley and Sons; 2000.

[123] Johnson NL, Kotz S, Kemp AW. *Univariate Discrete Distributions*. 2nd ed. New York: John Wiley and Sons; 1992.

[124] Hwang LC. A simple proof of the binomial theorem using differential calculus. Am Stat 2009; 63(1): 43–44.

[125] Johnson NL. A note on the mean deviation of the binomial distribution. Biometrika 1957; 44: 532–533.

[126] Kamat AR. A generalization of Johnson's property of the mean deviation for a class of distributions. Biometrika 1966; 53: 285–287.

[127] Bernoulli J. The MacTutor History of Mathematics archive, School of Mathematics and Statistics, University_of_St_Andrews"University of St Andrews, UK; 1713.

[128] Pascal B. Varia opera Mathematica. Tolossae: D. Petri de Fermat; 1679.

[129] Chattamvelli R, Jones MC. Recurrence relations for the noncentral density, distribution functions and inverse moments. J Stat Comput Simul 1995; 52: 289–299.

[130] De Moivre A. Approximatio ad summam terminorum binomii $(a+b)^n$ in seriem expansi" (self-published pamphlet); 1733. 7 p.

[131] Bertrand JF. *Calculus des Probabilites*. Paris: Gauthier-Villars; 1889.

[132] Diaconis P, Zabell S. Closed form summations for classical distributions: variations on a theme of De Moivre. Stat Sci 1991; 6: 284–302.

[133] Peizer DB, Pratt JW. A normal approximation to binomial, F, beta and other common related tail probabilities. J Am Stat Assoc 1968; 63: 1417–1456.

[134] Johnson NL, Kotz S, Balakrishnan N. *Continuous Univariate Distributions*. Volume 1. New York: John Wiley and Sons; 1994.

[135] Zelterman D. *Discrete Distributions—Applications in the Health Sciences*. New York: John Wiley and Sons; 2005.

[136] van der Geest PAG. The binomial distribution with dependent Bernoulli trials. J Stat Comput Simul 2005; 75(2): 141–154.

[137] Rice J. *Mathematical Statistics and Data Analysis*. Belmont (CA): Wadsworth; 2006.

[138] Ross SM. *Probability Models for Computer Science*. New York: Academic Press; 2002.

[139] Srivastava RC. Two characterizations of the geometric distribution. J Am Stat Assoc 1974; 69: 267–269.

[140] Boland PJ. *Statistical and Probabilistic Methods in Actuarial Sciences*. Boca Raton, FL: Chapman & Hall/CRC; 2007.

[141] Meyer A. Calcul des Probabilités, publié sur les manuscrits de l'auteur par F. Folie. Mem. Soc. R. Sci. Liége. German translation by E. Czuber: Vorlesungen über Wahrscheinlichkeitsrechnung von Dr A. Meyer. Teubner, Leipzig, 1879. References are to the German edition, 1874; 2(4): 1–458.

[142] Guenther WC. A note on the relation between incomplete beta and negative binomial. Metron 1973; 31: 349–351.

[143] Vellaisamy P, Upadhye NS. On the negative binomial distribution and its generalizations. Stat Probab Lett 2007; 77(2): 173–180.

[144] Withers Ch, Nadarajah S. Maximum of continuous versions of Poisson and negative binomial type distributions. Teor Veroyatnost i Primenen 2010; 55(3): 598–601.

[145] Bowman KO, Shenton LR. Skewness for maximum likelihood estimators of the negative binomial distribution. Far East J Theor Stat 2007; 22(1): 103–129.

[146] Fisher RA. The mathematical distributions used in the common tests of significance. Econometrika 1931; 3(4): 353–365.

[147] Anscombe FJ. The transformation of Poisson, binomial, and negative binomial data. Biometrika 1948; 35: 246–254.

[148] Freeman MF, Tukey JW. Transformations related to the angular and the square root. Ann Math Stat 1950; 21: 607–611.

[149] Shanmugam R. Incidence rate restricted Poissonness. Sankhya (Series B) 1991; 53: 191–201.

[150] Shanmugam R. Size biased incidence rate restricted Poissonness and its application in international terrorism. Appl Manage Sci 1993; 7: 41–49.

[151] Shanmugam R. Poisson distribution. In: Salkind NJ, editor. *Encyclopedia of Measurement and Statistics*. Thousand Oaks (CA): Sage Press; 2006. p 772–775.

[152] Shanmugam R. Spinned Poisson distribution with health management application. Healthc Manage Sci 2011; 14(4): 299–306.

[153] Rodríguez-Avi J, Conde-Sánchez A, Sáez-Castillo AJ, Olmo-Jiménez MJ. A generalization of the beta-binomial distribution. J R Stat Soc Ser C Appl Stat 2007; 56(1): 51–61.

[154] Le Gall F. Determination of the modes of a multinomial distribution. Stat Probab Lett 2003; 62(4): 325–333.

[155] Frisch R. Solution of a problem of the calculus of probabilities. Scand Actuar J 1924; 7: 153–174.

[156] Kamat AR. A property of the mean deviation for a class of continuous distributions. Biometrika 1965; 52: 288–289.

[157] Winkler P. On computability of the mean deviation. Inf Process Lett 1982; 15: 36–38.

[158] Babu GJ, Rao CR. Expansions for statistics involving the mean absolute deviation. Ann Inst Stat Math 1992; 44: 387–403.

[159] Pham-Gia T, Turkkan N, Duong QP. Using the mean deviation to determine the prior distribution. Stat Probab Lett 1992; 13: 373–381.

[160] Pham-Gia T, Hung TL. The mean and median absolute deviations. Math Comput Model 2001; 34(7): 921–936.

[161] Egorychev GP, Zima EV, Hazewinkel M. Integral representation and algorithms for closed form summation. Handb Algebra 2008; 5: 459–529.

[162] Kono H, Koshizuka T. Mean-absolute deviation model. IIE Trans 2005; 37: 893–900.

[163] Liu Y, Zhang Z. Mean-absolute deviation optimization model for hedging portfolio selection problems. IEEE Comput Soc 2009; 3: 76–79.

[164] Dong J. Mean deviation method for fuzzy multi-sensor object recognition, 2009. International Conference on Management of e-Commerce and e-Government. IEEE computer society; Nanchang, China; 2009. p 201–204.

[165] Erel E, Ghosh JB. Minimizing weighted mean absolute deviation of job completion times from their weighted mean. Appl Math Comput 2011; 217: 9340–9350.

[166] Jones MC, Balakrishnan N. How are moments and moments of spacings related to distribution functions? J Stat Plann Inference 2003; 103: 377–390.

[167] Leemis LM, McQueston JT. Univariate distribution relationships. Am Stat 2008; 62: 45–53.

[168] Gentle J, Hardle WK, Mori Y. *Handbook of Computational Statistics Concepts and Methods*. New York: Springer; 2012.

[169] Chattamvelli R. Power of the power-laws and an application to the PageRank metric. PMU J Comput Sci Eng 2010; 1(2): 1–7.

[170] Abramowitz M, Stegun IA, editors. *Handbook of Mathematical Functions*. New York: Dover publications; 1973.

[171] Jambunathan MV. Some properties of beta and gamma distributions. Ann Math Stat 1954; 25: 401–405.

[172] El-Saidi MA, Singh KP. Evaluating the cumulative distribution function and 100α percentage points of a family of generalized logistic distributions. Biom J 1991; 33: 865–873.

[173] Olapade AK. On extended Type I generalized logistic distribution. IJMMS 57; 2003. p. 3069–3074.

[174] Barreto-Souza W, Santos AHS, Cordeiro GM. The beta generalized exponential distribution. J Stat Comput Simul 2010; 80: 159–172.

[175] Krysicki W. On some new properties of the beta distribution. Stat Prob Lett 1999; 42: 131–137.

[176] Morgan EC, Lackner M, Vogel RM, Baise LG. Probability distributions for offshore wind speeds. Energy Convers Manage 2011;52:15–26.

[177] Glaser RE. A characterization of Bartlett's statistic involving incomplete beta functions. Biometrika 1980; 67(1): 53–58.

[178] Anderson TW. *An Introduction to Multivariate Statistical Analysis*. New York: John Wiley and Sons; 1971.

[179] Seber GAF. Linear hypotheses and induced tests. Biometrika 1965; 51: 41–47.

[180] Joshi PC, Lalitha S. Tests for two outliers in a linear model. Biometrika 1986; 73: 236–239.

[181] Haight FA. On the effect of removing persons with N or more accidents from an accident prone population. Biometrika 1956; 52: 298–300.

[182] Pearson K. On the applications of the double Bessel function $K_{\tau_1, \tau_2}(x)$ to statistical problems. Biometrika 1933; 25: 158–178.

[183] Wise ME. The incomplete beta function as a contour integral and a quickly converging series for its inverse. Biometrika 1950; 37: 208–218.

[184] Tio GG, Guttman I. The inverted Dirichlet distribution with applications. J Am Stat Assoc 1965; 60: 793–805.

[185] Benson D, Krishnamurthy K. Computing discrete mixtures of continuous distributions: noncentral chisquare, noncentral t and the distribution of the square of the sample multiple correlation coefficient. Comput Stat Data Anal 2003; 43(2): 249–267.

[186] Bock ME, Govindarajulu Z. A note on the noncentral chi-square distribution. Stat Probab Lett 1989; 7(2): 127–129.

[187] Cohen JD. Noncentral chi-square: some observations on recurrence. Am Stat 1988; 42(2): 120–122.

[188] Craig CC. Note on the distribution of noncentral t with an application. Ann Math Stat 1941; 17: 193–194.

[189] Farebrother RW. Algorithm AS 231: the distribution of a noncentral χ^2 variable with nonnegative degrees of freedom. Appl Stat 1987; 36: 402–405.

[190] Lenth RV. Cumulative distribution function of the noncentral t distribution. Appl Stat 1988; 38(1): 185–187.

[191] Ruben H. Noncentral chi-square and gamma revisited. Commun Stat 1974; 3(7): 607–637.

[192] Tang PC. The power function of the analysis of variance tests with tables and illustrations of their use. Stat Res Mem 1938; 2: 126–150.

[193] Ding CG, Bargmann RE. Sampling distribution of the square of the sample multiple correlation coefficient. Appl Stat 1991; 41: 478–482.

[194] Ding CG. On the computation of the distribution of the square of the sample multiple correlation coefficient. Comput Stat Data Anal 1996; 22: 345–350.

[195] Seber GAF. The noncentral chi-squared and beta distributions. Biometrika 1963; 50: 542–544.

[196] Chattamvelli R. Another derivation of two algorithms for the noncentral χ^2 and F distributions. J Stat Comput Simul 1995; 49: 207–214.

[197] Norton V. A simple algorithm for noncentral F distribution. Appl Stat JRSS 1983; 32: 84–85.

[198] Patnaik PB. The noncentral chi-square and F distributions and their applications. Biometrika 1949; 46: 202–232.

[199] Ding CG, Bargmann RE. Quantiles of the distribution of the square of the sample multiple-correlation coefficient. Appl Stat 1991; 40: 199–202.

[200] Gurland J. A relatively simple form of the distribution of the multiple correlation coefficient. J R Stat Soc Ser B 1968; 30(2): 276–283.

[201] L'Ecuyer P, Simard R. Inverting the symmetrical beta distribution. Trans Math Softw (TOMS) 2006; 32(4): 509–520.

[202] Saunders LR. An exact formula for the symmetrical incomplete beta function where the parameter is an integer or half-integer. Aust J Stat 1992; 34(2): 261–264.

[203] Aroian LA. Continued fractions for the incomplete beta function. Ann Math Stat 1941; 12: 218–223 (also 30, p. 1265).

[204] Özçag E, Ege I, Gürçay H. An extension of the incomplete beta function for negative integers. J Math Anal Appl 2008; 338(2): 984–992.

[205] Thompson DH. Approximate formulae for the percentage points of the incomplete beta function and the χ^2 distribution. Biometrika 1974; 34: 368–372.

[206] Soper HE. *The Numerical Evaluation of the Incomplete Beta Function, Tracts for Computers*. Cambridge: Cambridge University Press; 1921.

[207] Pearson K, editor. *Tables of the Incomplete Beta Function*. 2nd ed. Cambridge: Cambridge university press; 1968.

[208] Majumder KL, Bhattacharjee GP. Algorithm AS 63: the incomplete beta integral. Appl Stat 1973; 22: 409–411.

[209] Bosten NE, Battiste EL. Remark on algorithm AS 179: incomplete beta ratio. Commun ACM 1974; 17: 156–157.

[210] Alexander C, Cordeiro GM, Ortega EMM, Sarabia JM. *Generalized Beta-Generated Distributions, ICMA Centre - DP, 2011-05*. The University of Reading, UK; 2011.

[211] Adell JA, Jodrá P. Sharp estimates for the median of the $\Gamma(n + 1, 1)$ distribution. Stat Probab Lett 2005; 71: 185–191.

[212] Chen J, Rubin H. Bounds for the difference between median and mean of gamma and Poisson distributions. Stat Probab Lett 1986; 4: 281–283.

[213] Griffiths P, Hill ID, editors. *Applied Statistics Algorithms*. Royal Statistical Society. Chichester: Ellis Horwood; 1985.

[214] Köpf W. The gamma function. *Hypergeometric Summation: An Algorithmic Approach to Summation and Special Identities*. Chapter 1. Braunschweig: Vieweg; 1998. p 4–10.

[215] Pearson K, editor. *Tables of the Incomplete Γ-Function*. Cambridge: Cambridge University Press; 1934.

[216] Shea BL. Chi-square and incomplete gamma integral. Appl Stat R Stat Soc Ser C 1988; 37: 466–473.

[217] Shepp L. Normal functions of normal random variables. SIAM Rev 1964; 6: 459.

[218] Quine MP. A result of Shepp. Appl Math Lett 1994; 7(6): 33–34.

[219] Weibull M. The distributions of t- and F- statistics and of correlation and regression co-efficients in stratified samples from normal populations with different means. Skand Aktuar Tidskr 1953; 36: 9–106.

[220] Ellison BE. Two theorems for inference about the normal distribution with applications in acceptance sampling. J Am Stat Assoc (JASA) 1964; 64: 89–95.

[221] Patil GP, Kapadia CH, Owen DB. *Handbook of Statistical Distributions*. New York: Marcel Dekker; 1989.

[222] Box GEP, Muller ME. A note on the generation of random normal deviates. Ann Math Stat 1958; 29: 610–611.

[223] Kotlarski IL. An exercise involving Cauchy random variables. Am Math Monthly 1979; 86: 229.

[224] Arnold BC. Some characterizations of the Cauchy distribution. Aust J Stat 1979; 21(2): 166–169.

[225] Stuart A, Ord K. *Kendall's Advanced Theory of Statistics*. Volume 1, Distribution Theory. Oxford: Oxford University Press; 2009.

[226] Sahai H, Thompson WO. Comparison of approximations to the percentiles of the t, χ^2, and F distributions. J Stat Comput Simul 1974; 3: 81–93.

[227] Beaulieu NC, Abu-Dayya AA, McLane PJ. On approximating the distribution of a sum of independent lognormal random variables. IEEE Conference Proceedings, Atlanta, GA; 1993. 72–79.

[228] Filho JCSS, Cardieri P, Yacoub MD. Simple accurate lognormal approximation to log-normal sums. Electron Lett 2005; 41(18): 1016–1017.

[229] Daniel G, Gaston A. *Mathematical and Statistical Methods in Food Science and Technology*. New York: Wiley-Blackwell Press; 2014.

[230] Arnold BC. *Pareto Distribution*. Fairland (MD): International Cooperative Publishing House; 1983.

[231] Kleiber C, Kotz S. *Statistical Size Distributions in Economics and Actuarial Sciences*. New York: John Wiley and Sons; 2003.

[232] Michael H. Pareto's law. Math Intell 2010; 32(3): 38–43.

[233] Newman MEJ. Power laws, Pareto distributions and Zipf's law. Contemp Phys 2005; 46(5): 323–351.

[234] Lancaster HO. The median of χ^2. Aust J Stat 1968; 10: 83.

[235] Lancaster HO. *The Chi-Squared Distribution*. New York: John Wiley and Sons; 1969.

[236] Cheng SW, Fu JC. An algorithm to obtain the critical values of the t, χ^2 and F distributions. Stat Probab Lett 1983; 1(5): 223–227.

[237] Bienayme, **'Cahiers du Centre d'Analyse et de Mathématiques Sociales'**, n° 138, *Série Histoire du Calcul des Probabilités et de la Statistique*, n° 28, Paris, E.H.E.S.S.-C.N.R.S; 1852.

[238] Helmert FR. Ueber die Wahrscheinlichkeit der Potenzsummen der Beobachtungs-fehler und über einige damit im Zusam- menhange stehende Fragen. *Z. Math. Phys*, 1876;21:102–219.

[239] Zar JH. Approximations for the percentage points of the chi-squared distribution. Appl Stat 1978; 27: 280–290.

[240] Severo NC, Zelen M. Normal approximation to the chi-square and noncentral F probability functions. Biometrika 1960; 47(3): 411–416.

[241] Alam K, Rizvi HM. On noncentral chi-squared and noncentral F distributions. Am Stat 1967; 21(4): 21–22.

[242] Gabler S, Wolff C. A quick and easy approximation to the distribution of a sum of weighted chi-square variables. Stat Hefte 1987; 28: 317–323.

[243] Johnson NL. On an extension of the connection between Poisson and χ^2 distributions. Biometrika 1959; 46: 352–363.

[244] Jones MC. On the relationship between the Poisson-exponential model and the noncentral chi-squared distribution. Scand Actuar J 1987; 1987(1–2): 104–109.

[245] Knüsel L. Computation of the chi-square and Poisson distribution. SIAM J Sci Stat Comput 1986; 7(3): 1022–1036.

[246] Gosset WS. In: Pearson ES, Kendall MG, editors. *Studies in the History of Statistics and Probability*. London; 1970. p. 355–404.

[247] Jones MC, Faddy MJ. A skew extension of the T distribution, with applications. J R Stat Soc Ser B 2003; 65(1): 159–174.

[248] Dudewicz EJ, Dalal SR. On approximations to the t-distribution. J Qual Technol 1972; 4: 196–198.

[249] Fisher RA. Expansion of "Student's" integral in powers of n^{-1}. Metron 1925; 5: 109–112.

[250] Good IJ, Smith E. A power series for the tail-area probability of Student's T distribution. J Stat Comput Simul 1986; 23(3): 248–250.

[251] Good IJ. Very small tails of the t distribution, and significance tests for clustering. J Stat Comput Simul 1986; 23: 243–250.

[252] Ghosh BK. On the distribution of the difference of two t-variables. J Am Stat Assoc 1975; 70: 463–467.

[253] Kanji GK. *100 Statistical Tests*. London: Sage Publishers; 2006.

[254] Cacoullos T. A relation between t and F distributions. J Am Stat Assoc 1965; 60: 528–531 (also 60, p.1249).

[255] Ifram AF. On the characteristic functions of the F and t distributions. Sankhya-A 1970; 32: 350–352.

[256] Snedecor GW, Cochran WG. *Statistical Methods*. 8th ed. Ames (IA): Blackwell Publishing Professional; 1989.

[257] Siegel AF. The noncentral chi-squared distribution with zero degrees of freedom and testing for uniformity. Biometrika 1979; 66(2): 381–386.

[258] Phillips PCB. The true characteristic function of the F distribution. Biometrika 1982; 69: 261–264.

[259] Awad AM. Remark on the characteristic function of the F-distribution. Sankhya A 1980; 42: 128–129.

[260] Pestana D. Note on a paper of Ifram. Sankhya-A 1977; 39: 396–397.

[261] Davis WE, Khalil HM. Algorithm and series expansion for the F-distribution. Signum Newsl 1972; 7: 21–23.

[262] Mardia KV, Zemroch PJ. *Tables of the F and Related Distributions with Algorithms*. New York: Academic Press; 1978.

[263] Singh KP, Relyea GE. Computation of noncentral F probabilities: a computer program. Comput Stat Data Anal 1992; 43: 95–102.

[264] Schader M, Schmid F. Distribution function and percentage points for the central and noncentral F-distribution. Stat Hefte 1986; 27: 67–74.

[265] Fisher RA, Cornish EA. The percentile points of distributions having known cumulants. Technometrics 1960; 2: 209–225.

[266] Sengupta S. Some simple approximations for the doubly noncentral z distribution. Aust J Stat 1991; 33: 177–181.

[267] Aroian LA. Note on the cumulants of Fisher's Z distribution. Biometrika 1947; 34: 359–360.

[268] Wishart J. The cumulants of the z and of the logarithmic χ^2 and t distributions. Biometrika 1947; 34: 170–178.

[269] Fréchet, M. Sur la loi de probabilité de l'écart maximum. Ann Soc Polon Math 1927;6:93.

[270] Rosen P, Rammler E. The laws governing the fineness of powdered coal. J Inst Fuel 1933; 7: 29–36.

[271] Hribar L, Duka D. Weibull distribution in modeling component faults. IEEE Proceedings of ELMAR; Zadar; 2010. p 183–186.

[272] Murthy DNP, Xie M, Jiang R. *Weibull Models*. New York: John Wiley and Sons; 2004.

[273] Englehardt JD, Ruochen Li. The discrete Weibull distribution: an alternative for correlated counts with confirmation for microbial counts in water. Risk Anal 2011; 31(3): 370–381.

[274] Mihet-Popa L, Groza V. Annual wind and energy loss distribution for two variable speed wind turbine concepts of 3 MW. Instrumentation and Measurement Technology Conference (I2MTC), 2011. IEEE; Shanghai, China: 2011. p 1–5.

[275] Ueno T, Hirano S, Hirano Y, Oribe N, Nakamura I, Oda Y, Kanba S, Onitsuka T. Stability of the Rayleigh distribution. 4th International Congress on Image and Signal Processing (CISP), Volume 5; Shanghai, China; 2011. p 2376–2378.

[276] Salo J, El-Sallabi HM, Vainikainen P. The distribution of the product of independent Rayleigh random variables. IEEE Trans Antennas Propag 2006; 54(2): 639–643.

[277] Nadarajah S, Kotz S. Comments on "On the distribution of the product of independent Rayleigh random variables". IEEE Trans Antennas Propag 2006; 54(1–2): 3570–3571.

[278] Kostin A. Probability distribution of distance between pairs of nearest stations in wireless network. Electron Lett 2010; 46(18): 1299–1300.

[279] Hazewinkel M. *Probability Distribution*. Encyclopedia of Mathematics. New York: Springer; 2001.

[280] Rao CR. *Linear Statistical Inference and its Applications*. 2nd ed. New York: John Wiley and Sons; 1973.

[281] Aitchison J, Brown JAC. *The Lognormal Distribution*. Cambridge: Cambridge University Press; 1998.

[282] Giri NC. Approximations and tables of beta distributions. In: Gupta AK, Nadaraja S, editors. *Handbook of Beta Distribution*. New York: Marcel Dekker; 2004.

[283] Raja RB. A note on the moments of non-central chi-squared and F distributions. Metron 1986; 44: 121–130.

[284] Wilf HS. *Generatingfunctionology*. New York: Academic Press; 1994.

[285] Benjamin J, Cornell A. *Probability, Statistics, and Decision for Civil Engineers*. New York: Dover Press; 2014.

[286] Whittle P. *Probability via Expectation*. 4th ed. New York: Springer; 2000.

[287] Berberan-Santos MN. Expressing a probability density function in terms of another PDF: a generalized Gram-Charlier expansion. J Math Chem 2007; 42(3): 585–594.

[288] Shohat JA, Tamarkin JD. The problem of moments. Am Math Mon 1943; 76: 55–56.

[289] Smith PJ. A recursive formulation of the old problem of obtaining moments from cumulants and vice versa. Am Stat 1995; 49: 217–219.

[290] Dreier I, Kotz S. A note on the characteristic function of the t-distribution. Stat Probab Lett 2002; 57: 221–224.

[291] Montgomery DC, Runger GC, Hubele NF. *Engineering Statistics*. New York: John Wiley and Sons; 2011.

[292] Olofsson P. *Probability, Statistics, and Stochastic Processes*. New York: Wiley-Interscience; 2005.

[293] Zacks S. *Examples and Problems in Mathematical Statistics*. New York: John Wiley and Sons; 2014.

[294] Annavajjala R, Chockalingam A, Mohammed SK. On a ratio of functions of exponential random variables and some applications. IEEE Trans Commun 2010; 58(11): 3091–3097.

[295] Stewart WJ. *Probability, Markov Chains, Queues, and Simulation: The Mathematical Basis of Performance Modeling*. Princeton (NJ): Princeton University Press; 2011.

[296] Mathai AM. *Jacobians of Matrix Transformations, and Functions of Matrix Argument*. Singapore: World scientific; 1997.

[297] Bailey RW. Distributional identities of beta and χ^2 variates: a geometric interpretation. Am Stat 1992; 46(2): 117–120.

[298] Gradshteyn IS, Ryzhik IM. *Table of Integrals, Series and Products*. New York: Academic Press; 1980.

[299] Searle SR. *Matrix Algebra Useful for Statistics*. New York: John Wiley and Sons; 1982.

[300] Seber GAF. *A Matrix Handbook for Statisticians*. New York: John Wiley and Sons; 2008.

[301] Baringhaus L, Henze N, Morgenstern D. Some elementary proofs of the normality of $XY/(X^2 + Y^2)^{1/2}$ where X and Y are normal. Comput Math Appl 1988; 15: 943–944.

[302] Jones MC. Distributional relations arising from simple trigonometric formulas. Am Stat 1999; 53(2 Suppl 53, 393): 99–102.

[303] Bansal N, Hamedani GG, Key E, Volkmer H, Zhang H, Behboodian J. Some characterizations of the normal distribution. Stat Probab Lett 1999; 42: 393–400.

[304] Abdi H. Partial least square regression - a tutorial. Anal Chem Acta 2007; 35: 1–17.

[305] Pratt JW. A normal approximation for binomial, F, beta and other common related tail probabilities, II. J Am Stat Assoc 1968; 63: 1457–1483.

[306] Rohatgi VK, Saleh AKMdE. *Statistical Inference*. New York: John Wiley and Sons; 2001.

[307] Shanmugam R, Singh K. Testing of Poisson incidence rate restriction. Int J Reliab Appl 2001; 4: 263–268.

[308] Shanmugam R. An intervened Poisson distribution and its medical applications. Biometrics 1985; 41: 1025–1030.

INDEX

Statistics for Scientists and Engineers, First Edition. Ramalingam Shanmugam and Rajan Chattamvelli.
© 2015 John Wiley & Sons, Inc. Published 2015 by John Wiley & Sons, Inc.